CHEMICAL ECOLOGY
OF INSECTS

CHEMICAL ECOLOGY
OF INSECTS

William J. Bell

Professor of Entomology,
Physiology and Cell Biology
University of Kansas

AND

Ring T. Cardé

Professor of Entomology
University of Massachusetts

SINAUER ASSOCIATES, INC · PUBLISHERS
SUNDERLAND, MASSACHUSETTS

First published 1984 by
Chapman and Hall Ltd
11 New Fetter Lane, London EC4P 4EE
© *1984 William J. Bell and Ring T. Cardé*

Distributed in the USA by
Sinauer Associates, Inc., Publishers
Sunderland, MA 01375

Library of Congress Cataloging in Publication Data
Main entry under title:

Chemical ecology of insects.

Bibliography: p.
Includes index.
1. Insects—Physiology. 2. Insects—Ecology.
3. Chemical senses.   I. Bell, William J.
II. Cardé, Ring T.
QL495.C47   1984   595.7′05   83-20212
ISBN 0-87893-069-8
ISBN 0-87893-070-1 (pbk.)

Printed in Great Britain

Contents

PREDATORS, PARASITES AND PREY

CHEMICAL-MEDIATED SPACING

Preface

Our objective in compiling a series of chapters on the chemical ecology of insects has been to delineate the major concepts of this discipline. The fine line between presenting a few topics in great detail or many topics in veneer has been carefully drawn, such that the book contains sufficient diversity to cover the field and a few topics in some depth.

After the reader has penetrated the crust of what has been learned about chemical ecology of insects, the deficiencies in our understanding of this field should become evident. These deficiencies, to which no chapter topic is immune, indicate the youthful state of chemical ecology and the need for further investigations, especially those with potential for integrating elements that are presently isolated from each other. At the outset of this volume it becomes evident that, although we are beginning to decipher how receptor cells work, virtually nothing is known of how sensory information is coded to become relevant to the insect and to control the behavior of the insect. This problem is exacerbated by the state of our knowledge of how chemicals are distributed in nature, especially in complex habitats. And finally, we have been unable to understand the significance of orientation pathways of insects, in part because of the two previous problems: orientation seems to depend on patterns of distribution of chemicals, the coding of these patterns by the central nervous system, and the generation of motor output based on the resulting motor commands.

Studies of insects searching for food, mates, refugia and other resources point out the problem of investigating the modification of behavior by chemicals without providing potential cues of other modalities. It has been necessary to isolate the effects of chemicals on the insect's behavior, but further investigation should re-insert the other relevant factors, so that we can determine the extent to which insects rely solely on chemical signals and to what degree information derived through the other senses is important in locating resources or avoiding stress. Chapters on location of host plants and prey illustrate interesting parallels in the sequence of information from long-range cues that are utilized by an herbivore, predator or parasitoid to initiate search, to close-range information required to home in on the resource, to information that is derived from tasting and touching the potential resource. Studies are needed that examine in detail each step in this sequence. An important concept that is revealed in several places throughout the book is that the pattern of resources is related to the behavior of the insect searching for those resources. Some writers have gone further to suggest that the searching of insects is optimally related to these distributions, but authors of chapters in this volume have tempered this view and offered more mechanistic explanations for how insects obtain resources with minimum energy expenditure. Although equal time is not given to repellants and attractants, two chapters deal specifically with chemicals that reduce predation either directly or indirectly. Potential prey and hosts have evolved chemical means of signalling their inappropriateness. A third chapter discusses more subtle chemical signals that lead to partitioning of resources, a topic that easily could have been the theme of the book: when resources are limiting the success of a species depends on division of resources. The bark beetle chapter is a special case of complexity: intraspecific and interspecific chemical communication that encompasses every one of the concepts developed in the entire book. If the bark beetle example represents conceptual complexity, the social Hymenoptera offer the most complete catalogue of chemicals that can be produced and employed as informational signals. Evolution of sociality seemed to spawn a chemical language that is equivalent to the visual and auditory repertoire of higher vertebrates.

An important issue that is dealt with by several authors in this volume is that of developing appropriate terminology for chemical ecology. The consensus, with a few exceptions, seems to be toward more practical, functionally related terms, rather than classic Greek stems or teleogical terms. This emphasis, if widely adopted, should lead to more efficient communication among researchers. Nothing can inhibit progress in this field more effectively than terms that are teleological, poorly defined, non-mechanistic, non-probabilistic and difficult to spell.

Finally, chemical ecology as a discipline should not be allowed to detach itself from visual or auditory ecology. Already these categories look and sound artificial, and the danger is that the vast quantity of information available to insects through these latter modalities will be ignored or surpressed by the growing field of chemical ecology. Researchers in chemical ecology need to keep abreast of these homologous topics to profit optimally from their findings that relate chemicals to the ecology of insects.

William J. Bell, Ring T. Cardé

List of contributors

Anne L. Averill, *Department of Entomology, University of Massachusetts, Amherst, Massachusetts 01003, USA.*

Thomas C. Baker, *Department of Entomology, University of California, Riverside, California 92521, USA.*

William J. Bell, *Department of Entomology and Department of Physiology & Cell Biology, University of Kansas, Lawrence, Kansas 66045, USA.*

M. C. Birch, *Department of Zoology, Oxford University, South Parks Road, Oxford OX1 3PS, UK.*

S. Bradleigh Vinson, *Department of Entomology, Texas A & M University, College Station, Texas 77843, USA.*

J. W. S. Bradshaw, *Chemical Entomology Unit, Department of Biology, The University, Southampton, UK.*

Ring T. Cardé, *Department of Entomology, University of Massachusetts, Amherst, Massachusetts 01003, USA.*

Richard M. Duffield, *Department of Zoology, Howard University, Washington, DC 20059, USA.*

George C. Eickwort, *Department of Entomology, Cornell University, Ithaca, New York 14850, USA.*

Joseph S. Elkinton, *Department of Entomology, University of Massachusetts, Amherst, Massachusetts 01003, USA.*

P. E. Howse, *Chemical Entomology Unit, Department of Biology, The University, Southampton, UK.*

James E. Huheey, *Department of Chemistry, University of Maryland, College Park, Maryland 20742, USA.*

James R. Miller, *Department of Entomology and Pesticide Research Center, Michigan State University, East Lansing, Michigan 48824, USA.*

Hanna Mustaparta, *University of Trondheim, Zoological Institute, Rosenborg, 7000 Trondheim, Norway.*

L. R. Nault, *Department of Entomology, Ohio Agricultural Research and Development Center, Wooster, Ohio 44691, USA.*

P. L. Phelan, *Department of Entomology, University of California, Riverside, California 92521, USA.*

Ronald J. Prokopy, *Department of Entomology, University of Massachusetts, Amherst, Massachusetts 01003, USA.*

Bernard D. Roitberg, *Department of Entomology, University of Massachusetts, Amherst, Massachusetts 01003, USA.*

J. Mark Scriber, *Department of Entomology, University of Wisconsin, Madison, Wisconsin 53706, USA.*

Erich Städler, *Swiss Federal Research Station for Fruit-Growing, Viticulture and Horticulture, CH-8820 Waedenswil, Switzerland.*

Karen L. Strickler, *Department of Entomology and Pesticide Research Center, Michigan State University, East Lansing, Michigan 48824, USA.*

James W. Wheeler, *Department of Chemistry, Howard University, Washington, DC 20059, USA.*

Perceptual Mechanisms

1

Contact Chemoreception

Erich Städler

1.1 INTRODUCTION

In insects the chemistry of the environment is a dominant modality mediating adaptive behavior, including the choice of food and feeding, avoidance of danger, location of a sexual partner and the choice of a habitat for the progeny. However, it is important to recognize that other environmental stimuli such as visual, tactile, temperature and humidity also are part of the total sensory input to the central nervous system (CNS). These perceptual mechanisms together with the internal state of the CNS, as influenced by earlier sensory perception or developmental processes, determine the insect's behavior.

Chemoreception and its rôle in the ecology of insects have been reviewed by Chapman and Blaney (1979), Dethier (1970), Hansen (1978), Schoonhoven (1981) and Städler (1976, 1980). The topic of chemical perceptual mechanisms has been divided into contact chemoreception and olfaction, based on the characteristics of the stimuli (volatility), the transport medium (water versus air) and the morphology of the sensory organs. In insect chemoreception, this division is not as clear-cut as it may appear, as has been pointed out and substantiated by studies of smell in water and air in the diving beetle *Dytiscus marginalis* (Behrend, 1971). A relatively minor difference between taste and smell also is suggested by one discovery of 'gustatory' sensilla with receptor cells sensitive to different odors (Dethier, 1972; Städler and Hanson, 1975). However, these authors point out that the olfactory capability of 'gustatory' sensilla may not be a general phenomenon because other sensilla failed to react to odor stimulation.

The results of Städler and Hanson (1975) did not rule out the possibility that

Chemical Ecology of Insects. Edited by William J. Bell and Ring T. Cardé

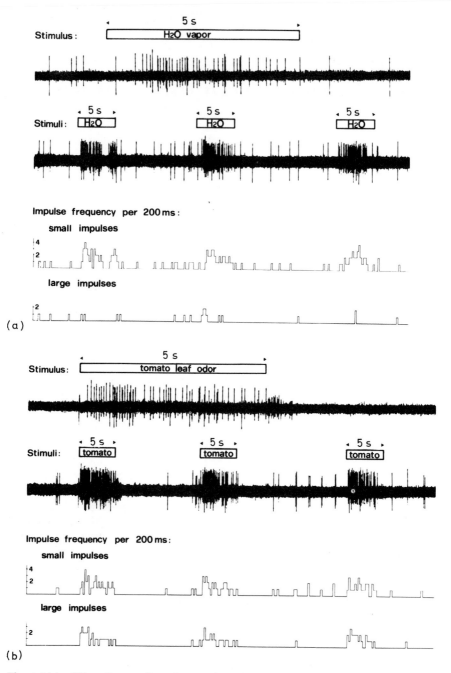

Fig. 1.1(a)   Sidewall recordings from a lateral sensillum styloconicum of a larva of *Manduca sexta* (tobacco hornworm). Stimulation with an airstream (0.8 m s^{-1}) carrying water vapor.
(b) Sidewall recording from the same sensillum. Stimulation with odor of a crushed tomato leaflet.

humidity or a cooling effect were responsible for the olfactory response of the lateral sensillum styloconicum of larval *Manduca sexta*, phenomena observed in other sensilla (Dethier and Schoonhoven, 1968; Schoonhoven, 1967). To resolve this question the lateral sensillum styloconicum was stimulated with an air stream $(0.8\,\mathrm{m\,s^{-1}})$ carrying water vapor or the odor of freshly crushed tomato leaf pieces. Examination of the data in Fig. 1.1(a), (b) suggests that the responses to tomato leaf odor and water vapor are identical. However, a close analysis of the frequencies of impulses with different amplitudes reveals that an additional receptor cell with larger impulses, not present in the water response, is firing in response to tomato. This shows that these 'gustatory' receptor cells can 'smell' at a close range (0.5 mm) (see Städler and Hanson, 1975). We propose to use the term 'contact chemoreception' to include taste or gustation, and close-range olfaction.

This review will stress the ecological point of view of contact chemo-reception. Morphological and strictly physiological aspects will be omitted if they are not related directly to the behavior and ecology of insects.

1.2 MORPHOLOGY OF SENSILLA

Contact chemoreceptive sensilla can have very different external shapes (Fig. 1.2). The common characteristics are a single pore at the tip of the sensillum (Altner, 1977; Altner and Prillinger, 1980; Zacharuk, 1980) and unbranched dendrites from 2 to 10 cells reaching to the tip. Often a mechanoreceptive cell is associated with the chemoreceptor cells, monitoring movements of the whole sensillum or its tip. Other cells are associated with these sensory neurons. The trichogen, tormogen and sheath cells play a role in the ontogeny of the sensillum, forming respectively the hairshaft, hairsocket and the sheath surrounding the dendrites. After the formation of the sensillum these associated cells have physiological functions: electrogenic pump, production of the fluid bathing the dendrites, and isolation and nutrition (?) of the receptor cells (for more details see Hansen, 1978; Thurm and Küppers, 1980; Zacharuk, 1980; Mustaparta, Chapter 2).

McIver *et al.* (1980) have recently observed a unique contact chemoreceptive sensillum, the bifurcate sensilla on the tarsi of the female black fly (*Simulidae*). This type of sensillum has characteristics of a contact chemoreceptor, although a sieve-like structure in the pore region, which increases the absorptive surface area at the tip, suggests a secondarily acquired olfactory function.

Contact chemoreceptive sensilla have been identified on the major parts of the body including the wings (Wolbarsht and Dethier, 1958) and the ovipositor of flies (Rice, 1976; Wallis, 1962; Wolbarsht and Dethier, 1958). This redundancy of receptors is probably a way to avoid unsuitable environments. Most sensilla are located on the proboscis, the tarsi and palpi. Contact chemoreceptors in the food channel have been described for flies (Stocker and Schorderet, 1981),

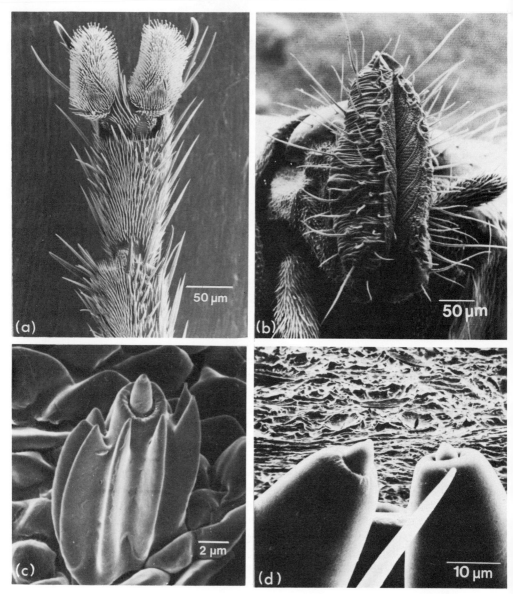

Fig. 1.2 Scanning electron micrographs of contact chemoreceptive organs and sensilla.
(a) Tip of the prothoracic tarsus of a female of *Delia brassicae* (cabbage root fly).
Contact chemoreceptive hairs are marked.
(b) Proboscis (Labellum) of *Psila rosae* (carrot rust fly) with sensory hairs which all
contains contact chemoreceptor cells.
(c) Sensillum styloconicum on the proboscis of the moth *Choristoneura fumiferana*
(spruce budworm moth).
(d) Pair of sensilla styloconicum (medial and lateral) of the larva of *Malacosoma
americana* (eastern tent caterpillar) in relation to a leaf surface (courtesy of Dr V. G.
Dethier, 1975).

caterpillars (De Boer *et al.*, 1977), moths (Eaton, 1979; Städler *et al.*, 1974), true bugs (Wensler and Filshie, 1969; Smith and Friend, 1972), cockroaches (Moulins and Noirot, 1972) and locusts (Louveaux, 1976). Many insects are known to 'inspect' food, conspecific insects and potential hosts with their antennae by contact, and although the antennae bear contact chemoreceptive sensilla, these organs have attracted only limited interest (Rence and Loher, 1977; Rüth, 1976).

The question of the phylogeny of contact chemoreceptive sensilla has been addressed by Altner and Prillinger (1980) and Bassemir and Hansen (1980). These authors studied four single-pored sensilla on the maxillary palp of a damselfly naiad which could be a phylogenetically old contact chemoreceptor sensillum. These sensilla lack any outer cuticular differentiation, the pore opening into an area of flat cuticle. These morphological studies are a beginning to the comparison of single-pored sensilla of insects of different phylogenetic orders.

1.3 METHODS OF INVESTIGATION

1.3.1 Ablation of sensory organs

An important method for ascertaining the role of specific contact chemical sensory input is the elimination of the sensory organ. Various techniques in conjunction with appropriate controls, such as unilaterally or sham-operated insects, have been used to monitor the possible negative effects of operations. Elimination of the sensory organs by removing sensilla or sensilla-bearing organs can be employed relatively easily, but many organs such as legs, ovipositor or proboscis obviously can not be removed without interfering with normal behavior. Several workers have employed acids to cauterize the sensory structures. For example, 5 N HCl effectively destroys the chemosensory cells in the tarsal sensilla of *Psila rosae* within five seconds (Städler, 1977) or individual sensilla can be destroyed with electro-coagulation (Blom, 1978; Clark, 1980). An elegant but difficult approach is to sever the sensory nerves as performed by Louveaux (1976) on *Locusta*.

1.3.2 Electrophysiology

Most investigations of contact chemoreceptive sensilla have used the tip-recording technique (Hodgson *et al.*, 1955; for a detailed description see Dethier, 1976). In this method, the stimulating solution in a glass capillary acting as an electrolytic bridge is used also as the recording electrode linked via an Ag/Ag Cl junction to the amplifier. An advantage of this technique is its simplicity. Since chemoreceptor responses are inherently variable due to differences between animals and sensilla, an important advantage of this technique is that it allows recording from different sensilla of a preparation in a relatively short period of time. The disadvantages of the technique are the following: (i) the nerve cell activity is recorded only during contact, which means that background activity before and after effects (rebound inhibition) following the removal of the stimulus can not be recorded; (ii) at the beginning of the recording artifacts can block the amplifier and, as a consequence, the first 100 ms or more of the recording may be lost. The first 100 ms of stimulation may contain enough information to trigger an adequate behavioral response (Dethier, 1968); (iii) the stimulating compounds have to be in a water phase, a serious limitation for the study of nonpolar chemicals (Blaney, 1974; Dethier, 1972); and (iv) both electrolytes and non-electrolytes influence the electrical conductivity of the stimulating and recording solution (Wolbarsht, 1958). This may result in a change in the amplitude of the nerve impulses which is a typical criterion for their identification (see examples in Fig. 1.1(a), (b)). When more than one cell of the sensillum is firing and when recordings of the stimulation with different compounds and extracts are compared this is especially troublesome.

Several solutions to the above difficulties have been proposed. To limit the contact artifact, special amplifiers have been designed (for example de Kramer and van der Molen, 1980a) and the use of a bucking voltage has been introduced by Wolbarsht (1958). Since chemoreceptors are sensitive to small electric currents, amplifiers with small bias currents should be used (<10 pA: de Kramer and van der Molen, 1977; Maes, 1977; Morita and Takeda, 1959; Wolbarsht, 1958).

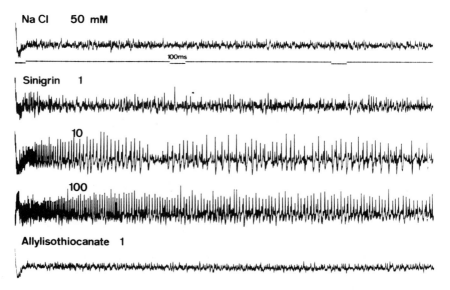

Fig. 1.3   Tip recordings from a tarsal D-hair of *Delia brassicae* (cabbage root fly) in response to stimulation with the glucosinolate sinigrin and its aglycon allylisothiocyanate which is not stimulatory.

To solve the problem of non-polar chemicals in water, Blaney (1975) used a suspension of leaf surface waxes in a 50 mM NaCl solution. Since our attempts to use this technique gave unclear results in the carrot rust fly, *Psila rosae*, we tested additional methods. The most promising results were obtained using 10% methanol extracts of leaf surfaces in a 100 mM NaCl solution as represented in Fig. 1.4 (Städler, unpublished). We found no indication of damage or stimulation by methanol in different fly species. These results are in agreement with those of Dethier (1951): the 50% behavioral rejection threshold for methanol is about 5 to 10 molar for the blowfly, *Phormia regina*, and the cricket, *Gryllus assimilis*.

Some of the recording and stimulating difficulties can be avoided by the side-wall recording technique (Dethier, 1976; Hanson, 1970; Morita and Yamashita, 1959; Rees, 1968). Examples of such recordings are given in Fig. 1.1. The problem with this method, which has prevented its wider use, is a technical one: (i) since normally only one sensillum of a preparation can be investigated, the variability between insects and sensilla cannot be separated; (ii) the technique itself creates inherent variability (Blaney, 1974).

The impulse activity of sensory neurons can be analyzed using computers. Different systems have been devised to simplify the task of discriminating between nerve impulses of different amplitudes and to produce spike frequency time histograms (van der Molen *et al.*, 1978). A simpler device modified from the design of McCook *et al.* (1975) was used to generate the histograms in Fig. 1.1. The question remains as to whether spike amplitude is the only useful parameter to discriminate between the nerve impulses of different cells (Frazier and Hanson, unpublished in Städler, 1982).

Because most receptor cells are not narrowly responsive to specific chemicals and more than one cell is located in a sensillum, the interpretation of recorded

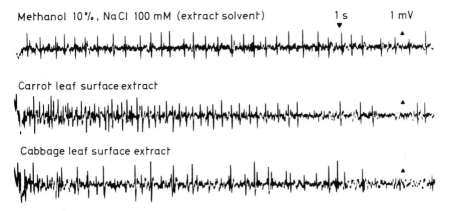

Fig. 1.4   Tip recordings from two tarsal D-hairs of *Psila rosae* (carrot rust fly) in response to methanolic leaf surface extracts (10% methanol in water with 100 mM NaCl).

responses, even to pure chemicals and especially to natural stimuli, is not always obvious. The differences in amplitude of the nerve impulses can sometimes, but not always, be used to distinguish between the activity of the different cells. Stimulation with pure compounds and mixtures of pure compounds have been used to identify different cells in terms of individual sensitivity spectra and to differentiate between them (for recent examples see Wieczorek, 1976; Städler, 1978). In addition, selective adaptation with one chemical followed by the stimulation with another compound (cross-adaptation) in a separate capillary can be used effectively as shown by Wieczorek (1976).

The use of interspike interval histograms and auto-correlations for the detection of temporal patterns in recorded responses of single cells has been described by Perkel *et al.* (1967) and has recently been applied by Dethier and Crnjar (1982) to the analysis of recordings from contact chemoreceptor cells of *Manduca sexta*.

1.4 PHYSIOLOGY

1.4.1 General sensory physiology

Our present knowledge of the physiological processes in chemosensory sensilla has been described in detail by Hansen (1978). In brief, the stimulating molecules reach the dendrites of sensory cells through the tip opening. This pore is filled with a viscous fluid which may also cover the tip of the sensillum. Bernays *et al.* (1975) found an acid mucopolysaccharide (hyaluronic acid) in the tip of the sensilla pores of the palps of *Locusta*. Both this viscous tip fluid (varying in quantity and quality) and presumed mechanical effects causing opening and closing of the pore (de Kramer and van der Molen, 1980b) may influence the diffusion of the stimulants to the dendritic membranes. The biological significance of these effects, which can be correlated with an increased electrical resistance and decrease of the sensory cell activity, was first shown by Bernays *et al.* (1972). Bernays and Chapman (1972) and Bernays and Mordue-Luntz (1973) showed that this mechanism causes a decrease of sensitivity lasting for up to 2 hours after feeding, is under hormonal control of the corpora cardiaca and is brought about by distension of the foregut. Changes of resistance and sensitivity in the blowfly (*Phormia regina*) sensory hairs owing to feeding with sucrose have also been reported by Angioy *et al.* (1979). However, Rachman (1979) found no changes in the sensitivity of labellar contact chemoreceptors or the behavioral threshold of the blowfly. Hall (1980) observed also a decrease in behavioral responsiveness after feeding but no decline in the peripheral sensitivity of the tarsi. Evidently, this sensitivity change is brought about by an inhibition of the feeding response in the CNS.

The dendritic outer segments are believed to contain receptor molecules which interact with appropriate stimulants. Attempts have been made to isolate the sugar receptor molecules (for recent progress see Hansen and Wieczorek, 1981; Kijima *et al.*, 1977; Tanimura and Shimada, 1981). Proteins with an enzymatic activity matching the characteristics of the sensitivity of the sugar-sensitive receptor cells are believed to be receptor proteins. The interaction of the stimulant with the receptor site(s) is currently thought to open ion channels (channel proteins), which allow a receptor current to flow as inferred from the measured receptor potentials (Hansen, 1978). The special cases of H_2O and salt receptor mechanisms are still open to debate (Wieczorek, 1980). The nerve impulse (spike) frequency produced by the generator, assumed to be located in or near the cell body, is a function of the receptor potential and thus of the stimulus strength (Fig. 1.3). The impulses are generated after a latency of a few milliseconds following the stimulus onset. The spike frequency reaches usually its maximum within the first 50–100 ms and declines rapidly to a constant discharge rate (Fig. 1.3). This adaptation to prolonged stimulation may last from several seconds to minutes after the stimulus has been removed, depending on the receptor cell type and the duration of the adaptation period (for further details see Thurm and Küppers, 1980).

The axons of the sensory receptor cells conduct the nerve impulses to the central nervous system. In the thoracic ganglion some second-order neurons connecting via synapses with the axons of tarsal receptor cells have been identified (Dethier, 1976; Rook *et al.*, 1980). The corresponding second-order neurons for the labellar sensory neurons are located in the suboesophagal ganglion (Dethier, 1976; van Mier *et al.*, 1980; Stocker and Schorderet, 1981).

1.4.2 Contact chemoreceptor perception of environmental chemicals

The receptor cells identified electrophysiologically (Table 1.1) can be compared with the chemical environments (sources of stimuli). The cells have been named according to the identified stimulus or stimuli to which the cell showed highest sensitivity. Such categorization is not without limitations because usually only a limited number of compounds can be tested on a single sensillum. Obviously, the narrower the sensitivity spectrum, the more precisely a cell can be labelled. The ideal description of the sensitivity of a cell would include not only the spectrum of compounds perceived but also a description of the stimulus-response or dose (concentration)-response curve (Wieczorek, 1976; Mitchell and Gregory, 1981). The interesting values are not only threshold concentrations but also the concentration for maximal receptor cell response, the concentration for half maximal response (K_b) and the slope of the curve at K_b (Hill coefficient). Such detailed descriptions of response curves are known for only a few chemoreceptor cells. Interesting examples have been provided by Wieczorek (1976) for the glucoside receptors of two different *Mamestra brassicae* strains.

Table 1.1 Contact chemoreceptive cells

Stimulus origin	Receptor sensitivity (perceived chemicals with low threshold)	Type of behavioral reaction*	Animal and sensory organ	Literature
Different (food sources):				
floral nectar	H$_2$O, salt, sugars	± feeding	Tarsi and labellum of Phomia regina	Dethier (1976); Shimada & Tanimura (1981); Gritsai (1978); Rachman (1980 a, b)
	Amino acids and small peptides by 'sugar' cell	+ feeding		
pollen	Salts of fatty acids by 'fifth or anion' cell	?		
honey dew	Albumin by 'sugar' and 'water' cell	+ feeding		
cadavers	Yeast extract by 'salt' cell	+ feeding	Tarsi and labellum of different Diptera	Angioy et al. (1978); Gothilf et al. (1971); Städler (1978); Tanimura & Shimada (1981)
	Sugars, salt	± feeding		
	Sugars, salt	± feeding	Tarsi, galea and labial palpi of Apis mellifera	Pappas & Larsen (1976, 1978; Whitehead & Larsen 1976; Whitehead (1978)
	Sugars	+ feeding	Galea of Choristoneura fumiferana	Städler & Seabrook (1975)
	Sugars, fatty acids, alcohols	+ feeding	Antennae of Periplaneta americana	Rüth (1976)
	Sugars	+ feeding	Maxillary palpi of Periplaneta americana	Wieczorek (1978)
Undamaged plant surfaces:				
Cruciferae	Mustard oil glucosides	+ oviposition	Tarsi of Pieris brassicae	Ma & Schoonhoven (1973)
Hyperiaceae	Mustard oil glucosides	+ oviposition	Tarsi of Delia brassicae	Städler (1978, Fig. 1.3)
	Hypericin (quinone)	+ feeding	Tarsi of Chrysomela brunsvicensis	Rees (1969)
	Salts, H$_2$O	− feeding		

Monocotyleoneae	Leafwaxes / Different deterrents	+ biting / – biting	Maxillary palpi of *Locusta migratoria*	Blaney (1975); Blaney & Duckett (1975) Blaney (1981)
Umbelliferae	Leaf surface extracts	+ oviposition	Tarsi of *Psila rosae*	Städler (unpubl., Fig. 1.4)
Plant interior (plant saps):				
Celastracede non-hosts	Dulcitol / Phloridzin, prunasin	+ feeding / – feeding	Maxilla of larval *Yponomeuta cagnagellus*	van Drongelen (1979, 1980)
Crassulaceae *Cruciferae*	Isocitric acid / Mustard oil glucosides	? / + biting, feeding	Maxilla of larval *Y. vigintipunctatus*	van Drongelen (1979)
	Sugars / Inositol, aminoacids	+ feeding / ?	Maxilla of larval *Pieris brassicae*	Ma (1972); Schoonhoven (1967, 1969)
+ different plants	Alkaloids (deterrents), salt	—	Maxilla and epipharyngeal organs of larval *Pieris brassicae*	Ma (1972)
Cruciferae	Mustard oil glucosides, salts, sugars	–, N feeding	Maxilla of larval *Mamestra brassicae*	Blom (1978); Wieczorek (1976)
+ different plants	Sugars and aminoacids by 'sugar' cell	+ feeding	Maxilla of larval	Mitchell & Gregory (1979)
Cruciferae		+ feeding	*Entomoscelis americana*	Albert (1980)
Gymnospermae	Sugars, proline	+ feeding	Maxilla of larval *Choristoneura fumiferana*	Ma (1977a); Ma & Kubo (1977)
Monocotyledonae	Sugars, adenosine	+ feeding	Maxilla of larval *Spodoptera exempta*	(Blaney (1981); (Winstanley & Blaney (1978); Mordue Luntz (1979)
Monocotyledonae + various plants	Sugars, alkaloids, acids, azadirachtin	+ feeding / – feeding	Maxillary palpi of adult *Locusta* and *Schistocerca*	van Drongelen (1979, 1980)
Rosaceae *Prunus* non-hosts	Prunasin, sorbitol / Phloridzin	N, + feeding / – feeding	Maxilla of larval *Y. evonymellus*	Schoonhoven (1981)
Malus	Sorbitol	? feeding	Maxilla of larval different Lepidoptera	

Table 1.1—continued

Stimulus origin	Receptor sensitivity (perceived chemicals with low threshold)	Type of behavioral reaction*	Animal and sensory organ	Literature
Solanaceae	Sugars, amino acids, chlorogenic acid	+ feeding	Maxilla of larval *Lepinotarsa decemlineata*	Mitchell (1974); Mitchell & Schoonhoven (1974)
Different plants	Aminoacids Alkaloids (deterrents)	+? feeding – feeding	Maxilla of larval Lepidoptera	Schoonhoven (1981); Clark (1981); Ma (1976b)
Insect hosts hemolymph	Unknown compound	+ oviposition	Ovipositor of *Pseudeucoila bochei*	Lammers (cited by van Lenteren 1981)
Mammalian hosts: skin, serum	Adenosine triphosphate	+ feeding	Labellum of *Glossina morsitans*	Galun (1976); Mitchell (1976)
+ nectar?	Salt, sugars, amino acids	?	Tarsi and proboscis of *Aedes aegypti*	Elizarov & Sinitsina (1974)
	Salt	? oviposition	Ovipositor of *Lucilia cuprina*	Rice (1976)
+ nectar?	Sugars	+ feeding	Tarsi and labellum of *Tabanus nigrovittatus*	Stoffolano (1980); Pietra & Stoffolano (1980)
Conspecific animals pheromones: eggs	Unknown compound(s)	– oviposition	Tarsi of *Pieris brassicae*	Behan & Schoonhoven (1978)
female markings	Unknown compound(s)	– oviposition	Tarsi of *Rhagoletis* species	Crnjar *et al.* (1978, 1982); Städler & Boller (unpubl.); Hurter *et al.* (1976)
			Tarsi of *Ceratitis capitata*	Städler & Boller (unpubl.)

* Explanations of behavioral reactions: + stimulation, – inhibition (deterrence), N no response at natural concentration

(a) Chemicals of general occurrence

Sugars

Sugar receptor cells have been identified in almost all contact-chemoreceptive sensilla. Exceptions are the tarsal hairs of different flies (Shiraishi and Tanabe, 1974; Städler, 1978) and *Pieris brassicae* (Ma and Schoonhoven, 1973). Another example is the epipharyngeal organ of *Manduca sexta* (De Boer *et al.*, 1977) which lacks a specific sugar receptor cell. However, in *Manduca* it has been shown that two cells sensitive to salts and presumably also to deterrents are inhibited by sucrose. In behavioral tests with maxillectomised larvae, the inhibition of the deterrent cells of the epipharyngeal organ correlated with an increased feeding response.

Hansen (1978) and Hansen and Wieczorek (1981) compared the sugar receptor cells of various insect species in detail. Most cells were found to be sensitive to glucose and sucrose (with the exception of lepidopteran larvae), and seem to have at least two receptor sites for pyranoses and furanoses. A closer examination of the sugar cells reveals remarkable differences between cells of different insects and even closely related species (Hansen and Wieczorek, 1981; Jakinovich *et al.*, 1981; Mitchell and Gregory, 1979). An explanation for these differences could lie in the adaptations to specific food sources and feeding behaviors. However, this is not evident in *Pieris rapae* when stimulatory activity and nutritive values are compared (Kusano and Sato, 1980; Mitchell, 1981). The classical blowfly water receptor also reacts to sugars. The receptor site in this cell is similar to the furanose site of the sugar receptor cell (Wieczorek, 1980).

Salts and water

Salt (cation)-receptive cells seem to be almost as common as sugar receptor cells (see Dethier, 1977a). The electrophysiological sensitivity thresholds to salt are usually about 100 mM, and in most insects higher concentrations are inhibitory. The question arises whether or not these cells (cation and anion receptor cells) are all really 'salt cells' in view of the fact that the rôle of salt recognition is not clear (Dethier, 1977a; Schoonhoven, 1968). Dethier (1977a) suggested that salt sensitivity may be an evolutionarily ancient common sensitivity for prevention of hypersalinity. Monitoring salt concentrations may also give information on the water content for plant feeding insects. High water content has proved advantageous for leaf-feeding lepidopteran larvae in terms of consumption and utilization of plant biomass (Scriber and Slansky, 1981; see also Scriber, Chapter 7). In mammalian parasites, salt perception may be involved in the localization of skin and initiation of probing.

A specific 'water' receptor cell has long been known to exist in the labellar hairs of different flies (Dethier, 1976). Although the activity of this cell releases drinking behavior in thirsty flies, the previously mentioned discoveries of Wieczorek (1980) question this designation.

Amino acids and proteins

As pointed out by Bernays and Chapman (1978), Dethier (1976), and McNeil and Southwood (1978) it has long been known that nitrogen is important for nutrition, especially for egg development. Thus it is surprising that only in phytophagous larvae have specific amino acid receptors been found. In contrast, in arthropods other than insects, receptor cells very sensitive to amino acids have been located (Bauer and Hatt, 1980). Specific protein receptors have not been identified, although Wallis (1961) recorded responses of labellar hairs of *Phormia regina* in response to hemoglobin and brain–heart infusions. The complex stimuli employed, however, do not allow a finite conclusion on the sensitivity to protein. Later Shiraishi and Kuwabara (1970) discovered that the classical sugar receptor cell is sensitive to some amino acids. Gritsai (1978) showed that the sugar cell in some sensilla responds to albumin. This protein also stimulated some of the 'water' receptor cells. These two cells enable *Musca domestica* to identify albumin as a protein source. In *Phormia regina*, the sensitivity of the water cell to water is inhibited by some amino acids (Gold-rich = Rachman, 1973) thus contributing to the fly's discriminatory ability. Rachman (1980a, b) recently showed that the salt (cation) cell responds to yeast extract and mediates a feeding response in flies fed only with sucrose which need protein for egg development. This correlation between feeding behavior and the chemosensory activity proves that highly specific receptor cells are not necessary for adapted feeding responses (protein intake).

Fatty acids and alcohols

In addition to the mechanoreceptor and the sugar-, water- and salt-sensitive cells, the chemoreceptive sensilla of the blowfly labellum contain a 'fifth' or 'anion' cell. The sensitivity spectrum of this cell is not well defined. Dethier and Hanson (1968) found that this cell responds to salts of fatty acids of C_5 to C_{14} chain length. Since natural foods like honey and apple also strongly stimulated this cell (Dethier, 1974a), it is evident that more investigation is required to characterize its sensitivity spectrum.

Some of the antennal contact chemoreceptive sensilla of *Periplaneta americana* contain cells sensitive to alcohols and fatty acids (Rüth, 1976); fruits stimulated the sugar and alcohol cell, whereas meat was perceived by the fatty acid cell.

(b) Undamaged plant surfaces

After phytophagous insects have located a potential host plant using visual and olfactory cues, contact with the plant surface seems to be essential before feeding or oviposition are initiated (Alfaro *et al.*, 1980; Blaney and Duckett, 1975; Calvert, 1974; Calvert and Hanson, 1981; Chapman, 1977; Ma and Schoonhoven, 1973; Röttger, 1978; Städler, 1976, 1977, 1980; see also Scriber, Chapter 7). Chemical analysis of the leaf surfaces with the exception of cuticular waxes (*n*-alkanes) are sparse (Chapman, 1977). Although the chemistry of the leaf surface is less complex than the interior, thus providing less information, many phytophagous insects can choose their host plants without the perception of the leaf interior. In host selection by phytophagous insects the nutritional quality and presence of secondary plant chemicals, as indicators of potential toxicity, are essential (Miller and Strickler, Chapter 6). It is unlikely that the plant surface contains enough information about the nutritional quality of the potential host plant, although there could be a correlation between the content of nutrients and allelochemicals. It seems likely, therefore, that the main chemical information derived from contact with the undamaged plant is the occurrence of secondary plant metabolites. This is exemplified by the tarsal chemoreceptor cells for mustard oil glucosides in *Pieris brassicae* and *Delia brassicae* (Fig. 1.3) and the receptor cell found in the tarsal hairs of the beetle *Chrysomela brunsvicensis* (Table 1.1). The special rôle of these receptor cells for host selection of the three species is indicated by Rees's (1969) finding that 'hypericine' cells do not occur in other *Chrysomela* species and that the male cabbage root fly in contrast to the female apparently has no receptor cell for mustard oil glucosides (Städler, 1978). In all three examples the sensitivity threshold is very low ($\sim 10^{-5}$M). Unfortunately, no data on the concentration of these chemicals on the leaf surfaces are available. Blaney's (1975) recordings of differential effects of leaf waxes from host and non-host plants are in accordance with the ability of locusts to differentiate during palpation.

Stürckow and Quadbeck (1958) and Stürckow (1959) recorded responses of contact chemoreceptor cells in tarsal hairs of the Colorado beetle, *Leptinotarsa decemlineata*, to host- and non-host plant extracts, alkaloid glucosides and salts. The observed reactions lack the typical phasic–tonic response which is, according to our present knowledge, typical of the functioning of most chemoreceptive cells. This raises the question whether or not the published recordings do show physiologically meaningful chemosensory reactions or if the observed volleys of nerve impulses occurring some time after contact were not signs of damage to the receptor cells.

Contact chemoreception of the dry plant surface is basically different from the well-studied perception of aqueous solutions. Since the stimulating compounds are in a dry state it may be that their distribution is not continuous and/or at a very low concentration. Thus not every contact between a sensillum tip and the plant would result in a stimulation. Although speculative, this may explain the specific behavior of various insect species after contacting plant leaves: two female flies (*P. rosae, D. brassicae*) perform so called 'oviposition runs,' on host leaves of about 20–30 s duration (Städler, 1977, 1978). Locusts make rapid vibrations (10–15 times per second) with the palps contacting the potential food plant surface. Blaney and Duckett (1975) showed that this palpation increases the amount of sensory input to the CNS by allowing disadaptation of chemoreceptors between contacts. This must also be true for the tarsal hairs of two flies mentioned above. Some butterflies use the fore tarsi to drum or tap the plant surface prior to oviposition (Ma and Schoonhoven, 1973;

Calvert, 1974; Calvert and Hanson, 1981; Feeny *et al.*, 1983). Again the function of this behavior may be analogous to palpation in the locust or 'running' in the flies.

Figure 1.4 shows examples of recordings from the tarsal, D-hairs of the carrot rust fly (*P. rosae*) in response to leaf surface extracts. The spike patterns show clear differences which may allow this fly to discriminate between its respective host and non-host plant. The same extracts also allow discrimination in an oviposition bioassay in the carrot rust fly (Städler, 1977). Although the non-host leaf extracts are not repellent or deterrent (compared with solvent), these extracts are also perceived. The different cell types observed in the recordings remain to be identified using isolated compounds.

(c) Plant interior (plant saps)

In contrast to the leaf surfaces much more is known about the chemistry of the leaf interior, which is probably considerably more complex. These compounds representing nutrients, toxic allelochemicals and allelochemicals with a host plant sign character, influence feeding behavior (see Scriber, Chapter 7). As can be seen from Table 1.1, contact chemoreceptors for all types of compounds have been identified in a variety of insect species. With the exception of the two beetle larvae and the locust, all of the examples are lepidopteran larvae mainly because the two pairs of sensilla styloconica on the galea (maxilla) of these larvae can be investigated easily (Schoonhoven and Dethier, 1966). The relatively small number of contact chemoreceptors (four sensilla with four neurons) and their importance for food plant discrimination (Hanson and Dethier, 1973) make them ideal for the study of contact chemoreception in relation to host plant selection.

Nutrients

The rôle of nutrients in inducing feeding is indicated by the presence of sugar and amino acid receptors. Receptor cells for lipids (sterols, phospholipids) have not yet been identified despite their rôle in nutrition (Beck and Schoonhoven, 1980; House, 1974) and host plant selection (Städler and Hanson, 1976, 1978). This is probably the result of the methodology (see Section 1.3) and not a general lack of sensitivity of contact chemoreceptors.

Nucleosides

A recently discovered specific receptor for the nucleoside adenosine and adenine by Ma (1977a) and Ma and Kubo (1977) was not unexpected, as Hsiao (1969) identified both compounds as feeding stimulants for the alfalfa weevil, *Hypera postica*. This receptor cell, identified in the lateral sensillum styloconicum of *Spodoptera exempta* larvae, perceives adenosine that occurs in host plants and its activity is correlated with feeding. Adenosine is known to occur in all living tissues and presumably *Spodoptera* larvae, as other insects, have no specific nutritional requirement for this compound. The question arises if this sensitivity is unique to *Spodoptera* larvae and why such a chemoreceptor has evolved. Since nucleotides and nucleosides occur at relatively high concentrations in growing tissue it may be speculated that such receptor cells could influence the preference for such tissue which is certainly also rich in nutrients.

Inositol
Another chemoreceptor cell with a sensitivity which cannot be easily explained is the inositol receptor found in many lepidopteran larvae (Schoonhoven, 1972a). Most insects appear to synthesize myo-inositol from glucose, thus having no nutritional requirement for this compound. One known exception is the larva of the moth, *Heliothis zea*, which is dependent upon dietary inositol (Chippendale, 1978). For some larvae inositol in pure form is a feeding stimulant (Blom, 1978, Städler and Hanson, 1978) but others such as *S. exempta* show no behavioral response, despite the fact that inositol occurs in its food plants and that it has a receptor cell for the compound (Ma, 1976a).

'Host-plant sign stimuli'
The first known example of a contact chemoreceptor cell sensitive to a sign or token stimulus was the mustard oil glucoside receptor cell of *P. brassicae* discovered by Schoonhoven (1967). These glucosides stimulate biting and synergize feeding stimulation with sucrose at a low concentration (Ma, 1972; Blom, 1978). Sensory and behavioral thresholds have been found to be in the same order of magnitude ($\sim 10^{-5}$M, sinigrin).

Receptor cells sensitive to host plant compounds that stimulate feeding have also been identified in several species of *Yponomeuta* moths. *Y. cagnagellus* feeding on *Euonymus* has one dulcitol (sugar)-sensitive cell in the medial and one in the lateral sensillum styloconicum (van Drongelen, 1979, 1980). Another species, *Y. evonymellus*, is also stimulated by dulcitol at the sensory and behavioral level even though the compound does not occur in its host plant. This contradictory result may be explained by the possible presence of dulcitol in an ancestral host plant (van Drongelen, 1979, 1980).

The correlation of sensivity to sorbitol (occurring in *Rosaceae*) and feeding on plants of the *Rosaceae* in the different *Yponomeuta* species suggests that this compound is also a feeding stimulant with a 'chemical host-plant sign stimulus' value (Table 1.1; Schoonhoven, 1981).

Deterrents from host plants
Inhibitory compounds may be at first unexpected in host plants, but such examples are known (Schoonhoven and Jermy, 1977; Schoonhoven, 1981). Städler and Hanson (1978) isolated fractions from tomato leaves which were deterrent to *M. sexta* although tomato is a preferred host plant. The compound or compounds for this deterrency and the corresponding receptors have not yet been identified. Further indirect evidence for inhibitory compounds occurring naturally in host plants may be found in examples of successful breeding of plants showing a non-preference type of resistance (Beck and Schoonhoven, 1980; Schoonhoven, 1981).

Some species possess contact chemoreceptors sensitive to specific host plant components which, at natural concentrations, release no behavioral reaction. For example, *Y. evonymellus* does not respond behaviorally to prunasin (cyanogenic glycoside) which occurs in its host plant. This glycoside can have a host-plant sign value, however, for other *Yponomeuta* species not feeding on *Prunus*, as they are both sensitive and deterred by this compound. Another example is the receptor cell of *M. brassicae* with a threshold close to the highest concentration of glucosinolates found in one of its host plants, the *Cruciferae*. No behavioral response occurred at the concentration found in most leaves of *Brassicas* (Blom, 1978). Since the same cell is sensitive to typical, toxic deterrents, the specifically low sensitivity to glucosinolates can contribute to some host-plant specificity in this species. The sensitivity to host plant deterrents may further allow phytophagous insects to choose leaves with a low content of these toxic allelochemics.

Deterrents from non-host plants
In many phytophagous insects the host plant spectrum is decisively determined by the botanical distribution of chemicals acting as feeding or oviposition deterrents (Dethier, 1980a; Jermy, 1966; Jermy and Szentesi, 1978; Schoonhoven and Jermy, 1977). Contact chemoreceptors sensitive to a wide variety of compounds have been identified in all species of Lepidoptera where such cells have been looked for (Schoonhoven, 1972a). Often more than one cell in various organs (sensilla styloconica, epipharyngeal organ) has been identified.

A special type of deterrent effect was shown by Ma (1977b). A sesquiterpene dialdehyde isolated from the warburgia plant blocks the sugar receptor cell of *S. exempta* for 10–20 min after two contacts of 3 min. This inhibitory effect on the sensory neuron has been correlated with a suppression of the feeding response to sucrose. Similarly, Kennedy and Halpern (1980) showed that ziziphin, a cyclic peptide alkaloid of *Rhamnaceae*, also temporarily inhibits the fly sugar (but not the salt) receptor.

The adaptive value of deterrent receptor cells in preventing toxication by allelochemics in non-host plants seem to be evident. Other examples which, in terms of our present knowledge, are more difficult to explain on the basis of selective value, are deterrent cells sensitive to allelochemics that do not occur in the insects' normal habitat (Schoonhoven, 1981). Possible explanations could be (i) a redundancy from earlier phylogeny; and, (ii) a still unknown chemical similarity of the used stimulating deterrent with compounds present in the insect's natural environment.

(d) Animal hosts

The few studies of contact chemoreceptors of arthropod parasites of mammals have led to the identification of receptors predicted from behavioral observations (reviewed by Galun, 1976). The most significant discovery in this regard was the localization of an ATP receptor in the labellar sensilla of the tsetse fly by Mitchell (1976). ATP, ADP and UTP have long been known to be a phago-stimulant for different hematophagous insects (Smith and Friend, 1972; Galun, 1976). Many behavioral studies revealed the importance of chemicals in the location and recognition of hosts or their environment. The rôle of contact chemoreceptors for the selection of mammalian hosts has not yet been investigated. Galun (1976) believes that the olfactory sense is much more important in this respect.

The chemoreceptors of insect parasites and predators have not as yet attracted much attention. The first chemoreceptors sensitive to the hemolymph of the host insect have been identified by Lammers (cited in van Lenteren, 1981) on the ovipositor of a parasitic wasp (Table 1.1). Many behavioral observations show in addition that contact chemoreceptors must be involved in the host selection of highly specialized parasites (Rotheray, 1981; further examples are given by Vinson, Chapter 8).

(e) Conspecific animals

Research on pheromone chemoreception in general has been dominated so far by the study of volatile signals (see Chapters 9, 12 and 13) and the corresponding receptors. It is only in recent years that the rôle of the non-volatile chemical 'markings' and contact sex pheromones (Huyton *et al.*, 1980; Rence and Loher, 1977; Schlein *et al.*, 1981) have received attention.

Non-volatile pheromones exist, but their chemical isolation and identification has largely just begun. Consequently, few specific receptor cells have been characterized. Using behaviorally active extracts, contact chemoreceptive sensilla with sensitive cells have been identified (Table 1.1). In Fig. 1.5, representative recordings from a tarsal D-hair of *Rhagoletis cerasi* are presented. Only artificial fruits which were contacted by females can stimulate a receptor cells, and thus activity due to food residues can be excluded. Similar recordings were also obtained from D-hairs of *Ceratitis capitata* (Städler and Boller, unpublished).

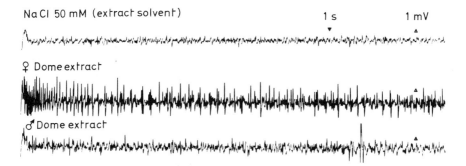

NaCl 50 mM (extract solvent)   1 s   1 mV

♀ Dome extract

♂ Dome extract

Fig. 1.5 Tip recordings from a tarsal D-hair of *Rhagoletis cerasi* (cherry fruit fly) in response to surface extracts of fruits with female pheromone markings and of fruits which had only contact with an equal number of males.

Prokopy *et al.* (1982) used both electrophysiological and behavioral techniques to identify the source of production of a fruit-marking pheromone in female *Rhagoletis pomonella*. This seems to be the first case in which electrophysiological techniques have been used to monitor the activity of different extracts of a non-volatile pheromone. Provided the correlations with the behavioral bioassays have been established, this approach seems very promising since the activity of small amount of extracts or fractions can be determined in a short time. This technique is analogous to the electroantennogram recordings performed in conjunction with gas–liquid chromatographic analysis of volatiles (see Mustaparta, Chapter 2).

Since the contact pheromones are spread as small droplets over the fruit surface, the contact chemoreception process is similar to the earlier mentioned perception of leaf surfaces. Females have also been observed running over the fruit surface for some time before oviposition occurs (Prokopy, 1981; see also Chapter 11). It can be assumed that this sampling of the dry surface does increase the sensory input to the CNS by the multiple contacts of the tarsal hairs during these inspection runs. Presumably, this type of multiple stimulation over time is comparable to the simultaneous perception of volatile pheromones with multiple sensory hairs on the antennae. Both the multiple stimulation of a limited number of sensilla and the multitude of sensitive sensilla will result in a low perception threshold for chemical signals occurring at a low concentration in the environment.

1.5 SENSORY CODING – CENTRAL PROCESSING

The generated spike activity of the different sensory cells in response to stimuli is decoded by the central nervous system (CNS). To investigate the sensory code of the receptor cells of a sensillum or several sensilla, different approaches can be used. The most direct method is to record from central neurons. In contrast to the investigation of central neurons responding to olfactory stimulations (Chapter 2), such recordings have been performed only recently from interneurons of the thoracic ganglion of the blowfly in response to the stimulation of tarsal contact chemoreceptive sensilla (Rook *et al.*, 1980).

A second possible method for investigating the decoding mechanisms of the CNS is the simultaneous recording from sensory and motor neurons or muscles. Using this technique with *P. regina*, Getting (1971) and Fredman (1975) studied the proboscis extension reflex and Pollack (1977) the labellar lobe spreading in response to the stimulation of the labellar hairs. The authors were able to confirm and extend earlier behavioral studies of the adaptation in the periphery and the CNS, the interaction (summation and inhibition) of the sensory input from different sensilla and receptor cells, and also the influence of the condition of the CNS (excitation, hunger; for more details see Dethier, 1976).

A third approach for elucidating the coding of sensory qualities and quantities is to correlate responses of sensory neurons and the behavioral reactions released by the same chemical stimuli. But correlations do not prove causative relations and they cannot replace studies of the CNS and its decoding mechanisms. Such attempts will have to be based on a thorough understanding of the inputs (sensory nervous activity) and the outputs (motor programs as units of behavioral reactions) of the central nervous structures.

Our understanding of sensory coding is far from complete. The main difficulties have been that the chemistry of natural stimuli remains unknown and thus the essential chemical components are unavailable. An additional problem is the variability of the sensory responses between animals and sensilla (Blaney, 1974; Nederstigt and van der Molen, 1980; Schoonhoven, 1977c). Successive stimulations of the same sensillum (Dethier, 1974b), should allow us and presumably the insect's CNS, to recognize sensory codes (Dethier and Crnjar, 1982). Additional sources of sensory variability such as age and feeding history have been identified and discussed by Dethier (1974b), Schoonhoven (1976, 1977a, b, c), Städler (1976) and Städler and Hanson (1976, 1978) (see also Section 1.3). The evidence for individually different sensory codes in animals of the same age and feeding history raises the question of the causal mechanisms (Schoonhoven, 1977b). The existence of strains with different sensitivities for glucosinolates in *M. brassicae* (Wieczorek, 1976) points to genetic determination. The genetics of chemoreceptive sensitivity may be a new insight into sensory physiology, as has been shown by crossing two *Yponomeuta* species with different sensitivities (van Drongelen and van Loon, 1980). Another promising approach seems to be the analysis of contact chemoreception in different strains and mutants of *Drosophila melanogaster* (Falk, 1979; Isono and Kikuchi, 1974; Siddiqi and Rodrigues, 1980; Tanimura and Shimada, 1981).

Boeckh (1980) and Dethier and Crnjar (1982) have discussed currently proposed and identified sensory coding mechanisms. Three basic types of systems have been distinguished: (i) labelled lines; (ii) temporal patterns, and (iii) across fiber patterns. Possibly combinations of these coding systems may exist as well.

Specific receptor cells with a narrow well-defined sensitivity spectrum are prerequisite for labelled lines. This spectrum of perceived compounds defines the quality of the sensory message. The quantity or intensity is coded in the firing rate of the sensory neuron. Impressive examples using feeding behavior to understand the sensory coding in labelled lines (deterrent, sugar, phagostimulant) have been documented in the blowfly by Dethier (review, 1976) and Rachman (= Goldrich, 1973; 1980a, b) and in caterpillars by Ma (1972), Blom (1978), and van Drongelen (1980). Further examples are represented by tarsal chemoreceptor cells that perceive identified compounds of host plants which stimulate oviposition (Ma and Schoonhoven, 1973; Rees, 1969; Städler, 1978). The identification of these 'labelled line' contact chemoreceptors has been based on pure chemicals and mixtures which are known to influence the behavior of the insect studied.

A remarkable difference between the sensory coding of individual olfactory and contact chemoreceptor cells is the much lower spontaneous activity in the unstimulated state of contact chemoreceptors (Fig. 1.1 a, b). As a consequence the contact chemoreceptor neurons can only increase the firing rate in response to a stimulus, whereas in olfactory neurons the firing rate can be both increased and inhibited by different chemicals (see also Mustaparta, Chapter 2). The reason for this difference is not known, but it may be that this limit in the coding ability of contact chemoreceptors is not as important as it may appear. In fact, evidence presented earlier for the inhibition of sugar receptor cells by deterrents and vice versa shows that a background activity could be generated by a universally occurring compound which could be modulated by other components of natural stimuli.

Temporal patterns within a single labelled line can be viewed as a stimulus-specific sequence of spike intervals during the tonic portion of the response of the sensory neuron (for further explanation see Perkel *et al.*, 1967). Dethier and Crnjar (1982) have presented evidence for such patterns in the response of a contact chemoreceptor cell of *M. sexta* larvae. An alternative to temporal coding may exist in a stimulus-specific length of the phasic portion of the spike train after stimulation, as has been suggested by Städler and Seabrook (1975) and Maes (1980). Since together with these temporal patterns other coding mechanisms have been discovered, it cannot be decided which of the identified codes are used by the CNS for the decoding process.

The concept of across-fiber patterns in sensory coding was first recognized by vertebrate sensory physiologists who found that, in many preparations, the responses of chemoreceptor cells were non-specific. This means that the individual neurons were found to have widely overlapping reaction spectra. Further, the activity of individual cells could not be readily linked to a defined behavioral reaction. Across-fiber patterns and the corresponding decoding process in the CNS can be viewed as an evolutionary solution to the problem of perceiving the very complicated chemical stimuli using a limited number of chemoreceptor cells. This coding mechanism is probably the physiological basis of the principle of 'Gestalt' perception as postulated long ago by ethologists (Tinbergen, 1951) for the integration of different environmental key stimuli in the CNS. Atema *et al.* (1977) used the appropriate term 'chemical search image' to describe chemoreception of food odors in fish. Schoonhoven (1977a) compared the across-fiber pattern with a complex multidimensional key and the central decoding mechanism with a complementary complex lock. Theories and experimental evidence for across-fiber patterns in insect contact chemoreception have been presented by Blaney (1975, 1981), and Blaney and Winstanley (1980) for locusts by Rüth (1976) in the cockroach and by Dethier (1973, 1974a, 1977b), Dethier and Crnjar (1982), van Drongelen *et al.* (1978), and Schoonhoven (1972a, b, 1977a) for lepidopteran larvae.

An important function of across-fiber patterns in lepidopteran larvae is the coding of quality Dethier and Crnjar (1982). Quality does not represent 'acceptable or unacceptable' but instead is a characteristic sensory impression of a particular plant sap. This has been concluded from the fact that the across-fiber patterns of different acceptable and unacceptable plant species are typical but not identical. The concentration (quantity) of the stimuli appeared to be of secondary importance since diluted saps evoked across-fiber patterns indistinguishable from those of the original plant saps.

The question arises whether the behavioral and physiological adaptations of insects to their chemical environment are reflected in different ways of processing sensory information. Part of this question has been answered by the attempt in Table 1.1 to classify the identified receptor cells according to the ecological environment. It seems that the more important a single pure chemical is, the more likely it is that a labelled line for this stimulus exists. If the important 'chemical message' is complex (as in leaf saps) it seems that across-fiber patterns prevail. However, this conclusion is tentative since we do not yet have enough data on both the chemistry of natural stimuli influencing insect behavior and the corresponding sensory physiology.

1.6 EVOLUTION OF CONTACT CHEMORECEPTION

The accumulated knowledge about insect chemoreceptors raises questions about the causes of the observed distribution and function of chemoreceptors. These questions focus on the problem of the evolution of the sensory organs and the corresponding central brain structures.

Hypotheses on possible evolutionary developments have been based on three aspects: (i) Chapman (1980) compared 25 insect species, using species of differing evolutionary age and known feeding habits; (ii) Dethier (1973, 1980b) and Dethier and Kuch (1971) compared insects with different and similar food preferences (phytophagous larvae and flies) with special reference to deterrence; and (iii) Schoonhoven (1981) discussed comparative investigations of host selection and chemoreception in lepidopteran larvae in general and the closely related *Yponomeuta* species in particular (Gerrits-Heybroek *et al.*, 1978; van Drongelen, 1979; van Drongelen and Povel, 1980).

Chapman (1980) found that there are fewer chemoreceptor sensilla and neurons in evolutionarily 'younger' orders of the *Endopterygota* and *Hemipteroidea* than in the 'older' orders of the *Apterygota* and *Orthopteroidea*. This reduction in numbers is probably also reflected in a reduction in the complexity of the neuronal architecture of the CNS. In the British *Orthopteroidea* relatively few species with specialized feeding habits (number of plant families) have been found. In contrast, 38–60% of the British *Hemipteroidea* and *Endopterygota* feed on a single genus. Further evidence for Chapman's supposition comes from studies of Blaney (1981) and Blaney and Winstanley (1980) on sensory coding in the contact chemoreceptors of locusts: labelled line coding seems to be largely lacking and a more quantitative type of coding seems to occur. This is remarkably different from what has been observed in the *Endopterygota* species.

Schoonhoven's conclusion can be viewed as a micro-evolutionary complement to macro-evolutionary traits as pointed out by Chapman (1980). Schoonhoven (1981) found that the diversity of chemoreceptors of closely related *Yponomeuta* species correlated with the shift in food plants, which can be an important step in speciation (Bush, 1975). Both Dethier (1980b) and Schoonhoven (1974) assume that the present chemoreceptor cells which are sensitive to a wide variety of compounds evolved from a basic type of chemoreceptor. It is tempting to believe that the early arthropods had one basic type of cell which, like unicellar organisms, had the ability to react both to attractive food components and to toxic deterrents. This would include the possibility that the same chemical could release opposite reactions depending on the concentration. If the epipharyngeal sensilla of caterpillars (*M. sexta* − de Boer *et al.*, 1977) and the sensilla on the palp of *Schistocerca* (Blaney, 1981) are viewed as primitive, evolutionarily old sensory organs, it could be inferred that the original receptors were excited (depolarization, increase of spike frequency) by deterrents and inhibited by sugars. Unfortunately, this assumption cannot readily explain how such cells could develop into specific sugar receptor cells which are excited with increasing sugar concentrations. However, the observation of Morita *et al.* (1977) that the fly sugar receptor cell is inhibited by the universal deterrent quinine seems to support the idea that this sugar receptor retained a sensitivity to deterrents during the evolution.

As pointed out by Schoonhoven (1981), it remains to be investigated to what extent the evolutionary changes in the discriminatory ability are based on the chemosensory organs and to what extent on the capability of the CNS. Answers to these questions may be found by further studies of the chemoreception of host races and comparative investigations on the genetics of behavior and receptor physiology. If the 'Gestalt' concepts for sensory perception is accepted, it has to be concluded that all types of sensory input, not only contact chemoreception, should be considered for such investigations. This emphasizes the need for a multidisciplinary and comparative approach to the study of the chemistry of natural stimuli, chemoreception and its evolution.

ACKNOWLEDGEMENTS

It is a pleasure to thank Drs William J. Bell, Elizabeth A. Bernays, Walter M. Blaney, Vincent G. Dethier, Kai Hansen, Frank E. Hanson, Louis M. Schoonhoven, John C. Stoffolano and Helmut Wieczorek for discussing and improving, and Michelle Flury for typing the manuscript.

REFERENCES

Albert, P. J. (1980) Morphology and innervation of mouthpart sensilla in larvae of the spruce budworm, *Choristoneura fumiferana* (Clem.) (*Lepidoptera: Tortricidae*). *Can. J. Zool.*, **58**, 842–51.

Alfaro, R. T., Pierce Jr., H. D., Borden, J. H. and Oehschlager, A. C. (1980) Role of volatile and nonvolatile components of Sitka spruce bark as feeding stimulants for *Pissodes strobi* (Peck) (*Coleoptera: Curculionidae*). *Can. J. Zool.*, **58**, 626–32.

Altner, H. (1977) Insect sensillum specificity and structure: An approach to a new typology. In: *Olfaction and Taste*, Vol. VI (Le Magnen, J. and Macleod, P., eds). pp. 295–303, Information Retrieval, London.

Altner, H. and Prillinger, L. (1980) Ultrastructure of invertebrate chemo-, thermo-, and hygroreceptors and its functional significance. *Int. Rev. Cytol.*, **67**, 69–139.

Angioy, A. M., Liscia, A. and Pietra, P. (1978) The electrophysiological response of labellar and tarsal hairs of *Dacus oleae* (Gemel.) to salt and sugar stimulation. *Boll. Soc. Ital. Biol. Sper.*, **54**, 2115–21.

Angioy, A. M., Liscia, A. and Pietra, P. (1979) Influence of feeding conditions on wing, labellar and tarsal hair resistance in *Phormia regina* (Meig.). *Experientia*, **35**, 60–61.

Atema, J., Holland, K. and Ikehara, W. (1980) Olfactory responses of yellowfin tuna (*Thunnus albacares*) to prey odors: Chemical search image. *J. Chem. Ecol.*, **6**, 457–65.

Bauer, U. and Hatt, H. (1980) Demonstration of three different types of chemosensitive units in the crayfish claw using a computerized evaluation. *Neurosci. Lett.*, **17**, 209–14.

Bassemir, U. and Hansen, K. (1980) Single-pore sensilla of damselfly-larvae: representatives of phylogenetically old contact chemoreceptors? *Cell Tissue Res.*, **207**, 307–20.

Beck, S. D. and Schoonhoven, L. M. (1980) Insect behavior and plant resistance. In: *Breeding Plants Resistant to Insects* (Maxwell, F. G. and Jennings, P. R., eds) pp. 115–35, John Wiley & Sons, New York.

Behan, M. and Schoonhoven, L. M. (1978) Chemoreception of an oviposition deterrent associated with eggs in *Pieris brassicae*. *Ent. exp. appl.*, **24**, 163–79.

Behrend, K. (1971) Riechen in Wasser und in Luft bei *Dytiscus marginalis* (L.). *Z. vergl. Physiol.*, **75**, 108–22.

Bernays, E. A., Blaney, W. M. and Chapman, R. F. (1972) Changes in chemoreceptor sensilla on the maxillary palps of *Locusta migratoria* in relation to feeding. *J. exp. Biol.*, **57**, 745–53.

Bernays, E. A. and Chapman, R. F. (1972) The control of changes in peripheral sensilla associated with feeding in *Locusta migratoria* (L.). *J. exp. Biol.*, **57**, 755–63.

Bernays, E. A. and Mordue-Luntz, A. J. (1973) Changes in the palp tip sensilla of *Locusta migratoria* in relation to feeding: The effects of different levels of hormone. *Comp. Biochem. Physiol.*, **45A**, 451–4.

Bernays, E. A., Blaney, W. M. and Chapman, R. F. (1975) The problems of perception of leaf-surface chemicals by locust contact chemoreceptors. In: *Olfaction and Taste*, Vol. V (Denton, D. A. and Coghlan, J. P., eds) pp. 227–9, Academic Press, New York.

Bernays, E. A. and Chapman, R. F. (1978) Plant chemistry and acridoid feeding behaviour. In: *Biochemical Aspects of Plant and Animal Coevolution* (Harborne, J. B., ed.), Academic Press, London, pp. 99–141.

Blaney, W. M. (1974) Electrophysiological responses of the terminal sensilla on the maxillary palps of *Locusta migratoria* (L.) to some electrolytes and non-electrolytes. *J. exp. Biol.*, **60**, 275–93.

Blaney, W. M. (1975) Behavioural and electrophysiological studies of taste discrimination by the maxillary palps of larvae of *Locusta migratoria* (L.). *J. exp. Biol.*, **62**, 555–69.

Blaney, W. M. (1981) Chemoreception and food selection in locusts. *Trends in Neurosciences*, February 1981, 35–8.

Blaney, W. M. and Duckett, A. M. (1975) The significance of palpation by the maxillary palps of *Locusta migratoria* (L.): An electro-physiological and behavioural study. *J. exp. Biol.*, **63**, 701–712.

Blaney, W. M. and Winstanley, C. (1980) Chemosensory mechanism of locusts in relation to feeding: the role of some secondary plant compounds. In: *Insect Neurobiology and Pesticide Action* (Neurotox 79), pp. 383–9. Chemical Industry, London.

Blom, F. (1978) Sensory activity and food intake: A study of input-output relationships in two phytophagous insects. *Netherlands J. Zool.*, **28**, 277–340.

Boeckh, J. (1980) Neural basis of coding of chemosensory quality at the receptor cell level. In: *Olfaction and Taste*, Vol. VII (van der Starre, H., ed.) pp. 113–22. Information Retrieval, London.

Boer de, G., Dethier, V. G. and Schoonhoven, L. M. (1977) Chemoreceptors in the preoral cavity of the tobacco hornworm, *Manduca sexta*, and their possible function in feeding behaviour. *Ent. exp. appl.*, **21**, 287–98.

Bush, G. L. (1975) Sympatric speciation in phytophagous parasitic insects. In: *Evolutionary Strategies of Parasitic Insects and Mites* (Price, P. W., ed.) pp. 187–206. Plenum Press, New York.

Calvert, W. H. (1974) The external morphology of foretarsal receptors involved with host discrimination by the nymphalid butterfly, *Chlosyne lacinia. Ann. ent. Soc. Am.*, **67**, 853–6.

Calvert, W. H. and Hanson, F. E. (1981) The role of sensory structures and preoviposition behavior in oviposition by the Texas Checkerspot butterfly, *Chlosyne lacinia* (Geyer) (*Lepidoptera: Nymphalidae*) (in preparation).

Chapman, R. F. (1977) The role of the leaf surface in food selection by Acridids and other insects. In: *Colloques Internationaux CNRS. Comportement des insectes et milieu trophique*, **265**, 133–49.

Chapman, R. F. (1980) The evolution of insect chemosensory systems and food selection behavior. In: *Olfaction and Taste*, Vol. VII (van der Starre, H., ed.) pp. 131–3. Information Retrieval, London.

Chapman, R. F. and Blaney, W. M. (1979) How animals perceive secondary compounds. In: *Herbivores. Their Interaction with Secondary Plant Metabolites* (Rosenthal, G. A. and Janzen, D. H., ed.) pp. 161–98. Academic Press, New York.

Chippendale, G. M. (1978) The functions of carbohydrates in insect life processes. In: *Biochemistry of Insects* (Rockstein, M., ed.) pp. 1–55. Academic Press, New York.

Clark, J. V. (1980) Changes in the feeding rate and receptor sensitivity over the last instar of the African armyworm, *Spodoptera exempta. Ent. exp. appl.*, **27**, 144–8.

Clark, J. V. (1981) Feeding deterrent receptors in the last instar African armyworm *Spodoptera exempta*: a study using salicin and caffein. *Ent. exp. appl.*, **29**, 189–97.

Crnjar, R. M., Prokopy, R. J. and Dethier, V. G. (1978) Electrophysiological identification of oviposition-deterring pheromone receptors in *Rhagoletis pomonella* (*Diptera: Tephritidae*). *J. N. Y. Ent. Soc.*, **86**, 283–4.

Crnjar, R. M. and Prokopy, R. J. (1982) Morphology and electrophysiological mapping of oviposition-deterring pheromone receptors in *Rhagoletis pomonella. J. Insect Physiol.* **28**, 393–400.

Dethier, V. G. (1951) The limiting mechanism in tarsal chemoreception. *J. gen. Physiol.*, **35**, 55–65.

Dethier, V. G. (1968) Chemosensory input and taste discrimination in the blowfly. *Science*, **161**, 389–91.

Dethier, V. G. (1970) Chemical Interactions between plants and insects. In: *Chemical Ecology* (Sondheimer, E. and Simeone, J. B., eds) pp. 83–102. Academic Press, New York.

Dethier, V. G. (1972) Sensitivity of the contact chemoreceptors of the blowfly to vapors. *Proc. Nat. Acad. Sci., U.S.A.*, **69**, 2189–92.

Dethier, V. G. (1973) Electrophysiological studies of gustation in lepidopterous larvae. II Taste spectra in relation to food plant discrimination. *J. comp. Physiol.*, **82**, 103–34.

Dethier, V. G. (1974a) The specificity of the labellar chemoreceptors of the glowfly and the response to natural foods. *J. Insect Physiol.*, **20**, 1859–69.

Dethier, V. G. (1974b) Sensory input and the inconstant fly. In: *Experimental Analysis of Insect Behaviour* (Barton Browne, L., ed.) pp. 21–31. Springer, Berlin.

Dethier, V. G. (1975) The monarch revisited. *J. Kansas Ent. Soc.*, **48**, 129–40.

Dethier, V. G. (1976) *The Hungry Fly*. Harvard University Press, Cambridge, Mass.

Dethier, V. G. (1977a) The taste of salt. *Am. Sci.*, **65**, 744–51.

Dethier, V. G. (1977b) Gustatory sensing of complex mixed stimuli by insects. In: *Olfaction and Taste*, Vol. VI (Le Magnen, J. and MacLeod, P., eds) pp. 323–31. Information Retrieval, London.

Dethier, V. G. (1980a) Food-aversion learning in two polyphagous caterpillars, *Diacrisia virginica* and *Estigmene congrua*. *Physiol. Ent.*, **5**, 321–5.

Dethier, V. G. (1980b) Evolution of receptor sensitivity to secondary plant substances with special reference to deterrents. *Am. Nat.*, **115**, 45–66.

Dethier, V. G. and Hanson, F. E. (1968) Electrophysiological responses of the chemoreceptors of the blowfly to sodium salts of fatty acids. *Proc. Nat. Acad. Sci. USA*, **60**, 1296–303.

Dethier, V. G. and Schoonhoven, L. M. (1968) Evaluation of evaporation by cold and humidity receptors in caterpillars. *J. Insect Physiol.*, **14**, 1049–54.

Dethier, V. G. and Kuch, J. H. (1971) Electrophysiological studies of gustation in Lepidopterous larvae. I Comparative sensitivity to sugars, amino acids, and glycosides. Z. vergl. Physiol., **72**, 343–63.

Dethier, V. G. and Crnjar, R. M. (1982) Candidate codes in the gustatory system of caterpillars. *J. gen. Physiol.*, **79**, 549–69.

Eaton, J. L. (1979) Chemoreceptors in the cibario-pharyngeal pump of the cabbage looper moth, *Trichoplusia ni* (*Lepidoptera: Noctuidae*). *J. Morphol.*, **160**, 7–16.

Elizarov, Yu. A. and Sinitsina, E. E. (1974) Contact chemoreceptors of the mosquito *Aedes aegypti* (*Diptera, Culicidae*). *Zool. J.*, **53**, 577–84.

Falk, R. (1979) Taste response of *Drosophila melanogaster*. *J. Insect Physiol.*, **25**, 87–91.

Feeny, P., Rosenberry, L. and Carter, M. (1983) Chemical aspects of oviposition behavior in butterflies. In: *Herbivore Insects: Host-seeking Behavior and Mechanisms* (Ahmad, S., ed.) pp. 27–76. Academic Press, New York.

Fredman, S. M. (1975) Peripheral and central interactions between sugar, water, and salt receptors of the blowfly, *Phormia regina*. *J. Insect Physiol.*, **21**, 265–80.

Galun, R. (1976) The physiology of hematophagous insect/animal host relationships. *Proc. XV Int. Congr. Ent.*, 257–65.

Gerrits-Heybroek, E. M., Herrebout, W. M., Ulenberg, S. A. and Wiebes, J. T. (1978). Host plant preference of five species of small ermine moths (*Lepidoptera: Yponomeutidae*). *Ent. exp. appl.*, **24**, 160–8.

Getting, P. A. (1971) The sensory control of motor output in fly proboscis extension. Z. Vergl. Physiol., **74**, 103–20.

Goldrich, N. R. (= Rachman, N. J.) (1973) Behavioral responses of *Phormia regina* (Meigen) to labellar stimulation with amino acids. *J. gen. Physiol.*, **61**, 74–88.

Gothilf, S., Galun, R. and Bar-Zeev, M. (1971) Taste reception in the Mediterranean fruit fly: Electrophysiological and behavioural studies. *J. Insect Physiol.*, **17**, 1371–84.

Gritsai, O. B. (1978) Protein-sensitive elements of the labellar sensilla of *Musca domestica*. *J. Evol. Biochem. Physiol.*, **14**, 142–9.

Hall, M. J. R. (1980) Central control of tarsal thresholds for proboscis extension in the blowfly. *Physiol. Ent.*, **5**, 17–24.

Hansen, K. (1978) Insect chemoreception. In: *Taxis and Behavior (Receptors and Recognition, Ser. B, Vol. 5)* (Hazelbauer, G. L., ed.) pp. 233–92. Chapman & Hall, London.

Hansen, K. and Wieczorek, H. (1961) Biochemical aspects of sugar reception in insects. In: *Biochemistry of Taste and Olfaction* (Cagan, R. H. and Kare, M. R., eds) pp. 139–162. Academic Press, New York.

Hanson, F. E. (1970) Sensory responses of phytophagus Lepidoptera to chemical and tactile stimuli. In: *Control of Insect Behavior by Natural Products* (Wood, D. L., Silverstein, R. M. and Nakajima, M., eds) pp. 81–91. Academic Press, New York.

Hanson, F. E. and Dethier, V. G. (1973) The role of gustation and olfaction in food plant discrimination and induction of preferences in the tobacco hornworm, *Manduca sexta*. *J. Insect Physiol.*, **19**, 1019–34.

Hodgson, E. S., Lettvin, J. Y. and Roeder, K. D. (1955) Physiology of a primary chemoreceptor unit. *Science*, **122**, 417–18.

House, H. L. (1974) Nutrition. In: *The Physiology of Insecta*, Vol. V (Rockstein, M., ed.) pp. 1–62. Academic Press, New York.

Hsiao, T. H. (1969) Adenine and related substances as potent feeding stimulants for the alfalfa weevil, *Hypera postica*. *J. Insect Physiol.*, **15**, 1785–90.

Hurter, J., Katsoyannos, B., Boller, E. F. and Wirz, P. (1976) Beitrag zur Anreicherung und teilweisen Reinigung des eiablageverhindernden Pheromons der Kirschenflige, *Rhagoletis cerasi* (L.). (*Dipt. Trypetidae*). *Z. angew. Ent.*, **80**, 50–6.

Huyton, P. M., Langley, P. A., Carlson, D. A. and Schwarz, M. (1980) Specificity of contact sex pheromones in tsetse flies, *Glossina spp. Physiol. Ent.*, **5**, 253–64.

Isono, K. and Kikuchi, T. (1974). Autosomal recessive mutation in sugar response of *Drosophila. Nature*, **248**, 243–4.

Jakinovich, Jr., W., Sugarman, D. and Vlahopoulos, V. (1981) Gustatory responses of the cockroach, house fly, and gerbils to methyl glycosides. *J. comp. Physiol.*, **141**, 297–301.

Jermy, T. (1966) Feeding inhibitors and food preference in chewing phytophagous insects. *Ent. exp. appl.*, **9**, 1–12.

Jermy, T. and Szentesi, A. (1978) The role of inhibitory stimuli in the choice of oviposition site by phytophagous insects. *Ent. exp. appl.*, **24**, 458–71.

Kennedy, L. M. and Halpern, B. P. (1980) Fly chemoreceptors: a model system for the taste modifier Ziziphin. *Physiol. and Behav.*, **24**, 135–43.

Kijima, H., Amakawa, T., Nakashima, M. and Morita, H. (1977) Properties of membrane-bound α-glucosidases: Possible sugar receptor protein of the blowfly, *Phormia regina. J. Insect Physiol.*, **23**, 469–79.

Kramer de, J. J. and Molen van der, J. N. (1977) The influence of very small currents on the reception and transduction of chemical stimuli in insect contact chemoreceptors. In: *Olfaction and Taste*, Vol. VI (Le Magnen, J. and MacLeod, P., eds) p. 355. Information Retrieval, London.

Kramer de, J. J. and Molen van der, J. N. (1980a) Special purpose amplifier to record spike trains of insect taste cells. *Med. Biol. Eng. Comp.*, **18**, 371–4.

Kramer de, J. J. and Molen van der, J. N. (1980b) The pore mechanism of the contact chemoreceptors of the blowfly, *Calliphora vicina*. In: *Olfaction and Taste*, Vol. VII (van der Starre, H., ed.) pp. 61–4. Information Retrieval, London.

Kusano, T. and Sato, H. (1980) The sensitivity of tarsal chemoreceptors for sugars in the cabbage butterfly, *Pieris rapae crucivora* (Boisduval). *Appl. ent. Zool.*, **15**, 385–91.

Louveaux, A. (1976) Répercussion sur la prise de nourriture de désafférences sensitives de la région péribuccale de la larve de Locusta. *J. Insect Physiol.*, **22**, 9–18.

Ma, W.-C. (1972) Dynamics of feeding responses in *Pieris brassicae* (Linn) as a function of chemosensory input: A behavioural, ultrastructural and electrophysiological study. *Mededelingen Landbouwhogeschool Wageningen*, **72**, 1–162.

Ma, W.-C. (1976a) Mouth parts and receptors involved in feeding behaviour and sugar perception in the African armyworm, *Spodoptera exampta* (*Lepidoptera, Noctuidae*). *Symp. Biol. Hung.*, **16**, 139–51.

Ma, W.-C. (1976b) Experimental observations of food-aversive responses in larvae of *Spodoptera exempta* (Wlk.) (*Lepidoptera, Noctuidae*). *Bull. Ent. Res.*, **66**, 87–96.

Ma, W.-C. (1977a) Electrophysiological evidence for chemosensitivity to adenosine, adenine and sugars in *Spodoptera exempta* and related species. *Experientia*, **33**, 356–8.

Ma, W.-C. (1977b) Alterations of chemoreceptor function in armyworm larvae (*Spodoptera exempta*) by a plant-derived sesquiterpenoid and by sulfhydryl reagents. *Physiol. Ent.*, **2**, 199–207.

Ma, W.-C. and Schoonhoven, L. M. (1973) Tarsal contact chemosensory hairs of the large white butterfly, *Pieris brassicae* and their possible role in oviposition behaviour. *Ent. exp. appl.*, **16**, 343–57.

Ma, W.-C. and Kubo, I. (1977) Phagostimulants for *Spodoptera exempta*: Identification of adenosine from *Zea mays*. *Ent. exp. appl.*, **22**, 107–112.

Maes, F. W. (1977) Simultaneous chemical and electrical stimulation of labellar taste hairs of the blowfly, *Calliphora vicina*. *J. Insect Physiol.*, **23**, 453–60.

Maes, F. W. (1980) Neural coding of salt quality in blowfly labellar taste organ. In: *Olfaction and Taste*, Vol. VII (van der Starre, H., ed.) pp. 123–6. Information Retrieval, London.

McCook, R. D., Gudewicz, D. and Wurster, R. D. (1975) A device for continuous determination of nerve spike frequency histograms. *J. Electrophysiol. Tech.*, **4**, 47–51.

McIver, S., Siemicki, R. and Sutcliffe, J. (1980) Bifurcate sensilla on the tarsi of female black flies, *Simulium venustum* (*Diptera: Simuliidae*): Contact chemosensilla adapted for olfaction. *J. Morphol.*, **165**, 1–12.

McNeil, S. and Southwood, T. R. E. (1978) The role of nitrogen in the development of insect plant relationships. In: *Biochemical Aspects of Plant and Animal Coevolution* (Harborne, J. B., ed.) pp. 77–98. Academic Press, London.

Mitchell, B. K. (1974) Behavioural and electrophysiological investigations on the responses of larvae of the colorado potato beetle (*Leptinotarsa decemlineata*) to amino acids. *Ent. exp. appl.*, **17**, 255–64.

Mitchell, B. K. (1976) Physiology of an ATP receptor in labellar sensilla of the tsetse fly, *Glossina morsitans morsitans* (Westw.) (*Diptera: Glossinidae*). *J. exp. Biol.*, **65**, 259–71.

Mitchell, B. K. and Schoonhoven, L. M. (1974) Taste receptors in colorado beetle larvae. *J. Insect Physiol.*, **20**, 1787–93.

Mitchell, B. K. and Gregory, P. (1979) Physiology of the maxillary sugar sensitive cell in the red turnip beetle, *Entomoscelis americana*. *J. comp. Physiol.*, **132**, 167–78.

Mitchell, B. K. and Gregory, P. (1981) Physiology of the lateral galeal sensillum in red turnip beetle larvae (Entomoscelis americana Brown): responses to NaCl, glucosinolates and other glucosides. *J. comp. Physiol.*, **144**, 495–501.

Mitchell, R. (1981) Insect behavior, resource exploitation, and fitness. *A. Rev. Ent.*, **26**, 373–96.

Mordue-Luntz, A. J. (1979) The role of the maxillary and labial palps in the feeding behaviour of *Schistocerca gregaria*. *Ent. exp. appl.*, **25**, 279–88.

Morita, H. and Takeda, K. (1959) Initiation of spike potentials in contact chemosensory hairs of insects. II. The effect of electric current on tarsal chemosensory hairs of *Vanessa*. *J. cell. comp. Physiol.*, **54**, 177–87.

Morita, H. and Yamashita, S. (1959) Generator potential of insect chemoreceptor. *Science*, **130**, 922.

Morita, H., Enomoto, K.-I., Nakashima, M., Shimada, I. and Kijima, H. (1977) The receptor site for sugars in chemoreception of the fleshfly and blowfly. In: *Olfaction and Taste*, Vol. VI (Le Magnen, J. and MacLeod, P., eds) pp. 39–46. Information Retrieval, London.

Moulins, M. and Noirot, Ch. (1972) Morphological features bearing on transduction and peripheral integration in insect gustatory organs. In: *Olfaction and Taste*, Vol. IV (Schneider, D., ed.) pp. 49–55. Information Retrieval, London.

Nederstigt, L. J. and Molen van der, J. N. (1980) On the variability in insect taste responses. In: *Olfaction and Taste*, Vol. VII (van der Starre, H., ed.) pp. 201. Information Retrieval, London.

Pappas, L. G. and Larsen, J. R. (1976) Gustatory hairs on the mosquito, *Culiseta inornata. J. exp. Zool.*, **196**, 351–60.

Pappas, L. G. and Larsen, J. R. (1978) Gustatory mechanisms and sugar-feeding in the mosquito, *Culiseta inornata. Physiol. Ent.*, **3**, 115–9.

Perkel, D. H. Gerstein, G. L. and Moore, G. P. (1967) Neuronal spike trains and stochastic point processes. I. The single spike train. *Biophy. J.*, **7**, 391–417.

Pietra, P. and Stoffolano Jr., J. G. (1980) Electrophysiological responses of the labellar and tarsal chemosensilla of *Tabanus nigrovittatus* (Macquart) (*Diptera: Tabanidae*) to salt and sugar stimulation. In: *Olfaction and Taste*, Vol. VII (van der Starre, H., ed.) p. 198. Information Retrieval, London.

Pollack, G. S. (1977) Labellar lobe spreading in the blowfly: regulation by taste and satiety. *J. comp. Physiol.*, **121**, 115–34.

Prokopy, R. J. (1981) Oviposition deterring pheromone system of apple maggot flies. In: *Management of Insect Pests With Semiochemicals* (Mitchell, E. R., ed.) pp. 477–494. Plenum Press, New York.

Prokopy, R. J., Averill, A. L., Bardinelly, C. M., Bowdan, E. S., Cooley, S. S., Crnjar, R. M., Spatcher, P. J. and Weeks, B. L. (1982) Site of oviposition-deterring pheromone production in *Rhagoletis pomonella* flies. *J. Insect Physiol.*, **28**, 1–10.

Rachman, N. J. (1979) The sensitivity of the labellar sugar receptors of *Phormia regina* in relation to feeding. *J. Insect Physiol.*, **25**, 733–9.

Rachman, N. J. (1980a) Physiology of feeding preference patterns of female black blowflies (*Phormia regina* Meigen) I. The role of carbohydrate reserves. *J. comp. Physiol.*, **139**, 59–66.

Rachman, N. J. (1980b) A possible role for the salt-sensitive chemosensory neurons in the recognition and acceptance of protein sources by female blowflies (*Phormia regina*). In: Olfaction and Taste, Vol. VII (van der Starre, H., ed.) p. 425. Information Retrieval, London.

Rees, C. J. C. (1968) The effect of aqueous solutions of some 1:1 electrolytes on the electrical response of the type 1 ('salt') chemoreceptor cell in the labella of *Phormia. J. Insect Physiol.*, **14**, 1331–64.

Rees, C. J. C. (1969) Chemoreceptor specificity associated with choice of feeding site by the beetle, *Chrysolina brunsvicensis* on its foodplant, *Hypericum hirsutum. Ent. exp. appl.*, **12**, 565–83.

Rence, B. and Loher, W. (1977) Contact chemoreceptive sex recognition in the male cricket, *Teleogryllus commodus. Physiol. Ent.*, **2**, 225–36.

Rice, M. J. (1976) Contact chemoreceptors on the ovipositor of *Lucilia cuprina* (Wied.), the Australian sheep blowfly. *Aust. J. Zool.*, **24**, 353–60.

Rook, M. B., Wolk van der, F. M. and Starre van der, H. (1980) Functional-anatomical mapping of taste-associated interneurons in the central nervous system of the blowfly, *Calliphora vicina* (R.-D.). In: *Olfaction and Taste*, Vol. VII (van der Starre, H., ed.) p. 281. Information Retrieval, London.

2.2 MORPHOLOGY

The olfactory receptor cells in insects are usually found on the antennae (Schneider, 1964). These primary sensory cells receive the information from the odor signals and transfer it via their own axons to the second order neurons in the antennal lobe (Fig. 2.1). The receiving part of the receptor cell is located within the olfactory organ or sensilla, whose outer structure can be seen on the

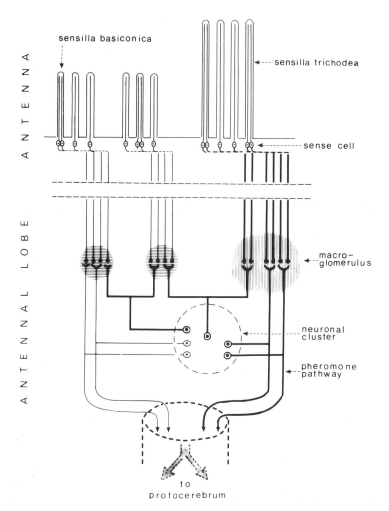

Fig. 2.1   Outline of the insect olfactory system, showing separate pathways for information from pheromones (thicker lines) and from other olfactory stimulants (thinner lines). As in moths, *s. trichodea* mediate the pheromone information and *s. basiconica* the information from other odors. Some second order neurons (local interneurons) make connections between glomeruli, while others (output neurons) project to the calyces and lateral lobe in protecerebrum (mofidied after Masson, 1981).

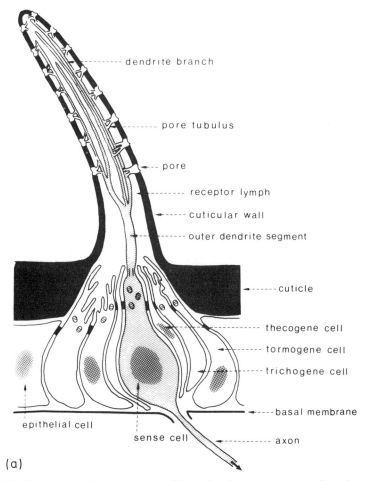

Fig. 2.2(a)   Scheme of a hair-shaped sensillum, showing one sensory cell, embraced by the three auxiliary cells. The dendrite of the sensory cell is divided into a branched outer segment and an unbranched inner segment (containing mitochondria). The conduction system for the odor molecules (pore and pore tubules) makes contact with the dendritic membrane.

surface of the antenna as hairs, plates or pores (cf. Kaissling, 1971). The information conveyed by the axons of the primary cells is transferred to second-order neurons in the antennal lobes via synapses which are confined to characteristic glomerular structures (Boeckh *et al.*, 1970; Pareto, 1972; Masson, 1973, 1977; Ernst *et al.*, 1977; Boeckh and Boeckh, 1979; Hildebrand *et al.*, 1980; Matsumoto and Hildebrand, 1981). After integration processes in the antennal lobe (involving convergence, divergence and connections via interneurons), the information is conveyed further via tractus olfactorio-globularis to higher

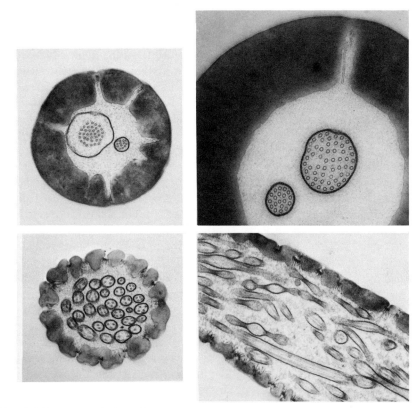

Fig. 2.2(b) Transmission electron micrographs of *Bombyx mori sensilla trichodea* (above, × 20 000 and 40 000) and *s. basiconica* (below, × 20 000). Fixation by freeze substitution results in almost round cross-sectional profiles of the dendrites, containing a regular array of microtubules. Pore tubules are well fixed in the outer parts of the hair (courtesy of Steinbrecht, 1980).

order neurons in the protocerebrum (Fig. 2.3). Terminations are found in the calyces of corpora pedunculata (mushroom bodies) and in the lateral protocerebral lobe (cf. Hanström, 1928; Jawlowski, 1958; Weiss, 1974; Strausfeld, 1976).

Morphological and physiological studies of the olfactory system in insects have so far mainly concerned the sensory organs and antennal lobe. The majority of these studies have been carried out on the sensory cells, which are relatively easy to approach. Nonetheless, results of investigations on the central processes of the antennal lobes are available, and further research in that area seems promising.

2.2.1 Antennae and olfactory sensilla

When considering antennae of different species, the variation in shape is striking. To understand the evolutionary development of variations in size and shape one must consider all of the different functions the antennae may serve (e.g. olfaction, taste, touch, perception of air-movement, heat and CO_2), which vary considerably between species (cf. Schneider, 1964). Antennae with featherlike shapes (e.g., in saturniids, lymantriids) seem most advanced as 'odor filters'. The surface is greatly increased by thousands of long vertically oriented hair sensilla on the antennal branches. The large surface area and characteristic geometry enable the antennae to sieve out odorous molecules effectively from the air passing through it (Kaissling, 1971).

Olfactory sensilla have been classified on the basis of size and shape as *sensilla trichodea*, *s. basiconica*, *s. coeloconica*, *s. placodea* (Schneider and Steinbrecht, 1968). Other types were later described in a number of species (Altner and Prillinger, 1980; Steinbrecht and Müller, 1976). The first two types, *s. trichodea* and *s. basiconica*, are hair-shaped sensilla. The *s. coeloconica* type, also called a pit-peg, consists of a peg embedded in a cavity of the cuticle which opens to the surface. *S. placodea*, or pore plates, consist of a circular cuticular structure which is perforated by pores. However, a classification of sensilla cannot be made exclusively on the basis of external morphology. Internal morphological properties revealed by studies of fine structure do not always correlate with the outer shape. For example, *s. trichodea* and *s. basiconica* usually differ in their long and short hairs respectively, but not always, and in this case the character of their fine structure is the principal difference between the two types (Schneider and Steinbrecht, 1968). Furthermore, intermingled with *s. trichodea* and *s. basiconica* in *B. mori* are found sensilla which show a fine structure of an intermediate type (Steinbrecht and Müller, 1971). The limited value of using the external morphology for classifying sensilla is discussed by Altner and Prillinger (1980).

One sensillum type for which the fine structure is well studied is *s. trichodea* of the silk moth *B. mori* (Steinbrecht and Müller, 1971; Steinbrecht, 1980). This hair-shaped sensillum (Fig. 2.2) can be characterized as a thin-walled (0.3 μm at half length) cuticular protrusion (10 μm long, diameter ca. 2 μm). The wall is penetrated by numerous pores (ca. 2500) over the entire surface, except at the most proximal part of the hair. The hair is innervated by one or more sensory cells, with dendrites penetrating the hair lumen and reaching out to the tip. Each dendrite is divided into an inner and outer segment, separated by a special ciliar structure which is short and 'neck-like' and contains the typical concentric doublets of tubules found in other sensory cilia. It is only the outer dendrite segment which resides within the hair lumen. The two segments also differ with respect to the contents of cell organelles. Whereas the outer segments mainly contain microtubules and no mitochondria (Fig. 2.2(b)), the latter are abundant within the inner segment. The outer segment is usually divided longitudinally into branches, relatively few in *s. trichodea*, but up to 50 in others, e.g. *s. basiconica* (Fig. 2.2(b)). The inner segment and the cell bodies are surrounded by three auxiliary cells (the thechogen, tormogen and trichogen) which form a sheath around the sensory cells. Thus an olfactory sensillum consists of at least four cells. The auxiliary cells, which carry out an important secretory function during the ontogenetic development of the sensillum, forming hair cuticle and the microtubules (Ernst, 1972), may have another important function during adult life. They apparently maintain the electrical potential (*ca.* +50 mV) between the haemolymph and the receptor lymph (Kaissling and Thorson, 1980). The receptor lymph fills the hair lumen around the outer segment of the dendrites and is separated from the haemolymph by tight junctions formed between the auxiliary cells just below the ciliary structure. The potential of the receptor lymph, which has relatively high K^+ content, is probably maintained by an electrogenic K^+-pump located in the folded membranes of the trichogen and tormogen cells. The membrane folds are only present on the side facing the receptor lymph. The sensory and auxiliary cells are located within the epithelium which is separated from the haemolymph by the basal membrane. An axon (0.3 μm) leading from a sensory cell to the brain is surrounded by glial cells, and does not interconnect with other axons, either by fusions or synapses, before reaching the brain (Steinbrecht, 1969).

The conduction system through which the odor molecules reach the dendrite membrane of the sensory cell occur in two principally different types (Steinbrecht and Schneider, 1980; Altner and Prillinger, 1980). The first type consists of a cuticular spoke channel, leading from longitudinal grooves on the hair surface to an inner cylinder that contains the dendrites (Steinbrecht and Müller, 1976). However, the conduction system has been mainly studied in the other type (Steinbrecht and Müller, 1971). Here, the molecules are adsorbed on the cuticle of the hair wall and diffuse through the pore openings (funnel 100 nm wide) which often lead into a fine (8 nm) and short (25 nm) pore channel. This widens into a cavity (50–100 nm). The inside of the cavity faces the receptor lymph of the liquor channel (200–400 nm in length) which penetrates the remainder of the hair wall. From the inside of the cavity 3–7 fine, pore tubules pass through the liquor channel and the hair lumen to the dendrite membrane. Electron microscopical studies have established that contact exists between these tubules and the dendrite membrane (Steinbrecht and Müller, 1971). Furthermore, Ernst (1969) has shown that tracer substances penetrate from the outside through the pores into the pore tubules, a route by which odorous molecules can evidently reach the dendrite membrane. Calculations by Adam and Delbrück (1968) suggest a two-dimensional surface diffusion of the odorous molecules through the pore-tubule conduction system. Pore tubules were earlier thought to be formed by the dendrite membrane, but were later shown to emanate from the trichogen cell (Ernst, 1972). This apparently excludes the possibility that receptor specificity is confined to the conduction system, as different receptor cells within the same sensillum (i.e., sharing the same conduction system) are usually keyed to different compounds (see Section 2.3.2).

2.2.2 Antennal lobe

Axons from olfactory receptor cells, forming a part of the antennal nerve, enter laterally or dorso-laterally into the ipsilateral antennal lobe and terminate in the glomeruli (Fig. 2.3). These synaptic structures consist of particularly fine and dense neuropil. The organization of the glomeruli in the cockroach, *Blaberus craniifer*, is uniform between individuals, both in the number per hemideuterum (109) and in contributions to the two sides (Chambille and Masson, 1980). Like the olfactory bulb in vertebrates, the glomerular structures occupy a large part of the lobes, surrounding a central region which in insects consists exclusively of coarser nerve fibers. The cell bodies are located cortically, usually in large clusters (Boeckh *et al.*, 1976). The low number of neurons in the antennal lobe compared to the number of receptor cells, suggests a strong convergence of primary axons onto these neurons (Boeckh and Boeckh, 1979).

Two main types of neurons have been identified in the antennal lobe of several insect species (Ernst *et al.*, 1977; Matsumoto and Hildebrand, 1981).

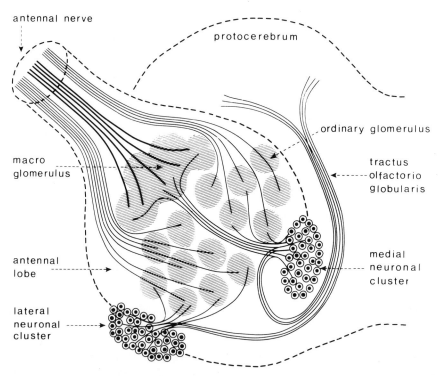

Fig. 2.3   Olfactory pathways in the antennal lobe. Axons from 'pheromome sense cells' terminate in the 'macroglomerulus' whereas axons from other olfactory cells project to numerous 'ordinary glomeruli'. Second-order cell bodies are located in the medial and lateral neuronal clusters of the antennal lobe. Axons from output neurons in these clusters form the tractus olfactorio globularis, which project to the calyces and lateral lobe of protocerebrum. The diagram is modified after Boeckh and Boeckh (1979) who found that second-order neurons in *Antherea sp.*, responding to pheromones, were located exclusively in the medial clusters. However, in *Manduca sexta*, Matsumoto and Hildebrand, have revealed that output neurons with dendritic arborizations in the 'macroglomerulus' are located in both clusters.

The local anaxonic interneuron (Fig. 2.1) is multiglomerular (makes connections between different glomeruli). The other type, the output neuron, is uniglomerular and sends one axon through the deutocerebral bundle of the tractus olfactorio-globularis to the protocerebrum. In *Manduca sexta*, the local interneurons have been further divided into (i) a male-specific type with dendritic arborizations in the 'macroglomerulus' (see below), (ii) a female-specific type with asymetrical arborizations, and (iii) a type found in both sexes with dendritic arborizations exclusively in 'ordinary glomeruli' (Matsumoto and Hildebrand, 1981). A further classification of antennal lobe neurons has been achieved by combined electrophysiological and morphological studies, as discussed in the following section.

Studies of the ultrastructure of glomeruli in *M. sexta* have revealed pre- and postsynaptic profiles, where one pre-synaptic region is commonly associated with several postsynaptic (Tolbert and Hildebrand 1981). Pre-synaptic profiles are not only present in terminals of primary axons. Identification of second-order neurons filled with horseradish peroxidase, have shown that local interneurons also possess pre-synaptic profiles. Similar indications that information is not only received by second-order neurons but also transmitted, were previously provided by degenerating studies in *Periplaneta americana* and *Calliphora erythrocephala* (Boeckh *et al.*, 1970). One of three types of pre-synaptic profiles is most common in *M. sexta* and seems to be the predominant type in the primary axon terminals (Tolbert and Hildebrand, 1981). These are characterized by their numerous small round and clear vesicles, resembling cholinergic terminals in vertebrates, that use acetylcholine as a transmitter (Sanes *et al.*, 1977).

2.3 ELECTROPHYSIOLOGY

2.3.1 Techniques

Electrophysiological recordings from receptor cells and from second-order neurons in the antennal lobe have led to a better understanding of the olfactory system. The first recordings from olfactory receptor cells in insects were performed by Schneider (1957) who obtained a slow, graded, summated receptor potential, the electro-antennogram or EAG. Besides being a valuable tool for studying the olfactory mechanisms, EAG-recordings have been used successfully in the identification of pheromones and host attractants. Fractions from female gland contents, separated by gas-liquid chromatography, have either been collected and tested on male antennae (Roelofs, 1979), or the EAG technique has been linked to the GLC-separation system by recording simultaneously from the gas chromatograph and the antenna detector (Moorhouse *et al.*, 1969; Arn *et al.*, 1975). By using both endogenous and synthetic pheromones and analogues, EAG recordings have also been used to suggest taxonomical relationships of pheromone–receptor systems between species of Lepidoptera (Priesner, 1979b).

In 1962, Boeckh reported the first quantitative recordings from single olfactory receptor cells in the carrion beetle, *Necrophorus vespilloides*. Tungsten micro-electrodes, with a tip diameter of about 1 μm, were inserted into the hair base, contacting the dendrites, thus picking up the nerve impulses extracellularly. This technique turned out to be well-suited for recordings from sensilla of short hairs of coeloconic and placodaic type in several species. Another technique, more suitable for recording from single cells of the long hair sensillum, *s. trichodea*, was introduced by Kaissling (1974). The tip of the hair was cut off and a glass capillary micro-electrode was brought into contact with the sensory cell by covering the cut end of the hair with the electrode tip. In that manner, it was possible to record nerve impulses and the slow receptor potential of the innervating cells. This is in principle the same technique used earlier for recording from contact chemosensilla where a natural pore opening exists at the tip of the hair (see Städler, Chapter 1). When recording from a single sensillum, impulses from one or from several cells may be obtained. In the latter case the spikes of the various cells are distinguished by differences in amplitudes. However, when three or more cells are present in one sensillum the size differences of the amplitudes may be unreliable for distinguishing among the cells. In these cases the spikes from different cells must be identified by using selective adaptation techniques (Vareschi, 1971; Kaissling, 1979; Priesner, 1979a, b). A cell is selectively adapted by stimulation with a 'key compound' for this cell; a second stimulus, acting on the same cell, applied immediately after the adaptation stimulus, will elicit a lower response than if it was applied alone. Any test compound which shows a reduced effect when applied immediately after the adaptation stimulus is thus assumed to have activated the same cell. However, if a compound when applied as a second stimulus, elicits the same response as when applied alone, it must have activated another cell which was unaffected by the adaptation stimulus. Thus far, intracellular recordings from olfactory receptor cells have not been reported.

Recordings from neurons in the antennal lobe were performed by Yamada (1968), who recorded extracellular nerve impulses from single neurons in *P. americana*. The neurons responded when odors were blown over the antennae. Later, similar recordings were carried out in other species, allowing the determination of second-order neuron response spectra in *P. americana* (Yamada, 1971; Boeckh, 1974; Boeckh *et al.*, 1976; Waldow, 1975), *Locusta migratoria* (Boeckh, 1974) and *Camponotus vagus* (Masson, 1973). Furthermore, extracellular recordings from neurons of the antennal lobes in two species of *Antherea* were performed during exposure to pheromone compounds of one of the species (Boeckh and Boeckh, 1979). Subsequent to the recordings, cobalt staining was performed which visualized the widely branching arborizations of certain pheromone-responding units in the 'macroglomerulus'. Recently, intracellular recordings from antennal lobe neurons were also performed in *M. sexta*, followed by intracellular injections of cobalt or horse-radish peroxidase (Matsumoto and Hildebrand, 1981).

The stimulation techniques chosen for studies of insect olfaction have to be considered with respect to the small quantities of pheromone compounds that are sometimes available. One technique, employed in studies of lepidopteran pheromones, is the use of cartridges (Kaissling and Priesner, 1970). A specific amount of a compound is applied on a piece of filter paper inside a glass tube (cartridge) through which a constant air stream (puff) is blown over the antennae. Another technique (employed in studies of bark beetle pheromones) involves the use of a syringe, inside of which is placed a vial, containing the odor solution (Kafka, 1970). After saturation, the piston of the syringe can be depressed by a motor-driven device at a constant velocity, while directed toward the antenna. In both cases a stream of pure air is blown over the antenna between each stimulation period. In other techniques the odor is introduced as a puff into an air stream, continually blowing over the antennae (Roelofs and Comeau, 1971).

2.3.2 Receptor cell responses

(a) General

The receptor cells are either activated, inhibited or unaffected when an odor is blown over the antennae. When activated, a negative slow deflection is recorded between the receptor lymph of a single sensillum and the haemolymph (Kaissling and Thorson, 1980). This deflection is assumed to be the summated receptor potential of the innervated olfactory cells of a particular sensillum. The sum of deflections from all olfactory sensilla is assumed to constitute the EAG recorded with relatively large diameter glass capillary electrodes. When recording from a single sensillum with glass capillary micro-electrodes during stimulation with a low pheromone concentration, small elementary potentials (0.5 mV) are observed 10–40 ms before a spike occurs. These potentials have been interpreted as the result of opening of single ion channels involved in the transduction process. The extracellularly recorded spikes consist of an initial positive deflection (2 ms) followed by a negative one (Kaissling and Thorson, 1980). In some cases, responses to odors may also be recorded as an inhibition of receptor cells, which then show a slow hyperpolarization and reduced firing rate (Boeckh, 1967a).

The specificity of an olfactory cell reflects the specificity of membrane receptors, also called acceptors, which are assumed to be located in the dendritic membranes. The possibility that the specificity might be confined to the conduction system, e.g., microtubules, was excluded by the observation that two sensory cells innervating the same sensillum, and sharing the same conduction system, could be activated by different 'key' substances. In activating a cell, the odor molecule is thought to bind to a membrane receptor protein

(Hansen, 1978; Städler, Chapter 1), whereby a conformational change to an active stage of the receptor occurs, leading to the opening of ion channels and depolarization of the cell membrane (Kaissling and Thorson, 1980). So far, different olfactory cells have been shown to possess varying degrees of specialization, some responding only to one of the test compounds, while others may respond to many different odors. Varying composition of different types of membrane receptors within each cell has been postulated as a model to explain the degree of specialization possessed by different olfactory cells (Kafka, 1974).

In an early study in the saturniid moth, *Antherea pernyi*, the term 'specialists' was introduced to describe cells responding most selectively to the pheromone compound, in contrast to the term 'generalists' for those cells responding less selectively to many different host compounds (Schneider *et al.*, 1964). So far, it seems that the cells activated by pheromones are specialized for one compound of a pheromone mixture (Kaissling, 1979; Priesner, 1979a; Den Otter, 1977; Mustaparta, 1979; Angst, 1981; Wadhams *et al.*, 1982). On the other hand, odors with other functions, such as food attractants, usually seem to activate cells which have broader response spectra. However, the term 'generalists' can be misleading since it may mean either cells which all respond differently (individual response spectra) or cells which belong to a certain response group, where the response spectra within a group is identical (Vareschi, 1971). Furthermore, these groups may either overlap or be non-overlapping (see Section 2.3.2(d)).

(b) Pheromones

Moths
In lepidopteran species the sex pheromones produced in the female abdominal gland specifically activate *s. trichodea* receptor cells on the male antennae, eliciting attraction and sexual behavior in the males (Kramer, 1978). The pheromone components of numerous lepidopteran species have been identified by chemical analysis, often in connection with electrophysiological tests and field observations. The molecular structures of these compounds are related in investigated species of *Noctuidae*, *Tortricidae*, and several other moth families. Most are straight-chain unsaturated hydrocarbons with acetate, aldehyde or alcohols as one terminal group. The availability of synthetic lepidopteran pheromones makes this stimulus—receptor system most suitable for studying the structure-activity relationship of the olfactory system (Kaissling, 1971, 1979; Priesner, 1979a, b; Schneider *et al.*, 1977; Den Otter, 1977) and for testing various hypotheses on molecular—receptor (acceptor) interaction (Kafka and Neuwirth, 1975) and transduction processes (Kaissling, 1974).

The number of *s. trichodea* sensory cells varies from one to five, depending on the lepidopteran species (Priesner, 1979a). In some species the cell generating the largest spike amplitude in each sensillum consistently belong to the same cell type, i.e. maximally responds to the same 'key' substance, which is the major pheromone component. This cell type (A) seems to be generally present in all *s. trichodea* of a species, while the other cell types, each keyed to a minor component, may be less frequently found. Since morphological studies in *B. mori* have revealed that *s. trichodea* contain two cells, one considerably larger than the other (Fig. 2.2(b)), it seems likely that the largest spikes, elicited by bombycol, originate from the large cell (Kaissling and Thorson, 1980). Similar relationships between spike amplitudes and specificities of *s. trichodea* sensory cells seem to exist in many Lepidoptera (Priesner, 1979a; Kaissling, 1979; Den Otter, 1977; O'Connell, 1975). Thus, even though most species use multicomponent pheromones, the receptor cells detecting these signals are 'specialists' for one 'key' substance.

Other questions of interest include the sensitivity and the specificity of the cells for their respective 'key' compounds. The pioneering experiment leading to the conclusion that one or two molecules of bombycol hitting the receptor cell is enough for eliciting a spike, was performed using electrophysiological recordings combined with measurements of adsorption of labelled bombycol on the antennae of *B. mori* (Kaissling and Priesner, 1970). In other species, the relative sensitivity of the cells to their respective 'key' compound has been determined by measuring the amount of the unlabelled compound needed to give a certain response at low levels. In that manner, it was found in *B. mori*, *A. polyphemus* (Kaissling, 1979) and various noctuids and tortricids (Priesner, 1979a; Den Otter, 1977) that approximately the same degree of response of the respective cells was elicited by equivalent amounts of their 'key' substances. Further determination of the specificity of a particular pheromone cell has been carried out by measuring the dose–response relationship for the 'key' substance and its derivatives. Studies of the pheromone cell specialized to *cis*-7, 8-epoxy-2-methyl-octadecane in *Lymantria dispar* showed that this cell also responds to the related derivatives, but only at higher concentrations (Kafka, 1974; Schneider *et al.*, 1977). The parallel dose–response curves which were obtained for the pheromones and several analogues indicate that a single type of membrane receptor is involved. In addition, the relative effects of numerous derivatives were determined by measuring the amount of compound necessary for eliciting equivalent responses. Thus the *cis*-isomer and the position of the epoxy group were found to be most important for the stimulatory effect. Recent studies of the effect of optical isomers of disparlure, have revealed that the two *s. trichodea* sense cells in *L. dispar* are each specialized to one of the enantiomers (Hansen, unpublished).

In several noctuids and tortricids, Priesner (1979a, b) determined the specificities of the various pheromone cells within *s. trichodea* by measuring their responses to four doses of each test compound, including the 'key' compound. In Table 2.1 examples are given for five cells (A–E) within the trichodeal sensillum of *Pseudaletia unipuncta*. The function of the components appears to be (A) main pheromone component, (B and D) interspecific inhibitors, (E) synergist, and (C) unknown (Hill and Roelofs, 1980; McDonough *et al.*, 1980; Farine *et al.*, 1981).

Table 2.1 Responses of five receptor cells (A to E) in male *sensilla trichodea* of *Pseudaletia unipuncta* to indicated amounts of test chemicals. Size of filled circles symbolizes frequency of nerve impulse response; open circles indicate lack of response (courtesy of Priesner.)

	Amount (µg)	Cell A	Cell B	Cell C	Cell D	Cell E
Z5–12:Ac	10	○	•	●	○	○
Z6–12:Ac	10		•	●		○
Z7–12:Ac	0.1			○		
	1	○		●		
	10	•	●	●	○	○
E7–12:Ac	10			●	○	○
Z8–12:Ac	10	•		●		
Z9–12:Ac	10	●	●	●	○	○
Z7–14:Ac	10	•	●	•	○	
Z8–14:Ac	10	•	●			
Z9–14:Ac	0.1		●	○		
	1	•	●	•		○
	10	●	●	•	○	○
E9–14:Ac	10	●	●	•	○	○
Z10–14:Ac	10	•	●		○	
Z11–14:Ac	10	•	•	•	○	○
Z9–16:Ac	10	●	•	•	○	○
Z10–16:Ac	10	●		○		
Z11–16:Ac	0.01	•				
	0.1	●	○			○
	1	●	•	○	○	○
	10	●	●	•	○	•
E11–16:Ac	10	●	•	○	○	○
Z12–16:Ac	10	●		○		
Z13–16:Ac	10	•	•	○	○	○
Z7–12:OH	10	○		•	○	•
Z9–14:OH	10	○	•	○	○	●
Z11–14:OH	10	○	○	○	○	●
Z9–16:OH	10	○	○			●
Z11–16:OH	0.1	○				●
	1	○			○	●
	10	•	○	○	•	●
E11–16:OH	10	•	○	○	○	●
Z12–16:OH	10		○		○	●
Z13–16:OH	10	○	○		○	●
Z7–12:Ald	10	○		○	•	○
Z9–14:Ald	10	○	○	○	●	○
Z11–14:Ald	10	○	○		●	
Z9–16:Ald	10	○			•	○
Z11–16:Ald	0.1				●	
	1	○			●	○
	10	•	○	○	●	•
E11–16:Ald	10	○	○	○	●	○
Z12–16:Ald	10	○			•	○
Z13–16:Ald	10	○			•	○

All cells having approximately the same threshold for their 'key' substances, also responded to other compounds, but only at higher concentrations. It is obvious from these data that the terminal group is most important; the acetate cells (A, B and C) were almost insensitive to aldehydes and alcohols, and the aldeyde cell (D) and alcohol cell (E) were almost insensitive to acetates. Furthermore, an increase or decrease of chain length from that of the 'key' substance, generally caused a decrease of the stimulatory effect. However, it does not seem to be the total chain length, but rather the length of a particular section of the chain, that is most important. With respect to this, it is of further interest that the receptors in the two families Noctuidae and Tortricidae differ; the part of the chain between the apolar alcyl group and the double bond being most critical for the stimulatory effect on receptor cells in noctuids and the other part of the chain most critical in tortricids (Priesner, 1979a, b). In this connection it is important to realize that the receptor specificity of a particular cell type generally remains consistent within closely related species. Thus, Priesner (1979a, b) found identical response spectra and dose-response relationships for cells keyed to the same compound within a species as well as in closely related species. This in contrast to the receptor cells of taxonomically distant species, e.g. noctuids and tortricids, for which the specificities differed as described above. These findings of constant specificities of pheromone cells belonging to one type, differ from the results of the early study of the red-banded leafroller moth, *Argyrothenia velutinana* (O'Connell, 1975). Here, two *s. trichodea* receptor cells were each found to be most sensitive to one of the pheromone components, $(Z)11-14$:Ac or $(E)11-14$:Ac, but individually the cells differed both in respect to absolute and relative sensitivities for the pheromones and other related compounds. In recent studies of the same species, however, the same cell types and two additional types of *s. trichodea* have each been found to possess the same sensitivities, response spectra and dose–response relationships, according to results of other tortricids (Priesner, unpublished).

These data provide substantial evidence that each type of pheromone cell in lepidopteran species possesses only one membrane receptor type, a specialist for the 'key' substance, and that pheromone constituents in these species can indeed be identified electrophysiologically be screening the effect of various derivatives on single trichodeal cells. Screening tests of this kind have been carried out in many species using single cell recordings as well as recordings of EAGs (Priesner, 1979a, b; Kaissling, 1979). Demonstration of these uniform receptor systems has stimulated studies of the physicochemical interaction between a stimulatory agent and the receptors. Based on electrophysiological data, a physicomathematical model has been proposed and tested, suggesting that three to five subsites within the pheromone bind to corresponding subsites in the membrane receptor (Kafka and Neuwirth, 1975). It appears that the spatial distribution of these subsites of high electron density (e.g., double bond, terminal groups, etc.) are important for the stimulatory effect of the chemical agent. This concept about the nature of the membrane receptor is interesting also in terms of the evolution of the pheromone receptors. Priesner (1977) has suggested that for certain related species the receptors have undergone a step-wise change. The idea was based mainly on electrophysiological studies of numerous moth families, demonstrating a gradual change in receptor responses (EAGs) of male antennae with the degree of taxonomic separation (Priesner, 1979a). In addition, the distinct difference in taxonomic neighbors (mentioned above for noctuids and tortricids) suggests evolutionary steps which have become preserved in these species. With respect to speculation on the evolution of the specific stimulus—receptor system, there appears to be substantial plasticity in both the produced compounds and the receptor cells. Thus, female pheromone glands contain compounds for which the conspecific male has no specific receptors, and the male antennae possesses specific receptor cells for which the conspecific female does not produce the 'key' compound (Priesner, 1979a). That kind of reserve capacity may be important in the evolution of the pheromone and its receptor.

Bark beetles

Several species of the genera *Ips*, *Scolytus* and *Dendroctonus* have been electro-physiologically investigated with respect to pheromone detection. Bark beetles use volatiles of the host tree as precursors for their pheromone components (Hughes, 1973, 1974). These are ingested when beetles bore into the bark and are at least in some instances modified by intestinal bacteria (Brand *et al.*, 1975) or fungi (Brand *et al.*, 1976). They are then released through the frass (boring dust and fecal material). The modified compounds have a strong attractive effect upon both sexes and are therefore called aggregation pheromones. In some cases the effect of the pheromones is increased when host tree volatiles are added. In *Dendroctonus*, the boring beetles are females and in *Ips* males, while in *Scolytus* the sex boring depends on the species (see Birch, Chapter 12 on bark beetles).

A few pheromones in various species (Silverstein, 1979; Blight *et al.*, 1979a, b) illustrate the characteristic differences between the pheromone structures used in the three genera (Fig. 2.4). In *Ips*, most components are monoterpenes of which ipsdienol, ipsenol, amitinol and perhaps also *cis*-verbenol are genus-specific. The prominent pheromones in species of *Dendroctonus* are brev-icomin and frontalin. In species of *Scolytus* two structures have been identified as pheromones, multistriatin and 4-methyl-3-heptanol. Structure—activity relationships have so far not been systematically investigated for the above pheromone compounds. Rather, in these insects, studies have mainly con-cerned the specificities of the receptor cells for known pheromone components, with the aim of elucidating the reception of the multicomponent aggregation pheromones. Electrophysiological studies have stressed the importance of optical forms (Fig. 2.4) as behavioral studies demonstrated that these are decisive for communication in bark beetles. All of the pheromone compounds mentioned are produced by the beetles as specific optical configurations, and

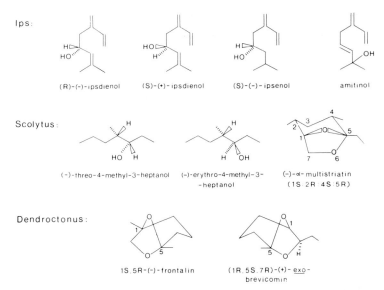

Fig. 2.4   Molecular structure of pheromones characteristic of three genera of bark beetles.

the receiving insects discriminate behaviorally between the enantiomers in the field (Lanier *et al.*, 1972; Lanier *et al.*, 1980; Birch *et al.*, 1980a, b; Blight *et al.*, 1979a; Light and Birch, 1979; Payne *et al.*, 1982).

Olfactory sensilla in bark beetles have been classified as *s. trichodea* and *s. basiconica* as in species of Lepidoptera. Because of high hair density, however, it has not been possible in most cases to identify the type of sensillum from which electrophysiological recordings have been made. The olfactory receptor cells responding to pheromones of *Ips* and *Scolytus* were shown to be specialized to one 'key' substance. Thus, in females of *Ips* species, separate groups of cells keyed to ipsenol, ipsdienol, amitinol, verbenone/verbenol, methylbutanol, phenylethanol (the two latter also isolated from males of *Ips*) and host volatiles were identified electrophysiologically (Mustaparta, 1979). In addition, various cell groups were shown to be optically specific. In fact, the optical isomers seems to act as independently as any other pheromone component. For example, the ipsdienol cells consist of two distinct types; one keyed to (+)-(Fig. 2.5) and the other to (−)-ipsdienol (Mustaparta *et al.*, 1980). On the other hand exclusively (−)-ipsenol cells are present in the investigated species of *Ips* (Mustaparta, 1980), which corresponds to their produced compound (Fish *et al.*, 1979).

It is difficult to compare the absolute sensitivities of the pheromone cells of bark beetles and moths on the basis of available experimental results, because different stimulation techniques have been used, and because the two classes of compounds have different vapor pressures. However, with cartridge-techniques,

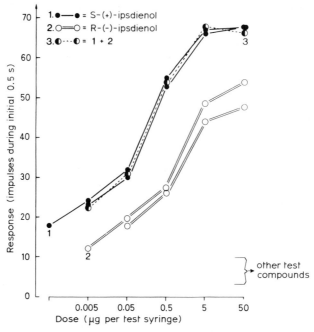

Fig. 2.5   Dose–response curves of a 'pheromone sense cell' in *Ips pini* (N.Y.) specialized to (S)-(+)-ipsdienol. The effect (ca. 10 times lower) of the opposite enantiomeric sample ((R)-(−)-ipsdienol) can be ascribed to the contamination (ca. 8%) of S-(+)-ipsdienol. Adding (R)-(−)- to S-(+)-ipsdienol did not increase the response to the latter, indicating no interaction at the receptor level.

the threshold dosage for the cells in *Ips* species was found to be in the range of 10^{-4}–10^{-3} μg of the 'key' substance on filter paper (Mustaparta *et al.*, 1979). This is roughly the same threshold found in moths for their 'key' substances. Comparisons of the number of molecules reaching the antennae in the two insect groups can, however, not be made. The specificities as well as the relative sensitivities of the receptor cells in *Ips* have been determined by measuring the dose-response relationship to six stimulation intensities using the syringe stimulation technique (Fig. 2.5). All cells with 'key' compounds of monoterpenes showed the same dose–response relationship. A linear increase of responses to logarithmic increase of the stimulation intensity occurred in the same concentration range. Above this level (about 100 μg inside the syringe) all dose–response curves levelled off, reaching a plateau, which probably represented maximal engagement of cell responsiveness (Mustaparta, 1980).

   Recordings of olfactory receptor cells in *Scolytus multistriatus* and *S. scolytus* revealed the same degree of specialization as that found for *Ips* species. In males of *S. multistriatus* (attracted to the three-component mixture, multilure) the majority of receptor cells were keyed to either of the two compounds produced by females, (−)-α-multistriatin and (−)-threo-4-methyl-3-heptanol (Angst, 1981).

However, no cells in the males were keyed to the third compound (−)-cubebene which is produced by the host tree. Both sexes of *S. scolytus* possess a majority of receptor cells keyed to either of the two male-produced compounds, (−)-threo-4-methyl-3-heptanol and the (−)-erythro-stereo-isomer, and in this species some cells were keyed also to (−)-*α*-multistriatin (Wadhams *et al.*, 1982). The relative sensitivities and specificities of these cells were determined in the same way as described for *Ips* species. Each cell was minimally activated by pheromone components other than its own 'key' substance. The functional types of olfactory receptor cells discussed here for *Scolytus*, as with the cells of *Ips*, apparently are all keyed to the specific isomer of the pheromone compounds. The cells specialized to (−)-*α*-multistriatin are 1000 times less sensitive to the geometrical isomer δ-multistriatin. However, the discrimination of the optical form of multistriatin seemed to be less pronounced. The substantially lower effect of the (+)-stereo-isomer (optical purity 99.5%), cannot be solely ascribed to contamination with the (−)-enantiomer. It implies that these cells in fact may also respond to (+)-multistriatin, although to a lesser extent. The cells keyed to methyl-heptanol are, however, highly optically specific in both species.

The findings in *Ips* and *Scolytus* are different from those obtained in species of *Dendroctonus*. Recordings of EAGs and responses of single cells (both combined with adaptation techniques) has led to the conclusion that reception of the multicomponent pheromones in species of *Dendroctonus* reflects interaction of different compounds on one and the same receptor cell, i.e., a single cell containing different types of membrane receptors, each interacting with a certain pheromone component. This idea is supported by observations that the prominent female-produced pheromone compound, frontalin, the synergistic host volatile, *α*-pinene, and the interspecific inhibitor, endobrevicomin, all activate the same receptor cell (Payne, 1975; Dickens and Payne, 1977). It has also been suggested that the receptor cells in *Dendroctonus* discriminate between optical configurations of frontalin by a similar mechanism, involving two types of optically specific membrane receptors located in the same cell (Payne *et al.*, 1982). However, what obscures the conclusion drawn from the experiments in *Dendroctonus* is that relatively high concentrations (100 μg) of the test substances were used. Cells of *Ips* and *Scolytus* exhibit maximal responses at this concentration level of a 'key' substance, and compounds other than the 'key' substance may elicit some response at this concentration. Therefore, a dose−response relationship should also be determined for the pheromone cells in species of *Dendroctonus* to substantiate that this species possesses single cells specifically responding to more than one 'key' substance.

Other insect species
Receptor cells specialized to pheromones have also been studied in other insect groups. In the honey bee, *Apis mellifera*, two separate cell types have been observed, one for queen substance and another for the alarm pheromone

(Vareschi, 1971). The response spectra of these cells do not overlap. Further-more, the cells keyed to the queen substance possess optical specificities (Kafka *et al.*, 1973). The ant, *Lasius fuliginosus*, has a highly specialized cell type for the alarm pheromone, undecane (Dumpert, 1972); the cell was 1000 times less sensitive to decane and to dodecane (Kafka, 1974).

The data presented above for most species investigated provide substantial evidence for specialized olfactory cells that receive information from a single component of a blend of sex pheromones, aggregation pheromones and others. This implies that the synergistic effects of the components in the mixtures upon behavioral responses are due to integration within the CNS of signals from separate primary axons, rather than to interaction of compounds on the receptor cells. It seems to hold true for most synergistic interactions, whether the substances are acting in sequence or together when eliciting identical behav-ioral responses. In a few cases, it has been proven electrophysiologically that behavioral synergists do not interact on the receptor cells. Thus the two syn-ergists, ipsenol and ipsdienol, were tested separately and combined on both 'ipsenol-' and 'ipsdienol cells' of *I. paraconfusus* (Mustaparta, 1979). The cell responses did not increase when the mixture was applied. Neither did en-antiomers, acting as behavioral synergists, interact on the receptor cells (Fig. 2.5).

(c) Inhibitors

Most compounds which interfere with a behavioral response elicited by another compound are interspecific interruptants and are also pheromone components of sympatric species (see Birch, Chapter 12; Cardé and Baker, Chapter 13). They interrupt the attraction of closely related sympatric species, which may otherwise be cross-attracted by other pheromone components or host volatiles. In some instances the interruption is mutual, (Light and Birch, 1979; Lanier *et al.*, 1980; Birch *et al.*, 1980a, b), and it might well be that this is a general phenomenon.

Two different mechanisms for interruption have been discussed: (i) inter-action of the inhibitor and the pheromone on the same receptor cell (com-petitive blocking), and (ii) activation of separate cells by the two compounds, causing interruption of response at a central level. The presence of specialized receptor cells for behavioral inhibitors has been demonstrated in many species by electrophysiological recordings from single receptor cells. Priesner (1979a, b) found that a particular *s. trichodea* receptor cell in some noctuids and tortricids is specialized for the compound causing interruption of pheromone attraction in these species, (exemplified in Table 2.1). Furthermore, in studies of the tortricid moth, *A. orana*, behavioral inhibitors for pheromone attraction did not show any significant inhibitory effect on the pheromone receptor cells (Den Otter, 1977). The mechanism of interspecific interruption was studied in more detail in bark beetles, where separate cells for inhibitors produced by a sympatric species were identified (Mustaparta *et al.*, 1977). Additional tests showed that the inhibitor does not influence the responses of pheromone receptor cells to their key substance. This applies to pheromone cells in *I. pini* and *I. paraconfusus*, where mutual interruption take place (Birch, Chapter 12, Fig. 12.2). In western *I. pini*, the interruptants (−)-ipsenol and (+)-ipsdienol produced by *I. paraconfusus* did not reduce the response of the pheromone cells keyed to (−)-ipsdienol (Mustaparta, 1979; Mustaparta *et al.*, 1980). Conversely, the response of (+)-ipsdienol cells in *I. paraconfusus* were not reduced by adding the interruptant (−)-ipsdienol. There was no observable difference between cells keyed to the same substance, whether they mediated inhibition in one species or attraction in the other. These results seem to justify the conclusion that the mechanism for interspecific interruption is not due to interaction of compounds on the receptor cells, but to activation of separate cells, the signals of which interfere with those from pheromone cells in the CNS. Such a system seems more efficient for blocking a behavioral response, as in the case of competitive blocking, the inhibitor would need to be present in a much higher concentration than the pheromone to prevent the latter from affecting the cell.

Competitive blocking of single membrane receptors is a well-established phenomenon (e.g., in synaptic and neuromuscular transmission), and it has been suggested that the same might hold true for the olfactory receptors of insects (Roelofs and Comeau, 1971). However, in the case of the established mechanisms for blocking effects, the chemical agent is disadvantageous to the 'receiving' individual and, therefore, specialized receptors for such an agent would not be predicted. The pheromones and interspecific inhibitors are important in conspecific mate finding. Here, the development of particular receptor cells for interspecific interruptants appear to be advantageous. No case is known where a compound is produced by one species with the sole function of blocking the attraction of another species to another volatile compound. If such a case exists the inhibitory mechanism is probably by competitive blocking. Other protective mechanisms are the production of repellents which may activate specific receptor cells, host plant compounds (e.g., geraniol) which hyperpolarize pheromone cells, and compounds with general irritating properties (defense compounds).

Inhibition also occurs between individuals of the same species. In *B. mori*, the secondary pheromone component, bombycal, acts as an inhibitor on the attraction caused by the main pheromone component, bombycol. For each component there are separate receptor cells as described above.

(d) Host volatiles

Although the reception of pheromones in many respects has now been elucidated, it remains to be shown conclusively how behaviorally important volatiles from host and food materials are discriminated by the receptor cells. There is probably not one single system for receptor cell specialization to these compounds. Volatiles released from host and food materials constitute a much more complex mixture than do pheromone blends. Various kinds of host and food materials have been classified by Städler (1980), as decaying organic materials, stored products, plant roots, undamaged plant surfaces, flowers, plant interior and animal hosts. In this section studies of the specificities of single receptor cells are discussed suggesting innate abilities of receptor cells to discriminate volatiles originating from these kinds of odors.

Pheromones and odors of other environmental sources are generally received by separate receptor populations. Thus, in most investigated species, cells keyed to pheromones are not activated by food or host odors, and as a rule the reverse holds true for cells strongly activated by host volatiles with the exception of *Dendroctonus* (see above). The only cross-effect sometimes observed is inhibition. Geraniol, a common substance in plants, inhibits the responses of cells keyed to the pheromone in two species of moths (Schneider *et al.*, 1964; Den Otter *et al.*, 1978). Similarly, inhibition of the spontaneous discharge of 'pheromone cells' in bark beetles of *Ips* is common when stimulating with vapor of bark from the host tree (Mustaparta, unpublished). In moths, the cells activated by pheromones and host odors are located in different sensilla *s. trichodea* and *s. basiconica/s. coeloconica*, respectively (Schneider *et al.*, 1964; Den Otter *et al.*, 1978; Van der Pers, 1978; Van der Pers and Den Otter, 1978). Similarly, in species of Coleoptera possessing *s. basiconica* and *s. trichodea*, the receptor cells of the former respond to host odors, whereas those of *s. trichodea* seem to be pheromone specialists (Mustaparta, 1975; Ma and Visser, 1978).

One of the main problems in studies of host volatiles is to find the important components of the mixture which elicit the specific behavior, such as attraction of the particular insect species. An interesting approach in the search for important host compounds is the investigation of the head space odors of carrot leaves, using linked GC-separation technique and EAG-recordings from the carrot fly, *Psila rosae* (Guerin *et al.*, 1983). Preliminary observations suggest that the antennae respond mainly to methylisoengenol and asarone, giving no signal or only a minor signal in the GC-flamedetector. Hence, the results obtained so far, illustrate the difficulties in finding the effective host and food volatiles, problems also adherent to the identification of pheromones.

In several electrophysiological studies of single receptor cells important compounds of host and food materials could be predicted and were in some instances shown to elicit behavioral responses. The results of these studies have demonstrated a large variety in the degree of receptor cell specialization to host and food volatiles. In bark beetle species *Ips* the vapor of bark was used as an indication stimulus for finding cells tuned to host odors (Mustaparta, 1979). Some cells, strongly activated by this vapor, responded maximally to a single component present in the host. In studies of other species a variety of compounds have been tested, some naturally occurring in the respective hosts and some with structural similarities. On the basis of the compounds to which the cells respond (response spectrum) and dose—response relationships, attempts have been made to classify the cells. Thus, studies of 'grass odor' receptor cells of *s. coeloconicae* in the grasshopper, *L. migratoria* (Boeckh, 1967b; Kafka, 1970), 'food receptor' cells of *s. basiconicae* in the cockroach, *P. americana* (Sass, 1978) and 'host odor' receptor cells of *s. placodea* in the honey bee, *A. melifera* (Vareschi, 1971), have revealed different systems for the reception of the respective test compounds. The grass odor receptor cells, strongly activated by short-chained (C_5-C_7) fatty acids, alcohols and aldehydes, could not strictly be placed in groups, with respect either to the 'key' substance or to dose—response relationships (Kafka, 1970). Thus, these cells seem to possess a specialization to certain hydrocarbonesters, but with individually specificities, suggesting that several membrane receptors are involved in the same cell. In contrast to these results, cell groupings could be made in the two other studies. In *P. americana* the cells of a particular 'food receptor type' were classified according to the most potent compound as pentanol, hexanol, octanol and butyric acid types. Furthermore, the cells within a group showed identical specificities. Between groups, however, there was consistent overlap for the same substances. In the honey bee, the host odor receptor cells were grouped in distinct 'reaction groups', without overlap between the groups. On the other hand, cells within one group, i.e., responding to the same substances, did not show identical dose—response relationships to these compounds. Further conditioning experiments showed that bees discriminate behaviorally among odors which differentially activate the same cells (i.e., odors belonging to the same reaction group). However, the frequency of errors was higher for these compounds than for those of different groups.

Studies of two species, the carrion beetle, *Necrophorus vespilloides* (Boeckh, 1962; Waldow, 1973) and the blowfly, *Calliphora vicina* (Kaib, 1974), both dependent on the odor of meat for finding food and breeding material, demonstrate variations in receptor cell specificities for the same odor. Both species possess numerous receptor cells mediating these chemical messages and, in *Calliphora*, these were separate cells from those mediating messages from flowers. Although the quality of meat-specific volatiles are largely unknown, certain known components activated particular cells as strongly as meat odor. Responses of these cells in *Necrophorus* and *Calliphora* were different as were cells within each species. In *Calliphora*, six groups of cells differ in their responses to various stages of meat decomposition, which apparently enables the fly to detect the condition of meat. One group, most sensitive to short-chained hydrocarbon esters, gradually increased its response in relation to the age of meat (from 1 to 20 days). Like the grass odor cells, in *Locusta* the meat odor cells in *Calliphora* showed individual specificities.

The results of these studies suggest that the receptor cells responding to host and food volatiles possess a relatively low degree of specialization, and thus respond to many compounds and with individual differences of their relative sensitivities. However, in some instances a higher degree of specialization, including identical cells, seems to be present. It may even be that some host compounds should be regarded as specific 'key' substances for olfactory cells. Further studies involving screening of the effects of all components of host and food volatiles may provide further insight into this possibility.

(e) Kairomonal action of pheromones

A variety of chemical signals may directly or indirectly mediate important messages between predator and prey or parasite and host (Vincent, Chapter 8). Pheromones of one species may function directly as kairomones: odors of prey or host species may be attractive to the predator or parasite respectively. Conversely, recognition of the pheromones of enemies by the prey or host insect has also been demonstrated. For example, the attractant mixture of ipsdienol, *cis*-verbenol and 2,3,2-methylbutenol, which is used for trapping *I. typographus*, is also attractive to its predator, *Thanassimus formicarius* (Bakke and Kvamme, 1978). EAG responses showed that the predator possesses olfactory receptor cells responding to bark beetle pheromones; by comparing the dose–response relationships (EAG responses obtained for different pheromone dosages in predator and prey), it was found that the curves roughly covered the same dosages in the two species (Hansen, 1983). This suggests that the predator, *T. formicarius*, possesses receptor cells with the same sensitivities as those of the prey to the prey's pheromones, and therefore may recognize these compounds (kairomones) at the same concentrations that the prey uses for aggregation. In recent studies of single olfactory cells in *T. formicarius* and *I. typographus* it was found that the olfactory receptor cells of the predator and prey, responding to the prey's pheromones, possess the same specificities as those described above for other *Ips* species (Tømmerås and Mustaparta, unpublished). The two distant relative species of beetles have apparently developed specialized receptor cells with the same specificities for compounds which, however, give different messages (food volatiles vs. pheromones) for the two insect groups.

Another kairomonal action of pheromones is suggested to exist between bark beetles of different genera (Byers and Wood, 1981; Tømmerås *et al.*, 1983). Here, a pheromone component of one genus may be used as a cue by beetles of the other genus for finding suitable host materials. Tømmerås *et al.*, found that *I. typographus* and *Dendroctonus micans* have specific receptor cells for pheromones of the other genus.

2.3.3 Antennal lobe responses

The results discusssed above suggest that information about pheromones and about host- and food volatiles is generally detected and transmitted via different primary neurons to the CNS, and in principle by the use of two different mechanisms. In most insect species, it appears that pheromone receptor cells, each containing one type of membrane receptor, transfer information about a particular compound to CNS via a 'labeled line system' (Boeckh, 1980). This means that information from different pheromone components is transmitted via separate, non-overlapping primary axons. Thus, the presence of activity in any single primary axon in this system encodes for (is labeled to) the quality of one particular compound. Different pheromone compounds acting synergistically, or pheromone and inhibitors acting antagonistically, elicit activation in differently labeled lines. In the first step of integration, terminals of equally or differently labeled lines may converge on identical second-order neurons, either directly or via interneurons.

In the case of host and food volatiles another mechanism (across-fiber pattern) is suggested (Boeckh, 1980). Here, the 'key' information about the quality of a single host volatile is apparently transmitted to the CNS via over-lapping lines. However, it is also possible that information about certain host volatiles (as about pheromones) may be mediated via specialized cells and the labeled line mechanism. In addition to these two mechanisms, temporal response pattern may also modify the coding process (Boeckh, 1980).

The axons of sex pheromone receptor cells terminate in one specific glomerulus, the macroglomerulus, which is significantly larger than the other ordinary glomeruli (Figs 2.1 and 2.3). The morphology of the macroglomerulus is described in males of several species, and demonstrates the sexual dimorphism of the antennal lobes (Jawlowski, 1958; Boeckh *et al.*, 1977; Boeckh and Boeckh, 1979; Chambille and Masson, 1980; Matsumoto and Hildebrand, 1981). The termination of pheromone cells at the macroglomerulus has further been verified by staining techniques following electrophysiological recordings in some lepidopteran species. The use of extracellular cobalt staining in males of *A. pernyi* and *A. polyphemus* revealed that antennal lobe neurons, responding to pheromones, have arborizations in the macroglomerulus which is located dorsally at the entrance of the antennal nerve (Boeckh and Boeckh, 1979). In *M. sexta*, intracellular recordings from neurons in the antennal lobe followed by staining with cobalt and horseradish peroxidase, demonstrated that those neurons which respond to sex pheromones all have arborizations in the macroglomerulus (Matsumoto and Hildebrand, 1981), whereas neurons which did not respond to pheromones projected to ordinary glomeruli. In both species the cell bodies of antennal lobe neurons are located in two main cortical clusters, a larger lateral and a smaller medial one (Fig. 2.3). In *M. sexta*, a few neurons were also found, concentrated in a third small cluster anterior to the glomerular neuropil.

Recordings from antennal lobe neurons of *Antherea sp.* demonstrated that neurons responding to pheromones were located in the medial cluster, while no neurons studied in the lateral cluster responded to these compounds (Boeckh and Boeckh, 1979). Responses to pheromones were exclusively excitatory with a phasic–tonic pattern imposed upon a low resting activity (ca 1 imps^{-1}). Responses were elicited exclusively by stimulation with sex pheromones on the ipsilateral antennae. All central neurons responding to pheromones were activated by both components of their respective pheromone blends in the two *Antherea* species. Only a quantitative difference in the effect on the neurons was observed between the major and minor pheromone components. In *A. polyphemus* the major component (E)-6-(Z)-11-hexadecadienyl acetate had a threshold effect at $10^{-7}\,\mu g$ while the aldehyde had to be applied at a much higher concentration, 10^{-5} or $10^{-4}\,\mu g$, before eliciting a response. Corresponding results were obtained in *A. pernyi*. Comparison between the sensitivities of receptor cells and central neurons showed that the central neurons had a lower threshold for the major component than the receptor cells, whereas the opposite was the case for the minor component. In addition, the major component was found to dominate the responses of the central neurons. Adding various amounts of the minor component to the major one at low concentrations, did not cause increased response. Neither did the (behaviorally) most important mixture 9:1 (major:minor) show a higher activity than the major component alone. These findings clearly demonstrate that receptor cells, keyed to different pheromone components, converge on the same second order of antennal lobe neurons. However, the studies do not reveal how the minor component, via central pathways, increases the behavioral response to the major component, and how the pheromone 'profile' (i.e., simultaneous information from all of the pheromone receptor cells) is coded. Boeckh and Boeckh (1979) have suggested that additional neurons in the antennal lobe, responding in a different way to the two pheromone components, might also be involved. Discrimination between two other pheromone components by most antennal lobe neurons have, however, been demonstrated in recent studies with *B. mori* (R. Ohlberg, unpublished). Some neurons were activated by the major component, bombycol, and inhibited by the minor component, bombycal whereas other neurons were activated only when both components were applied. In addition, other excitatory and inhibitory types of response to these pheromone components were observed. Most of the antennal lobe neurons responding to pheromones were also influenced by odorless air puffs, which indicated that information about wind mediated by mechanosensitive receptor cells is also integrated at this level (Waldow, 1975; Pareto, 1972).

As mentioned in Section 2.2, histological studies have revealed the presence of two types of neurons, anaxonic interneurons and output neurons in the antennal lobe of several insect species (Fig. 2.1). Intracellular recordings followed by staining have resulted in a more detailed classification in *M. sexta* (Matsumoto and Hildebrand, 1981). One type of local interneuron, with arborizations in the macroglomerulus, responded to pheromones in two distinct patterns, either a longlasting excitation ('on' response) or an inhibition followed by excitation ('off–on' response). Cells exhibiting the latter response were also activated by the 'green odor', (E)-2-hexenal, which elicited a simple 'on' response. Thus the first integrative step of lines mediating pheromone and 'green odor' reception seems to take place in local interneurons making connections between the macroglomerulus and ordinary glomeruli. This integration seems to result in increasing the contrast between the two classes of stimuli, pheromone and 'green odor'. In *M. sexta*, recordings were obtained more frequently from local interneurons than from output neurons (about 8 : 1) and, to date, output neurons responding to pheromones have not been reported. It is hoped that future electrophysiological characterization of output neurons with arborizations in the macroglomerulus will elucidate the coding processes of pheromones and other odors in *M. sexta*, as well as in other species.

Responses of antennal lobe neurons to food and host volatiles have been studied in several insect species (Boeckh *et al.*, 1976). Except in one ant species (Masson, 1973), the neurons were found to respond only to stimulation of the ipsilateral antennae and to react with both excitatory ('on') and inhibitory ('off') responses. In a more recent study on *M. sexta* (Matsumoto and Hildebrand, 1981), the local interneurons reacting exclusively to 'green odors' were found to exhibit either an 'off–on' response or an 'on' response. Output neurons, reacting to the 'green odor', showed either a long-lasting excitation ('on' response) or an inhibition ('off' response). Both the local interneurons and the output neurons have dendritic arborizations in 'ordinary' glomeruli. Special attention has been paid in two species (*L. migratoria* and *P. americana*) to the importance of across-fiber pattern in central coding. Here, physiological data on specificities of receptor cells (Boeckh, 1967b; Sass, 1978) and of central neurons (Boeckh, 1974) were compared with morphological data (Ernst *et al.*, 1977). Recordings from central neurons demonstrated both convergence and divergence of receptor cells on to the antennal lobe neurons. Studies of the cockroach indicate that no response spectrum of a central neuron simply reflects the activity of one receptor cell type. Different receptor cells all respond to a series of fruit odors (apple, banana, lemon, orange) but to a different degree (Sass, 1978). The central neurons, however, respond either to lemon or to orange or to both, but not to the whole series of fruit odors (Boeckh, 1975). As suggested by Ernst *et al.* (1977), this implies that the processes involved are not simply convergence or divergence of receptor cells into second-order neurons, but a rather more complicated network of integration.

2.4 CONCLUDING REMARKS

Present knowledge about insect olfaction is largely based upon the integrated results of behavioral, morphological, and electrophysiological investigations performed over the past 25 years. All-important for the progress of these investigations have been the concomitant isolation, identification, and synthetization of relevant chemical signals. Two general conclusions can be drawn concerning the olfactory mechanisms underlying behavioral responses, based on the majority of electrophysiological investigations discussed in this chapter:

(i) Pheromones as a rule are perceived via 'specialist-type' receptor cells and via a labeled line mechanism. Consequently, behavioral synergism caused by different pheromone compounds reflects the interaction at the CNS level of information carried via different pheromone-labeled lines. In contrast, host and food volatiles are perceived via less specific receptor cells and via an across-fiber mechanism. However, the possibility remains that the latter mechanism is utilized also for pheromone perception in some species (*Dentroctonus* sp. and *A. velutiana*).

(ii) Interspecific interruption is due to convergence of information reaching the CNS via different (pheromone vs. inhibitor) labeled lines. This concept is founded on demonstrations of 'specialist-type' receptor cells for interspecific interruptants, the stimulation of which does not affect the response of the pheromone receptor cells.

The studies discussed in this chapter do not completely explain why the ratio of compounds in a pheromonone blend is of such critical importance for behavioral responses as shown in many species of moths. However, the demonstration of 'specialist-type' receptor cells for the single pheromone components, and the observed lack of interaction of the components at the receptor level, implies that it is the ratio of activity between differently labeled lines that informs about the quality of the pheromone blend and triggers the behavioral response. Furthermore, differences in activity along a particular labeled line provide an explanation for the fact that a single compound may act as a synergist at a low concentration and as an interspecific interruptant at a high concentration (Byers and Wood, 1981). Thus, increased activity in one labeled line may change the behavioral meaning of the information from attraction to no attraction. This may also explain the seeming paradox that the only electro-physiological effect observed in response to some compounds reducing catches of insects in traps is an activation of pheromone receptor cells (Priesner, 1979a). Here, it appears likely that the additional stimulus changes the ratio of labeled line activity elicited by the pheromone blend in a manner reducing the attraction. The ultrastructure of the antennal lobe leaves many possibilities open as to how the CNS may integrate the information carried via differently labeled lines. Further investigations are needed before any firm conclusions can be drawn about the manner in which the insect CNS transforms semiochemical information into purposeful behavioral responses.

Finally, pheromone communication may be important in the evolution of insect species (see Section 2.3.2(b)). In two species the responsiveness of the pheromone receptor cells varies between populations according to variations in the produced pheromone compounds (Priesner, 1979a, Mustaparta *et al.*, 1980). Hybrids of the respective populations possess either three phenotypes of receptor cells (the two parental and one intermediate) (Priesner, 1979a), or exclusively the two parental types (Mustaparta *et al.*, unpublished). Further investigations along these lines may provide valuable information concerning evolutionary events in insects.

ACKNOWLEDGEMENTS

I wish to express my sincere thanks to Drs William J. Bell, Department of Entomology, University of Kansas; John Hildebrand, Department of Biological

Sciences, Columbia University and Ernst Priesner, Max-Planck-Institute, See-wiesen for valuable suggestions and criticism during the preparation of this chapter.

REFERENCES

Adam, G. and Delbrück, M. (1968) Reduction of dimensionality in biological diffusion processes. In: *Structural Chemical and Molecular Biology* (Rich, A. and Davidson, N., eds) pp. 198–215. Freeman, San Francisco.

Altner, H. and Prillinger, L. (1980) Ultrastructure of invertebrate chemo-, thermo-, and hygroreceptor and its functional significance. *Int. Rev. Cytol.*, **67**, 69–139.

Angst, M. (1981) Sinnesphysiologische Untersuchungen über die Pheromon-perzeption von *Scolytus*-Arten, besonders *Scolytus multistriatus*: Feldversuche, Electroantenno-gramme und Eizelzellableitungen von den Antennen. Thesis. ETH. Zürich.

Arn, H., Städler, E. and Rauscher, S. (1975) The electroantennographic detector – a selective and sensitive tool in the gas chromatographic analysis of insect pheromones. *Z. Naturforsch.*, **30**, 722–5.

Bakke, A. and Kvamme, T. (1978) Kairomone response by the predators *Thanasimus formicarius* and *Thanasimus rufipes* to the synthetic pheromones of *Ips typographus*. *Norw. J. Ent.*, **25**, 41–3.

Birch, M. C., Light, D. M., Wood, D. L., Browne, L. E., Silverstein, R. M., Bergot, B. J., Ohloff, G., West, J. R. and Young, J. C. (1980a) Pheromonal attraction and allomonal interruption of *Ips pini* in California by the two enantiomers of ipsdienol. *J. Chem. Ecol.*, **6**, 703–17.

Birch, M. C., Svihra, P., Paine, T. D. and Miller, J. C. (1980b) Influence of chemically mediated behavior on host tree colonization by four cohabiting species of bark beetles. *J. Chem. Ecol.*, **6**, 395–414.

Blight, M. M., Wadhams, L. J., Wenham, M. J. and King, C. J. (1979a) Field attraction of *Scolytus scolytus* (F.) to the enantiomers of 4-methyl-3-heptanol, the major component of the aggregation pheromone. *Forestry*, **52**, 83–90.

Blight, M. M., Wadhams, L. J. and Wenham, M. J. (1979b) The stereo-isomeric composition of the 4-methyl-3-heptanol produced by *S. scolytus* and the preparation and biological activity of the four synthetic stereo-isomers. *Insect Biochem.*, **9**, 525–33.

Blum, M. (1978) Biochemical defences of insects. In: *Biochemistry of Insects* (Rockstein, M., ed.) pp. 465–513. Academic Press, New York.

Boeckh, J. (1962) Electrophysiologische Untersuchungen an einzelnen Geruchsrezep-toren auf den Antennen des Totengräbers (*Necrophorus*, Coleoptera). *Z. vergl. Physiol.*, **46**, 212–48.

Boeckh, J. (1967a) Inhibition and excitation of single insect olfactory receptors, and their role as a primary sensory code. In: *Olfaction and Taste*, Vol. II (Zotterman, Y., ed.) pp. 721–35. Pergamon Press, New York.

Boeckh, J. (1967b) Reaktionsschwelle, Arbeitsbereich und Spezifität eines Geruchs-rezeptors auf der Hauschreckenantenne. *Z. vergl. Physiol.*, **55**, 378–406.

Boeckh, J. (1974) Die Reaktionen olfactorischer Neurone im Deutocerebrum von Insekten im Vergleich zu den Antwortmustern der Geruchssinneszellen. *J. comp. Physiol.*, **90**, 183–205.

Boeckh, J. (1975) Nervous mechanisms in Insect Olfaction. In: *Pheromones and Defensive Secretions in Social Insects* (Noirot, C., Howse, P. E. and Le Masne, G., eds) pp. 155–171. Symp. Int. Union for the Study of Social Insects, Dijon.

Boeckh, J. (1980) Neural basis of coding of chemosensory quality at the receptor cell level. *Olfaction and Taste*, Vol. VII (van der Starre, H., ed.). Information Retrieval, London.

Boeckh, J., Sandri, C. and Akert, K. (1970) Sensorische Eingänge und synaptische Verbindungen im Zentralnervensystem von Insekten. *Z. Zellforsch.*, **103**, 429–46.

Boeckh, J., Ernst, K-D., Sass, H. and Waldow, U. (1976) Zur nervösen Organisation antennaler Sinneseingänge bei Insekten unter besonderer Berücksichtigung der Riechbahn. *Verh. Dtsch. Zool. Ges.*, pp. 123–39. Gustav Fischer, Stuttgart.

Boeckh, J., Boeckh, V. and Kühn, A. (1977) Further data on the topography and physiology of central olfactory neurons in insects. In: *Olfaction and Taste*, Vol. VI (Le Magnen, J. and MacLeod, P., eds). Information Retrieval, London.

Boeckh, J. and Boeckh, V. (1979) Threshold and odor specificity of pheromone-sensitive neurons in the deutocerebrum of *Antherea pernyi* and *A. polyphemus* (Saturnidae). *J. comp. Physiol.*, **132**, 235–42.

Boppré, M., Petty, R. L., Schneider, D. and Meinwald, J. (1978) Behaviorally mediated contacts between scent organs: another prerequisite for pheromone production in *Danaus chrysippus* males (Lepidoptera). *J. comp. Physiol.*, **126**, 97–103.

Brand, J. M., Bracke, J. W., Markowetz, A. J., Wood, D. L. and Browne, L. E. (1975) Production of verbenol pheromone by a bacteria isolated from bark beetles. *Nature*, **254**, 136–7.

Brand, J. M., Bracke, J. W., Britton, L. N., Markowetz, A. J. and Barras, S. J. (1976) Bark beetle pheromones: Production of verbenone by a mycangial fungus of *Dendroctonus frontalis*. *J. Chem. Ecol.*, **2**, 195–9.

Byers, J. A. and Wood, D. L. (1981) Interspecific effects of pheromones on the attraction of bark beetles, *Dendroctonus brevicomis* and *Ips paraconfusus* in the laboratory. *J. Chem. Ecol.*, **7**, 9–17.

Cardé, R. T., Baker, T. C. and Roelofs, W. L. (1975) Behavioral role of individual components of a multichemical attractant system in the oriental fruit Moth. *Nature*, **253**, 348–9.

Chambille, I. and Masson, C. (1980) The deutocerebrum of the cockroach *Blaberus Cranifer* Burm. Spatial organization of the sensory glomeruli. *J. Neurobiol.*, **11**, 135–57.

Den Otter, C. T. (1977) Single sensillum responses in the male moth *Adoxophyes orana* (F.v.R.) to female sex pheromone components and their geometrical Isomers. *J. comp. Physiol.*, **121**, 205–22.

Den Otter, C. J., Schuil, H. A. and Sander-Van Oosten, A. (1978) Reception of host-plant odors and female sex pheromones in *Adoxophyes orana* (Lepidoptera: Tortricidae): Electrophysiology and Morphology, *Ent. exp. appl.*, **24**, 370–8.

Dickens, J. C. and Payne, T. L. (1977) Bark beetle olfaction: Pheromone receptor system in *Dendroctonus frontalis*. *J. insect. Physiol.*, **23**, 481–9.

Dumpert, K. (1972) Alarmstoffrezeptoren auf der Antenne von *Lasius fuliginosus* (Latr.) (Hymenoptera, Formicidae). *Z. vergl. Physiol.*, **76**, 403–25.

Eberhard, W. G. (1977) Aggressive chemical mimicry by a bolas spider. *Science*, **198**, 1173–5.

Eisner, T. (1970) Chemical defense against predation in arthropods. In: *Chemical Ecology* (Sonderheimer, E. and Simeone, J. B., eds) pp. 157–217. Academic Press, New York.

Ernst, K.-D. (1969) Die Feinstruktur von Riechsensillen auf der Antenne des Aaskäfers *Necrophorus* (Coleoptera). *Z. Zellforsch.*, **94**, 72–102.

Ernst, K.-D. (1972) Die Ontogenie der basiconische Riechsensillen auf der Antenne von *Necrophorus* (Coleoptera). *Z. Zellforsch.*, **129**, 217–36.

Ernst, K.-D., Boeckh, J. and Boeckh, V. (1977) A neuroanatomical study on the organization of the central antennal pathways in insects. *Cell Tiss. Res.*, **176**, 285–308.

Farine, J.-P. Frérot, B. and Isart, J. (1981) Physiologie des Invertébrés – Facteurs d'isolement chimique dans la sécrétion phéromonale de deux Noctuelles Hadeninae: *Mamestra brassicae* (L) et *Pseudaletia unipuncta* (Haw.). *CR Acad. Sci. Paris.*, **292**, 101–4.

Fish, R. J., Brown, L. E., Wood, D. L. and Hendry, L. B. (1979) Pheromone bio-synthetic pathways: conversions of deuterium-labelled ipsdienol with sexual and enantio selectivity in *Ips paraconfusus* Lanier. *Tetrahedron Lett.*, **17**, 1465–8.

Guerin, P. M., Städler, E. and Buser, H. R. (1983) Identification of host plant attractants for the carrot fly, *Psila rosae*. *J. Chem. Ecol.*, **9**, 843–61.

Hansen, K. (1978) Insect Chemoreception. In: *Taxis and Behavior*. (Hazelbauer, G. L., ed.) pp. 233–92. Chapman and Hall, London.

Hansen, K. (1983) Reception of bark beetle pheromone in the predaceous clerid beetle, *Thanasimus formicarius* (Coleoptera: Cleridae) *J. Comp. Physiol.*, **150**, 371–8.

Hanström, B. (1928) *Vergleichende Anatomie des Nervensystems der Wirbellosen Tiere unter Berücksichtigung seiner Funktion*. Springer Verlag, Berlin.

Hildebrand, J. G., Matsumoto, S. G., Camazine, S. M., Tolbert, L. P., Blank, S., Ferguson, H. and Ecker, V. (1980) Organisation and physiology of antenna centres in the brain of the moth *Manduca sexta*. In: *Insect Neurobiology and Pesticide Action* (Neurotox 79) pp. 375–82. Society for Chemical Industry, London.

Hill, A. S. and Roelofs, W. L. (1980) A female-produced sex pheromone component and attractant for males in the armyworm moth, *Pseudalitia unipuncta*. *Environ. Ent.*, **9**, 408–11.

Hughes, P. R. (1973) *Dendroctonus*: Production of pheromones and related compounds in response to host monoterpenes. *Z. Angew. Ent.*, **73**, 294–312.

Hughes, P. R. (1974) Myrcene: A precursor of pheromones in *Ips* beetles. *J. Insect Physiol.*, **20**, 1271–5.

Jawlowski, H. (1958) Nerve tracts in the bee (*Apis mellifera*) running from the sight and antennal organs to the brain. *Ann. Univ. M. Curic-Slodowska* (C), **12**, 307–23.

Kaib, M. (1974) Die Fleisch- und Blumenduftrezeptoren auf der Antenne der Schmeiss-fliege *Calliphora vicina*. *J. comp. Physiol.*, **95**, 105–121.

Kaissling, K. E. (1971) Insect Olfaction. In: *Handbook of Sensory Physiology*, Vol. IV, *Chemical senses 1. Olfaction* (Beidler, L. M., ed.) pp. 351–431. Springer, Berlin.

Kaissling, K. E. (1974) Sensory transduction in insect olfactory receptors. In: *Biochemistry of Sensory Functions* 25. Colloquium der Gesellsch. Biologie, Chemie, Mosbach (Jaenecke, J., ed.) pp. 243–73. Springer Verlag, Heidelberg.

Kaissling, K. E. (1979) Recognition of pheromones by moths, especially in *Saturniids* and *Bombyx mori*. In: *Chemical Ecology: Odor Communication in Animals* (Ritter, F. J., ed.) pp. 43–56. Elsevier/N. Holland, Amsterdam.

Kaissling, K. E. and Priesner, E. (1970) Die Riechschwelle des Seidenspinners. *Natur-wiss.*, **1**, 23–8.

Kaissling, K. E. and Thorson, J. (1980) Insect olfactory sensilla: structural, chemical and electrical aspects of the functional organization. In: *Receptors for Neurotransmitters, Hormones and Pheromones in Insects* (Satelle, D. B., Hall, L. M. and Hildebrand, J. G., eds). Elsevier/N. Holland, Amsterdam.

Kafka, W. A. (1970) Molekulare Wechselwirkungen bei der Erregung einselner Riech-zellen, *Z. vergl. Physiol.*, **70**, 105–43.

Kafka, W. A. (1974) Physiochemical aspects of odor reception in insects, *Ann. N. Y. Acad. Sci.*, **237**, 115–28.

Kafka, W. A., Ohloff, G., Schneider, D. and Vareschi, E. (1973) Olfactory discrimi-nation of two enantiomers of 4-methyl-hexanoic acid by the migratory locust and the honeybee. *J. comp. Physiol.*, **87**, 277–84.

Kafka, W. A. and Neuwirth, J. (1975) A model of pheromone moleculeacceptor inter-action. *Z. Naturforsch.*, **30**, 278–82.

Kramer, E. (1978) Insect pheromones. In: *Taxis and Behavior* (Hazelbauer, G. L., ed.) pp. 205–29. Chapman & Hall, London.

Kullenberg, B. and Bergström, G. (1974) The pollination of *Ophrys* orchids. *Nobel Symposium 25, Chemistry in Botanical Classification*, pp. 253–8.

Lanier, G. N., Birch, M. C., Schmidt, R. F. and Furniss, M. M. (1972) Pheromone of *Ips pini* (Coleoptera: Scolytidae): Variation in response among three populations. *Can. Ent.*, **104**, 1917–23.

Lanier, G. N. and Wood, D. L. (1975) Specificity of response to pheromones in the Genus *Ips* (Coleoptera: Scolytidae). *J. Chem. Ecol.*, **1**, 9–23.

Lanier, G. N., Claesson, A., Stewart, T., Piston, J. J. and Silverstein, R. M. (1980) *Ips pini*: The basis for interpopulational differences in pheromone biology. *J. Chem. Ecol.*, **6**, 677–87.

Light, D. M. and Birch, M. C. (1979) Inhibition of the attractive pheromone response in *Ips paraconfusus* by (R)-(−)-ipsdienol. *KOM Nawi*, **726**, 1.

Ma, Wei-Chun and Visser, J. H. (1978) Single unit analysis of odor quality coding by the olfactory antennal receptor system of the colorado beetle. *Ent. exp. appl.*, **24**, 520–33.

McDonough, L. M., Kamm, J. A. and Bierl-Leonhardt, B. A. (1980) Sex pheromone of the armyworm, *Psudaletia unipuncta* (Haworth) (Lepidoptera: Noctuidae). *J. Chem. Ecol.*, **6**, 566–72.

Masson, C. (1973) *Contribution a l'Etude du Systeme Antennaire chez les Fourmis.* These, Fac. Sci. Nat. Univ. Provence, Marseilles.

Masson, C. and Strambi, C. (1977) Sensory antennal organization in an ant and a wasp. *J. Neurobiol.*, **8**, 537–48.

Masson, C. (1981) La communication chimique chez les insects. *La Recherche*, **12**, 406–16.

Matsumoto, S. G. and Hildebrand, J. G. (1981) Olfactory mechanisms in the moth *Manduca sexta*: Response characteristics and morphology of central neurons in the antennal lobes. *Proc. R. Soc. Lond. Ser. B*, **213**, 249–77.

Moorhouse, J. E., Yeadon, R., Beevor, P. S. and Nesbitt, B. F. (1969) Method for use in studies of insect chemical communication. *Nature*, **223**, 1174–5.

Mustaparta, H. (1975) Responses of single olfactory cells in the pine weevil *Hylobius abietis* L. (Col: Curculionidae). *J. comp. Physiol.*, **97**, 271–90.

Mustaparta, H. (1979) Chemoreception in bark beetles of the genus *Ips*: synergism, inhibition and discrimination of enantiomers. In: *Chemical Ecology: Odor Communication in Animals* (Ritter, F. J., ed.) Elsevier/N. Holland, Amsterdam.

Mustaparta, H. (1980) Olfactory receptor specificities for multicomponent chemical signals. In: *Receptors for Neurotransmitters, Hormones and Pheromones in Insects* (Satelle, D. B., Hall, L. M. and Hildebrand, J. G., eds) Elsevier/N. Holland, Amsterdam.

Mustaparta, H., Angst, M. E. and Lanier, G. N. (1977) Responses of single receptor cells in the pine engraver beetle *Ips pini* (Say) (Coleoptera: Scolytidae) to its aggregation pheromone ipsdienol and the aggregation inhibitor, ipsenol. *J. comp. Physiol.*, **121**, 343–7.

Mustaparta, H., Angst, M. E. and Lanier, G. N. (1979) Specialization of olfactory cells to insect- and host produced volatiles in the bark beetle *Ips pini* (Say). *J. Chem. Ecol.*, **5**, 109–23.

Mustaparta, H., Angst, M. E. and Lanier, G. N. (1980) Receptor discrimination of enantiomers of the aggregation pheromone ipsdienol, in two species of *Ips*. *J. Chem. Ecol.*, **6**, 689–701.

O'Connell, R. J. (1975) Olfactory receptor responses to sex pheromone components in the redbanded leafroller moth. *J. gen. Physiol.*, **65**, 179–205.

Pareto, A. (1972) Die zentrale Verteilung der Fühlerafferenz bei Arbeiterinnen der Honigbienen, *Apis mellifera* L. *Z. Zellforsch.*, **131**, 109–40.

Payne, T. L. (1975) Bark Beetle olfaction III. Antennal olfactory responsiveness of *Dendroctonus frontalis*, Zimmermann and *D. brevicomis*, Le Conte (Coleoptera: Scolytidae) to aggregation pheromones and host tree terpenes hydrocarbons. *J. Chem. Ecol.*, **1**, 233–42.

Payne, T. L., Richerson, J. V., Dickens, J. C., Berisford, C. W., Hedden, R. L., Mori,

K., Vité, J. P. and Blum, M. S. (1982) Southern pine beetle: olfactory receptor and behavior discrimination of enantiomers of the attractant pheromone frontalin. *J. Chem. Ecol.* (in press).

Priesner, E. (1977) Evolutionary potential of sepecialized olfactory receptors. In: *Olfaction and Taste*, Vol. VI (Le Magnen, J. and MacLeod, P., eds). Information Retrieval, London.

Priesner, E. (1979a) Specificity studies on pheromone receptors of noctuid and tortricid Lepidoptera. In: *Chemical Ecology: Odor Communication in Animals* (Ritter, F. J., ed.). Elsevier/N. Holland, Amsterdam.

Priesner, E. (1979b) Progress in the analysis of pheromone receptor systems. *Ann. Zool. Ecol. Anim.*, **11**, 533–46.

Roelofs, W. L. and Comeau, A. (1971) Sex pheromone perception: Synergists and inhibitors for the red-banded leafroller attractant. *J. Insect Physiol.*, **17**, 435–48.

Roelofs, W. L. (1979) Electroantennograms. *Chemtech.*, **9**, 222–7.

Sanes, J. R., Prescott, D. J. and Hildebrand, J. G. (1977) Cholinergic neurochemical development of normal and deafferented antennal lobes during metamorphosis of the moth, *Manduca sexta. Brain Res.*, **199**, 389–402.

Sass, H. (1978) Olfactory receptors on the antenna of *Periplaneta*: Response constellations that encode food odors. *J. comp. Physiol.*, **128**, 227–33.

Schneider, D. (1957) Elektrophysiologische Untersuchungen von Chemo- und Mechanorezeptoren der Antenne des Seidenspinners *Bombyx mori* L. *Z. vergl. Physiol.*, **40**, 8–41.

Schneider, D. (1964) Insect Antennae. *A. Rev. Ent.*, **9**, 103–22.

Schneider, D., Lacher, V. and Kaissling, K. E. (1964) Die Reaktionsweisen und das Reaktionsspektrum von Riechzellen bei *Antherea pernyi* (Lepidoptera, Saturniidae). *Z. vergl. Physiol.*, **48**, 632–62.

Schneider, D. and Steinbrecht, R. A. (1968) Checklist of insect olfactory sensilla. *Symp. Zool. Soc. Lond.*, **23**, 279–97.

Schneider, D., Boppré, M., Schneider, H., Thompson, W. R., Boriack, C. J., Petty, R. L. and Meinwald, J. (1975) A pheromone precursor and its uptake in male *Danaus* butterflies. *J. comp. Physiol.*, **97**, 245–6.

Schneider, D., Kafka, W. A., Beroza, M. and Bierl, B. A. (1977) Odor receptor responses of male gypsy and nun Moth (Lepidoptera, Lymandriidae) to disparlure and its analogues. *J. comp. Physiol.*, **113**, 1–15.

Silverstein, R. M. (1979) Enantiomeric composition and bioactivity of chiral semiochemicals in insects. In: *Chemical Ecology: Odor Communications in Animals* (Ritter, F. J., ed.). Elsevier/N. Holland, Amsterdam.

Städler, E. (1980) Chemoreception in Arthropods: sensory physiology and ecological chemistry. In: *Animals and Environmental Fitness* (Gilles, R., ed.) pp. 223–41. Pergamon Press, London.

Steinbrecht, R. A. (1969) On the question of nervous syncytia: Lack of axon fusion in two insect nerves. *J. Cell. Sci.*, **4**, 39–53.

Steinbrecht, R. A. (1980) Cryofixation without cryoprotectants. Freeze substitution and freeze etching of an insect olfactory receptor. *Tissue and Cell*, **12**, 73–100.

Steinbrecht, R. A. and Müller, B. (1971) On the stimulus conducting structures in insect olfactory receptors. *Z. Zellforsch.*, **117**, 570–75.

Steinbrecht, R. A. and Müller, B. (1976) Fine structure of the antennal receptors of the bed bug, *Cimex léctularius* L. *Tissue and Cell*, **8**, 615–36.

Steinbrecht, R. A. and Schneider, D. (1980) Pheromone communication in moths sensory physiology and behavior. In: *Insect Biology in the Future "VBW 80"* (Locke, M. and Smith, D. S., eds). Academic Press, New York.

Strausfeld, N. J. (1976) *Atlas of Insect Brain*. Springer Verlag, Berlin.

Tolbert, L. P. and Hildebrand, J. G. (1981) Organization and synaptic ultrastructure of glomeruli in the antennal lobes of the moth *Manduca sexta*: a study using thin section and freeze-fracture. *Proc. R. Soc. Lond. Ser. B*, **213**, 279–301.

Tømmerås, B. Å., Mustaparta, H. and Gregoire, J.-Cl. (1983) Receptor cells in *Ips typographus* and *Dendroctonus micans* to pheromones of the reciprocal genus. *J. Chem. Ecol.* (in press).

Van Der Pers, J. C. N. (1978) Responses from olfactory receptors in females of three species of small eremine moths (Lepidoptera: Yponomeutidae) to plant odors. *Ent. exp. appl.*, **24**, 394–8.

Van Der Pers, J. N. C. and Den Otter, C. J. (1978) Single cell response from olfactory receptors of small eremine moths to sex-attractants. *J. Insect Physiol.*, **24**, 337–43.

Vareschi, E. (1971) Luftunterscheidung bei der Honigbiene – Einzelzell Ableitungen und Verhaltensreaktionen. *Z. vergl. Physiol.*, **75**, 143–73.

Wadhams, L. J., Angst, M. E. and Blight, M. M. (1982) Olfactory responses of *Scolytus scolytus* (F) (Coleoptera: Scolytidae) to the stereo-isomers of 4-methyl-3-heptanol. *J. Chem. Ecol.*, **8**, 1982.

Waldow, U. (1973) Electrophysiologie eines neuen Aasgeruchrezeptors und seine Bedeutung für das Verhalten des Totengräbers (*Necrophorus*). *J. comp. Physiol.*, **83**, 415–24.

Waldow, U. (1975) Multimodale Neurone im Dentocerebrum von *Periplaneta americana*. *J. comp. Physiol.*, **101**, 329–41.

Weaver, N. (1978) Chemical control of behavior – interspecific. In: *Biochemistry of Insects* (Rockstein, M., ed.). Academic Press, New York.

Weiss, M. J. (1974) Neuronal connections and the functions of the corpora pedunculata in the brain of the American cockroach. *Periplaneta americana* (L). *J. Morph.*, **142**, 21–70.

Yamada, M. (1968) Extracellular recording from single neurons in the olfactory centre of the cockroach. *Nature*, **217**, 778–9.

Yamada, M. (1971) A search for odor encoding in the olfactory lobe. *J. Physiol.*, **214**, 127–43.

Odor Dispersion and Chemo-orientation Mechanisms

3

Odor Dispersion

Joseph S. Elkinton and Ring T. Cardé

3.1 INTRODUCTION

The transfer of chemical information from an emitting to a receiving organism can occur directly by way of contact chemoreception or by dispersion through a transport medium. In this chapter we focus on transport of odor signals in the air. The various mathematical models discussed here have been used to describe the dispersion of pheromones in still air (Bossert and Wilson, 1963; Mankin *et al.*, 1980a) or moving air (Wright, 1958; Bossert and Wilson, 1963; Aylor *et al.*, 1976; Miksad and Kittredge, 1979; Fares *et al.*, 1980). They can be applied to the dispersion of any airborne odor such as plant volatiles inducing host finding. An understanding of odor dispersion is requisite for an accurate interpretation of odor-induced behaviors. (See Bell, Chapter 4, and Cardé, Chapter 5).

As odor molecules disperse from a source the odor concentration in the air declines with increasing distance. The active space is defined as the volume of air inside which the odor concentration is above threshold, i.e., the level sufficient to produce a behavioral reaction in the receiving organism. The active space concept may be refined to include an upper response threshold as well. If odor concentrations close to the source are sufficiently high to suppress the behavior (as could be produced by artificial sources) then the active space will occur at distances further from the source (Baker and Roelofs, 1981).

The various dispersion models discussed below have been used to estimate the size of the active space and the maximum distance of communication. Such descriptions depend upon an accurate knowledge of the odor emission rate and the behavioral threshold. Alternatively, the emission rate and maximum

Chemical Ecology of Insects. Edited by William J. Bell and Ring T. Cardé
© 1984 Chapman and Hall Ltd.

communication distances are measured and the dispersion models used to estimate the behavioral threshold. Unfortunately, neither emission rates nor behavioral thresholds are well defined for most organisms. The emission rates of pheromone from natural sources have been measured in relatively few species.

The behavioral threshold is certainly a more complicated phenomenon than that implied by the usual practice of assigning it a single and constant value (K). The same odor compounds may elicit different behaviors in the same organism at different concentrations (e.g., Rust, 1976; Baker and Cardé, 1979), or the threshold for a particular behavior may vary with the ratio of the compounds that comprise the odor (Roelofs, 1978). Furthermore, the process by which an odor signal elicits a behavioral response involves temporal summation of the signal over a short time interval at sensory receptor surfaces and within the central nervous system. Consequently, odor concentrations insufficient to produce a response upon brief exposure may elicit the response when exposure is sustained (Cardé and Hagaman, 1979). The response latency (the time interval between the arrival of the stimulus and response) increases with decreasing concentration. On the other hand, prolonged exposure to odor may produce either sensory adaptation or habituation, thus increasing the threshold level of the organism in subsequent exposures (Shorey 1976). Pulsed signals in theory may have a lower threshold than continuous signals because of an improvement in the signal to noise ratio, which comes about as the number of comparisons of signal to background is elevated (Murlis and Jones, 1981; Cardé *et al.*, 1983). Ambient temperatures influence the responsiveness of organisms (Mankin *et al.*, 1980b; Cardé and Hagaman, 1983) with attendant changes in the size of the active space (Baker and Roelofs, 1981). Thresholds for each species may vary with diel and annual fluctuations in responsiveness, and substantial differences in responsiveness may occur between individuals. Finally, threshold is usually expressed in terms of concentration in the atmosphere (e.g., $g\,cm^{-3}$). However, perception of the odor signal is governed by the rate of adsorption of odor molecules upon the olfactory sense organs. In moving air the rate of adsorption may relate more closely to odor flux ($g\,cm^{-2}\,s^{-1}$) than odor concentration.

The theoretical problems of defining thresholds are compounded by the practical difficulties of measuring them. In general, thresholds occur at concentrations or fluxes too low to measure directly by sampling the air and by quantifying the odor compounds collected. Instead, concentrations must be inferred from known emission rates (or dilutions thereof) and the assumptions of a particular dispersion model. The behavioral thresholds to pheromones have thus been *estimated* for only a few insects (Kaissling, 1971; Sower *et al.*, 1971; Shapas and Burkholder, 1978; Aylor *et al.*, 1976; Mankin *et al.*, 1980b; Hagaman and Cardé, 1983). Despite the theoretical and practical difficulties involved, the concept of a behavioral threshold is nevertheless a useful and necessary prerequisite for delineating the size and shape of an active space.

Bossert and Wilson (1963) showed that the ratio of the natural pheromone release rate to the behavioral threshold (Q/K) is a fundamental characteristic of different communication systems. In some communication functions such as with alarm pheromones it is imperative that the signal fade quickly. Organisms producing alarm pheromones generally have a high value of Q/K. Other chemical communication systems such as sex pheromones utilize a relatively persistent signal, characterized by low Q/K ratios. Q/K ratios in different organisms and communication systems are summarized by Matthews and Matthews (1978).

3.2 DISPERSION IN STILL AIR

In completely still air, odor dispersion is governed by molecular diffusion. The flux or rate of transport of a gas through a reference plane ($g\,cm^{-2}\,s^{-1}$) perpendicular to a given direction is equal to the product of (D), the diffusion coefficient ($cm^2\,s^{-1}$), and the concentration gradient. The diffusion coefficient is a property of the gas at a given temperature and is a function of molecular weight and intermolecular forces. Heavier molecules move at slower speeds at a given temperature and thus have lower diffusion coefficients than lighter molecules. Most pheromone molecules have diffusion coefficients in the range of $0.03-0.07\,cm^2\,s^{-1}$ (Wilson et al., 1969; Mankin et al., 1980a). However the effective values of diffusion coefficients are often greatly increased by the presence of air currents that exist even in most enclosed environments. The rate of diffusion driven by air currents is independent of the molecular species.

Bossert and Wilson (1963) were the first to use diffusion equations to estimate pheromone concentrations and active space dimensions in still air. They derived equations for three still air situations: (i) an instantaneous puff, (ii) a point source emitting at a continuous rate, and (iii) a moving point source such as an ant depositing a trail pheromone. For an instantaneous puff on the reflecting plane surface, such as an ant releasing a momentary burst of alarm pheromone, the active space is a half sphere with a radius (R) at (t) seconds after release of the puff given by:

$$R(t) = \left[4Dt\log\left(\frac{2Q}{K(4\pi Dt)^{3/2}} \right) \right]^{1/2}$$
$$\text{for } 0 \leqslant t \leqslant \frac{1}{4\pi D}\left(\frac{2Q}{K} \right)^{2/3} \qquad (3.1)$$
$$= 0 \text{ otherwise,}$$

where Q is the amount released (g), K is the behavioral threshold and D the diffusion coefficient. For a continuously emitting source the radius tends to a maximum (R_{max}) given by:

$$R_{max} = \frac{Q}{2K\pi D} \tag{3.2}$$

These equations combined with known values for Q and D provide a simple way to estimate the behavioral threshold for any organism based upon the time it takes the active space to attain a given radius and elicit the response in a still air environment (Wilson *et al.*, 1969).

Bossert and Wilson's model applies to a source resting on a non-adsorbing plane surface with no adjacent boundaries. Mankin *et al.* (1980a) present a still air dispersion model expanded to include the effects of boundaries with various degrees of odor adsorption at various distances from the source.

3.3 DISPERSION IN THE WIND

3.3.1 Sutton model

In nature, most odor signals are carried downwind and are dispersed by turbulent eddies. The transport of airborne material has been studied for many years, particularly with reference to dispersion of chemical warfare agents and more recently for air pollutants. The theoretical treatments have proceeded along several lines including atmospheric analogues of the equations for molecular diffusion (e.g., Roberts, 1923). However, most treatments of dispersion stem from the 'statistical approach' originating in the work of Taylor (1921) which focuses on the average trajectory of individual particles moving in the wind. Long-term studies conducted by O. G. Sutton (1947, 1953) culminated in his well-known equation for dispersion from a single, continuously emitting point source:

$$C_{(x,y,z)} = \frac{2Q}{\pi C_y C_z \bar{u} x^{2-n}} \cdot \exp\left[-x^{n-2}\left(\frac{y^2}{C_y^2} + \frac{z^2}{C_z^2}\right)\right] \tag{3.3}$$

$C_{(x,y,z)}$ is the concentration at any point (x,y,z) with the axes aligned so that x is the mean downwind direction, y is crosswind, z is vertical and with the source located at the point $(0,0,0)$ at the surface of a reflecting plane. The constant Q is the release rate, $\bar{u}$ is the mean wind speed, n is an index $(0 < n < 1)$ that varies with the vertical wind speed profile and C_y and C_z are the respective horizontal and vertical dispersion coefficients. The dispersion coefficients are measures of the amount of turbulence and are functions of n, wind speed, and 'macroviscosity' which is a

measure of the roughness of the surface (Sutton, 1953; Wright, 1958). However, Sutton suggests (1953, p. 292) that the following 'typical' values of
$C_y = 0.4\ C_z = 0.2$ and $n = 0.25$ can be used to estimate dispersion in a moderate
wind over level ground under neutral atmospheric conditions.

The Sutton equation belongs to a class known as Gaussian dispersion models
in which the concentration of odor along any axis perpendicular to the downwind (x) direction is assumed to follow a normal or Gaussian distribution
(Fig. 3.1). The dispersion coefficients determine the width of the plume and
thus are related to the standard deviation of the concentration along the crosswind and vertical axes.

Wright (1958) and Bossert and Wilson (1963) independently introduced the
use of Sutton's equation to analyze pheromone dispersion in the wind. Bossert
and Wilson solved the equation for the maximum distance of communication:

$$X_{\max} = \left[\frac{2Q}{K\pi C_y C_z \bar{u}}\right]^{1/(2-n)} \tag{3.4}$$

They estimated a Q/K for the gypsy moth (*Lymantria dispar*) using this
equation with the 'typical values' suggested by Sutton (1953 p. 292) combined
with an approximation of X_{max} based on a report by Collins and Potts (1932)

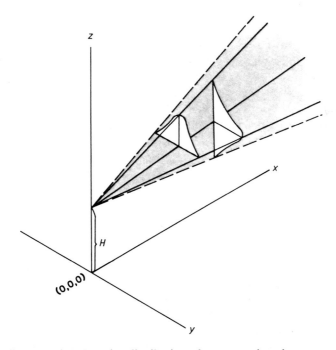

Fig. 3.1   The normal or Gaussian distribution of concentration along any axis perpendicular to the mean wind direction assumed by all plume models of the Gaussian type.

that gypsy moth females could attract males over a distance of 4 km. Unfortunately, there is no way of knowing what proportion of the 4 km distance was flown by the males prior to contacting the pheromone.

The work of Wright (1958) and Bossert and Wilson (1963) implanted the Sutton equation in the pheromone literature, despite the fact that most atmospheric scientists have long since adopted other Gaussian models (Mason, 1973; Fares *et al.*; 1980) and despite the theoretical objections to the use of time average models (Wright, 1958; Aylor, 1976; Aylor *et al.*; 1976). Furthermore, the 'typical values' for n, C_y and C_z have been used, regardless of the fact that the conditions under which they apply seldom prevail.

Various investigators have used Bossert and Wilson's equation (3.4) to estimate the maximum distance of communication or the behavioral threshold. Sower *et al.* (1971, 1973) used previous estimates of threshold and emission rate to calculate the maximum distance of communication for the noctuid moth *Trichoplusia ni*. Shapas and Burkholder (1978) used the same equation to estimate the theoretical maximum distance of communication of the dermestid beetle *Trogoderma glabrum* in a wind tunnel using known rates of release and thresholds estimated from experiments conducted in enclosed laboratory containers. Discrepancies between the theoretical and observed values of X_{max} prompted them to suggest modified values for C_y and C_z in a wind tunnel. Baker and Roelofs (1981) used equation (3.4) to estimate a lower *and upper* threshold for upwind flight in the Oriental fruit moth, *Grapholitha molesta*. The equation predicted a 3000-fold concentration difference between these two thresholds.

Hartstack *et al.* (1976) utilized the Sutton equation (3.3) to estimate the size of a pheromone active space as part of an effort to simulate the capture of moths in pheromone traps. Nakamura (1976) and Nakamura and Kawasaki (1977) estimated the size of the active space of the noctuid *Spodoptera litura* using Sutton's equation in conjunction with a function that modified the release rate to account for pheromone 'deposition' or adsorption to the surface. Hirooka and Suwani (1976) derived an equation similar to that of Sutton for diffusion of the pheromone of the arctiid moth *Hyphantria cunea* using a variable called 'gustiness' in place of the dispersion coefficients. They report that the size of the 'pheromone-effective sphere' defined by the maximum downwind distance of communication does not change very much with changes in wind velocity and 'gustiness'. However the *width* of their plume changes drastically with gustiness as it does with other Gaussian models.

3.3.2 Gaussian plume models

The typical values suggested by Sutton for 'n' and the dispersion coefficients in equation (3.3) only apply under conditions of neutral atmospheric stability. Neutral stability occurs when air temperature decreases with height above ground at a rate close to 1 °C per 100 m, which is known as the adiabatic lapse rate. However, atmospheric conditions are often not neutral. Unstable or super-adiabatic conditions predominate on sunny days when the radiant heat absorbed by the ground is transmitted to the air at ground level. The warm air rises into the cooler, denser air above causing a maximum amount of turbulence and mixing of airborne odors. At night, particularly on clear nights, radiant heat loss from the ground cools the air at the surface relative to the warmer air above, establishing inversion or stable conditions in which turbulence and dispersion rates are at a minimum. This general pattern is modified by cloud cover which increases stability during the day and reduces stability at night. In forests with a closed canopy, sunlight absorbed by the foliage may cause an inversion with cooler layers of air beneath the canopy.

Sutton's model can be extended to stable or unstable conditions by using different values for 'n' and different dispersion coefficients (Sutton, 1953, p. 291). However, theoretical objections to Sutton's model (see Gifford, 1968, p. 88) have led atmospheric scientists to utilize the following more general Gaussian plume equation:

$$C_{(x,y,z,H)} = \frac{Q}{2\pi\sigma_y\sigma_z\bar{u}} \exp\left[-\tfrac{1}{2}\left(\frac{y^2}{\sigma_y^2} + \frac{(z-H)^2}{\sigma_z^2}\right)\right] \tag{3.5}$$

where C_{xyz}, Q, and $\bar{u}$ are defined as in the Sutton equation and 'H' is the height of the source above ground. The dispersion coefficients σ_y and σ_z are equal to the standard deviation of the Gaussian distribution of odor concentration along any horizontal (σ_y) or vertical (σ_z) axis perpendicular to the mean downwind direction. The dispersion coefficients depend upon the atmospheric stability and they vary with the downwind distance (x) in contrast to the Sutton equation in which the dispersion coefficients (C_y, C_z) are constants at any distance under a given set of terrain and atmospheric conditions. In the Gaussian plume equation, the change in concentration over downwind distance depends on the changing value of the dispersion coefficients. The downwind distance (x) does not appear explicitly in the equation as it does in Sutton's model. Differences between various applications of the Gaussian plume model depend on the values chosen for the dispersion coefficients which are derived empirically from tracer experiments, in contrast to the theoretical derivation attempted by Sutton.

Fares *et al.* (1980) were the first to apply a general Gaussian plume model (3.5) to pheromone dispersion and the first to emphasize the importance of atmospheric stability on pheromone communication. They used dispersion coefficients derived from tracer experiments conducted in a pine forest. They hypothesized that the diurnal pattern of bark beetle responses to pheromone may relate to diurnal changes in stability. (For further discussion of the selective forces influencing diurnal activity patterns see Cardé and Baker, Chapter 12.)

Elkinton *et al.* (1984) used probit analysis to test the predictions of the Gaussian plume model using the dispersion coefficients suggested by Pasquill

(1961); Gifford (1968, p. 102); Mason (1973) and Fares *et al.* (1980) as well as the Sutton model. Model predictions were compared against observed gypsy moth active spaces which were mapped downwind of a pheromone source using male wing-fanning activity as an assay (Fig. 3.2). None of the models gave statistically significant predictions as to where gypsy moth wing-fanning would occur in relation to the mean wind direction. A peculiar feature of the Fares model was that it predicted higher concentrations of pheromone at 80 m than at 20 m along directions offset by more than a few degrees from the mean wind direction.

3.3.3 Instantaneous versus time-average dispersion models

The Gaussian dispersion models discussed so far all predict the average concentration of odor in space over a fixed time interval. The models are derived from

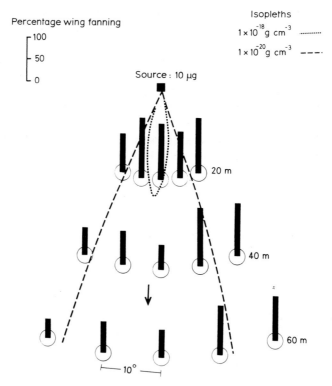

Fig. 3.2   Predicted concentration isopleths of airborne material utilizing the Gaussian plume model and the Pasquill 'Prairie Grass' dispersion coefficients versus the occurrence of male gypsy moth wing-fanning behavior over a 10-minute interval following pheromone release from a point source. The 1×10^{-18} isopleth approximates the minimum concentration that produces a similar wing-fanning response in the wind tunnel.

experiments in which tracer chemicals are released from a source and adsorbed by collectors arrayed downwind for the duration of that interval. The width or standard deviation of the Gaussian plume predicted from such experiments increases and the odor concentration at any point in space decreases with the duration of the sample interval (Fig. 3.3). Sutton's model, for instance, applies to a 3-minute sample interval. The dispersion coefficients utilized by Pasquill (1961; Gifford, 1968 p. 102) were derived from studies with a 10-minute sampling interval. The models predict average odor concentrations only for the same sample intervals as those from which the dispersion coefficients are derived. For example, to test the Sutton model, Elkinton *et al.* (1984) examined the wing-fanning behavior of male gypsy moths *during a 3-minute sample interval*. Pheromone biologists have generally used these models without considering the sample interval to which they apply.

Wright (1958), Shorey (1976), Aylor (1976) and Aylor *et al.* (1976) observed that the concentration of pheromone perceived at any instant in time is quite different from the concentration predicted by a time-average model. The instantaneous plume is a narrow swath of disjunct odor filaments that meanders downwind with the large-scale turbulent eddies. The time-average concentrations are obtained from collectors that experience little or no odor most of the time interspersed by bursts of high concentration. Consequently, an insect located downwind of an odor source will experience bursts of odor at concentrations well above the predicted average concentrations. According to Gifford (1960) the ratio of peak/average concentrations is small (1–5) along the mean plume centerline but increases to values above 100 away from the plume axis. Aylor *et al.* (1976) estimated a peak/average ratio of the order of 25 within

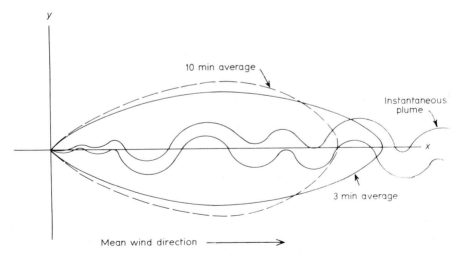

Fig. 3.3 Diagrammatic representation of a concentration isopleth of an instantaneous plume and time-average Gaussian plumes with 3-minute and 10-minute averaging times (redrawn from Slade, 1968).

a few meters of the pheromone source. They distinguished between the small peak/average ratio (2–3) that occurs within the instantaneous plume due to its disjunct, filamentous nature and the large peak/average ratio (*ca.* 50) that is caused by plume meander. The determination of peak/average concentrations depends upon the sampling time used to determine the average *and* the peak concentrations (Gifford, 1960). Odor receptors can respond to odors within a fraction of a second, an interval shorter than the sampling time used to measure peak concentrations in most experiments. However, using an ion generator Murlis and Jones (1981) measured ion concentrations at locations 2–15 m downwind with very short sample intervals (a few ms). At these distances they found peak/average concentration ratios around 20.

The occurrence of peak pheromone concentrations at levels well above those predicted by time average models may explain why Elkinton *et al.* (1984) observed wing-fanning responses at locations well outside the range predicted by the Gaussian models they examined. The median concentration predicted by these models for locations where a wing-fanning response occurred was several orders of magnitude below the threshold known to produce such wing-fanning in a wind tunnel (Hagaman and Cardé, 1983).

The work of Murlis and Jones (1981) illustrates the discontinuous, filamentous nature of the instantaneous odor plume. The continuous stream of ions they released arrived at samplers downwind in discrete bursts. The disjunct nature of the plume is also evident in the 'instantaneous' (1 minute) pattern of wing-fanning onset in the data of Elkinton *et al.* (1984, unpublished) along a continuous cross-section of the plume 20 m downwind of the source (Fig. 3.4).

The turbulent eddies that cause an odor plume to disperse occur simultaneously over a range of sizes in the atmosphere. Energy derived from the largest eddies is passed down into smaller and smaller eddies until it dissipates as heat. As pointed out by Slade (1968), Mason (1973), Aylor (1976) and Aylor *et al.* (1976) the dispersion of an instantaneous odor plume is driven by eddies that are about the same size as the plume. Eddies that are smaller than the plume redistribute the pheromone within the plume whereas the larger eddies cause the plume to meander downwind intact. At the small end of the spectrum there exists a lower eddy size limit below which molecular viscosity causes the eddies to break down and dissipates their energy as heat. Typically, the minimum eddy size is in the order of 1 mm.

Aylor (1976) and Aylor *et al.* (1976) adapted a model of the instantaneous odor plume originally developed by Batchelor (1952) based upon the notion of dispersion relative to the plume centerline which meanders downwind with the large-scale eddies. As the plume expands, the rate of expansion changes, depending on the size of the plume in relation to the abundance of eddies that exist in that size range. There are three phases of plume growth in Aylor's model. During the first phase, which lasts at most a few cm, the plume is smaller than the smallest eddies and expands slowly by molecular diffusion alone. The second phase begins when the plume attains a diameter equal to that

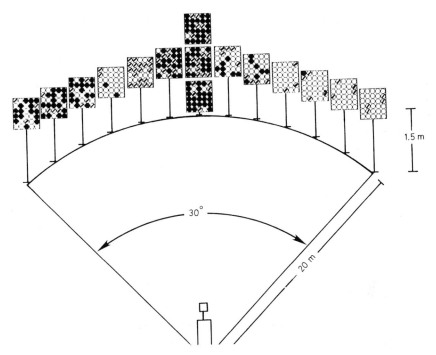

Fig. 3.4 The onset of wing-fanning behavior (solid circles) over a 1-minute period following pheromone release among 390 male gypsy moths deployed in wire mesh cages along a continuous 30° arc at a distance of 20 m from a pheromone (+disparlure) source. The height of the source was 1.5 m. Moths that were already fanning at the time of pheromone release and therefore could not initiate fanning are indicated by circles with a slash.

of the smallest eddies and the expansion of the plume accelerates rapidly. During the final phase when the plume has expanded to the size of the large eddies the rate of expansion slows.

Miksad and Kittredge (1979) proposed an instantaneous plume model similar in many respects to that of Aylor *et al*. The principal difference is that Miksad and Kittredge believe that a pheromone plume is initially much smaller than the smallest eddies and the initial molecular diffusion stage lasts much longer, producing a narrow 'filament' that is carried downwind over a distance of many meters. The length of the molecular diffusion phase depends upon the size of the pheromone source, its height above ground and the wind speed. During a second phase of molecular diffusion the smallest eddies begin to chop the filament into fragments, each of which continues to expand by molecular diffusion. The third and fourth stages of plume expansion correspond to the last two stages of the model of Aylor *et al*.: an explosive phase of plume growth during which the plume is torn apart by eddies in the 'inertial subrange' (see

Pasquill 1974, p. 46) followed by a final phase of slower growth. In contrast to Aylor's model, the explosive growth phase in Miksad and Kittredge's model occurs many meters downwind of the source. The length of the initial filament stage (up to 40 m) rests upon the assumption that the size of the typical phero-mone source is much smaller than the smallest eddies. The size of the typical pheromone source they cited (0.1 to 1 mm) is smaller than that typical for most insects. Consequently, the initial filament stage may only be a few cm as in Aylor *et al.* (1976).

Attempts to verify these instantaneous plume models or apply them to predicted pheromone active spaces in the field are hampered by the fact that they consider only dispersion relative to the plume centerline. The meandering centerline would cause even the narrow filament stage of Miksad and Kittredge to cut a wide swath within a few seconds. Aylor *et al.*, (1976) used a wing-fanning assay of caged male gypsy moths, attempting to keep the moths within the meandering plume by moving them continuously in the track of a visual tracer. For unexplained reasons they failed to observe a wing-fanning response beyond 2.5 m. The intermittent occurrence of odor at samplers fixed in space as in the experiments of Elkinton *et al.* (1984) and Murlis and Jones (1981) is caused by a combination of plume meander and the disjunct occurrence of odor within the plume. Only fixed samplers with extremely fast response times such as those utilized by Murlis and Jones (1981) could distinguish between these two sources of odor discontinuity.

Predictions of instantaneous odor concentration at locations fixed in space using a model for dispersion relative to the meandering plume centerline require that we describe plume meander as well. Miksad and Kittredge (1979), for example derived a probability density function that specifies the probability that a given location will be within the plume. Similar probabilities can be derived from the ion concentration studies of Murlis and Jones (1981). An entirely different approach is suggested in David *et al.* (1982) who adapted the ideas of Davidson and Halitsky (1958) to odor dispersion. They suggest that over short distances the speed and direction of any individual 'parcel' of air will be approximately constant even though the speed and direction of sequential 'parcels' as monitored by a windvane is changing continuously. If we conceive of the odor plume as a continuous series of puffs, each puff will thus travel in an approximate straight line at constant speed after it leaves the source. By monitoring the changing wind speed and direction at the source we could predict the location of each puff at each subsequent instant and thus map the plume meander. Davidson and Halitsky (1958) applied this simple model to dis-persion in the vertical plane under strong winds from sources elevated 100 m above ground. Whether this idea can be usefully coupled with models that describe plume expansion relative to the meandering centerline to provide an accurate description of odor dispersion near the ground remains to be seen. Undoubtedly vegetation or objects on the ground will affect the trajectory of odor puffs (see Wall *et al.*, 1981).

3.4 EFFECTS OF WIND SPEED AND AIR TEMPERATURE ON DISPERSION

3.4.1 Wind speed

A major prediction of the Sutton equation and subsequent Gaussian models is that the size of the odor active space and the maximum distance of communication decrease with increasing wind speed. The instantaneous plume model of Aylor *et al.* (1976) and Miksad and Kittredge (1979) also predict shrinking active spaces at higher wind speeds. The principal cause of this effect is that, at higher wind speeds, a given amount of odor is entrained into a larger 'initial volume' (e.g., the volume attained by the plume in the first second after release) thereby diluting the plume at all subsequent locations downwind. In addition, the 'turbulent intensity' as measured by the standard deviation of the wind speed increases at higher wind speeds (R. Shaw, personal communication) causing more rapid expansion of the odor puff. However, the puff is also being transported more rapidly downwind at higher wind speeds. This usually results in a narrower plume as indicated by a smaller value of the dispersion coefficients (σ_y, σ_z) at higher wind speeds which can be related to smaller values for the standard deviation of wind direction at higher wind speeds.

Various studies have documented a decline in the communication distance at increasing wind speeds. Nakamura (1976) reported that the active space of *S. litura*, as defined by male wing-fanning responses, decreased at higher wind speeds but also at the lowest wind speeds as well. He attributed the latter trend to increased pheromone 'deposition' at the lowest speeds.

The effect of wind speed on the size of the active space is compounded by possible effects on the rate of odor release. The rate of odor release from many synthetic or natural sources is governed by the rate of transport to the surface of the releasing substrate and would therefore be independent of wind speed. Hirooka and Suwani (1976) argued on theoretical grounds that the release rate from any small dispenser would be unaffected by wind speed. However, Elkinton *et al.* (1984) demonstrated that release rates of disparlure from cotton wick dispensers increase with increasing wind speed.

The dispersion models discussed in this chapter are all expressed in terms of odor concentration in space ($g\,cm^{-3}$). Odor receptors, however, respond to the rate of odor adsorption per unit time which is more directly related to the flux ($g\,cm^{-2}\,s^{-1}$) of odor molecules over the receptor organ than to odor concentration. The rate of flux through a stationary receptor increases in direct proportion to the wind speed offsetting the inverse decline in concentration. The efficiency of adsorption, however, may decrease at higher wind speeds.

Wind speed may also directly affect the response and pheromone release behavior of organisms. For example, Kaae and Shorey (1972) reported that the persistence of the calling behavior of *T. ni* was greatest at intermediate wind speeds which they suggest were optimal for successful mate finding. Various studies have shown that higher wind speeds suppress the flight behavior of various male Lepidoptera (e.g., Sower *et al.*, 1973). Wind speeds vary with time of day and thus may affect the diel pattern of trap catch or pheromone responsiveness (See Cardé and Baker, Chapter 12).

3.4.2 Air temperature

Air temperature may affect the size of the active space in several ways. As discussed above the vertical gradient in air temperature determines the stability of the atmosphere and thus the rate of dispersion. Air temperature will affect the rate of odor release from synthetic dispensers and natural odor sources. It may affect the 'calling behavior' of pheromone-producing organisms (see Cardé and Baker, Chapter 12) and determine the response threshold of odor-receiving organisms (Cardé and Hagaman, 1983). Baker and Roelofs (1981) document the changing size of an active space as a function of air temperature.

3.5 DEPOSITION AND VERTICAL DISTRIBUTION OF ODORS

The belief that pheromone molecules sink as they leave the source because they are heavier than air has been invoked to explain the common observance that many insects follow odor plumes near ground level or that they approach an odor source from below. However, this belief is a misconception because the gravitational fall of individual molecules in the air is overwhelmed by collisions with other molecules. The pheromone released from a source will only sink if the pheromone-laden air as a whole is significantly more dense than surrounding air. Mankin *et al.* (1980a) argued convincingly that under most conditions pheromone-laden air will not sink because the pheromone concentrations are sufficiently low to cause a negligible increase in air density. The observed ground-level flight of insects orienting to an odor source is more likely to be caused by the requirements of the optomotor orientation mechanism (Kennedy, 1940; Kuenen and Baker, 1982; Elkinton and Cardé, 1983; Cardé, Chapter 5).

Nevertheless, various odor dispersion models predict that highest concentrations occur at ground level. This prediction depends upon the height of the odor source and assumptions concerning the reflection or adsorption of odors at the surface. The simplest case, as in the model of Bossert and Wilson (1963), occurs when the source is located on the surface and odor adsorption on the surface is negligible. Complete reflection of the plume results in a simple doubling of the concentration at all points downwind which accounts for the factor '2' in the numerator of the first term. The same result occurs in the Gaussian plume model for ground level sources and no adsorption. For a source above ground level an additional term is added to the Gaussian plume equation (Gifford, 1968 p. 99; Fares *et al.*, 1980):

$$C_{(x,y,z,H)} = \frac{Q}{2\pi\sigma_y\sigma_z\bar{u}} \exp\left[-\frac{y^2}{2\sigma_y^2}\right] \left[\exp\left(-\frac{(z-H)^2}{2\sigma_z^2}\right) + \alpha \exp\left(-\frac{(z+H)^2}{2\sigma_y^2}\right)\right], \quad (3.6)$$

where α is a constant that depends upon the degree of adsorption of pheromone at the surface. The physical analogy of this third term is a mirror image of the source an equal distance below the surface (Fig. 3.5). When adsorption at the surface is complete equation (3.6) reduces to (3.5). When the source is located on a surface with complete reflection, equation (3.6) reduces to the Gaussian plume analogue of the Sutton equation (3.3). At any rate, all these models predict that, given a certain amount of reflection from the surface, maximum concentrations occur at ground level beginning at some point downwind from an elevated source. The same prediction is obtained by Miksad and Kittredge (1979) from their instantaneous plume model. These results occur despite the fact that pheromone-laden air does not sink. It is doubtful that a simple doubling of concentration is a sufficient increase to alter behavior.

The results of the male gypsy moth wing-fanning experiments of Elkinton *et al.* (1984) did not support the conclusion that pheromone concentrations were highest at ground level downwind from an elevated point source. At a distance of 20 m from the source they found that a significantly higher proportion of males initiated wing-fanning at the same height as the source (1.6 m) compared to moths near ground level (0.5 m). The same trend was evident at 60 m although the difference between the proportions responding at the two heights was not statistically significant. At both distances the response of moths near ground level was delayed *ca.* 48 s at 20 m and 19 s at 60 m compared to the

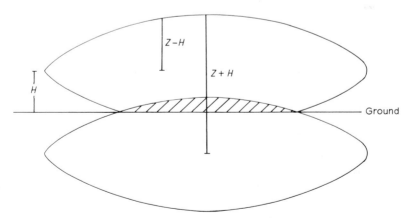

Fig. 3.5   Diagram of the mirror image analogy incorporated in the Gaussian plume model to account for the reflection of airborne material from the ground surface after release from an elevated source.

moths at the source height indicative of slower wind speeds near ground level. Wind direction data from a bi-directional wind vane at the site gave no evidence for net vertical movements of the wind.

Very little work has been done on the rate of pheromone adsorption and re-entrainment by vegetation and other objects in nature. Various studies have documented reduced trap catch of insects to pheromones in heavy vegetation. These results, however, may have easily been caused by disruption of the flight behavior by the vegetation rather than pheromone adsorption. Adsorption to vegetation may be substantial (Wall *et al.*, 1981).

An additional aspect of the vertical distribution of pheromone has been considered by Schal (1982). The odor dispersion models discussed in this chapter all assume that the mean wind flow is horizontal. The occurrence of buoyant or convective transport of odors is treated by increasing the rate of dispersion to that characteristic of unstable conditions. On the relatively small scale of odor communication this description may not be adequate. Schal demonstrated that air currents at night beneath a Costa Rican rain forest move predominantly upward in response to a substantial lapse rate. He found that male cockroaches of various species positioned themselves in trees at levels above that of pheromone producing females of the same species.

3.6 CONCLUSION

Except for odor dispersion in still air which has been treated theoretically by Bossert and Wilson (1963) and Mankin *et al.* (1980a), the analysis of odor dispersion has been dominated by the use of the Sutton equation ever since it was introduced to the pheromone literature by Wright (1958) and Bossert and Wilson (1963). It has been widely used to calculate behavioral thresholds and maximum distances of communication of various pheromone emitting organisms. Atmospheric scientists have turned to the more general Gaussian plume model to describe the average concentration of airborne material dispensing from a point source. This latter model has the advantage that it can be applied to a range of atmospheric stabilities and depends upon dispersion coefficients that are determined experimentally for a given situation.

The Sutton model has persisted despite the fact, first observed by Wright (1958), that the odor plume encountered by an organism at any instant in time is very different from the plume predicted by the Sutton equation or any other time-average model. The validation experiments of Elkinton *et al.* (1984) have underscored the inadequacies of various time-average Gaussian models for predicting the dimensions of the active space. Aylor *et al.* (1976) and Miksad and Kittredge (1979) developed instantaneous plume models for dispersion relative to an instantaneous plume centerline which meanders downwind. It is difficult to use these models to predict concentrations at fixed locations downwind because they do not predict the meander of the plume centerline. Indeed a major reason for the continued use of the Sutton equation is the ease of the calculations involved. Perhaps the best approach for a replacement for the Sutton equation will be a model which couples instantaneous dispersion relative to a meandering plume centerline (Aylor *et al.*, 1976) with a description of the plume meander (Davidson and Halitsky, 1958; David *et al.*, 1982) or with a probability density function that assigns a likelihood that a fixed point downwind of a source will at a given moment lie within the instantaneous plume (Miksad and Kittredge, 1979; Murlis and Jones, 1981). The key to either approach will be understanding the meandering of the plume under various atmospheric and terrain conditions.

ACKNOWLEDGEMENTS

Our research on pheromone dispersion has been supported by NSF Grant PMC 7912014 and by a grant from the USDA Expanded Gypsy Moth Program. We are greatly indebted to Conrad Mason, John Murlis, Roger Shaw and Ralph Charlton for invaluable discussion and review of this paper.

REFERENCES

Aylor, D. E. (1976) Estimating peak concentrations of pheromones in the forest. In: *Perspectives in Forest Entomology* (Anderson, J. F. and Kaya, M. K., eds) pp. 177–88. Academic Press, New York.

Aylor, D. E., Parlange, J.-Y. and Grannett, J. (1976) Turbulent dispersion of disparlure in the forest and male gypsy moth response. *Env. Ent.*, **5**, 1026–32.

Baker, T. C. and Cardé, R. T. (1979) Analysis of pheromone-mediated behaviors in male *Grapholitha molesta*, the Oriental fruit moth (Lepidoptera: Tortricidae). *Env. Ent.*, **8**, 956–68.

Baker, T. C. and Roelofs, W. L. (1981) Initiation and termination of Oriental fruit moth male response to pheromone concentration in the field. *Env. Ent.*, **10**, 211–18.

Batchelor, G. K. (1952) Diffusion in a field of homogeneous turbulence: II. The relative motion of particles. *Proc. Cambr. Phil. Soc.*, **48**, 345–62.

Bossert, W. H. and Wilson, E. O. (1963) The analysis of olfactory communication among animals. *J. theoret. Biol.*, **5**, 443–69.

Cardé, R. T. and Hagaman, T. E. (1979) Behavioral responses of the gypsy moth in a wind tunnel to air-borne enantiomers of disparlure. *Env. Ent.*, **8**, 475–84.

Cardé, R. T. and Hagaman, T. E. (1983) Influence of ambient and thoracic temperatures upon sexual behaviour of gypsy moth. *Physiol. Ent.*, **8**, 7–14.

Cardé, R. T., Dindonis, L. L., Agar, B. and Foss, J. (1983) Apparency of pulsed and continuous pheromone to male gypsy moths. *J. Chem. Ecol.* (in press).

Collins, C. W. and Potts, S. F. (1932) Attractants for the flying gipsy moths as an aid in locating new infestations. *U.S.D.A. Tech. Bull.* No. 336.

David, C. T., Kennedy, J. S., Ludlow, A. R., Perry, J. N. and Wall, C. (1982). A re-appraisal of insect flight towards a distant, point source of wind-borne odor. *J. Chem. Ecol.*, **8**, 1207–15.

Davidson, B. and Halitsky, J. (1958) A method of estimating the field of instantaneous ground concentration from tower bivane data. *J. Air Poll. Cont. Ass.*, **7**, 316–19.

Elkinton, J. S. and Cardé, R. T. (1983) Appetitive flight behavior of male gypsy moths (Lepidoptera: Lymantriidae). *Env. Ent.* (in press).

Elkinton, J. S., Cardé, R. T. and Mason, C. J. (1984) Evaluation of time-average dispersion models for estimating pheromone concentration in a deciduous forest. *J. Chem. Ecol.* (in press).

Fares, Y., Sharpe, P. J. H. and Magnuson, C. E. (1980) Pheromone dispersion in forests. *J. Theoret. Biol.*, **84**, 335–59.

Gifford, F. A. Jr. (1960) Peak to average concentration ratios according to a fluctuating plume dispersion model. *Int. J. Air Poll.*, **3**, 253–60.

Gifford, F. A. Jr. (1968) An outline of theories of diffusion in the lower layers of the atmosphere. In: *Meteorology and Atomic Energy* (Slade, D. H., ed.) pp. 65–116. US Atomic Energy Commission, Oak Ridge, Tennessee.

Hagaman, T. E. and Cardé, R. T. (1983) Effect of pheromone concentration on the organization of pre-flight behaviors of the male gypsy moth, *Lymantria dispar* (L.). *J. Chem. Ecol.* (in press).

Hartstack, A. W., Witz, J. A., Hollingsworth, J. P. and Bull, D. L. (1976) SPERM – a sex pheromone emission and response model. *Trans. Am. Soc. Agr. Eng.*, **19**, 1170–80.

Hirooka, Y. and Suwani, M. (1976) Role of insect sex pheromone in mating behavior. I. Theoretical consideration on release and diffusion of sex pheromone in the air. *Appl. ent. Zool.*, **11**, 126–32.

Kaae, R. S. and Shorey, H. H. (1972) Sex pheromones of noctuid moths. XXVII. Influence of wind velocity on sex pheromone releasing behavior of *Trichoplusia ni* females. *Ann. Ent. Soc. Am.*, **65**, 436–40.

Kaissling, K-E. (1971) Insect olfaction. In: *Handbook of Sensory Physiology IV Chemical Senses 1. Olfaction* (Beidler, L. M., ed.) pp. 351–431. Springer-Verlag, New York.

Kennedy, J. S. (1940) The visual responses of flying mosquitoes. *Proc. Zool. Soc. Lond., Ser. A.*, **109**, 221–42.

Kuenen, L. P. S. and Baker, T. C. (1982) Optomotor regulation of ground velocity in moths during flight to sex pheromone at different heights. *Physiol. Ent.*, **7**, 193–202.

Mankin, R. W., Vick, K. W., Mayer, M. S., Coffelt, J. A. and Callahan, P. S. (1980a) Models for dispersal of vapors in open and confined spaces: Application to sex pheromone trapping in a warehouse. *J. Chem. Ecol.*, **6**, 929–50.

Mankin, R. W., Vick, K. W., Mayer, M. S. and Coffelt, J. A. (1980b) Anemotactic response threshold of the Indian meal moth, *Plodia interpunctella* (Hübner) (Lepidoptera: Pyralidae) to its sex pheromone. *J. Chem. Ecol.*, **6**, 919–28.

Mason, C. J. (1973) Meteorological dispersion models and their application to aerobiological problems. Presented at: Workshop/Conference III, Ecological Systems Approaches to Aerobiology, US/IBP Aerobiology Program, University of Michigan, Ann Arbor, Michigan.

Matthews, R. W. and Matthews, J. R. (1978) *Insect Behavior*. Wiley-Interscience, New York.

Miksad, R. W. and Kittredge, J. (1979) Pheromone aerial dispersion: A filament model. *14th Conf. Agric. and For. Met., Am. Met. Soc.*, pp. 238–43.

Murlis, J. and Jones, C. D. (1981) Fine-scale structure of odour plumes in relation to insect orientation to distant pheromone and other attractant sources. *Physiol. Ent.*, **6**, 71–86.

Nakamura, K. (1976) The effect of wind velocity on the diffusion of *Spodoptera litura* (F.) sex pheromone. *Appl. ent. Zool.*, **11**, 312–19.

Nakamura, K. and Kawasaki, F. (1977) The active space of the *Spodoptera litura* (F.) sex pheromone and the pheromone component determining this space. *Appl. ent. Zool.*, **12**, 162–77.

Pasquill, F. (1961) The estimation of the dispersion of wind-borne material. *Met. Mag.*, **90**, 33–49.

Pasquill, F. (1974) *Atmospheric Diffusion*. Halsted Press. John Wiley, New York.

Roberts, O. F. T. (1923) The theoretical scattering of smoke in turbulent atmosphere. *Proc. Roy. Soc. A.*, **104**, 640–54.

Roelofs, W. L. (1978) The threshold hypothesis for pheromone perception. *J. Chem. Ecol.*, **4**, 685–99.

Rust, M. K. (1976) Quantitative analyses of male responses released by female sex pheromone in *Periplaneta americana. Anim. Behav.*, **24**, 681–5.

Schal, C. (1982) Intraspecific vertical stratification as a mate-finding mechanism in tropical cockroaches. *Science*, *215*, 1405–7.

Shapas, T. J. and Burkholder, W. E. (1978) Patterns of sex pheromone release from adult females, and effects of air velocity and pheromone release rates on theoretical communication distances in *Trogoderma glabrum. J. Chem. Ecol.*, **4**, 395–408.

Shorey, H. H. (1976) *Animal Communication by Pheromones*. Academic Press, New York.

Slade, D. H. (ed.) (1968) *Meterorology and Atomic Energy*. US Atomic Energy Commission, Oat Ridge, Tennessee.

Sower, L. L., Gaston, L. K. and Shorey, H. H. (1971) Sex pheromones of noctuid moths. XXVI. Female release rate, male response threshold, and communication distance for *Trichoplusia ni. Ann. Ent. Soc. Am.*, **64**, 1448–56.

Sower, L. L., Kaae, R. S. and Shorey, H. H. (1973) Sex pheromones of Lepidoptera. XLI. Factors limiting potential distance of sex pheromone communication in *Trichoplusia ni. Ann. Ent. Soc. Am.*, **66**, 1121–2.

Sutton, O. G. (1947) The problem of diffusion in the lower atmosphere. *Q. J. R. Met. Soc.*, **73**, 257–81.

Sutton, O. G. (1953) *Micrometeorology*. McGraw-Hill, New York.

Taylor, G. I. (1921) Diffusion by continuous movements. *Proc. Lond. Math. Soc., Ser. 2*, **20**, 196–212.

Wall, C., Sturgeon, D. M., Greenway, A. R. and Perry, J. N. (1981) Contamination of vegetation with synthetic sex-attractant released from traps for the pea moth, *Cydia nigricana. Ent. exp. Appl.*, **30**, 111–15.

Wilson, E. O., Bossert, W. H. and Regnier, F. E. (1969) A general method for estimating threshold concentrations of ordorant molecules. *J. Insect Physiol.*, **15**, 597–610.

Wright, R. H. (1958) The olfactory guidance of flying insects. *Cand. Ent.*, **90**, 81–9.

4

Chemo-orientation in Walking Insects

William J. Bell

4.1 INTRODUCTION

This section on walking insects, together with the following on flying insects, defines and illustrates mechanisms by which insects utilize chemical information available to them for purposes of locating mates, food and other resources or for avoiding repellents or stress sources. This discussion follows Städler, Chapter 1, and Mustaparta, Chapter 2 on the acquisition and processing of chemical information through peripheral receptors and the central nervous system, Elkinton and Cardé, Chapter 3, on airborne dispersal of chemicals, and is an introduction to the remaining chapters in this volume on ecological implications of resource localization and stress avoidance.

4.2 CLASSIFICATION OF ORIENTATION MECHANISMS

The first attempt to treat animal orientation in mechanistic terms was published by Loeb (1918). This was followed in 1919 by Kühn's comprehensive framework of animal orientation, and then much later by the compilation of Fraenkel and Gunn (1940). The following definitions of orientation mechanisms are from Kühn (1919) and Fraenkel and Gunn (1940).

Orthokinesis: Speed or frequency of locomotion is dependent on the intensity of stimulation.
Klinokinesis: Responses in which the rate of random turning, or angular velocity, depends on the intensity of stimulation.
Klinotaxis: Directed orientation made possible by means of regular deviations and

Chemical Ecology of Insects. Edited by William J. Bell and Ring T. Cardé
© 1984 Chapman and Hall Ltd.

involving comparison of intensities at successive points in time. Later, Ewer and Bursell (1950) expanded klinotaxis to include longitudinal temporal processing along the axis of movement, as well as transverse (lateral) deviations to the axis of movement.

Tropotaxis: Symmetrical orientation. The animal turns so that symmetrically positioned lateral sense organs are equally stimulated; asymmetry of stimulation leads to turning toward one of the two symmetrical positions.

Telotaxis: Goal orientation. Locomotion along the line joining the animal to a point source of stimulation; maintenance of a certain part of the field of stimulation on a particular point of the sensory apparatus − the fixation point.

Menotaxis: Maintenance of a given direction of the body axis by preserving a certain distribution of stimulation over the sensory surface, using compensatory movements (=compass orientation).

The problems with this system of classification have recently been summarized by Bell and Tobin (1982): (i) the terms are inconsistent, some based on sources of external sensory input or mechanisms of information processing, and some on the geometrical structure of locomotory pathways, (ii) integration of different mechanisms is neglected, even though it is theoretically possible, for example, for an animal to utilize klinotaxis and tropotaxis at the same time and to select the strongest difference obtained, (iii) the role of internally stored directional information is not considered, thus mistakenly referring to orientation that is made relative to the previous direction as 'random' orientation, (iv) data are not always sufficient to use these definitions, leading investigators to simply pick the most likely one instead of publishing an accurate description of the behavior observed; for example, most citations of 'chemotaxis' would be more accurate labeled as 'chemo-orientation' until critical tests are performed to determine which type of mechanism is involved, (v) the system, with a few exceptions, does not account for orientation based on more than one sensory modality, even though orientation in the real world is undoubtedly based on more than one modality.

To utilize the data available on chemo-orientation for a better understanding of principal internal mechanisms involved, Bell and Tobin (1982) have suggested that researchers describe, instead of or in addition to invoking the above terms, (i) the *information available* to an organism, (ii) the type of *information processing* employed, (iii) the *motor output* (e.g., search patterns) elicited by a stimulating chemical, and (iv) the nature of the *guidance system*, based on processing of the available sensory information. These components of orientation are expanded in the following paragraphs.

4.2.1 Information available

There are two general categories of information available to an organism for chemo-orientation: (i) *Externally derived sensory information* (=*allothetic*, Mittelstaedt and Mittelstaedt, 1973; =*exokinetic*, Jander, 1970, 1975), and (ii) *internally-derived information*, termed *endokinetic* if genetic or otherwise

Table 4.1   Types of external chemosensory information

Contact chemoreception	*Example*
Single contact chemical stimulus	Feeding deterrent
Trails or patches on a substrate	Nectar patch
Olfactory chemoreception	
Single olfactory chemical stimulus	Alarm pheromone pulse
Airborne trail or patch	Prey odor patch
Gradient	Sex pheromone

stored (Jander, 1970, 1975) or *idiothetic* if stored through proprioreceptors or otherwise learned (Mittelstaedt and Mittelstaedt, 1973).

Chemical sensory information falls into two broad, sometimes overlapping categories of mediation by contact and olfactory chemoreception. As shown in Table 4.1, these categories can be subdivided into single events, as when an insect discovers a single resource, and multiple events as when an insect enters a continuum such as an odor trail, patch or gradient.

Internally derived or stored orientation information is as necessary as externally derived information (Table 4.2). The sign (positive or negative) of the response to a chemical that is perceived is often learned, but with informational chemicals produced by the insects themselves or their prey, predator or host plant, the sign of response is usually genetically fixed. For example, alarm pheromones elicit a negative response and sex pheromones elicit a positive response. Spatial information can be genetically specified, as with components of search patterns, one form of endokinetic orientation (Jander, 1970), or derived from proprioreceptors and stored in memory as a kind of locomotory history (track memory) for use in subsequent idiothetic orientation (Burger, 1972). When an insect is caused to turn by an external directional stimulus, for example, it will probably swing back toward its original direction when the stimulus is removed. Such compensatory responses, which in a general context are sometimes called counterturning or alternation behavior, are controlled by internally stored information about the previous course direction of the insect.

Table 4.2   Types of internally stored orientation information

Short-term memory
    Track memory (proprioreception)
Long-term memory
    Sign of response (+ or −)
    Turn angle selection (reduction in circular variance)
    Direction (left or right)
    Directional constancy (generating loops or zig-zags)
    Pattern generator (turning rate, stops schedule)
    Terminator (timer in energy units, steps, sensory adaptation)
    Sequence (of patterns)
    Noise (variation)

4.2.2 Information processing

At the level of the chemoreceptor, differences in stimulus intensity result in differences in the rate of impulses transmitted to the central nervous system. As an insect moves in space or remains in one place, this information alone can potentially provide for a *temporal comparison*, and thus an interpretation of the stimulus pattern that is changing in its vicinity or that it is passing through. If an insect has a multiple receptor system with receptors spaced at sufficient distances to perceive a detectable difference in stimulus intensities, an *instantaneous comparison* can be made to interpret the nature of the stimulus pattern at one point in time. As summarized by Städler, Chapter 1 and Mustaparta, Chapter 2, nothing is known as yet about the neural mechanisms responsible for integration through temporal and instantaneous comparisons.

4.2.3 Motor output patterns

Triggered by chemical stimuli, an insect may terminate active (moving) or passive (non-moving) ranging, during which chemical stimuli are unavailable, and initiate species-specific motor output, which can be referred to as *guided search*. Locomotory patterns of guided search can fit anywhere in a continuum from seemingly random locomotion to highly structured patterns of repeating loops, zig-zags, spirals or sine waves. In other cases the pattern is referred to a second modality (e.g., wind currents). Where a second modality is not involved, an insect must rely on internally-stored information; where a second modality is involved, orientation is based on externally-derived non-chemical sensory information.

4.2.4 Guidance systems

Motor output, controlled by internally stored information, can be guided by changes in chemosensory information in at least three distinct ways: (i) the basic motor pattern persists, but with *modulation of one specific parameter*, such as changes in turning velocity, zig-zag amplitude or turn dimension, (ii) the basic pattern persists, but with a *change in* the *directional vector*, (iii) one pattern of motor output is *replaced by another*, as in a shift from circling to zig-zags to straight locomotion. Two useful descriptive terms pertaining to guidance are *direct* and *indirect orientation*. Turns executed during guided search are directly related to the polarity of the chemical gradient in direct orientation. Turns are not necessarily related to the chemical gradient in indirect orientation, and the chemical source is localized indirectly through a pattern of turning, looping, zig-zagging, or through non-chemical directional information, such as wind currents.

4.3 INDIRECT CHEMO-ORIENTATION BASED ON INTERNALLY STORED INFORMATION

Insects probably 'range' as do other organisms in a relatively straight line; 'ranging' switches to 'searching' when the odor of a resource is detected but cannot be localized and when one resource item is discovered and others are sought (Jander, 1975). Some species range by following a menotactic angle over relatively long distances with respect to a visual stimulus or to the direction of a wind current, thereby maintaining a straight walking course (e.g., Linsenmair, 1970, 1973; Bell and Kramer, 1979); when a pheromone is perceived, the animal may initiate upwind, zigzag orientation. Other species 'perch' while seeming to 'wait' for information about resources in the vegetation with only the antennae moving, prior to searching if a relevant odor is perceived (Schal *et al.*, 1983a).

A common type of indirect chemo-orientation is a species-specific search pattern that is released by perception of an odor, but where the insect cannot obtain sufficient information to execute orientation relative to a boundary or to a gradient. A walking insect can potentially control its rate of locomotion, turn direction constancy (left or right), frequency of discrete turns, dimensions of discrete turns or turn velocity (e.g., degree s^{-1}) in response to a perceived chemical stimulus. These alterations in locomotory mode, which generate a search pattern, need not be directly related to the spatial configuration of the stimulus pattern. Search patterns were first pointed out as such by Dethier (1957) in blowflies fed on a drop of sucrose solution (Fig. 4.1(a)). Subsequently, Murdie and Hassell (1973) revealed how the search pattern of the hungry fly correlates with localization of other sucrose drops (resources) in a patch. A similar example is the search pattern of the enticed cockroach *Blattella germanica*, stimulated by contact sex pheromone (Fig. 4.1(b)). The cockroach exhibits courtship turning (Bell and Schal, 1980), and if no female stimuli are encountered, it engages in local search (Schal *et al.*, 1983b). Search patterns of this kind rely to a large extent on genetically stored information that dictates the turning velocity, frequency of changes in direction and duration of search. As depicted in Fig. 4.1(a) and (b), paths are never identical, indicating the injection of variation (noise) at some level in the CNS and that search patterns are not dictated as are fixed action patterns. The paths of many parasitoid species searching for prey consist of a series of components, each serving to contract the area searched (Vinson, Chapter 8). This mechanism is exactly opposite to that described above for flies and cockroaches that locate a resource and then search for another or for one they have lost contact with. Another example is the specific response to an alarm pheromone, as elaborated in ants (Bradshaw and Howse, Section 15.3.2) and aphids (Nault and Phelan, Chapter 9). In all cases, however, a search pattern is *released* by external sensory information, but *controlled* by internally stored orientation information.

Fig. 4.2 illustrates two examples of another simple type of search pattern that utilizes 'boundary information' at the border of a resource patch. In the first

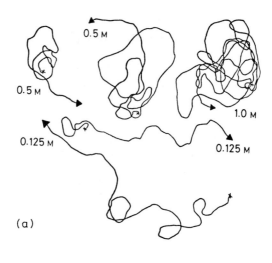

0.5 M

0.5 M

1.0 M

0.125 M

0.125 M

(a)

(b)

10 cm

Fig. 4.1(a)   Search pathways of blowflies after feeding on a drop of sucrose solution; 'x' shows position of drop; numbers refer to sucrose concentration (Dethier, 1957).
(b) Search pathways of the male German cockroach after antennal contact with (non-volatile) female sex pheromone; small dot designates point of contact; arrows show direction of movement (Schal *et al.*, 1983b).

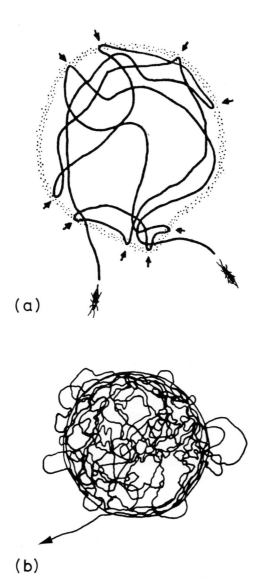

(a)

(b)

Fig. 4.2(a)　Path of a female *Nemeritis canescens* on a glass plate bearing a patch of host secretion. Stippling marks the edge of the patch; arrows indicate turns made at the patch edge (Waage, 1978).
(b)　Path of a sugar-deprived *Phormia regina* on a 5-cm patch of 1.0 M sucrose (Nelson, 1977).

case, the parasitic wasp, *Nemeritis canescens*, becomes 'locked into' a resource patch by executing turn angles exceeding 90° relative to its track direction each time it approaches the outer limits of substrate prey odor (Fig. 4.2(a)) (Waage, 1978). The second example shows a similar kind of turning when *Phormia regina* reaches the border of a sugar patch that it has been walking on (Fig. 4.2(b)) (Nelson, 1977). Recently, Havukkala (1980) showed that temporal comparison is employed by *Tenebrio molitor* to process boundary information of the type leading to large abrupt turns. This was accomplished by running beetles on a screen through which air of selected relative humidities could flow; a temporal, but not spatial change, was effected by switching abruptly from high to low or low to high relative humidity. Havukkala (1980) points out that the modification in turning is simply a reduction in the circular variance of turn angle dimensions generated by the insect (Fig. 4.3). In this mechanism the abrupt, large-dimension turn that is *released* by a change in the chemical stimulus, is *controlled* by internally stored orientation information.

Arrestment and escape, as reviewed recently by Kennedy (1977, 1978), can be accomplished by modulating locomotory rate and turning velocity. For example, increased turning velocity and decreased locomotory rate allows an insect to remain in a resource patch (Gunn and Pielou, 1940), whereas decreased turning velocity and increased locomotory rate allow an insect to escape from a stress source. Temporal comparison seems most likely to be the mode used to process gradient information that leads to output in terms of alterations in locomotory and turning rate. Contrary to some notations in the literature, the spatial-temporal stimulus pattern or at least thresholds must be monitored by the animal in order for these mechanisms to work.

4.4 INDIRECT CHEMO-ORIENTATION BASED ON A SECOND MODALITY

Among walking, swimming and flying arthropods the use of a non-chemical modality for obtaining directional information is well documented (see also Cardé, Chapter 5). Most examples refer to orientation in a wind current, as described below, but there is also evidence for odor-stimulated visual and gravity orientation in walking insects. For example, when the walking aphid, *Aphis fabae*, enters a steep gradient of aggregation pheromone it turns up the gradient; the pheromone also increases the probability of the insect turning toward another aphid or a visual mimic (Kay, 1976). Tropical species of cockroaches perch on vegetation such that males are located above females; upward-moving air currents at night probably carry pheromones from females to males (Schal, 1982). Laboratory studies indicate that upward-moving sex pheromone stimulates male *Periplaneta americana* to engage in walking search behavior (Schal, 1982) and downward oriented walking movements by males (Silverman and Bell, 1976). Visual cues and chemical trail information are

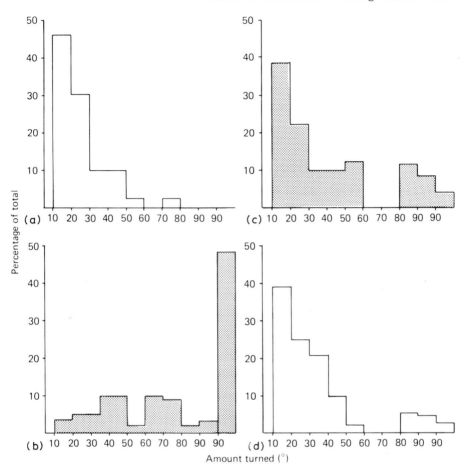

Fig. 4.3   Distributions of angles turned each second by *Tenebrio molitor* adults walking on a horizontal arena and exposed to a vertical air-stream of $10 \, \text{cm} \, \text{s}^{-1}$ at 25°C. The duration of the initial humidity condition was 5 s. (a and b) dry-acclimitized beetles: (a) 50% R.H. during seconds 1–5 and 85% R.H. during seconds 6–10. (c and d) moist-acclimitized beetles: (c), 85% R.H. during seconds 1–5 and 50% R.H. during seconds 6–10 (Havukkala, 1980).

integrated in foraging orientation in the harvester ant, *Pheidole militicida* (Hölldobler and Möglich, 1980). Investigators properly attempt to isolate the behavior released by chemicals by restricting potential information input through other sensory modalities. Thus insects probably make greater use of non-chemical information than the literature suggests.

The direction of a wind current is relatively stable in many habitats and it is easy to see that movement upwind would bring a walking insect closer to an upwind odor source, and movement downwind, the reverse. It is only necessary for an animal to be equipped with sensory organs that can detect the presence of

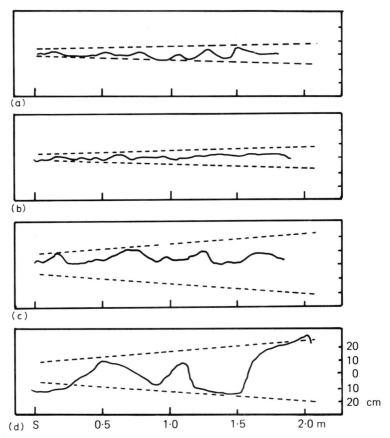

Fig. 4.4   Upwind orientation by the male cockroach, *Periplaneta americana*, in a plume of synthetic female sex pheromone, periplanone B. Plume widths at 1.0 m from the pheromone source: (a and b) 8 cm; (c and d) 28 cm. Wind direction is from left to right in the diagram; wind speed is 22 cm s^{-1}. S, position of the pheromone source (Tobin, 1981).

the current and sensory organs to maintain mechanical contact with the substratum so as to control speed of movement in the current. Most studies indicate that walking insects determine the direction of low velocity wind currents through antennal receptors or by the deflection of the antennae (e.g., Linsenmair, 1970; Bell and Kramer, 1979). No experiments to date have tested the possibility that walking insects make use of the apparent ground pattern movement, as do flying insects, but this is a viable possibility. Insect species that normally fly have been shown in the laboratory to walk upwind when an air current carries host odor (see examples in Miller and Strickler, Chapter 6 and Birch, Chapter 12). In their natural habitat, these species probably fly upwind in response to odors and then land and complete their search for an odor source with or without wind orientation (see Cardé, Chapter 5). For example, the cabbage fly, *Delia brassicae*, flies upwind in response to crucifer volatiles, lands and faces upwind, and then walks or flies upwind to the source (Hawkes, 1974; Hawkes and Coaker, 1979). An exception is *Bombyx mori*, a non-flying moth in which the male walks upwind in response to female sex pheromone (Schwinck, 1954, 1955; Kramer, 1975). The mechanism for maintaining an upwind course in an odor-plume has been investigated in a variety of walking insects and flying insects made to walk. The available evidence cited below suggests that mechanisms of walking walkers and walking flyers are similar, although complicated somewhat by a failure to differentiate in the literature between experiments that simulate walking conditions normally encountered by flying insects and conditions devised that merely test possible orientation mechanisms of flying insects made to walk.

It has now been demonstrated that walking insects locate an upwind odor source by at least three mechanisms. First, perception of an attractant chemical releases upwind orientation in certain species of flies (*Drosophila* − Flügge, 1934), moths (*B. mori* − Schwinck, 1954, 1955), beetles (*Ips confusus* − Wood and Bushing, 1963; *Dendroctonus frontalis* − Payne *et al.*, 1976; *Leptinotarsa desemlineata* − De Wilde *et al.*, 1969; Visser and Nielsen, 1977; *Trogoderma variable* − Shapas and Burkholder, 1978), locusts (*Schistocerca gregaria* − Kennedy and Moorhouse, 1969), cockroaches (*P. americana* − Rust *et al.*, 1976) and ants (*Novomessor* − Hölldobler *et al.*, 1978).

Second, perception of the plume boundary, i.e. a marked decrease in odor concentration, releases a turn to the opposite tack (counterturn), keeping a walking insect within a plume (Fig. 4.4) (*B. mori* − Kramer, 1975; *P. americana* − Tobin, 1981). These results show that olfactory information processing is involved to a certain extent in regulating orientation within an odor plume. Since cockroaches with only one antenna turn into the plume when they reach a boundary, we can assume that temporal comparisons are employed, at least by

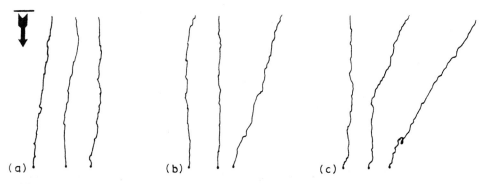

Fig. 4.5 Upwind orientation by the male cockroach, *Periplaneta americana*, in an airstream (22 cm s⁻¹) permeated with female sex pheromone extract (no plume boundaries). (a) 10^{-1}; (b) 10^{-2}; (c) 10^{-3} µg sex pheromone. Wind direction (arrow) is from top to bottom of page; bar shows 1.0 m. Recorded on a servosphere apparatus (Bell and Kramer, 1980).

these individuals (Tobin, 1981). Third, zig-zag turns are executed by cockroaches walking well within the boundaries, as well as at the boundaries, of a wide plume (Fig. 4.4) (Tobin, 1981) or by cockroaches and moths walking in air evenly permeated with odor (Fig. 4.5) (Bell and Kramer, 1980; Kramer, 1975). These turning patterns support the concept of an internal control system which sequentially generates a series of zigzag counterturns within the boundaries of a plume and the concept of counterturning induced by olfactory perception of a plume boundary (Tobin, 1981). Walking cockroaches may be more strongly influenced by the exact pheromone boundary of a plume than are flying moths, since cockroaches consistently remain with the plume, whereas moths often fly relatively far out of a narrow plume (see Cardé, Chapter 5). Overflying of moths may be due, however, to 'slippage' rather than longer lag times in responding to the plume boundary.

4.5 DIRECT CHEMO-ORIENTATION

That insects use longitudinal temporal processing for directional orientation is inferred from studies in which only temporal information (non-directional changes in stimulus intensity) is made available (e.g., Havukkula, 1980) and from experiments in which insects with one antenna are still able to locate an odor source (e.g., Martin, 1964). Because an animal using longitudinal temporal comparison must move from place to place, sequentially testing for changes in concentration, to detect the polarity of a gradient, the resulting turns are not as closely correlated with the polarity of the gradient as would be predicted for instantaneous comparison. An insect that crosses through an odor field and walks down the gradient will turn with an equal probability to the left and right because at that point it probably has no information to indicate which direction will lead up the gradient. Its subsequent turns, guided by temporal comparison can lead, over a loop, back toward the odor source (Bell and Tobin, 1981).

One type of experiment that has yielded direct evidence for instantaneous processing in insects is that of Martin (1964) who placed small tubes over the antennae of honeybees and measured the strength of turning tendency with different concentrations of odor solutions placed in the tubes. These manipulations restricted bees to instantaneous comparison when orienting in the simulated 'odor field'. Martin found that the 'minimum separable', or ratio of concentrations that could be distinguished between the antennae of honeybees was $1:10$ at high absolute concentrations and $1:2.5$ at low absolute concentrations. Kramer (1978) reported that male *B. mori* detect concentration differences of bombykol greater than $1:10$ through instantaneous comparison while walking in a two-channel wind stream.

Instantaneous comparison has also been demonstrated by crossing and fixing the antennae in place. An insect walks in the opposite direction to a zone containing the higher concentration of an attractant. This technique of inverting chemical inputs has been used in honeybees orienting to odors in the air (Martin, 1964), the ant *Lasius fuliginosus* and the termite *Schedorhinotermes lamanianus* orienting to trail pheromones on the substratum (Hangartner, 1967; Kaiss and Leuthold, 1975) and the cockroach *P. americana* orienting to trails of aggregation pheromone (Bell *et al.*, 1973). On the other hand, the termite *Hodotermes mossambicus* runs in wide serpentine routes on a trail. This behavior and the fact that the pattern is unchanged by unilateral antennectomy, suggests that temporal comparison is utilized exclusively by *H. mossambicus* (Leuthold, 1975).

Two other types of experiments, ablation of one antenna and the statistical analysis of turn directions, have been used to distinguish between temporal and instantaneous comparisons. There are numerous examples of antennal ablation (cf. Fraenkel and Gunn, 1940) in which operated insects orient in a more circuitous pathway toward an odor source, go only toward the direction of the intact antenna (circus movements), or perform just as well with one antenna as with two. The rationale put forward by Fraenkel and Gunn (1940) is that an animal normally uses instantaneous comparison if it exhibits circus movements when one antenna is removed; if circus movements do not result, then the animal is thought to normally use temporal comparison. The evidence is now sufficient to criticize this theory on several grounds. First, although circus movements may occur shortly after ablation, often the animal regains its ability to orient effectively and does not simply move toward the intact antenna (Rust *et al.*, 1976; Bell and Tobin, 1981), suggesting that in some cases the traumatic effects of ablation are reversible. Second, Martin (1964) showed that unilaterally antennectomized honeybees apparently orient as well as intact honeybees, and only execute circus movements when the one remaining antenna was fixed in place. As we know that honeybees can use instantaneous comparison in processing olfactory information, circus movements should undoubtedly result from ablating one antenna. Thus the single case that could firmly support the circus movement theory of Fraenkel and Gunn (1940) fails to do so. Instead it seems clear that the loss of one antenna in honeybees leads to temporal comparison in which the remaining antenna is moved transversely to detect the gradient polarity; this ability was negated when the remaining antenna was fixed. The evidence suggests that removing one antenna simply creates a temporal comparison animal, given that the animal already has the ability to engage in this process.

The last method for inferring instantaneous comparison and direct chemo-orientation is through statistical analysis of discrete turn angles relative to the chemical gradient. For example, the male cockroach, *P. americana*, turns more directly toward a sex pheromone source (in still air) within 35 cm of the source then when it is more distant from the source. Whereas the turn accuracy is the same for unilaterally antennectomized males as for intact males between 35 and 70 cm of the source, turns within 35 cm are more accurate in intact than one-antenna males (Bell and Tobin, 1981). Discrete turns up the concentration gradient can apparently be made more accurately when individuals have the capability for instantaneous comparison from the pair of antennae. At points distant from the odor source, temporal comparison would be required to decipher the more shallow gradient. Males with either one or two antennae apparently utilize this potential ability equally, and although males with one antenna exhibit more circuitous routes than intact males, they located the odor source in the same length of time as intact males. These results suggest switching between temporal and instantaneous comparison, depending on the steepness of the chemical gradient.

4.6 CONCLUSIONS

The importance of multiple modality stimuli in host plant, prey and mate location is emphasized in several chapters of this book. In cases where laboratory experiments have included two or more modalities, the results indicate an integration of these types of information in guiding orientation. Visual and gravity information, wind currents and probably acoustical cues are employed in chemo-orientation. Because the pattern of dispersed chemicals is at best irregular, unstable and relatively short-lived, the extractable information is in many ways imprecise as compared with that offered by other modalities. When studies of chemo-orientation have progressed further it is likely to be revealed that given the availability of other orientation cues, insects rely on chemicals primarily to initiate search and for gross boundary information. When only chemical information is available to an insect, the literature indicates that the information can be employed through temporal and instantaneous comparisons to locate an odor source; the accuracy of chemo-orientation depends to a large extent on the stability and slope of the chemical gradient.

ACKNOWLEDGEMENTS

The author appreciates the comments of Drs T. R. Tobin and R. T. Cardé. Research on cockroaches reported here was supported by the Psychobiology Program, National Science Foundation (BNS 82–03986).

REFERENCES

Bell, W. J., Burk, T. and Sams, G. R. (1973) Cockroach aggregation pheromone: directional orientation. *Behav. Biol.*, **9**, 251–5.
Bell, W. J. and Kramer, E. (1979) Search and anemotactic orientation of cockroaches. *J. Insect Physiol.*, **25**, 631–40.
Bell, W. J. and Kramer, E. (1980) Sex pheromone-stimulated orientation of the American cockroach on a servosphere apparatus. *J. Chem. Ecol.*, **6**, 287–95.

Bell, W. J. and Schal, C. (1980) Patterns of turning in courtship orientation of the male German cockroach. *Anim. Behav.*, **28**, 86–94.

Bell, W. J. and Tobin, T. R. (1981) Orientation to sex pheromone in the American cockroach: analysis of chemo-orientation mechanisms. *J. Insect Physiol.*, **27**, 501–8.

Bell, W. J. and Tobin, T. R. (1982) Chemo-orientation. *Biol. Rev. Cambr. Phil. Soc.*, **57**, 219–60.

Burger, M.-L. (1972) Der Anteil der propriozeptiven Erregung an der Kurskontrolle bei Arthopoden (Diplopoden und Insekten). *Verhandlungen der Deutschen Zoologischen Gessellschaft.*, **65**, 220–5.

Dethier, V. G. (1957) Communication by insects: physiology of dancing. *Science*, **125**, 331–6.

Ewer, D. W. and Bursell, E. (1950) A note on the classification of elementary behaviour patterns. *Behaviour*, **3**, 40–7.

Flügge, C. (1934) Geruchliche Raumorientierung von *Drosophila melanogaster*. *Z. vergl. Physiol.*, **20**, 463–500.

Fraenkel, G. S. and Gunn, D. L. (1940) *The Orientation of Animals.* Oxford University Press, Oxford: Revised edition (1961), Dover Publications, New York.

Gunn, D. L. and Pielou, D. P. (1940) The humidity behaviour of the mealworm beetle, *Tenebrio molitor* L. III. The mechanisms of the reaction. *J. Exper. Biol.*, **17**, 307–16.

Hangartner, W. (1967) Spezifität und Inaktivierung des Spurpheromons von *Lasius fuliginosus* Latr. und Orientierung der Arbeiterinnen im Duftfeld. *Z. vergl. Physiol.*, **57**, 103–36.

Havukkula, I. (1980) Klinokinetic and klinotactic humidity reactions of beetles *Hylobius abietis* and *Tenebrio molitor*. *Physiol. Ent.*, **5**, 133–40.

Hawkes, C. (1974) Dispersal of adult cabbage root fly (*Erioischia brassicae* (Bouche)) in relation to a brassica crop. *J. appl. Ecol.*, **11**, 83–93.

Hawkes, C. and Coaker, T. H. (1979) Factors affecting the behavioral responses of the adult cabbage root fly, *Delia brassicae*, to host plant odour. *Ent. exp. appl.*, **25**, 45–58.

Hölldobler, B., Stanton, R. C. and Markl, H. (1978). Recruitment and food-retrieving behaviour in *Novomessor* (Formicidae, Hymenoptera). I. Chemical signals. *Behav. Ecol. Sociobiol.*, **4**, 163–81.

Hölldobler, B. and Möglich, M. (1980) The foraging system of *Pheidole militicida* (Hymenoptera: Formicidae). *Insectes Sociaux*, **27**, 237–64.

Jander, R. (1970) Ein Ansatz zur modernen Elementarbeschreibung der Orientierungshandlung. *Z. Tierpsychol.*, **27**, 771–8.

Jander, R. (1975) Ecological aspects of spatial orientation. *A. Rev. Syst. Ecol.*, **6**, 171–88.

Kaiss, M. and Leuthold, R. H. (1975) Mechanisms of chemical orientation in *Hodothermes mossambicus* and *Schedorhinotermes lamanianus*. Orientation mediated by pheromones in social insects. In: *Pheromones and Defensive Secretions in Social Insects* (Noirot, Ch., Howse, P. E. and Le Masne, G., eds) pp. 197–211, *IUSSI*, Dijon.

Kay, R. H. (1976) Behavioural components of pheromonal aggregation in *Aphis fabae*. *Physiol. Ent.*, **1**, 249–54.

Kennedy, J. S. (1977) Olfactory responses to distant plants and other odor sources. In: *Chemical Control of Insect Behaviour* (Shorey, H. H. and McKelvey Jr., J. J., eds) pp. 67–91, Wiley-Interscience, New York.

Kennedy, J. S. (1978) The concepts of olfactory 'arrestement' and 'attraction'. *Physiol. Ent.*, **3**, 91–8.

Kennedy, J. S. and Moorhouse, J. E. (1969) Laboratory observations on locust responses to wind-borne grass odour. *Ent. exp. appl.*, **12**, 487–503.

Kramer, E. (1975) Orientation of the male silkmoth to the sex attractant bombykol. In: *Olfaction and Taste*, Vol. 5 (Denton, D. and Goghlan, J. D., eds) pp. 329–35. Academic Press, New York.

Kramer, E. (1978) Insect pheromones. In: *Taxis and Behaviour* (Hazelbauer, G. L., ed.) pp. 205–29. Chapman & Hall, London.

Kühn, A. (1919) *Die Orientierung der Tiere im Raum*. Fischer, Jena.

Leuthold, R. H. (1975) Orientation mediated by pheromones in social insects. In: *Pheromones and Defensive Secretions in Social Insects* (Noirot, Ch., Howse, P. E. and Le Masne, G., eds) pp. 197–211. *IUSSI*, Dijon.

Linsenmair, K. E. (1970) Die Interaktion der paarigen antennalen sinnesorgane dei der windorientierung laufender käfer (Insecta, Coleoptera). *Z. vergl. Physiol.*, **70**, 247–77.

Linsenmair, K. E. (1973) Die windorientierung laufener insekten. *Fortschr. Zool.*, **21**, 59–79.

Loeb, J. (1918) *Forced Movements, Tropisms and Animal Conduct*. Lippincott, Philadelphia.

Martin, H. (1964) Zur Nahorientirung der Biene im Duftfeld zugleich ein Nachweis fuer die Osmotropotaxis bei Insekten. *Z. vergl. Physiol.*, **48**, 481–533.

Mittelstaedt, H. and Mittelstaedt, M.-L. (1973) Mechanismen der Orientierung ohne richtende Aussenreize. *Fortschr. Zool.*, **21**, 45–58.

Murdie, G. and Hassel, M. P. (1973) Food distribution, searching success and predator–prey models. In: *The Mathematical Theory of the Dynamics of Biological Populations* (Hiorns, R. W., ed.) pp. 87–101. Academic Press, New York.

Nelson, M. C. (1977) The blowflies dance: role in the regulation of food intake. *J. Insect Physiol.*, **23**, 603–12.

Payne, T. L., Hart, E. R., Edson, L. J., McCarty, F. A., Billings, F. M. and Coster, J. E. (1976) Olfactometer for assay of behavioral chemicals for the southern pine beetle, *Dendroctonus frontalis* (Coleoptera: Scolytidae). *J. Chem. Ecol.*, **2**, 411–9.

Rust, M. K., Burk, T. and Bell, W. J. (1976) Pheromone-stimulated locomotory and orientation responses in the American cockroach *Periplaneta americana*. *Anim. Behav.*, **24**, 52–67.

Schal, C. (1982) Intraspecific vertical stratification as a mate-finding mechanism in tropical cockroaches. *Science*, **215**, 1405–7.

Schal, C., Gautier, J.-Y. and Bell, W. J. (1983a) Behavioral ecology of cockroaches. *Biol. Rev. Cambr. Phil. Soc.* (in press).

Schal, C., Surber, J. L., Vogel, G., Tobin, T. R., Tourtellot, M. K., Leban, R. A. and Bell. W. J. (1983b) Search strategy of sex pheromone-stimulated male German cockroaches. *J. Insect Physiol.*, **29**, 575–9.

Schwink, I. (1954) Experimentelle Untersuchungen über Geruchssinn und Stroemungswahrnehmung in der Orientierung bei Nachtschmetterlingen. *Z. vergl. Physiol.*, **37**, 19–56.

Schwink, I. (1955) Weitere Untersuchungen zur Frage der Geruchsorientierung der Nachtschmetterlinge: Partielle Fuehleramputation bei Spinnermaennchen, Insbesondere am Seidenspinner *Bombyx mori*. *Z. Vergl. Physiol.*, **37**, 439–58.

Shapas, T. J. and Burkholder, W. E. (1978) Patterns of sex pheromone release from adult females, and effects of air velocity and pheromone release rates on theoretical communication distances in *Trogoderma glabrum*. *J. Chem. Ecol.*, **4**, 395–408.

Shorey, H. H. and Farkas, S. R. (1973). Sex pheromones of Lepidoptera. 42. Terrestrial odor trail following by pheromone-stimulated males of *Trichoplusia ni*. *Ann. ent. Soc. Am.*, **66**, 1213–4.

Silverman, J. M. and Bell, W. J. (1976) Role of strato and horizontal object orientation on mate finding and predator avoidance by the American cockroach. *Anim. Behav.*, **27**, 652–57.

Tobin, T. R. (1981) Pheromone orientation: the role of internal control mechanisms. *Science*, **214**, 1147–9.

Visser, J. H. and Nielson, H. K. (1977) Specificity in the olfactory orientation of the colorado beetle, *Leptinotarsa decemlineata. Ent. exp. appl.*, **21**, 14–22.

Waage, J. K. (1978) Arrestment responses of the parasitoid *Nemeritis canescens* to a contact chemical produced by its host, *Plodia interpunctella. Physiol. Ent.*, **3**, 135–46.

Wilde, J. De, Hille Ris Lambers-Suverkopp, K. and Tol. A. van (1969) Responses to airborne plant odour in the Colorado beetle. *Neth. J. Pl. Pathol.*, **75**, 53–7.

Wood, D. L. and Bushing, R. W. (1963). The olfactory response of *Ips confusus* (Le Conte) (Coleoptera: Scolytidae) to the secondary attraction in the laboratory. *Can. Ent.*, **95**, 1066–78.

Wood, D. L., Browne, L. E., Silverstein, R. M. and Rodin, J. O. (1966) Sex pheromones of bark beetles. I. Mass production, bioassay, source, and isolation of the sex pheromones of *Ips confusus* (Le C.). *J. Insect Physiol.*, **12**, 523–36.

5

Chemo-orientation in Flying Insects

Ring T. Cardé

Dedicated to Prof. J. S. Kennedy in recognition of more than 40 years of contributions to innovative experimental methods and incisive analysis of insect chemo-orientation

5.1 INTRODUCTION

Our perceptions of the maneuvers and sensory inputs that animals employ to locate a stimulus have undergone continued refinement and reformulation following the efforts of Loeb (1918) and Kühn (1919) to classify orientation mechanisms. The adduction that the principal mechanism of 'long-distance' flying orientation to an airborne chemical stimulus in the wind is an optomotor-guided, chemically-induced, upwind orientation (or *anemotaxis*) has gained wide acceptance as a considerable body of experimental evidence in support of this mechanism has accumulated, principally with moth attraction to upwind pheromone sources. Besides optomotor anemotaxis, a number of alternative tactics have been proposed; these strategies have posited the existence of either spatial or temporal distributions of chemical stimulus that, if recognized, could serve as cues to guide the responder toward the chemical source or, if not to supply cues as to its direction, at least provide information as to its proximity. But our current understanding of the 'instantaneous' structure of a chemical stimulus emanating from a point source in continual or intermittent wind is imperfectly developed (see Elkinton and Cardé, Chapter 3) and this limits our ability to hypothesize intelligently on alternatives to the upwind anemotaxis paradigm.

The almost certainly independent evolution of upwind orientation to chemicals in different phylogenetic groups (at least at the ordinal level) and the inherent variability in environmental (particularly wind) conditions in various habitats, ranging from open grassland to dense tropical forests, may have dictated disparate as well as multiple solutions to the problem of discovering the

Chemical Ecology of Insects. Edited by William J. Bell and Ring T. Cardé
© 1984 Chapman and Hall Ltd.

location of a potential mate or host by its emitted chemicals. Most experimental studies have focused upon male moth orientation to either female-emitted or synthetic pheromone in a wind tunnel. Under natural conditions this process, based upon the recapture of released, male saturniid moths, is reputed to occur over distances as great as 11 km (e.g., Mell, 1922; Bossert and Wilson, 1963).

This chapter follows analyses of odor dispersal (Elkinton and Cardé, Chapter 3) and chemo-orientation in walking insects (Bell, Chapter 4) and refers to the definitions and concepts discussed therein. Important reviews of flying orientation to chemical sources include those of Farkas and Shorey (1974), Kennedy (1977, 1982), Kennedy et al. (1981), Bell and Tobin (1982), and the discussion of the concepts of 'attraction' and 'arrestment' by Kennedy (1978).

The present review re-examines the terminology, and the evidence supporting the various mechanisms, placing some emphasis on the nature of the sensory cues available in the heterogeneous natural environment in which the insect must negotiate its course, rather than the finely stratified wind milieu and artificial visual setting created by the experimenter in a wind tunnel. The available sensory cues and potential guidance mechanisms will be considered separately for chemo-orientation in moving and still air.

5.2 ORIENTATION IN THE WIND

5.2.1 Appetitive strategy

The probability of contacting an active space of a semiochemical obviously depends upon the dispersion pattern of the chemical stimulus and the position of the responding insect.* In theory the optimal searching or 'appetitive'† strategy of insects flying in wind prior to entering an active space would appear to be crosswind (Cardé, 1981), because a crosswind bias in the flight path decreases the 'searching' time and the distance travelled (an energetic advantage) before encountering a plume. However, field observations of male gypsy moth (*Lymantria dispar*) appetitive flight have not demonstrated a *tendency* toward crosswind flight (Elkinton and Cardé, 1983). Of course, the critical information necessary in such field observations is an accurate, simultaneous assessment of wind and insect movement; such measurements generally have not been attempted. Thus, although flight with crosswind bias appears to constitute an optimal strategy for encountering an active space in the wind,

---

* The proximate environmental cues and underlying circadian rhythms of responsiveness which are involved in the *initiation* of 'searching' behavior are beyond the scope of this review. Some aspects of this interesting topic are covered in Chapter 13.

† Searching or 'appetitive' strategy is used here to delineate behavior patterns that are characterized generally by increased locomotion and lowered thresholds to stimuli inducing the next behavior, in this case orientation toward a chemical. Appetitive behavior is thus defined teleologically, and as a consequence currently is avoided by most behaviorists. But it has heuristic value and I do not know of a suitable, non-teleological replacement term.

validation of this hypothesis as a generic strategy common to many flying insects remains to be established.

Many flying insects may 'perch' prior to receiving the chemical stimulus, and again, perching strategies that maximize the probability of encountering the active space should be expected. Schal (1982) found Costa Rican cockroaches exhibit a vertical stratification of perching sites on trees by species and within a species by sex, evidently in part to maximize the opportunity of encountering pheromone from a conspecific. When the pheromone-emitting and responding cockroaches are perched on different trees, the responder flys toward the emitter. Similarly, some moths 'perch' (rather than fly) during at least part of their daily cycle of pheromone responsiveness.

5.2.2 Optomotor response

A flight vector that is aligned either preferentially or systematically in some fashion with wind direction, as in the proposed crosswind flight strategy for appetitive flight or flight upwind toward a chemical stimulus, obviously implies perception of the direction of wind flow. Once airborne, an organism cannot monitor the direction of wind movement directly by anemoreceptors. Instead, visual reference to the deflection of the flight trajectory by the wind supplies navigational cues, a point strongly emphasized by Kennedy in 1940. This mechanism is analogous to that presumed to be used by walking insects: detecting the direction of wind flow by the deflection of the antennae (or possibly other anemoreceptors) relative to the walking direction (see Chapter 4). Thus, in upwind flight the degree of drift relative to a straight-ahead path (the head-body axis direction) is appraised visually by comparison of the course heading and the apparent direction of ground movement along the flight track. Compensatory maneuvers correct the track toward the upwind course. The mechanism for orienting upwind can undoubtedly be somewhat more complicated, as will be noted below, because the overall flight path in wind, at least in several moths which have been observed closely, appears to assume a zigzag path in the horizontal plane; nonetheless, the general principle of perception of wind flow holds. Another mechanism of upwind movement would include organisms that have determined the direction of wind flow while in contact with a substrate. After becoming airborne such flyers presumably could maintain an upwind heading for a short time by employing fixed optical cues. Eventual loss of the scent might induce landing or possibly a 'casting' flight (cf. Kennedy and Marsh, 1974) until the scent is detected again. Such a mechanism, while simpler than the optomotor anemotaxis discussed, would appear more energy and time-consuming, and therefore a less efficient strategy. But it might well explain host-finding of some phytophagous species which fly directly upwind in response to host-released chemicals (Hawkes *et al.*, 1978; Dindonis and Miller, 1980) and in which rapid location of a chemical source may not be of great advantage (cf. mate location).

5.2.3 Upwind anemotaxis

If wind flow is of a sufficient magnitude for a flying organism to detect its direction, then one obvious tactic for locating a chemical source in such winds is to proceed upwind. This process has been studied in wind tunnels in several moth species (Traynier, 1968; Farkas and Shorey, 1972, Farkas *et al.*, 1974; Kennedy and Marsh, 1974; Kennedy *et al.*, 1980, 1981; Marsh *et al.*, 1978; Cardé and Hagaman, 1979; Sanders *et al.*, 1981; Baker and Kuenen, 1982; Kuenen and Baker, 1982; Cardé and Crankshaw, 1983; Cardé *et al.*, 1983). In most of these examples the flight track along the axis of the pheromone plume has been characterized as a zigzag (successive lateral reversals) along an upwind trajectory.

The explanation advanced by Kennedy and Marsh (1974) for the zigzag behavior was that the successive turns were initiated *after* the loss of the scent; the turn carried the moth back towards the direction where the scent was last sensed. (The *direction* towards the source, however, would be detected by the optomotor response discussed previously.) Complementary evidence that such turns are initiated by the *loss* of the plume come from Traynier's (1968) wind tunnel observations with the moth *Anagusta kuhniella* in homogeneous pheromone clouds; moths were reputed to fly upwind in a straight path (see also the end of this section).

Tests designed to compare the flight track of males in narrow discrete plumes vs. homogenous clouds of pheromone in *Adoxophes orana* (Kennedy *et al.*, 1980, 1981, see Fig. 5.1) and *L. dispar* (Cardé and Crankshaw, 1983), unlike

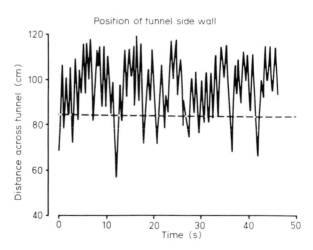

Fig. 5.1   The lateral (crosswind) excursions of a male *Adoxophes orana* tacking upwind in a wind tunnel over time. The broken line is the approximate (time average) demarcation between the even cloud of pheromone (top) and pheromone-free air bottom. Note the frequent track reversals occurring within the homogeneous cloud (after Kennedy *et al.*, 1980).

Trannier's 1968 observations, have shown that the zigzag path occurs under both conditions, in contradiction to the mechanism proposed by Kennedy and Marsh (1974). Lateral reversals, then, seem to occur in the presence of wind-borne pheromone and thus represent an internally generated pattern of flight.

Movement upwind, as noted in Section 5.2.2, requires visual appraisal of how the apparent movement of ground pattern (along the track direction) deviates from the body axis (the course heading, see Fig. 5.2). That the path is zigzag suggests that this tactic offers a navigational advantage over a straight-ahead course. The wide lateral excursions characteristic of the zigzag subject an airborne organism to increased angular deflections of path relative to the upwind direction over the deflection of track that would be incurred when the heading is aimed continually upwind. (Some walking insects also follow a sequentially generated zigzag path upwind and this navigational hypothesis may apply to these cases as well, except that antennal deflection rather than optical cues would be used to appraise the direction upwind. The situation may be quite complex, involving the comparison of the degree of antennal deflection on each leg of the zigzag (see Chapter 4)). A zigzag flight track offers an alter-nating series of comparisons of deflection and elevates the navigational infor-mation available to the insect. The course upwind could be set by adjusting the heading between reversals to maintain a flow of the ground pattern that alter-nates on the right and left legs of the zigzag at the same drift angle between the body axis (course heading) and the track.

In the gypsy moth the position of the moth as it flies along the zigzag is such that the moth's body axis is aligned with the course heading. The advantage of this strategem, as contrasted with maintaining an upwind body axis heading along the zigzag course, may be related to an improvement in flight efficiency.

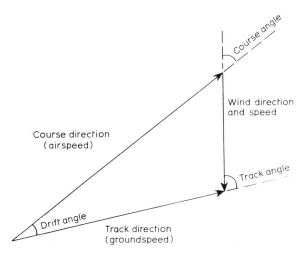

Fig. 5.2   The triangle of velocities for an insect tacking to the right of the wind heading (after Marsh *et al.*, 1978).

The visual effects generated in both situations are similar in terms of the amount of angular deflection, although the path of the visual field across the eyes would be quite different. If the insect's body is aligned with the course direction, the flow of the visual field beneath the eyes alternates at an angle relative to the head-body axis that is equal to the drift angle (Marsh *et al.*, 1978).

Another explanation for the zigzag path is that it allows the moth to fly up the plume by sequential sampling of the pheromone plume's spatial structure when the wind drops below the anemotactic threshold. This strategy will be explored in Section 5.3.2. Another consideration is that a zigzag path (noting both the lateral and the vertical components) would increase the probability re-contacting chemical stimulus after loss of the scent (Kennedy, 1978; David *et al.*, 1982; see Section 5.2.6).

The effects of increases in pheromone concentration upon the zigzag flight of *Grapholitha molesta* are to increase (i) the frequency of turns, (ii) decrease the airspeed (and therefore the distance across the plume's axis between turns.) The angles between turns, however, remain relatively constant (Kuenen and Baker, 1982). The resultant effect is to provide a narrower zigzag path and slow the speed of progress up the plume as the moth approaches the pheromone source, as covered in Section 5.2.4. The precision and success of *G. molesta* orientation to pheromone is quite dependent on a narrow range of optimal pheromone concentration (Cardé *et al.*, 1975; Baker and Cardé, 1979; Baker and Roelofs, 1981) and this facet of their behavior differs from most other moths such as *L. dispar* which orient to a rather broad range of stimulus concentrations.

Observations of *L. dispar* in a narrow pheromone plume in a wind tunnel (Cardé and Crankshaw, 1983) show some similarities with *G. molesta*: as pheromone concentration increases, the distance travelled on the directly upwind component of the flight track on each zigzag decreases as does the ground speed. In contrast to *G. molesta*, in the gypsy moth the interleg angles narrow slightly at increased pheromone levels and the distance across the plume between the turns remains relatively constant.

When the plumes are removed suddenly, and gypsy moths allowed to cast into a wide, homogeneous cloud of pheromone, the zigzagging, upwind flight resumes, but with some alterations: the width of the zigzag path is appreciably increased, as is the velocity along each leg, compared to a narrow pheromone plume. These trends in flight behavior in plumes vs. clouds remain similar over roughly 100-fold pheromone ranges and so seem attributable to the *spatial structure* of the chemical stimulus rather than pheromone flux. These observations compliment those with *A. orana* (Kennedy *et al.*, 1981) in that zigzagging is an internally generated behavior ('pre-programmed'), but, in addition, they show that the zigzag path of *L. dispar* is influenced by the external information about the spatial structure of the pheromone stimulus.

As noted at the beginning of this section, Traynier (1968) reported that the pyralid moth *A. kuhniella*, in a supposed even cloud of pheromone, flies directly upwind in an essentially straight (or at least not zigzag) path over at least a meter. Kellogg *et al.* (1962) stated that *Drosophila* exposed to a homogeneous cloud of banana odor also fly straight upwind. Both of these observations argue against a zigzag path as a generic component of anemotaxis, but the subject deserves study, particularly in groups of insects other than moths and at low stimulus concentrations such as would occur at some distance from the chemical source.

Bark beetles (Birch, Chapter 12) and many other Coleoptera (Cardé and Baker, Chapter 13) are attracted to aggregation and sex pheromones, but the mechanisms these insects use to fly to these sources are poorly documented and understood. The elm bark beetle, *Scolytus multistriatus*, in a wind tunnel, has been shown to fly horizontally *upwind* over a meter in the presence of synthetic pheromone, an anemotactic response (Choudhury and Kennedy, 1980). A major feature of bark bettle host colonization and mate recruitment is mass attack; this dictates that the pheromone emanates from a relatively large source such as a section of a tree; beetles may also orient to the visual cues from the tree trunk.

5.2.4 Regulation of flight velocity by optomotor cues and pheromone flux

The visual environment is important to regulation of flight velocity. An airborne organism cannot detect the magnitude of wind velocity any more than its direction, except by reference to movement of its body position relative to the visual field. This mechanism leads to the prediction that a constant *airspeed* would be difficult to maintain in differing wind speeds, but apparent ground velocity (i.e., the rate at which the ground appears to move or the so-called retinal velocity) could be regulated. The constant ground speed must come about because of compensatory changes in course heading and airspeed, because of the triangle of velocities (Fig. 5.3, see Marsh *et al.*, 1978). Wind

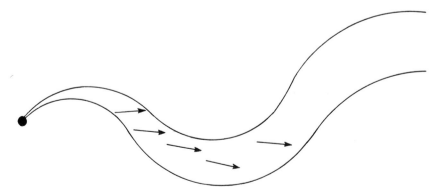

Fig. 5.3   Top view of a meandering chemical plume. The arrows indicate the wind (and towards the source) direction within the plume (after David *et al.*, 1982).

tunnel observations with day- and night-flying moths show that wind velocities ranging up to nearly 2 m s^{-1} produce widely disparate moth *airspeeds*, but the same apparent rates of movement relative to the ground pattern (Marsh *et al.*, 1978; Cardé and Hagaman, 1979; Kuenen and Baker, 1982). In a wind tunnel, floor patterns of stripes or large spots produce the optomotor effect quite readily. Yet a moving pattern directly below a flying moth is not the only visual input that can modulate flight velocity: visual features from a ceiling pattern influence the optomotor reaction in *Christoneura fumiferana*, the spruce bud-worm, more strongly than ventral cues (Sanders *et al.*, 1981). Obviously neither of these situations stimulates the visual fields available in nature. There the optomotor response will be dependent upon what ground and other patterns (e.g. trees) are resolvable and the relative importance of stimulating various regions of the eyes (David, 1979).

As noted previously, stimulus intensity can also exert a powerful effect on flight velocity; in moths responding to pheromone, increases in pheromone concentration (or flux) decrease the rate of upwind movement. In *L. dispar*, the flight track upwind follows a zigzag course and a major difference in the flight pattern, accounting for different average ground speeds along the plume's *axis*, is the change in the course heading on legs of the zigzag at various concentrations (Cardé and Crankshaw, 1983). *G. molesta* also decreases its ground velocity as pheromone concentration is elevated (Kuenen and Baker, 1982). In *Petinophora gosspiella*, the pink bollworm moth, Farkas *et al.* (1974) reported that variations in flight speed at various concentrations of pheromone were effected by changes in wing beat frequency. The decreases in ground velocity that have been associated with increases in pheromone concentration suggest that as the moth nears the chemical stimulus, flight speed would decrease prior to transition to other 'close-range' behaviors such as landing.

5.2.5 Fine-scale distribution of the stimulus

Other mechanisms for detection of the distance to the source have been proposed. Wright (1958) suggested that the frequency of encountering packets of relatively high concentrations of pheromone, as generated by turbulent dispersion, would increase as the source was approached. Murlis and Jones (1981), using data from an ion source and a detector 15 m downwind, maintain that a chemical stimulus would arrive downwind in a series of discrete bursts typically 0.1 s long and 0.5 s apart. A mechanism relying on the overall plume dimensions, proposed by Miksad and Kittridge (1979), contends that within many meters of the source the plume's diameter is very narrow ($<$ 1 cm) and that the plume's diameter could supply cues as to the emitter's proximity, but there are several objections to this aspect of their odor dispersion model (see Chapter 3). Alternatively, distance to the source might be related to pulses of pheromone generated in those insects which 'call' by rhythmically protruding and retracting their pheromone gland (Cardé *et al.*, 1983). Detection of such pulses (Conner *et al.*, 1980) at close range, prior to their dispersal by turbulence, might provide such cues. However, no evidence of 'pulsing' a pheromone plume (at 1-s on and 2-s off intervals) upon the upwind flight path was detected in the gypsy moth (Cardé *et al.*, 1983).

Kennedy *et al.* (1980) reported with *A. orana* that *sustained* upwind flight (> 2 sec.) does not occur in homogeneous clouds of pheromone, in contrast to persistent flight along discrete plumes. They advanced the explanation that discontinuities (increases and decreases in pheromone concentration), as might be generated by atmospheric turbulence or perceived during zigzag flight along a narrow plume, are requisite for *sustained* upwind flight. In this view, the continued presence of a relatively even flux of pheromone causes 'habituation' of upwind flight (Kennedy, 1981). However, similar wind tunnel tests with the gypsy moth in which the male is allowed to fly along a discrete plume, the pheromone stimulus is removed, and the moth casts into a homogeneous cloud of pheromone, have shown that these males maintain upwind progress in an evidently uniform cloud of pheromone (Cardé and Crankshaw, 1983). The importance of 'fine-scale' structure of the pheromone plume as well as its 'instantaneous boundaries,' also crucial to sampling along the longitudinal axis remains a fertile area for investigation.

5.2.6 Strategies after loss of the scent

In wind tunnel assays the sudden removal of pheromone causes moths to begin progressively widening 'casts,' similar to the upwind zigzag flight but without movement upwind relative to the visual field (Kennedy and Marsh, 1974). Such casting and 'station keeping' could obviously enhance the probability of regaining contact with the lost scent. David *et al.* (1982) proposed that loss of the scent occurs routinely in nature because the direction *upwind* and the direction *along the plume's axis* are not always coincident (Fig. 5.3, see Chapter 3 for an explanation). Thus, casting and station keeping would be crucial to following a plume to its source, because these maneuvers would place the flying organism in an optimal position to contact the next meander of the plume. The conclusion that there are substantial disparities between the heading upwind and the axis of the plume (David *et al.*, 1982) obviously implies that the mechanism of upwind anemotaxis, as outlined in the previous sections, without casting and station keeping, will not routinely lead a flying organism to a chemical stimulus. To a large extent the probability of a flying organism locating the source of a chemical stimulus in wind will be dependent on the frequency and magnitude of these disparities and on the success of the casting and station keeping maneuvers in optimizing the likelihood of recontacting the plume The integration of these meteorological and behavioural events require further detailed study.

5.3 ORIENTATION IN STILL AIR

5.3.1 Dispersion pattern of the stimulus

Still air can be viewed in two ways. One somewhat catholic definition would include air velocity below a level detectable by the optomotor reaction, thus embracing discrete plume structures similar to those generated in air speeds which can be monitored by the optomotor reaction. A second and more obvious definition would encompass chemical stimuli generated in still air. Dispersion in completely still air would be dominated by molecular diffusion, rather that turbulent diffusion in an airflow, but Mankin *et al.* (1980) has pointed out that in most 'still' air situations turbulent diffusion will overwhelm molecular diffusion (see Chapter 3). The spatial distributions of chemical stimulus in these two cases will obviously be quite different.

5.3.2 Orientation mechanisms in still air

There appear to be two mechanisms by which an organism could fly along a plume toward a chemical source without benefit of anemotaxis. First, it could alter its rate of turning, relative to its previous track, in response to perceived changes in stimulus intensity. Second, it could alter the turn angles relative to its last heading, again in response to perceived changes in stimulus intensity. Both of these mechanisms (self-steering klinokinesis and self-steering klinotaxis, respectively, in Kennedy's 1978 terminology) involve stored (idothetic) information on its previous position and orientation maneuvers as a guide to changing its rate of turning or the magnitude of the turn angles. The external chemical information (the odor gradient) necessary for self-steering would be compared over time either along the path of movement (longitudinal sampling) or along each side of the organism (transverse sampling).

Orientation to a chemical stimulus by sensing differences in concentration was invoked by Farkas and Shorey (1972, 1974) as an alternative to anemotaxis in location of a chemical source, although *detectable* differences in concentration along a plume's centerline or within the spherical active space (generated by molecular diffusion in a windless environment) only exist within several cm of the source. The Farkas and Shorey wind tunnel experiments with *P. gossyiella* showed that males in wind, having set course anemotactically along a pheromone plume, could continue a zigzag path along the plume for about a meter after the wind flow was abruptly terminated. Farkas and Shorey (1972) concluded that such 'chemical trail-following did not necessitate wind cues'; the zigzag path was attributed to a re-iterative loss of scent and a turn back toward the direction in which scent was *last* detected. However, such moths had initiated flight along the plume in wind and maintenance of the path over as short a distance as a meter could be sustained by adhering to the course setting determined by the idothetic information, including the stored information on the *previous* visual field. The responding organism could steer its course up an odor plume by turning about 180° after sensing a decrease in odor concentration (Shorey, 1973).

The same phenomenon has been studied in considerably greater detail by Baker and Kuenen (1982) in male *G. molesta* attraction to pheromone in a wind tunnel. As with *P. gossypiella*, moths continue flight up a plume over 1–2 m when wind flow ceases. The considerable similarities in *G. molesta*, however, between flight patterns (particularly in distances along each tack of the zigzag) in plumes of essentially identical structure but in a wind velocity of 0.4 m s^{-1} vs. a windless condition immediately following cessation of a 0.4 m s^{-1} wind suggests that males were detecting the plume structure. Additional evidence supporting longitudinal klinotaxis was provided by flying the moths in wind in the tunnel at nearly a zero *airspeed*. This was accomplished by moving the stripped wind tunnel floor downwind at a speed sufficient (using the optomotor response) to synchronize the moth's airspeed with the wind flow. Again, the zigzag tracks of the moths were similar under 'windless' and wind conditions. We can conclude that these two moth species, having started upwind in a pheromone plume, can continue along the plume at least under windless conditions over 1–2 m by steering their track along the plume's path. Given the inevitable occurrence of windless (below detectable levels) conditions in many environments — but particularly within the vegetative canopy — such a mechanism will undoubtedly be found among many moths and in other insect groups.

5.4 CONCLUSION

Optomotor-guided, upwind anemotaxis has been verified experimentally as a major mechanism used by flying insects to discover the location of a chemical stimulus. In several moth cases where detailed analysis in a wind tunnel has been possible, the anemotactic reaction to pheromone involves an internally generated, zigzag path upwind (using stored spatial information), a course-setting maneuver that may facilitate upwind progress, especially in low wind velocities, by allowing more effective comparisons of drift. Pheromone concentration may dictate the rate of net upwind progress and some features of the zigzag path may be modified by the plume's dimensions.

Should the wind cease (or drop below the threshold requisite for the anemotactic response) then another system, a self-steering, zigzag flight along the plume (longitudinal klinotaxis), appears to allow continued progress toward the pheromone source in two moth species in which this phenomenon has been studied. The importance of visual cues (rather than the plume structure) in setting this non-anemotactic course, the distance over which it might be sustained, and its generic nature as a mechanism in other insects, all remain to be determined. Other strategies (e.g., 'klinotaxis' initiated in still air without previous course setting in the wind) remain plausible, but await validation in appropriate experiments. That anemotaxis is a universal solution to flight toward a chemical stimulus in the wind needs verification in groups of insects other than moths.

Terms such as anemotaxis and klinotaxis, while seemingly defining a reaction precisely, tend to promote the notion of 'single-solution' orientation systems. Yet insects may well process the available information necessary for several mechanisms, selecting the anemotactic reaction, for example, only when there are sufficient anemo and visual cues and flight directly upwind provides continued contact with the chemical stimulus. Such terms also tend to camouflage the importance of non-chemical and idothetic inputs.

Future studies should define the role that internally-stored information on position (particularly as defined by visual cues) plays in maintenance of a course setting and gauging the congruence of heading and wind direction. It is doubtful that the range of orientation mechanisms used by flying insects in attraction to chemicals will be understood fully until the visual and wind cues and, most especially, the structure of the pheromone plume present in natural environments are adequately defined and duplicated experimentally.

ACKNOWLEDGEMENTS

The critical reviews of Drs W. J. Bell, J. S. Elkinton, J. S. Kennedy and T. R. Tobin are greatly appreciated. This research was supported by grants from the USDA and NSF Grant PMC 79-12014.

REFERENCES

Baker, T. C. and Cardé, R. T. (1979) Analysis of pheromone-mediated behaviors in male *Grapholitha molesta*, the Oriental fruit moth (Lepidoptera: Tortricidae). *Env. Ent.*, **8**, 956–68.

Baker, T. C. and Roelofs, W. L. (1981) Initiation and termination of Oriental fruit moth male response to pheromone concentrations in the field. *Env. Ent.*, **10**, 211–8.

Baker, T. C. and Kuenen, L. P. S. (1982) Pheromone source location in flying moths: A supplementary non-anemotactic mechanism. *Science*, **216**, 424–7.

Bell, W. J. and Tobin, T. R. (1982) Chemo-orientation. *Biol. Rev.*, **57**, 219–60.

Bossert, W. H. and Wilson, E. O. (1963) The analysis of olfactory communication among animals. *J. Theoret. Biol.*, **5**, 443–69.

Cardé, R. T. (1981) Precopulatory sexual behavior of the adult gypsy moth. In: *The Gypsy Moth: Research Toward Integrated Pest Management* (Doane, C. C. and McManus, M. L., eds) pp. 572–87. *USDA Tech. Bull*, No. 1584.

Cardé, R. T., Baker, T. C. and Roelofs, W. L. (1975) Ethological function of components of a sex attractant system for Oriental fruit moth males, *Grapholitha molesta* (Lepidoptera: Tortricidae). *J. Chem. Ecol.*, **1**, 475–91.

Cardé, R. T. and Hagaman, T. E. (1979) Behavioral responses of the gypsy moth in a wind tunnel to air-borne enantiomers of disparlure. *Env. Ent.*, **8**, 475–84.

Cardé, R. T. and Crankshaw, O. (1983) Effect of pheromone plume structure upon upwind flight of the gypsy moth. *J. Chem. Ecol.* (unpublished).

Cardé, R. T., Dindonis, L. L., Agar, B., and Foss, J. (1983) Apparency of pulsed and continuous pheromone to male gypsy moths. *J. Chem. Ecol.* (in press).

Choudhury, J. H. and Kennedy, J. S. (1980) Light versus pheromone-bearing wind in the control of flight direction by bark beetles, *Scolytus multistriatus. Physiol. Ent.*, **5**, 207–14.

Conner, W. E., Eisner, T., Vander Meer, R. K., Guerrero, A., Ghiringelli, D. and Meinwald, J. (1980) Sex attractant of an arctiid moth (*Utethesia ornatrix*): A pulsed chemical signal. *Behav. Ecol. Sociobiol.*, **7**, 55–63.

David, C. T. (1979) Optomotor control of speed and height by free-flying *Drosophila. J. Exp. Biol.*, **92**, 389–92.

David, C. T., Kennedy, J. S., Ludlow, A. R., Perry, J. N. and Wall, C. (1982) A reappraisal of insect flight towards a distant point source of wind-borne odor. *J. Chem. Ecol.*, **8**, 1207–15.

Dindonis, L. L. and Miller, J. R. (1980) Host-finding behavior of onion flies, *Hylemya antiqua. Env. Ent.*, **9**, 769–72.

Elkinton, J. S. and Cardé, R. T. (1983) Appetitive flight behavior of male gypsy moths. *Env. Ent.* (in press).

Farkas, S. R. and Shorey, H. H. (1972) Chemical trail-following by flying insects: A mechanism for orientation to a distant odor source. *Science*, **178**, 67–8.

Farkas, S. R., Shorey, H. H. and Gaston, L. K. (1974) Sex pheromones of Lepidoptera. Influence of pheromone concentration and visual cues on aerial odor trail following by males of *Pectinophera gossypiella. Ann. Ent. Soc. Am.*, **67**, 633–5.

Farkas, S. R. and Shorey, H. H. (1974) Mechanism of orientation to a distant pheromone source. In: *Pheromones* (Birch, M. C., ed.) pp. 81–95. North Holland, Amsterdam.

Hawkes, C., Patton, S. and Coaker, T. H. (1978) Mechanisms of host finding in adult cabbage root fly, *Delia brassicae. Ent. Exp. Appl.*, **24**, 210.

Kellogg, F. E., Frizel, D. E. and Wright, R. H. (1962) The olfactory guidance of flying insects. IV. *Drosophila. Can. Ent.*, **94**, 884–8.

Kennedy, J. S. (1940) The visual responses of flying mosquitoes. *Proc. Zool. Soc. Lond. Ser. A*, **109**, 221–42.

Kennedy, J. S. and Marsh, D. (1974) Pheromone-regulated anemotaxis in flying moths. *Science*, **184**, 999–1001.

Kennedy, J. S. (1977) Olfactory responses to distant plant and other odor sources. In: *Chemical Control of Insect Behavior* (Shorey, H. H. and McKelvey, J. J., eds) pp. 67–91. Wiley Interscience, New York.

Kennedy, J. S. (1978) The concepts of olfactory 'arrestment' and 'attraction.' *Phys. Ent.*, **3**, 91–8.

Kennedy, J. S. (1982) Mechanism of moth sex attraction: A modified view based on wind-tunnel experiments with flying male *Adoxophyes. Les Médiateurs chimiques* INTRA Symp. **7**, 189–92.

Kennedy, J. S., Ludlow, A. R. and Sanders, C. J. (1980) Guidance system used in moth sex attraction. *Nature*, **288**, 475–77.

Kennedy, J. S., Ludlow, A. R. and Sanders, C. J. (1981) Guidance of flying male moths by wind-borne sex pheromone. *Physiol. Ent.*, **6**, 395–412.

Kuenen, L. P. S. and Baker, T. C. (1982) Optomotor regulation of ground velocity during flight to sex pheromone at different heights. *Physiol. Ent.*, **7**, 193–202.

Kühn, A. (1919) *Die Orientierung der Tiere im Raum*. Fischer, Jena.

Loeb, J. (1918) *Forced Movements, Tropisms and Animal Conduct*. Lippincott, Philadelphia.

Mankin, R. W., Vick, K. W., Mayer, M. S., Coffelt, J. A. and Callahan, P. S. (1980) Models for dispersal of vapors in open and confined spaces: Application to sex pheromone trapping in a warehouse. *J. Chem. Ecol.*, **6**, 929–50.

Marsh, D., Kennedy, J. S. and Ludlow, A. R. (1978) An analysis of anemotactic zigzagging flight in male moths stimulated by pheromone. *Physiol. Ent.*, **3**, 221–40.

Mell, R. (1922) *Biologie und Systematik der chinesischen Sphingiden.* Friedlander, Berlin.

Miksad, R. W. and Kittredge, J. (1979) Pheromone aerial dispersion: A filament model. *14th Conf. Agric. For. Met. 4th Conf. Biomet.*, pp. 238–43.

Murlis, J. and Bettany, B. W. (1977) Night flight towards a sex pheromone source by *Spodoptera littoralis* (Boisd.) (Lepidoptera, Noctuidae). *Nature*, **268**, 433–5.

Murlis, J. and Jones, C. D. (1981) Fine-scale structure of odour plumes in relation to insect orientation to distant pheromone and other attractant sources. *Physiol. Ent.*, **6**, 71–86.

Sanders, C. J., Lucuik, G. S. and Fletcher, R. M. (1981) Responses of male spruce budworm (Lepidoptera: Tortricidae) to different concentrations of sex pheromone as measured in a sustained-flight wind tunnel. *Can. Ent.*, **113**, 943–8.

Schal, C. (1982) Intraspecific vertical stratification as a mate-finding mechanism in tropical cockroaches. *Science*, **215**, 1405–7.

Shorey, H. H. (1973) Behavioral responses to insect pheromones. *A. Rev. Ent.*, **18**, 349–80.

Traynier, R. M. M. (1968) Sex attraction in the Mediterranean flour moth, *Anagasta kuhniella*: Location of the female by the male. *Can. Ent.*, **100**, 5–10.

Wright, R. H. (1958) The olfactory guidance of flying insects. *Can. Ent.*, **80**, 81–9.

Plant-herbivore relationships

6

Finding and Accepting Host Plants

James R. Miller and Karen L. Strickler

6.1 INTRODUCTION

The plant-based resource units insects use for refugia, mating sites, and food vary in two fundamental ways. First, their distributions in time and space are usually patchy, as plants themselves assume distributions dictated by the patchiness of suitable soils and growing conditions. Second, the resources offered by plants vary qualitatively as judged by relative contributions to the fitness of the individual user. As one example, trees with rough rather than smooth bark promote higher survival of gypsy moth larvae (*Lymantria dispar*) (Campbell *et al.*, 1975; Barbosa, 1978), while larvae fed on white and red oak foliage experience greater survival, develop more rapidly and produce more fecund adults than do those fed on the foliage of most other tree species (Hough and Pimentel, 1978; Barbosa, 1978). The quality of resources offered by a given plant varies widely with factors such as plant tissue, growth stage, and plant nutritional states (see Scriber, Chapter 7).

The question of what mechanisms and strategies shape the patterns of resource use by insect herbivores in a heterogeneous world can be addressed in two parts: (i) what factors control behaviors that lead to investment of time, energy and gametes? and (ii) what are the physiological consequences of these investment behaviors? Questions (i) and (ii) are the focus of this and the following Chapter, respectively. This chapter is intended both to clarify some of the principles underpinning insect host finding and consuming, and to provide selective examples of the role chemicals play in mediating these processes for some insects.

Chemical Ecology of Insects. Edited by William J. Bell and Ring T. Cardé
© 1984 Chapman and Hall Ltd.

6.2 FUNDAMENTAL CONCEPTS AND DEFINITIONS

6.2.1 Significance of differential peripheral sensitivity and differential CNS responsiveness

Minimal components necessary to explain how animals in a heterogeneous world use some available resources more than others are: (i) a control mechanism capable of initiating and terminating the investment of time, energy, and other fitness correlates, and (ii) cues to trigger the initiation and termination of these actions. In the insect/plant system, the control mechanism is readily identified as the central nervous system (CNS) of the insect. The cues mediating investment are the chemical and physical attributes of the plant that are sensed by the various peripheral receptors. These receptors transmit the sensory correlates of plant attributes to the CNS in the form of trains of action potentials (e.g., chemoreceptors) or as graded receptor potentials (e.g., photoreceptors). Pivotal to being a non-random investor should be: (i) having peripheral receptors that generate discernably *different* patterns of electrical potential when exposed to different sets of stimuli, and (ii) behaving so as to invest when stimulated by cues from favorable resources *and* to withhold investment when stimulated by cues from unfavorable resources. Those insects whose neuronal wiring leads to better correlations between host-plant cues and host-plant suitability should be favored evolutionarily.

How insects stimulated by a given set of cues settle on certain behavioral options rather than others (i.e., make decisions) is a puzzle for which few pieces have been assembled. The simple lock and key model of insect decision-making, in which a select few 'token stimuli' (Fraenkel, 1959) trigger stereotyped host-investment behaviors, is not upheld by accumulating facts. A more dynamic and interactive model, incorporating factors from both the external and internal milieu, has been articulated and partially substantiated by Dethier (1982; *et ante*). Although this model concerns decisions by insects accepting or rejecting foods, it can help in conceptualizing other resource investment behaviors discussed herein.

Dethier envisions that 'all foods contain positive factors . . . which, if presented alone, would stimulate feeding, and negative factors which, if presented alone, would prevent feeding.' Acceptability of a given food (plant) is a function of the ratio of positive to negative factors. Additionally, internal factors modulate response to any ratio of stimulatory and inhibitory external inputs. For example, when an animal has food in the digestive tract,

'internal negativity adds to the negative factors provided by a plant that is only marginally acceptable, and that plant is refused. In a state of deprivation, when there is little or no internal inhibition, the ratio of sensory input from the marginally acceptable plant is biased in a positive direction, and the plant is accepted. This meal will be small because as partial satiety approaches, internal negativity will cause the plus/minus ratio to become smaller. For this reason, feeding will also be intermittent'   (Dethier, 1982)

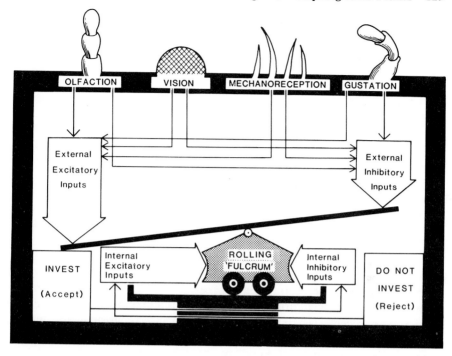

Fig. 6.1   Mechanical analogue of Dethier's 1982 model for the influence of external and internal factors on insect investment behaviors.

A mechanical analog of Dethier's model, with some embellishment, is presented in Fig. 6.1. The downward position of the central bar indicates whether or not the insect engages in investment behaviors. A strong influence tipping the bar one direction or the other is the balance between external excitatory and inhibitory inputs from the peripheral receptors (Fig. 6.1, top). However, the fulcrum of the balancing device rolls depending on the relative strength of counteracting excitatory and inhibitory internal inputs. This feature simulates changes in insect behavioral thresholds known to vary widely according to nutritional and reproductive status as well as diurnal cycles. Hence, the amount of external excitatory input needed to balance a given amount of external inhibitory input varies. Engaging in accepting or rejecting behaviors may raise the opposing and lower the contributing internal inputs (arrows at bottom of Fig. 6.1) and thus allow for homeostasis.

6.2.2 Overview of the host-finding and accepting process

Given that insects possess differential peripheral sensitivity to cues from potential resources and differential responsiveness mediated by the CNS (see Städler and Mustaparta, Chapters 1 and 2), we turn to how these abilities are

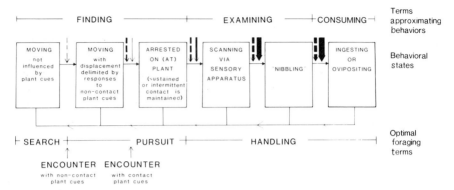

Fig. 6.2   Composite overview of behavioral events leading a generalized insect to feed or oviposit on a host plant. Broken and solid downward arrows indicate non-contact and contact sensory cues, respectively.

applied in the host-finding and accepting process. The composite overview (Fig. 6.2) of behavioral events leading a generalized insect to feed or oviposit on a host plant is provided to orient the reader (terms and complexities are treated below). This scheme is arbitrarily restricted to insects already moving and having a low threshold for response to host stimuli. Presumably, the insect has already passed through dispersal or migratory phases and is primed physiologically (Kennedy, 1958) to use host plants. Consistent with classical views (Thorsteinson, 1960; Kennedy, 1965; *et ante*), Fig. 6.2 suggests that host use results from a chain of behavioral responses. An insect can progress through the steps, remain for an extended time at a step, or regress to any prior step in the sequence if conditions do not stimulate progression, i.e., the option to move is always open. Rather than emphasizing the uniqueness of each set of stimuli eliciting a given response, Fig. 6.2 suggests considerable overlap in the stimuli eliciting the range of behaviors. Moving to the right in this scheme, there is a proposed increase in: (i) the number of sensory modalities stimulated, (ii) the number of cues sensed, and (iii) stimulus strength. In addition, behavioral thresholds will be altered by interactive effects (Kennedy, 1974).

Although Figs 6.1 and 6.2 conveniently organize the various aspects of host-finding and accepting treatment below, caution is advised in their interpretation. Certainly not all insects progress through every step shown in Fig. 6.2; the scheme must be collapsed for some and expanded for others. More important, there is danger in giving the impression that the essentials of insect foraging are captured accurately. However comforting our schemes, insects know nothing of them.

6.2.3 Perspective on terms

Once rooted in convention, certain terms continue to flourish even though they may connote mechanisms inaccurately and may actually impede clear communication and thinking. Despite considerable attention paid to terms (e.g., Dethier *et al.*, 1960; Beck, 1965; Kennedy, 1965; de Ponti, 1977; Singer, 1982), this problem remains acute in the field of insect/plant interactions. Terms like host-plant 'preference' and 'selection' have for years elicited an obvious degree of intellectual squirming (e.g., Thorsteinson, 1960, p. 193) and perhaps some of these difficulties should be sidestepped no longer. Attention paid to terms in the following discussion is intended to clarify concepts and promote mechanistic accuracy.

6.2.4 Finding

The term *find* has been used (e.g., Thorsteinson, 1960; Beck, 1965; *et ante*) to describe the phenomenon of insects arriving near or on a resource. More precisely, we use *find* to mean to behave so as to establish and maintain proximity with something, sensed by the finder's nervous system, that was previously apart and of undetermined location. Find spans the mechanistic range from pure chance events to 'purposeful' search. In contrast, the term *locate* favors the latter mechanism.

The process of finding entails spatial displacement. Since plants are relatively stationary or only intermittently mobile, insects must move to find plants. Such movement can occur in several ways (Table 6.1). From the standpoint of finding resources, more important than whether an organism moves actively or passively is whether and how it controls starting and stopping. The mover potentially can regulate the quality of its environments by starting to move when resources are inadequate or sparse and stopping at or confining movement close to rich resources. This deceptively simple principle is basic to all foraging behaviors. Required, of course, is differential peripheral sensitivity and CNS responsiveness.

Table 6.1 Examples of ways a finder can move

Type of movement	Method of propulsion	Examples	
		Without control over stopping*	With control over stopping
Passive	Displacement primarily due to being swept along by external medium	Ballooning insect larvae (McManus, 1979) and mites (van de Vrie *et al.*, 1972)	Dispersing aphids (Kennedy and Fosbrooke, 1972) and leafhoppers (Nickipatrick, 1965)
Active	Displacement primarily due to organism's own locomotory devices	?†	Most insects

* Since the overwhelming majority of insects control movement initiation, starting was not included in this treatment
† No known examples

The essential features of the finding process (Fig. 6.2 left) are: (i) finder moves, (ii) finder contacts cues from the resource, and (iii) finder responds to the cues by adjusting locomotion to increase proximity to the resource (arrestment, see Kennedy, 1978). Movement during finding can occur at various levels of sophistication. At the simplest level, an insect may contact the resource totally by chance and be arrested entirely via contact cues. At the next level (that shown in Fig. 6.2), an insect may be stimulated by volatile-chemical or visual cues at a distance from the resource and adjust its movements so that the probability of contact with the source is increased. One strategy for doing so is for movement to be randomly directed (kinesis) while effecting changes in the rate of displacement (orthokinesis) or the rate of turning (klinokinesis) (see Kennedy, 1978, and Chapters 4 and 5, this volume, for further explanations of these terms and processes). In a more sophisticated strategy, insects direct their movements toward the source (taxes), either by moving up stimulus gradients (e.g., positive chemotaxis) or moving in a particular direction with respect to environmental cues when stimulated by non-contact resource cues. For example, plume-following positive anemotaxis may occur in response to chemicals dispersed in moving air (see Bell and Cardé, Chapters 4 and 5).

Another set of terms needs to be dealt with in discussing resource finding. Workers developing theories of optimal foraging often categorize events leading to resource use as: search, encounter, pursuit, and handling; these terms are now firmly entrenched in ecological literature (MacArthur and Pianka, 1966; Schoener, 1971; Pyke *et al.*, 1977; Stanton, 1980). Optimal foraging terms (Fig. 6.2, bottom) can be integrated with those favored by behavioral researchers (Fig. 6.2, top) by linking each to the behavioral states (Fig. 6.2, middle) they approximate. *Search* is used in optimal foraging to describe animal movements before any contact is made with the resource or its cues. Although it is often used more broadly and has inescapable teleological connotations, search might properly apply to responsive animals moving in ways that increase the probability of first detecting stimuli from resources. *Encounter* can be viewed as a point where sensory information from the resource is first received. Given that there are multiple types of resource cues involved, there may be multiple encounters in a host-finding sequence (Fig. 6.2). This ambiguity is often not recognized in ecological literature where encounter is used variously to refer to reception of non-contact or contact cues (pre- and post-alighting cues; Rausher, 1983). *Pursuit* may be used to describe movements stimulated by cues from the resource that increase the probability of establishing and maintaining contact with the source. The change from search to pursuit is often observable by altered rates of displacement and turning, or other such behaviors.

No matter what terms are used to describe the events leading to host use, each should be defined behaviorally. Thus, behaviors rather than terms are the primary focal points of Fig. 6.2.

6.2.5 Examining

There should be strong selective pressure on an insect herbivore to possess effective sensory apparatus and to behave so that the sensing system is maximally engaged when critical decisions are to be made about consuming a potential resource. Certainly, intensive information gathering should precede oviposition or sustained feeding on any plant if an insect's tolerance for hosts is narrow.

In the behavioral sequence of Fig. 6.2 leading to plant use, the term *examining* is used to encompass behaviors indicative of sensory sampling. We use examining to mean inspect closely via the sensory apparatus. That some insects do sample plants intensively before using them for food or oviposition has been documented and examples will be presented later. The occurrence and significance of examining behaviors could go unappreciated if they are viewed only as part of the mechanics of getting food into the gut or eggs on (in) a substrate. Two behavioral states are shown under the examining phase (Fig. 6.2). Insects may be said to be *scanning* when the sensory apparatus is engaged more than fleetingly and sampling is extended to multiple points in space. An additional step in plant examining we call *nibbling*. For feeding, this means intermittently taking small amounts of food into the mouthparts and buccal cavity and sometimes swallowing. Ovipositor insertion without egg deposition would be analogous behavior in ovipositing. These behaviors represent the ultimate form of sensory sampling since most of the peripheral receptors excitable by cues from the plant are fully stimulated. This maximal level of stimulation sets the stage for the decision on whether to proceed in consuming the host plant. In optimal foraging terms, *handling* encompasses all events included under examining and consuming.

6.2.6 Consuming

If after nibbling, the insect continues ingesting or ovipositing, we say it is consuming. The rates and duration of consuming greatly influence the outcome of insect/plant interactions from either's perspective. On a given resource, consumptive rates and durations depend largely on the waxing and waning internal inhibitory and excitatory inputs experienced as the insect consumes (see Fig. 6.1). Once a consumptive bout has terminated, the probability of initiating another on the same site depends on whether the insect remains arrested or begins moving. The degree of consumption is most readily quantified by measuring time spent consuming while on plants and amounts ingested or numbers of eggs laid.

6.2.7 Clarification on preferring, selecting and accepting

When the foraging behaviors of insect herbivores operate over time, certain plants in the local environment are consumed to a greater degree than would be expected from their relative abundance. These plants are said to be 'preferred' (Hassell and Southwood, 1978), and often preferred plants are said to be 'selected' for consumption over those less preferred. Although these terms are convenient, we agree with Hassell and Southwood (1978) that their simplistic definitions or lack of definitions belie the complex behaviors mediating the end result. Furthermore, these terms perpetuate misconceptions concerning behavioral mechanisms. Strictly speaking, *prefer* means to put before something else in one's liking or positive responsiveness, while *select* means to pick out from among others on the basis of distinguishing qualities. Strongly implied by these terms is the anthropomorphic idea of weighing alternatives when considering a given item. There is a pitfall in assuming that insects consider alternatives as they respond to a given plant. In contrast, the term *accept* does not specify whether or not alternatives are weighed but simply indicates that consumatory behaviors occur (Thorsteinson, 1958). What many writers mean by host preference is host acceptability, and this substitution is easily made. We suggest that prefer and select be reserved for the special cases of accepting where weighing of alternatives is known to occur. Clearly, they apply whenever multiple items come into an insect's sensory field simultaneously and a given item is consistently taken. Selection after non-simultaneous or successive sampling of alternatives is more complex because it requires memory. For insects, it is difficult to prove experimentally that alternatives, existing only in memory, feature in the decision-making process; but this does not mean they do not (e.g., Mulkern, 1969, p. 521).

In practice, there may be no clear line of distinction between acceptance and rejection, particularly when only minimal consumption occurs. In such cases, it may be difficult to distinguish between 'nibbling' primarily for obtaining sensory information from the resource, and a low level of consumption. In other cases, sustained feeding or the deposition of a full complement of eggs clearly indicate that the examined plant has been accepted. Judging whether hosts are accepted or rejected is probably less helpful to our understanding of insect/plant interactions than quantifying degrees of consumption on various plants and determining how this influences fitness.

The distinction between accepting irrespective of alternatives and selecting is more than trivial, for the latter represents a foraging strategy much more sophisticated than the former. Selectors can avoid suffering negative consequences from investing heavily in a poor resource when a better one was 'just around the corner.' Potentially, selectors could explore the environment, finding and examining potential resources while deferring consumption. Then they could re-find those resources 'judged best' or items like them and use these predominantly. The advantages for improved resource utilization are clear.

It should be pointed out, however, that patterns in foraging behavior by herbivorous insects appear to be largely explicable via collective behaviors and mechanisms encompassed by Figs 6.1 and 6.2 and which do not invoke memory. Thorsteinson's (1960) assessment remains apropos: foraging by herbivorous insects is more akin to a series of take-it-or-leave-it-situations (with the durations of staying being variable) than 'comparison shopping.' Whether insect herbivores can and do comparison shop is a point highly worthy of investigation.

6.3 WHEN AND WHY USE CHEMICALS IN HOST-PLANT FINDING AND ACCEPTING

6.3.1 Evidence for the involvement of multiple sensory modalities

The pendulum of opinion on the amount and diversity of information used by insects during host-finding and accepting has been swinging over the years. As early as 1920, Brues suggested inferentially that insect 'botanical instinct' was based on responsiveness to the complex of chemical and physical stimuli emanating from plants. By the 1950's, Fraenkel (1953) had moved to the opposite extreme. In apparent awe of the prospect that many plants were laden with 'secondary chemicals' not participating in primary plant metabolism but in defense against herbivores and pathogens, Fraenkel championed the position that these allelochemicals (Whittaker and Feeny, 1971) were primarily responsible for mediating foraging behavior of insect herbivores. In contradiction to facts not then known, Fraenkel (1953) argued that all insects had similar nutritional requirements, and presence of required nutrients in plants was fairly universal. Hence, nutrients were viewed as unimportant in influencing foraging behavior when compared to the allelochemicals that vary distinctively among plant taxa. These 'odd' chemicals were termed 'token stimuli' and their role in governing insect behavior was exalted. After a heated debate and the procurement of additional facts, the significant role of plant nutrients as well as allelochemicals in stimulating host acceptance was recognized (Thorsteinson, 1960). The importance of both factors was reestablished in Kennedy's (1958) 'dual discrimination theory' that suggests both may play a significant role even in one insect.

Ecumenicity did not proceed much beyond the bounds of one sensory modality, however, and chemoreception continued to be viewed as paramount in determining host acceptance. One of the powerful influences for this bias was the isolation of distinctive chemical compounds from plants and the demonstration that these released oviposition or feeding behavior in adapted insect herbivores, e.g., mustard oil glycosides for cabbage butterflies, *Pieris rapae* and *P. brassicae* (Verschaeffelt, 1910) and the diamondback moth, *Plutella maculipennis* (Thorsteinson, 1953). Such examples suggested that acceptance was explicable simply by a few token stimuli and nutrients. Another major influence was the early interpretation of electrophysiological experiments on insects which suggested that many chemoreceptors were specialized, possibly to the point of one receptor for each important compound. This interpretation demanded that relatively few stimuli be involved, and thus generated simplistic explanations of the physiology of host acceptance.

These views have yielded to a more complex interpretation of insect perception of chemicals. Many of the so-called specialist receptors were found to be sensitive to a variety of chemicals (Dethier, 1975). Rather than responding via 'hard wiring' or 'labeled lines' (Dethier, 1971) to a few key stimuli, insects are now envisioned as responding to a summation of inputs from the various chemoreceptors. This 'gestalt' is believed to arise via across-fiber patterning processes. The full turn in thinking on the amounts of chemosensory information potentially useful to insects is reflected in Dethier's (1980) arguments that insects are at a selective advantage if they are capable of receiving high amounts of information from their foodstuffs. Presence of this information should set the stage for evolution of adaptive behavioral responses to complex sets of stimuli whose subtle variations reflect ecologically significant differences.

Extending this line of reasoning, insects finding and examining hosts should not restrict themselves to olfactory and gustatory information. Although work on mechano- and photoreception in host-finding and acceptance may be overshadowed by work on chemoreception, examples for oviposition are readily available (see reviews by Thorsteinson, 1960; Beck, 1965; Chew and Robbins, 1982; Prokopy and Owens, 1983). An example of the involvement of mechanoreception and its interaction with chemoreception in acceptance of oviposition sites is provided by the diamondback moth, *P. maculipennis*, which lays eggs in the concavities of plant leaves and stems (Gupta and Thorsteinson, 1960). Significantly more eggs are laid on dimpled than smooth plastic surfaces. A synergistic increase in egg deposition results when dimpled plastic is combined with allyl isothiocyanate, a constituent of crucifer sap.

The role of vision in host-finding and accepting is becoming increasingly appreciated (see Prokopy and Owens, 1983, for a recent review). Examples are the responses of: aphids to yellow and green objects (Moericke, 1969); the apple maggot fly, *Rhagoletis pomonella*, to red spheres (Prokopy, 1968); and butterflies to green surrogate leaves (Ilse, 1937; David and Gardiner, 1962). Beck's 1965 assessment was that 'color alone probably does not account for the host specificity of any insect species, but color frequently exerts an influence on the early stages of orientation.' Thorsteinson (1960) likewise saw shape, size, and color as being 'too variable and lacking the identifiable uniqueness required to explain the obvious discriminatory power of insects.'

Opinions that visual stimuli merely supplement the more important chemical cues may require some re-evaluation. For example, maximal oviposition by caged gravid onion flies, *Delia antiqua*, is released by the synergistic interplay of onion-produced alkyl sulfidés, a smooth yellow cylinder positioned vertically, and moist sand (Harris and Miller, 1982). Chemicals without cylinders (or cylinders without chemicals) released little oviposition from flies that had been ovipositing regularly. Moreover, when chemical and mechanoreceptory cues were held constant and only the color of surrogate host plants varied, more flies oviposited in dishes containing yellow than blue or grey stems. The greatest behavioral difference was not the rate of visitation, but the degree to which the differently-colored stems promoted stem runs and probing with the ovipositor followed by egg deposition (Harris and Miller, unpublished). In some cases, color and shape may be very influential even in the latter stages of foraging behaviors.

6.3.2 Relative advantages and disadvantages of using the different sensory modalities

Given that insects finding and consuming host plants can draw upon olfaction, gustation, vision, mechanoreception, and perhaps additional signals, the relative advantages and disadvantages of using each modality are of interest (see Marler and Hamilton, 1966, for general comparison of sensory modalities). Certainly, there is a great difference in the distance from the source over which these four senses operate. Mechanoreception and gustation require direct contact with the source, whereas olfaction and vision are thought to operate at ranges up to at least decimeters from the source (Finch, 1980; Prokopy and Owens, 1983). Although both olfactory and visual cues transmit information across distances, their physical properties differ vastly, and this affects spatial distribution of the signals. Because photons are minute, self-propelled, and move at extremely high velocity, their distribution on leaving an object is instantaneously established and relatively independent of air movement. Light from an evenly illuminated object is reflected in all directions including upward, provided there are no physical obstructions. Hence, visual cues generated under daylight are effectively multi-directional and not acutely variable with small changes in distance from the source.

Volatile molecules, by comparison, are large and slow-moving. In still air they move by diffusion and assume distributions essentially omni-directional but highly variable with a modest change in distance; concentration increases sharply as the source is approached. However air in nature is usually moving at speeds in excess of the linear diffusion velocity (ca. 3 cm min^{-1}; Levine, 1978) for the organic molecules commonly serving as olfactory signals for insects. Hence, the resultant signal from a source of volatile chemicals in moving air is a quasiconical plume having a rather flat concentration gradient along its windward axis, but sharp boundaries laterally (see Cardé and Elkinton, Chapter 3). Thus, the active space for olfactory signals in still or moving air is far more limited in direction than that for visual signals, assuming there are few obstructions. It is unclear whether there are pronounced differences in the distances over which vision and olfaction operate in host-finding by insect herbivores (Prokopy and Owens, 1983; Finch, 1980). Visual responses to host plants might be particularly favored for high-flying insects as chemical plumes are highly dispersed with increasing altitude (Wanta and Lowry, 1976).

Compared with vision and olfaction distant from the source, mechano-reception and gustation occur in direct contact with the plant where signal levels are high and noise levels should be low. Nearly every bit of signal encountered should have arisen from the plant and hence be relevant information. Likewise, vision and olfaction when in contact with a plant should be relatively free of noise.

Not all sensory modalities are equally subject to disruption by changes in environmental conditions. The effectiveness of olfaction in long-range resource finding by insects depends on wind velocity and direction (as outlined in Chapter 3) and the effectiveness of vision depends upon light intensity (Goldsmith and Bernard, 1974). In contrast, olfaction when in contact with the source as well as gustation and mechanoreception are not subject to these disruptions.

One of the most important criteria in judging the relative merits of using these four sensory modalities in host examining and accepting is their effectiveness in generating different patterns of sensory input for better versus poorer plant resources. The more characteristics that interact with the sensory system to shape the resultant reading, the higher the probability that it will be unique and constitute useful information for cueing consuming behaviors. Undoubtedly, gustation yields the most information on plant composition, especially when receptors contact plant sap. Any in the multitude of soluble constituents transportable to the sensillum liquor of taste receptors and participating in modifying the subsequent firing patterns will contribute information on chemical composition. Olfaction, on the other hand, is restricted to chemicals that evaporate at rates high enough to load insect antennae at rates modifying receptor firing. Vision can also provide information on differences in composition of plant compounds that modify incident light sufficiently to permit wavelength (color) differentiation.

Plant form is best detected by vision. Diversity of visual image is well exemplified in plants and is generated by combinations and permutations of overall dimensions, proportions, complexities in edge designs, etc. Mechanoreceptors permit textural sampling of particular sites they contact and may generate information on surface conformations indistinguishable by vision, e.g., subterranean oviposition substrates.

When modalities are considered separately, it is likely that gustation and vision would most frequently be the best indicators of plant composition and form, respectively. However, this generalization need not always hold. In some cases olfaction or color might reflect the greatest and most consistent compositional differences between good and poor plants. Likewise, mechanoreception, keying in on features like trichomes, might at times be more critical in cueing host examining and consuming than is overall form.

Undoubtedly, there would be great advantage in combining information across all available sensory modalities. The probability of making correct decisions should increase if promoted by favorable and deterred by unfavorable gustatory x visual x tactile cues. Additionally, if any one of these cues (including chemicals) were weak or lacking, insects using multiple modalities might fare satisfactorily by relying on other modalities in their sensory repertoire. These views represent steps away from the paradigm that chemical stimuli are the predominating regulars of host plant acceptance and the other modalities are only ancillary.

6.4 EXAMPLES OF THE INFLUENCE OF CHEMICALS IN HOST-PLANT FINDING AND ACCEPTING

6.4.1 Finding

Although the idea is intriguing, most insects do not find host plants by responding to volatiles with the same speed and pinpoint efficiency seemingly exhibited by responders to sex pheromones. Commonly, plants may be contacted by insects moving without apparent influence from odors. For example, the grasshopper, *Melanoplus sanguinipes*, not in contact with a suitable plant is stimulated into random movements (search) by hunger, favorable temperatures, and sufficient illumination (Mulkern, 1969). These movements eventually bring the grasshopper into visual range (encounter) of a vertical object to which it orients, approaches, and climbs (pursues). Chemical cues encountered upon contact with or in the immediate vicinity of the plant may arrest movement and induce examining. If not so stimulated, the hungry grasshopper leaves the vertical object and continues 'random movement.' Likewise, moth larvae (*Echaetias egle*), dislodged from their milkweed hosts, orient and proceed to any nearby plant by using visual cues and then examine the plant while ascending (Dethier, 1937). Aphids settling after dispersal flights apparently make little use of volatiles in their approaches to potential hosts (Van Emden, 1972), being cued primarily by vision. This is true also for some ovipositing butterflies approaching host plants (Rausher, 1983). In some of these cases, volatiles might increase responsiveness to visual cues (Kennedy, 1974). Although chemicals may sometimes play little or no role in approach to plants, they probably do exert an influence during the arrestment phase of most insects.

Plant finding can be accidental. Certainly this is true for ballooning insects or mites raining down upon potential hosts (Table 6.1). Also, insects having established contact with random plants for reasons other than consuming may (due to shifting behavioral thresholds) later find them potential food items and begin examining.

When hungry or gravid insects move within centimeters of potential host plants and encounter their volatiles, displacement is often delimited by olfactory responses. Monarch butterfly larvae (*Danaus plexippus*) not able to contact host and non-host foliage but moving on a screen immediately over the plants, turn and double back more frequently to milkweed foliage than to non-host foliage (Dethier, 1937). In similar tests (de Wilde, 1958), Colorado potato beetle larvae, *Leptinotarsa decemlineata*, frequently stop when immediately over solanaceous leaves. Larvae of the carrot root fly (*Psila rosae*) and mustard beetle (*Phaedon cochleariae*) usually stop when first contacting an odor plume, then raise and wave the head to and fro prior to and while moving toward the odor source (Jones and Coaker, 1979; Tanton, 1977). Numerous walking insects, e.g., some grasshoppers (Williams, 1954), and the vegetable weevil, *Listroderes obliquis* (Matsumoto, 1962), reportedly face the source and move directly to it. These and other examples from walking or crawling insects (Kennedy, 1977; Jones and Coaker, 1978) suggest that, at short range, plant odors can evoke ortho- and klinokineses as well as klino- and tropotaxes (see Bell, Chapter 4). These behaviors can also probably be exhibited by many insects flying near a source of plant volatiles, but documented examples are less common (see Kennedy, 1977).

Surprisingly, given the interest, well-documented examples of host-plant finding from distances greater than a meter or so (long-range in the sense of Kennedy, 1977) remain few. The earliest substantiated cases involved *Drosophila* (Kalmus, 1942; *et ante*) and locusts (Haskell *et al.*, 1962; Kennedy and Moorhouse, 1969). These insects turn and walk upwind upon detecting odors from rotting fruit and grass, respectively, and arrive at the source. The actual chemical effectors are unknown.

Long-range attraction has probably been more thoroughly explored in the cabbage fly (*Delia brassicae*)/crucifer interaction than in any other insect/plant relationship. Based on behavioral observations in the field, Hawkes (1974) concluded that mated gravid females moved upwind to crucifer crops from distances up to 15 meters. In a series of wind tunnel experiments (Hawkes and Coaker, 1979; *et ante*), crucifer volatiles or synthetic allylisothiocyanate (ANCS) stimulated *D. brassicae* to land in the odor plume, face into the wind, and then walk or fly upwind in a series of flights with a mean distance of 0.5 m. A zigzagging path characteristic of flying insect responses to sex pheromones was not evident. Although long-range, chemically mediated host finding is possible in *D. brassicae*, the concentration of ANCS required for optimal upwind movement was incredibly high (*ca*. 1 μg per liter of air) and required a release rate in the tunnel of 2 mg min^{-1}. Since the total amount of volatile aglucone present in the average moderate-sized crucifer is *ca*. 1 mg (Finch, 1978) and this would never be released instantaneously, it is unclear whether the wind-tunnel responses to ANCS represent 'normal' behavior.

Likewise, very high release rates of ANCS are required to effect significant *D. brassicae* catches in monitoring traps (Finch, 1980). Significantly, volatile aglucone *blends* released at milligram rates effect catches equivalent to gram rates of pure ANCS. However, no level of any mixture is much more attractive than the optimal release rate of ANCS (Finch and Skinner, 1982). None of these synthetic chemical preparations nor even crucifer extracts seem to deserve the title of 'potent attractants' for *D. brassicae* since the best catches are only *ca*. 3–5-fold greater than those of unbaited yellow traps. Although this is one of the best-studied examples of chemical attraction to host plants, it is not clear whether anemotactic responses to crucifer chemicals reflect the entire host-finding repertoire of cabbage flies in the field.

The behaviors of onion flies (*D. antiqua*) responding to onion (*Allium cepa*) volatiles are similar to those reported above for *D. brassicae*. Flies, entering plumes of volatiles from damaged onions several meters downwind of the source, alight on the soil, turn into the wind, and then move upwind toward the source in a series of short non-zigzagging flights punctuated with landing and walking (Dindonis and Miller, 1980a). It is not known how far downwind these behaviors are initiated. Both *D. antiqua* males and females respond in this manner to onion odors regardless of reproductive development, suggesting damaged onions may serve needs other than just oviposition sites, as was previously supposed (Dindonis and Miller, 1980b).

The most important releasers of these finding behaviors in *D. antiqua* are thought to be *n*-dipropyl disulfide and other related alkyl sulfides (Vernon *et al.*, 1981) emitted when onion cells are disrupted (Whittaker, 1976). Traps baited with synthetics catch up to 10 times more flies than do unbaited traps (Dindonis and Miller, 1981; Vernon *et al.*, 1981); however, these synthetics are not consistently attractive (Miller and Haarer, 1981). Optimizing release rates (Dindonis and Miller, 1981) or employing blends of onion alkyl sulfides (Vernon *et al.*, 1981) lead to only modest increases in attraction. The best synthetic alkyl sulfide treatments do not consistently reproduce the activity of freshly chopped onion, suggesting additional components may be involved.

Onions damaged and beginning to decompose in the field become up to 5 times more attractive than freshly damaged material and 50 times more attractive than unbaited traps (Miller *et al.*, unpublished). This potent onion fly attractant is thought to be a mixture of volatiles from onion plus volatile metabolites from the soft-rotting bacteria attacking damaged onions (Ishikawa *et al.*, 1981). Flies feed heavily upon decomposing onions as they do on enzymatically hydrolyzed yeast, which is also highly attractive (Miller and Haarer, 1981). Thus, in commercial onion fields, food sources appear to be stronger attractants of *D. antiqua* than oviposition attractants.

Involvement of volatile chemicals in host finding by the apple maggot fly, *R. pomonella*, was suggested by the observation (Prokopy *et al.*, 1973) that volatiles from bagged apples increased trap catches several fold in trees devoid of fruit. Indeed, extracts of apple volatiles released from red spheres in a wind tunnel elicited a 2–3 fold increase in visitation over spheres alone (Fein *et al.*, 1982). Flies stimulated by apple odors turned into the wind and walked steadily or took 'short hopping flights' toward the source. The activity was systematically traced to a GLC fraction containing: hexyl acetate (34%), (*E*)-2-hexen-l-yl acetate (< 2%), butyl 2-methylbutanoate (8%), propyl hexanoate (12%), hexyl propanoate (5%), butyl hexanoate (28%), and hexyl butanoate (10%). In a wind tunnel, this mixture of volatiles caught as many flies as the original extract. None of the components elicited full activity when presented alone.

Host finding in apple flies is apparently mediated by a combination of olfactory and visual responses.

Walking Colorado potato beetle adults, *L. decemlineata*, also exhibit positive odor-conditioned anemotaxis when stimulated by host odors in a wind tunnel (de Wilde *et al.*, 1969; Visser and Ave, 1978). In contrast, non-solanaceous plants are neutral or elicit downwind movement. Considerable effort was expended in identifying the chemicals responsible for attraction (Visser *et al.*, 1979). Several of the multitude of constituents present in the vapor head space over cut potato leaves (Visser, 1979) have been identified as (Z)-3-hexen-l-ol, (Z)-3-hexen-l-yl acetate, (E)-2-hexenal, and (E)-2-hexen-l-ol. None of these compounds, which collectively constitute the general 'green leaf odor' of plants (Visser *et al.*, 1979), was significantly attractive when tested singly. However, when released along with the volatiles from intact potato plants, the chemicals disrupted attraction (Visser and Ave, 1978). These workers suggest that precise blends of potato volatiles are needed to attract *L. decemlineata*. However, no defined mixture of synthetics, (each known to contribute to attraction) has been reported as an attractant for *L. decemlineata*. Moreover, the responses of flying Colorado potato beetles to solanaceous volatiles are yet to be investigated. This and the other examples discussed above point to the inescapable conclusion that we still know little about the role of chemicals in long-range host finding by insect herbivores.

6.4.2 Examining and consuming

Once insects are arrested on prospective host plants, there is considerable commonality in the ensuing examining and consuming behaviors irrespective of ingestive apparatus, taxonomic position, or developmental stage. To give a complete view of the examining and consuming process, we present a composite of examples, each emphasizing different aspects of the behavioral sequence in Fig. 6.2. Some of the chemicals acting as external excitatory and inhibitory inputs (Fig. 6.1) are identified.

(a) Feeding

Upon reaching plants, the *Locusta* and *Schistocerca* grasshoppers studied by Williams (1954) usually touch a leaf with the antennae, then lower the head until the labrum touches the leaf. The tips of the maxillary palpi are next brought into contact with the leaf 'a number of times and in rapid succession.' Finally, these grasshoppers close the mandibles on the edge of the blade, taking a small bite. *Melanoplus* spp. grasshoppers exhibit similar pre-swallowing behaviors (Mulkern, 1969), although the biting response frequently begins before the antennae contact the food. Odors elicit biting on objects such as glass dishes, wood sticks, paper, etc. Although discriminate feeders, these grasshoppers are indiscriminate biters, particularly when starved. After several small 'exploratory bites' (Williams, 1954) of material are taken into the bucal cavity and thoroughly chewed, grasshoppers either begin sustained feeding on a plant or move away. In the studies of Williams (1954) and Mulkern (1969), the initiation of swallowing and the duration of a feeding bout were dependent on the presence of phagostimulants and the absence of feeding deterrents, un-specified chemically. Harley and Thorsteinson (1967) report that secondary chemicals like digitonin and lupinine reduce grasshopper feeding on artificial diets whereas sucrose and lecithin are feeding stimulants.

Similar behaviors are exhibited by vegetable weevil adults, *L. obliquis.* Nearing a source of plant odor (Matsumoto, 1962, 1967), these weevils wave their antennae and begin tapping them 'excitedly' on the substrate near the source. When the antennal clubs touch the source, or with sufficient antennal tactile stimulation alone, the maxillary palpi move outward from the mouth and vibrate strongly while the other mouthparts remain stationary. Palpus vibration, suggested to be equivalent to a 'sniff' (lick) in mammals (Matsu-moto, 1967), continues as long as the antennae drum; frequency of vibration during this scanning behavior appears to be independent of stimulus strength. Tactile stimuli delivered to either the left or right antennal club evoke maxillary palpus extension on both sides but vibration only on the side of the stimulated club. Although various stimuli readily elicit behaviors up to this point, a progressively more restricted set of stimuli is required for biting, swallowing, and continued feeding. Chemical preparations stimulating feeding for longer than 10 minutes when applied to filter papers were Chinese cabbage juice, sucrose, and sodium chloride. Feeding on papers with citric acid solution lasted less than 1 minute whereas quinine-HCL solution elicited only a 'brief nibble' (Matsumoto, 1967). A number of compounds such as coumarin at higher concentrations inhibit consuming (Matsumoto, 1962).

Analogous behaviors are exhibited by sucking insects. Usually within seconds after alighting on a potential host, aphids insert their stylets and sample plant contents for a brief period, then repeat the process at nearby sites. These 'test probes' (Van Emden, 1972) are prime examples of scanning and 'nibbling,' as the proboscis and bucal cavity are endowed with chemoreceptors. Aphids may leave unsuitable plants after a few brief probes. On suitable hosts, they assume a characteristic feeding position and stance and begin sustained ingesting. Aphids on more suitable plants spend more time imbibing and less time changing positions than those on less suitable hosts. Sinigrin is one of the external excitatory inputs for some crucifer-feeding aphids (Nault and Styler, 1972) and an inhibitory input for some non-crucifer feeders.

Colorado potato beetle larvae bite various substrates emitting odors (de Wilde, 1958). Larvae persist in biting a solanaceous plant until the epidermis is penetrated and direct contact is gained with leaf sap. Depending upon the chemicals encountered, the telescoping antennae are retracted and ingesting commences. On the susceptible potato, *Solanum tuberosum*, larvae spend 90% of their time feeding and little time resting and wandering. However, on a resistant *S. demissum*, only 26% of the time was spent feeding while 76% was spent resting (de Wilde, 1958). After 30 minutes, larvae on the former plant had settled into undisrupted feeding, while those on the latter fed for stretches no longer than three minutes. Different cues from the two *Solanum* species effect different time budgets on the plants and different degrees of consumption.

The classic example of the lock and key model of insect feeding behavior comes from work on silkworm larvae, *Bombyx mori* (Hamamura *et al.*, 1962, *et ante*). In this view, feeding occurs in three distinct steps, each mediated by a discrete set of chemical stimulants. Originally only a few chemicals were thought to be necessary at each step; however, the list necessary to approach the phagostimulation of mulberry leaves grew to: attractants − citral, terpinylacetate, linalylacetate, linalol, and β-γ-hexenol; biting factors − sitosterol and isoquercitrin or morin; and swallowing factors − cellulose along with the 'cofactors' sucrose, inositol, inorganic phosphate and silica. The model of sequential unlocking of feeding behaviors for *B. mori* by key chemicals appeared well-supported when Hamamura *et al.* (1962) reported that larvae fed almost as well on a formulation of eight of the above substances incorporated into agar jelly as they did on mulberry leaves. However, further examination suggests *B. mori* feeding behavior is complex. Ishikawa *et al.* (1969) report that ratios of feeding stimulants had very pronounced effects on the rates of *B. mori* feeding. These authors conclude that 'the interacting effects of a mixture of chemical stimuli is far more important than individual stimuli as a determinant of the palatability of the diet.' Furthermore, having unique sets of chemicals stimulating biting as opposed to swallowing has not emerged as a general pattern in insects.

Attempts have been made to isolate and identify the dominant chemicals responsible for feeding specificity of a number of insect herbivores. An extensive effort for the cotton boll weevil, *Anthonomus grandis*, led to the conclusion that there were no dominant compounds in this case (Temple *et al.*, 1968). On the contrary, these workers concluded that full feeding responses probably require a complicated profile of components. Few active components were identified because of the difficulties with reformulating active mixtures.

Even in cases where dominant compounds are known, they alone do not account for host discrimination even for insect specialists. For example, Nielsen *et al.* (1979) concluded that monophagy in flea beetles like *Phyllotreta armoraciae* is not due to presence of glucosinolates alone, nor their precise ratios, although these factors are important. Glucosinolates must be presented in the context of other stimuli, also important determinants of consuming (e.g., cucurbitacins and cardenolides are potent feeding inhibitors of flea beetles). This conclusion is supported for other crucifer feeders where glucosinolates strongly influence consuming (Chew, 1980; Feeny *et al.*, 1970; Slansky and Feeny 1977).

(b) Oviposition

The oviposition behavior of the cabbage fly (*D. brassicae*), elucidated in detail by Zohren (1968), serves as an exemplary case study. Gravid females, held in small glass observation arenas, reach potted *Brassica* plants either by walking or flying onto the leaves. After a latency period ($\bar{x} = 4.5$ min) during which flies intermittently walk or fly to various sites on the leaves, they initiate continuous and randomly directed runs ('leaf-surface-run') on the surface of host leaves and no longer fly. On contacting leaf veins, flies follow them to the plant stem and descend rapidly to the base ('stem-run') where they move sideways and 'crablike' around the stem, head downward ('revolving-run'). Females next move onto the soil at the stem base and take short walks, changing directions frequently. Generally, they then return to the stem and climb it ('climbing'). By this time, extension of the ovipositor is noticeable. It is used to probe the soil when the flies move back to the substrate. Before eggs are laid, flies usually interrupt soil probing with additional climbing. When the ovipositor is re-inserted in the soil, flies often rake the soil particles with their metathoracic legs and finally lay some eggs. Thereafter, the chain of behaviors is often re-initiated at the leaf-surface-running phase and more eggs are ultimately deposited. In multiple bouts during 75 minutes, a total of 7 eggs was produced per female.

Although Zohren (1968) offered little explanation of the biological significance of these intricate behaviors, we suggest many of them exemplify examining. During the leaf-surface-run, stem-run, and climbing phases, females probably are receiving strong olfactory stimulation via the antennae and gustatory stimulation via taste receptors on the tarsi as well as on the mouthparts which sometimes contacted the plant (Zohren, 1968; Nair and McEwen, 1976). Additionally, visual scanning could be occurring during these phases. The revolving-run behaviors might yield information on resource size and chemical composition of the stem, particularly if the epidermis is scratched during crablike sidling. Soil composition is presumably sampled via antennae, tarsi, mouthparts, and ovipositor. Extensive probing of the soil with the ovipositor and raking with the legs may be scanning behaviors contributing to signal intensification and a representative reading. Mechanoreceptors on the *D. brassicae* ovipositor measure soil texture (Traynier, 1967). Although the electrophysiological input resulting from each behavior has not been elucidated, this example is a convincing case in which oviposition is preceded by extensive examining.

The dominant chemicals thought to act as the external excitatory cues releasing oviposition by *D. brassicae* are the glucosinolates characteristic of crucifers. Sinigrin applied to a fava bean plant stimulates oviposition (Zohren, 1968). Furthermore, correlations can be shown between levels of certain crucifer aglycones (e.g., 4-methylthio-3-butenyl isothiocyanate) and eggs laid (Ellis *et al.*, 1980). As was true for phagostimulation, however, glucosinolates appear to be only part of the story. Nair and McEwen (1976) suggest optimal release of oviposition depends upon both glucosinolate patterns and absence of inhibitory compounds.

The onion fly, *D. antiqua*, exhibits oviposition behaviors similar to *D. brassicae*

(Harris and Miller, unpublished). The best synthetic chemical releasers of onion fly oviposition are alkyl sulfides (like *n*-dipropyl disulfide) containing a propylthiol moiety (Ishikawa *et al.*, 1978; Vernon *et al.*, 1978); these are chemicals that also mediate host plant finding. However, even when the release rate is carefully optimized, *n*-dipropyl disulfide alone elicits only ca. 20% of the oviposition on authentic onion tissue in a side-by-side comparison (Miller and Keller, unpublished). These results again suggest the involvement of additional chemicals. *D. antiqua* ovipositional response to dipropyl disulfide is strongly inhibited by minute quantities of diallyl disulfide, characteristic of unsuitable hosts like garlic, *A. sativum* (Harris and Miller, unpublished).

Some butterflies such as the female cabbage white, *P. brassicae*, exhibit a 'drumming reaction' before ovipositing (Ilse, 1937). This consists of very quick alternating movements of the first pair of legs and results in forceful contact between foretarsi and plant tissues. Diverse explanations have been advanced for this phenomenon. Bell (1909) suggested that this behavior drives away spiders and parasites, whereas Vaidya (1956) astonishingly suggested it is nothing but an external manifestation of the 'labor pains' suffered by the pre-ovipositional female. The suggestion that this behavior is a method of sampling the physical and chemical properties of the plant is currently favored (Ilse, 1956; Chew and Robbins, 1982) and is supported by the finding that the foretarsi of drumming female butterflies are richly endowed with chemo-receptors sensitive to constituents of plant saps liberated during drumming (Ma and Schoonhaven, 1973). Drumming appears to be an exaggerated case of pre-ovipositional chemosensory sampling. Perhaps examining behaviors of other insects have been misinterpreted by researchers not expecting them.

6.5 ECOLOGICAL CONSIDERATIONS

6.5.1 Trade-offs between suitability and findability

The ultimate purpose of finding, examining, and consuming behaviors is, of course, to increase the probability that the insect (or its offspring) consumes hosts on which fitness, i.e., developmental rate, survivorship and/or fecundity, are maximized (Jaenike, 1978; Mitchell, 1981; Rausher, 1983). These fitness components are influenced by such plant characteristics as nutritional quality, presence or absence of toxins, and exposure to natural enemies, which collectively determine the host plant's suitability. Although documented cases are yet few, it is generally hypothesized that the acceptability of plant cues is positively correlated with suitability.

Plant suitability is not the only factor that determines plant use by herbivores. Plant *findability* − the probability that a plant will be found − is also important. The concept of findability is similar to plant apparency (Feeny, 1976), except that findability is defined in terms of the behavior and sensory capabilities of the insect (see Section 6.2.4 for definition of *find*), rather than general plant characteristics. When the most suitable host is infrequently found because it is rare or cryptic, the herbivore's fitness may be higher if it expands consumption to include less suitable but more findable hosts rather than deferring all consumption until the most suitable plant is found. Optimal diet models (Schoener, 1971; Pyke *et al.*, 1977; Jaenike, 1978) have been developed to explore these trade-offs between findability and suitability. A postulate of these models is that foragers should be capable of expanding and shrinking their dietary range as resource availabilities change. Furthermore, the order in which potential resources become acceptable should correspond to the rank order of host suitability. The models predict that fitness is maximized when successive potential hosts are added to those being consumed as long as the net fitness gain after adding the host is greater than the net fitness gain without the addition.

Few proximate mechanisms to explain how host findability influences host acceptability have been advanced. Memory and weighing of alternatives after sampling host availabilities may play an important role for some insects. However, activity- or time-driven threshold shifts (e.g., Kennedy and Booth, 1963; Singer, 1982) may explain much of the available data. For example, Singer (1982) shows that individuals of the butterfly, *Euphydras editha*, become increasingly polyphagous with time after oviposition if they are prevented from actually ovipositing. Immediately after laying an egg cluster, no plants are accepted. Over the next several hours the number of acceptable host species increases as various potential hosts are presented intermittently (but oviposition is not permitted), until the maximal range is accepted. Notably, the rank order of times that plants become acceptable reflects the same acceptability ranking found in the field. The data suggest a steady and continuous decrease in the acceptance threshold (becoming more accepting) until limits are reached. The degree of polyphagy displayed by *E. editha* populations in the field is directly correlated with the rate at which individuals from these populations become more accepting in the manipulative experiments.

These results are congruent with the model of Fig. 6.1. We suggest that *E. editha*, having just oviposited, has low internal excitatory inputs and perhaps elevated internal inhibitory inputs; the fulcrum is thus far to the left where cues from even the best host trigger no oviposition behavior. As time passes and more eggs move into the oviducts, internal excitatory inputs wax and/or internal inhibitory inputs wane so the fulcrum rolls to the right. The thresholds for accepting host plants should be reached in the order determined by the relative balance between external excitatory and inhibitory inputs for each plant, which in turn should be correlated with suitability. The tendency to be monophagous or polyphagous depends on the rate at which acceptance thresholds change, i.e., the rate at which the fulcrum moves to the right, and on the limits to movement in either direction.

The same decrease in acceptance thresholds as was observed by Singer (1982) with caged animals presumably occurs as an insect moves through the environment. These changes in threshold can be measured by time elapsed (or energy expended) until a resource becomes acceptable. Abundance, distribution, and crypticity directly determine the rate at which first encounters with cues from suitable host plants take place, and thus the degree to which the rolling fulcrum has shifted due to changing internal inputs.

For animals that search randomly, optimal foraging models predict that the time or energy expenditure at which a resource becomes acceptable depends not on the abundance of that plant, but on the abundance of more suitable plants (Pyke *et al.*, 1977; Jaenike 1978). The most suitable plant should always be consumed when encountered, no matter how rare; if sufficiently abundant, it should preclude the use of less suitable plants, even if these are also abundant. This mathematically derived prediction is also consistent with the concept of changing acceptance thresholds. When the highest ranked plants are very abundant, the fulcrum remains at the left, thus acceptance thresholds are high and all consumption takes place only on the most suitable host. If less abundant, the waxing internal excitatory inputs drive the fulcrum progressively to the right so that cues from lower-ranked hosts become acceptable.

The prediction that abundance of more suitable hosts affects the acceptability of less suitable ones is borne out by data on apple maggot flies (Roitberg *et al.*, 1982). Apples that have never received an egg are thought to be more suitable due to lack of intraspecific competition than are apples containing an egg. In trees with high rather than low densities of unused apples, flies are less likely to reach a behavioral state where used apples stimulate oviposition (see also Prokopy *et al.*, Chapter 11).

6.5.2 Foraging in a patchy environment

Like many resources, plants are frequently distributed in patches. Such patches may be discrete, like the apple trees in which apple maggot flies forage for fruits (Roitberg *et al.*, 1982). Alternatively, patches may be diffuse and hard to define, e.g., areas of a meadow where the probability of re-encountering a legume species is greater than in the surrounding vegetation (Stanton, 1982a). Rather than search randomly, insects may take advantage of such patchy distributions to increase encounter probabilities with plants. Encounter with and successful consumption of one suitable resource item may lead to an increased rate of turning, shorter flights, or more time in a patch than is observed after encounters with unsuitable host cues (Stanton 1982a; Roitberg *et al.*, 1982). When such behaviors occur, the probability of acceptance of an item may be affected by that item's abundance, in contrast to the situation for foragers that search randomly. For example, sulfur butterfly females (*Colias philodice eriphyle*) in a montane habitat fly shorter distances after landing on an acceptable plant in a dense patch than in a sparse patch (Stanton, 1982a). In dense patches, the probability of re-encountering the acceptable plant is higher. Thus, the butterfly biases its search toward dense patches and this leads to runs of foraging on one plant species. In an extreme case, pest populations of *Colias* in alfalfa fields rarely move between fields during their entire lifetime, unlike conspecifics in the montane habitat which move much farther (Tabashnik, 1979).

Like the decision to consume a specific host plant, the decision to search in a resource patch can be visualized as a process involving the balancing of external and internal excitatory and inhibitory cues. Waage (1979), for example, presents a model in which insects experience a continuously waning arrestment response to host odors in a patch until the threshold for moving is reached. Successful consumption, however, pushes the arrestment response to its upper level or at least to some increment greater than it was before consumption. Waage's model can be visualized in terms of Fig. 6.1. In this case, 'invest' refers to remaining in a patch. The continuously waning arrestment response is a decreasing internal excitatory (or increasing inhibitory) input that eventually pushes the fulcrum to the left, at which time the patch is rejected. Successful consumption of a host increases the internal excitatory input and pushes the fulcrum to the right. Thus, the insect remains in the patch longer than it would have without consumption.

For patch finding, we are beginning to determine the rate at which the fulcrum rolls, and how far it is pushed after successful consumption. Roitberg *et al.* (1982) show that apple maggot flies search a tree devoid of fruit for approximately four minutes. If introduced into a tree while ovipositing on a hand-held apple (simulating a successful oviposition in the tree), the female searches the tree for about nine minutes.

Patches of different size and density may differ in findability for insects. For example, dense patches of plants should have longer odor plumes than do sparse patches of equal area. Assuming that an herbivore uses primarily chemical cues when searching for plant patches, Stanton (1982b) predicts from theoretical models of odor-plume geometry that the active space of a dense patch is more likely to be encountered than is the active space of a sparse patch. She points out, however, that an increase in patch density should lead to a less than proportional increase in the active space of the odor plume from the patch. Thus, the probability that a given plant is encountered in the denser patch is less than the probability that a given plant is encountered in the sparse patch. This prediction has yet to be tested experimentally.

Visual cues from patches of different density have not been analyzed in the same theoretical fashion. However, if color is the most important cue used in patch finding, dense patches should be more easily found than sparse patches because increased plant surface area increases saturation. If pattern or form are more important cues, then dense patches may be more difficult to find because these cues may be more difficult to resolve. The probability of encountering patches of different density depends on the degree to which various visual and/or chemical cues are involved in patch finding.

Similarly, the plant species surrounding a potential host may affect the probability that it will be encountered by a searching insect (see review by Stanton, 1982b). For specialist insects, the presence of other plant species may act as noise that increases the crypticity of the host-plant. Generalist insects, on the other hand, might be more attracted to a patch with several potential host species than to a patch with only one host species. This difference between specialist and generalist insects may explain in part the differences in results of experimental studies of the effects of patch diversity on herbivore distributions (Stanton, 1982b). The possibility of using other plant species to increase the crypticity of crop plants to insects has been examined at numerous times with varying results (Bach, 1980; Stanton, 1982b; Risch *et al.*, 1983), but it is not clear in most cases what cues are used in long distance host-plant finding by the herbivores in question, and how these interact with the cues from nearby plants.

6.5.3 Caveat

The expectation that acceptable plants are most suitable for the insects that consume them has been challenged on numerous occasions by reports of oviposition on toxic plants (e.g., Chew, 1977; Wiklund 1981), greater acceptability of less suitable plants (e.g., Stanton, 1980; Courtney, 1981), and lack of acceptance of suitable plants (e.g., Waldbauer 1962; Harley and Thorsteinson, 1967; Smiley, 1978; see also Chew and Robbins, 1982, for further examples). These apparent 'errors' may be explained by factors less tangible than nutritional quality and allelochemicals that also affect plant suitability, e.g., risk of predation (Price *et al.*, 1980). However, lack of time to evolve in response to recent plant introductions (Chew, 1977; Tabashnik, 1982) and plant adaptations, or inherent unpredictability of plant resources (Stanton, 1980) may also be involved. We should, therefore, beware of trying too hard to fit the behaviors of insects to our adaptationist schemes (Gould and Lewontin, 1979). The insect's environment is constantly changing, and it would be surprising if there was no lag time in the adjustment of insect acceptability rankings to fitness on potential host plants.

6.6 SUMMARY AND CONCLUSIONS

The process whereby insects in a heterogeneous world invest in suitable hosts has been dissected into finding, examining, and consuming; and a composite picture of corresponding behaviors has been drawn. Although chemicals might not always play the dominant role in mediating each of these phases, chemicals are usually important mediators of the behavior during examining and consuming. The emergent view from the examples above and other reviews (Hedin *et al.*, 1974; Finch, 1980) is that chemical mediation of host finding and consuming generally occurs via mixtures of chemicals; some stimulate finding and consuming, while others inhibit these and stimulate antagonistic behaviors. The complexity of some of these chemical signals suggests that a myriad of compounds would be encountered if all plant chemicals known to influence host finding and consuming were added to the 1974 catalogue of Hedin *et al.* Indeed the list could grow nearly as long as the botanists' lists of phytochemicals.

We make no attempt to characterize these chemicals beyond the generalizations that: (i) wide ranges of both 'primary' and 'secondary' plant substances act as excitatory stimuli; (ii) inhibition of investment behaviors is triggered mainly by secondary substances but sometimes by unfavorable balances of primary nutrients; and (iii) because they are relatively non-volatile and effectively compartmentalized, many phytochemicals generating the inhibitory inputs influence insect behavior only during or after the examining phase when direct contact has been established. Readers are referred to Hedin *et al.* (1974) for analysis of behaviorally active phytochemicals by chemical class.

The view that chemical signals mediating host finding and consuming are accurately characterized as *aromas* and *flavors* does not imply that all chemicals are equally influential. Those cues most closely correlated with physiological or ecological detriment should be the most potent inhibitors of investing. Chemicals contributing most to the distinctiveness of aromas and flavors from the most suitable host plants might be most excitatory. Along the continuum between the extremes lie the chemical co-participants responsible for much of the fine tuning. Those researchers attempting to isolate and identify the chemicals mediating finding, feeding, and ovipositing must be prepared to deal with complexity. Perhaps our field can benefit from the technical expertise and experimental approaches of food scientists who are more familiar with the problems in dealing with flavors and aromas.

Presently, the need for highly quantitative work on the effects of chemicals mediating the behavior of insect herbivores seems more urgent than the need for expanding the lists of chemicals influencing behavior. Areas needing emphasis are: assessing the purity of chemical stimuli, quantifying rates of delivery to the insect, standardizing internal states of the insect, quantifying acceptance thresholds, and analyzing behavioral responses in detail. In addition, it is important that the effects of host-plant chemicals are, at some point, examined in the context of naturally accompanying stimuli. Experimental designs should include the authentic plant as a positive control against which to judge the activity of plant stimuli. Such positive controls have been conspicuously absent from many experiments. Relying only on negative controls (untreated filter papers, etc.) may lead to an overestimate of stimulus effectiveness, especially when no attention is paid to shifting acceptance thresholds. The behaviors exhibited at the extremes of an insect's threshold range are particularly worthy of quantification (Singer, 1982) because they are vital parameters influencing host range and the evolution of insect/plant relationships.

A mechanistic view of the influence of external and internal factors on insect investing has been provided by Dethier (1982). We have used this model to integrate information on the physiological, behavioral, and ecological aspects of foraging by insect herbivores. This model should serve as a stimulus for experiments leading to a clearer view of the mechanisms governing insect herbivory.

ACKNOWLEDGEMENTS

Support for this work was provided by National Science Foundation grant no. PCM-8110995 to J. R. Miller and United States Department of Agriculture Competitive grant no. 5901-2261-0-1-455-0 to K. L. Strickler. We thank B. Tabashnik, and M. Harris for helpful discussions on this manuscript, Michigan State Agricultural Experiment Station Journal Article no. 10774.

REFERENCES

Bach, C. E. (1980) Effects of plant density and diversity on the population dynamics of a specialist herbivore, the striped cucumber beetle, *Acalymma vittata* (Fab.). *Ecology*, **61**, 1515–30.

Barbosa, P. (1978) Host plant exploitation by the gypsy moth, *Lymantria dispar*. *Ent. exp. appl.*, **24**, 228–37.

Beck, S. D. (1965) Resistance of plants to insects. *A. Rev. Ent.*, **10**, 207–32.

Bell, T. R. (1909) Common butterflies of the plains of India. *J. Bombay Nat. Hist. Soc.*, **19**, 3.

Brues, C. T. (1920) The selection of food-plants by insects, with special reference to lepidopterous larvae. *Am. Nat.*, **54**, 313–32.

Campbell, R. W., Hubbard, D. L. and Sloan, R. L. (1975) Location of gypsy moth pupae and subsequent pupal survival in sparse, stable populations. *Env. Ent.*, **4**, 597–600.

Chew, F. S. (1977) Coevolution of pierid butterflies and their cruciferous food plants. II. The distribution of eggs on potential food plants. *Evolution*, **31**, 568–79.

Chew, F. S. (1980) Foodplant preferences of *Pieris* caterpillars (Lepidoptera). *Oecologia*, **46**, 347–53.

Chew, F. S. and Robbins, R. K. (1982) Egg-laying in butterflies. *Symp. R. Ent. Soc. Lond.*, Vol. II. (in press).

Courtney, S. (1981) Coevolution of pierid butterflies and their cruciferous foodplants. III. *Anthocharis cardamines* (L.) survival, development and oviposition on different plants. *Oecologia*, **51**, 91–6.

David, W. A. L. and Gardiner, B. O. C. (1962) Oviposition and the hatching of eggs of *Pieris brassicae* (L.) in a laboratory colony. *Bull. Ent. Res.*, **53**, 91–109.

Dethier, V. G. (1937) Gustation and olfaction in lepidopterous larvae. *Biol. Bull., Woods Hole*, **72**, 7–23.

Dethier, V. G. (1971) A surfeit of stimuli: a paucity of receptors. *Am. Sci.*, **59**, 706–15.

Dethier, V. G. (1975) The monarch revisited. *J. Kansas Ent. Soc.*, **48**, 129–40.

Dethier, V. G. (1980) Evolution of receptor sensitivity to secondary plant substances with special reference to deterrents. *Am. Nat.*, **115**, 45–66.

Dethier, V. G. (1982) Mechanisms of host-plant recognition. *Ent. exp. appl.*, **31**, 49–56.

Dethier, V. G., Browne, L. B. and Smith, C. N. (1960) The designation of chemicals in terms of the responses they elicit from insects. *J. Econ. Ent.*, **53**, 134–6.

Dindonis, L. L. and Miller, J. R. (1980a) Host-finding behavior of onion flies, *Hylemya antiqua. Env. Ent.*, **9**, 769–72.

Dindonis, L. L. and Miller, J. R. (1980b) Host-finding responses of onion and seedcorn flies to healthy and decomposing onions and several synthetic constituents of onion. *Env. Ent.*, **9**, 467–72.

Dindonis, L. L. and Miller, J. R. (1981) Onion fly trap catch as affected by release rates of *n*-dipropyl disulfide from polyethlene enclosures. *J. Chem. Ecol.*, **7**, 411–8.

Ellis, P. R., Cole, R., Crisp, P. and Hardman, J. A. (1980) The relationship between cabbage root fly egg-laying and volatile hydrolysis products of radish. *Ann. appl. Biol.*, **95**, 283–9.

Feeny, P. (1976) Plant apparency and chemical defense. In: *Biochemical Interactions Between Plants and Insects*, Vol. 10 (Wallace, J. W. and Mansell, R. L., eds) pp. 1–40. *Recent Advances in Phytochemistry*. Plenum Press, New York.

Feeny, P., Paawwe, K. L. and Demong, N. J. (1970) Flea beetles and mustard oils: host plant specificity of *Phyllotreta cruciferae* and *P. striolata* adults (Coleoptera: Chrysomelidae). *Ann. ent. Soc. Am.*, **63**, 832–41.

Fein, B. L., Reissig, W. H. and Roelofs, W. L. (1982) Identification of apple volatiles attractive to the apple maggot, *Rhagoletis pomonella. J. Chem. Ecol.* **8**, 1473–87.

Finch, S. (1978) Volatile plant chemicals and their effect on host plant finding by the cabbage root fly (*Delia brassicae*). *Ent. exp. appl.*, **24**, 350–9.

Finch, S. (1980) Chemical attraction of plant-feeding insects to plants. In: *Applied Biology*, Vol. 5 (Coaker, T. H., ed.) pp. 67–143. Academic Press, New York.

Finch, S. and Skinner, G. (1982) Trapping cabbage root flies in traps baited with plant extracts and with natural synthetic isothiocyanates. *Ent. exp. appl.*, **31**, 133–9.

Fraenkel, G. (1953) The nutritional value of green plants for insects. *Trans. Int. Cong. Ent.*, 9th, Amsterdam 2, 90–100.

Fraenkel, G. S. (1959) The *raison d'etre* of secondary plant substances. *Science*, **129**, 1466–70.

Goldsmith, T. H. and Bernard, G. D. (1974) The visual system of insects. In: *The Physiology of Insecta*, Vol. 2 (Rockstein, M., ed.) pp. 165–272. Academic Press, New York.

Gould, S. J. and Lewontin, R. C. (1979) The spandrels of San Marco and the Panglossian paradigm: a critique of the adaptationist programme. *Proc. R. Soc. Lond. Ser. B*, **205**, 581–98.

Gupta, P. D. and Thorsteinson, A. J. (1960) Food plant relationships of the diamondback moth (*Plutella maculipennis* (Curt.)) II. Sensory regulation of oviposition of the adult female. *Ent. exp. appl.*, **3**, 305–14.

Hamamura, Y., Hayashiya, K., Naito, K., Matsumura, K. and Nishida, J. (1962) Food selection by silkworm larvae. *Nature*, **194**, 754–5.

Harley, K. L. S. and Thorsteinson, A. J. (1967) The influence of plant chemicals on the feeding behavior, development, and survival of the two-striped grasshopper, *Melanoplus bivittatus* (Say), Acrididae: Orthoptera. *Can. J. Zool.*, **45**, 305–19.

Harris, M. O. and Miller, J. R. (1982) Synergism of visual and chemical stimuli in the oviposition behavior of *Delia antiqua* (Meigen) (Diptera: Anthomyiidae). *Proc. 5th Int. Symp. Insect–Plant Relationships, Pudoc Wagenigen.*

Haskell, P. T., Paskin, M. W. and Moorhouse, J. E. (1962) Laboratory observations on factors affecting the movements of hoppers of the desert locust. *J. Insect Physiol.*, **8**, 53–78.

Hassell, M. P. and Southwood, T. R. E. (1978) Foraging strategies of insects. *A. Rev. Ecol. Syst.*, **9**, 75–98.

Hawkes, C. (1974) Dispersal of adult cabbage root fly (*Erioischia brassicae* (Bouche)) in relation to a brassica crop. *J. Appl. Ecol.*, **11**, 83–93.

Hawkes, C. and Coaker, T. H. (1979) Factors affecting the behavioral responses of the adult cabbage root fly, *Delia brassicae*, to host plant odour. *Ent. exp. appl.*, **25**, 45–58.

Hedin, P. A., Maxwell, F. G. and Jenkins, J. N. (1974) Insect plant attractants, feeding stimulants; repellents, deterrents, and other related factors affecting insect behavior. In: *Proceedings of the Summer Institute on Biological Control of Plant Insects and Diseases* (Maxwell, F. G. and Harris, F. A., eds) pp. 494–527. University Press of Mississippi, Jackson.

Hough, J. A. and Pimentel, D. (1978) Influence of host foliage on development, survival, and fecundity of the gypsy moth. *Env. Ent.*, **7**, 97–102.

Ilse, D. (1937) New observations on responses to colour in egg-laying butterflies. *Nature*, **140**, 544–5.

Ilse, D. (1956) Behavior of butterflies before ovipositing. *J. Bombay Nat. Hist. Soc.*, **53**, 486–8.

Ishikawa, S., Hvias, T. and Arai, N. (1969). Chemosensory basis of host plant selection in the silkworm, *Ent. exp. appl.* **12**, 544–54.

Ishikawa V., Ikeshoji, T. and Matsumoto, Y. (1978) A propylthio moiety essential to the oviposition attractant and stimulant of the onion fly, *Hylemya antiqua* Meigen. *Appl. Ent. Zool.*, **13**, 115–22.

Ishikawa, Y., Ikeshoji, T. and Matsumoto, Y. (1981) Field trapping of the onion and seed-corn flies with baits of fresh and aged onion pulp. *Appl. Ent. Zool.*, **16**, 490–3.

Jaenike, J. (1978) On optimal oviposition behavior in phytophagous insects. *Theor. Pop. Biol.*, **14**, 350–6.

Jones, O. T. and Coaker, T. H. (1978) A basis for host plant finding in phytophageous larvae. *Ent. exp. appl.*, **24**, 272–84.

Jones, O. T. and Coaker, T. H. (1979) Responses of carrot fly larvae, *Psila rosae*, to the ordorous and contact-chemostimulatory metabolites of host and non-host plants. *Physiol. Ent.*, **4**, 353–60.

Kalmus, H. (1942) Anemotaxis in *Drosophila*. *Nature*, **50**, 405.

Kennedy, J. S. (1958) Physiological condition of the host-plant and susceptibility to aphid attack. *Ent. exp. appl.*, **1**, 49–65.

Kennedy, J. S. (1965) Mechanisms of host plant selection. *Ann. Appl. Biol.*, **56**, 317–22.

Kennedy, J. S. (1974) Changes of responsiveness in the patterning of behavioral sequences. In: *Experimental Analysis of Insect Behavior* (Barton Brown, L., ed.) pp. 1–6. Springer-Verlag, New York.

Kennedy, J. S. (1977) Olfactory responses to distant plants and other odor sources. In: *Chemical Control of Insect Behavior – Theory and Application* (Shorey, H. H. and McKelvey, Jr, J. J., eds) pp. 67–92. John Wiley & Sons, New York.

Kennedy, J. S. (1978) The concepts of olfactory 'arrestment' and 'attraction'. *Physiol. Ent.*, **3**, 91–8.

Kennedy, J. S. and Booth, C. O. (1963) Co-ordination of successive activities in an aphid. The effect of flight on the settling responses. *J. Exp. Biol.*, **40**, 351–69.

Kennedy, J. S. and Moorhouse, J. E. (1969) Laboratory observations on locust responses to wind-borne grass odour. *Ent. exp. appl.*, **12**, 487–503.

Kennedy, J. S. and Fosbrooke, I. H. M. (1972) The plant in the life of an aphid. In: *Insect/Plant Relationships* (van Emden, H. F., ed.) pp. 129–40. Royal Entomological Society, Blackwell Scientific Publications, London.

Levine, I. N. (1978) *Physical Chemistry*. McGraw-Hill Book Company, New York.

Ma, W. C. and Schoonhoven, L. M. (1973) Tarsal chemosensory hairs of the large white butterfly, *Pieris brassicae*, and their possible role in oviposition behavior. *Ent. exp. appl.*, **16**, 343–57.

MacArthur, R. H. and Pianka, E. R. (1966) On optimal use of a patchy environment. *Am Nat.*, **100**, 603–9.

McManus, M. L. (1979) Sources and characteristics of airborne materials: insects and other microfauna. In: *Aerobiology: Ecological Systems Approach* (Edmonds, R. L., ed.) pp. 54–70. Dowden, Hutchinson & Ross, Stroudsburg, PA.

Marler, P. and Hamilton, W. J. (1966) *Mechanisms of Animal Behavior*. John Wiley & Sons, Inc., New York.

Matsumoto, Y. (1962) A dual effect of coumarin, olfactory attraction and feeding inhibition, on the vegetable weevil adult, in relation to the uneatability of sweet clover leaves. *Jap. J. Appl. Ent. Zool.*, **6**, 141–9.

Matsumoto, Y. (1967) Vibration mechanisms of maxillary palpi in the vegetable weevil adult, *Listroderes obliquus* Klug (Coleoptera: Curculionidae), during food searching. *Appl. Ent. Zool.*, **2**, 31–8.

Miller, J. R. and Haarer, B. K. (1981) Yeast and corn hydrolysates and other nutritious materials as attractants for onion and seed flies. *J. Chem. Ecol.*, **7**, 555–62.

Mitchell, R. (1981) Insect behavior, resource exploitation and fitness. *A. Rev. Ent.*, **26**, 373–96.

Moericke, V. (1969) Host plant specific color behavior by *Hyalopterus pruni* (Aphididae). *Ent. exp. appl.*, **12**, 524–34.

Mulkern, G. B. (1969) Behavioral influences on food selection in grasshoppers (Orthoptera: Acrididae). *Ent. exp. appl.*, **12**, 509–23.

Nair, K. S. S. and McEwen, F. L. (1976) Host selection by the adult cabbage maggot, *Hylemya brassicae* (Diptera: Anthomyiidae): effects of glucosino-lates and common nutrients on oviposition. *Can. Ent.*, **108**, 1021–30.

Nault, L. R. and Styler, W. E. (1972) Effects of sinigrin on host selection by aphids. *Ent. exp. appl.*, **15**, 423–37.

Nickipatrick, W. (1965) The aerial migration of the six-spotted leafhopper and the spread of the virus disease aster yellows. *Int. J. Bioclimat. Biomet.*, **9**, 219–27.

Nielsen, J. K., Dalgaard, L., Larsen, L. M. and Sorensen, H. (1979) Host plant selection of the horse-radish flea beetle *Phyllotreta amoraciae* (Coleoptera: Chrysomelidae): feeding responses to glucosinolates from several crucifers. *Ent. exp. appl.*, **25**, 227–39.

Ponti, O. M. B. de. (1977) Resistance in *Cucumis sativus* L. to *Tetranychus urticae* Konch. 1. The role of plant breeding in integrated control. *Euphytica*, **26**, 633–40.

Price, P. W., Bouton, C. E., Gross, P., McPherson, B. A., Thompson, J. N. and Weis, A. E. (1980) Interactions among three trophic levels: Influence of plants on interactions between insect herbivores and natural enemies. *A. Rev. Ecol. Syst.*, **11**, 41–65.

Prokopy, R. J. (1968) Sticky spheres for estimating apple maggot adult abundance. *J. Econ. Ent.*, **61**, 1082–5.

Prokopy, R. J., Moericke, V. and Bush, G. L. (1973) Attraction of apple maggot flies to odor of apple. *Env. Ent.*, **2**, 743–9.

Prokopy, R. J. and Owens, E. D. (1983) Visual detection of plants by herbivorous insects. *A. Rev. Ent.*, **28**, 337–64.

Pyke, G. H., Pulliam, H. R. and Charnov, E. L. (1977) Optimal Foraging: A selective review of theory and tests. *Q. Rev. Biol.*, **52**, 137–54.

Rausher, M. D. (1983) The ecology of host selection behavior in phytophagous insects. In: *Impact of Variable Host Quality on Herbivorous Insects* (Denno, R. F. and McClure, M. S., eds) Academic Press, New York (in press).

Risch, S. J., Andow, D. and Altieri, M. (1983) Agroecosystem diversity and pest control: Data, tentative conclusions, and research directions. *Env. Ent.*, **12**, 625–9.

Roitberg, B. D., van Lenteren, J. C., van Alphen, J. J. M., Galis, F. and Prokopy, R. J. (1982) Foraging behavior of *Rhagoletis pomonella*, a parasite of hawthorn (*Crataegus viridis*) in nature. *J. Anim. Ecol.*, **51**, 307–25.

Schoener, T. W. (1971) Theory of feeding strategies. *A. Rev. Ecol. Syst.*, **11**, 369–404.

Singer, M. C. (1982) Quantification of host preference by manipulation of oviposition behavior in the butterfly *Euphydryas editha*. *Oecologia*, **52**, 224–9.

Slansky, F., Jr. and Feeny, P. (1977) Stabilization of the rate of nitrogen assimilation by larvae of the cabbage butterfly on wild and cultivated food plants. *Ecol. Monogr.*, **47**, 209–228.

Smiley, J. (1978) Plant chemistry and the evolution of host specificity: New evidence from *Heliconius* and *Passiflora*. *Science*, **201**, 745–7.

Stanton, M. L. (1980) The dynamics of search: Food plant choice in *Colias* butterflies. Ph.D. Dissertation. Harvard University, Cambridge, Massachusetts, USA.

Stanton, M. L. (1982a) Searching in a patchy environment: Food plant selection by *Colias p. eriphyle* butterflies. *Ecology*, **63**, 839–53.

Stanton, M. L. (1982b) Spatial patterns in the plant community and their effects upon insect search. In: *Herbivore Insects: host-seeking behavior and mechanisms* (Ahmad, S., ed.) Academic Press, New York (in press).

Tabashnik, B. E. (1979) Population structure of Pierid butterflies. III. Pest populations of *Colias philodice eriphyle*. *Oecologia*, **47**, 175–83.

Tabashnik, B. E. (1982) Host range evolution: The shift from native legume hosts to alfalfa by the butterfly, *Colias philodice eriphyle*. *Evolution* **371**, 150–62.

Tanton, M. T. (1977) Response to food plant stimuli by larvae of the mustard beetle, *Phaedon cochleariae*. *Ent. exp. appl.*, **22**, 113–22.

Temple, C., Jr., Roberts, E. C., Frye, J., Struck, R. F., Shealy, Y. F., Thompson, A. C., Minyard, J. P. and Hedin, P. A. (1968) Constituents of the cotton bud. XIII. Further studies on a nonpolar feeding stimulant for the boll weevil. *J. Econ. Ent.*, **61**, 1388–93.

Thorsteinson, A. J. (1953) The chemotactic basis of host plant selection in an oligophagous insect (*Plutella maculipennis* (Curt.)). *Can. J. Zool.*, **31**, 52–72.

Thorsteinson, A. J. (1958) The chemotactic influence of plant constituents on feeding by phytophagous insects. *Ent. exp. appl.*, **1**, 23–7.

Thorsteinson, A. J. (1960) Host selection in phytophagous insects. *A. Rev. Ent.*, **5**, 193–218.

Traynier, R. M. M. (1967) Stimulation of oviposition by the cabbage root fly, *Erioischia brassicae*. *Ent. exp. appl.*, **10**, 401–12.

Vaidya, V. G. (1956) On the phenomenon of drumming in egg-laying butterflies. *J. Bombay Nat. Hist. Soc.*, **54**, 216–17.

Van de Vrie, M., McMurtry, J. A. and Huffaker, C. B. (1972) Ecology of tetranychid mites and their natural enemies. III. Biology, ecology, and pest status and host-plant relations of tetranychids. *Hilgardia*, **41**, 343–432.

Van Emden, H. F. (1972) Aphids as phytochemists. In: *Phytochemical Ecology* (Harborne, J. B., ed.) pp. 25–43. Academic Press, London, New York.

Vernon, R. S., Pierce, H. D., Jr., Borden, J. H. and Oehlschlager, A. C. (1978) Host selection by *Hylemya antiqua*: identification of oviposition stimulants based on proposed active thioalkane moieties. *Env. Ent.*, **7**, 728–31.

Vernon, R. S., Jubb, G. J. R., Borden, J. H., Pierce, H. D., Jr. and Oehlschlager, A. C. (1981) Attraction of *Hylemya antiqua* (Meigen) (Diptera: Anthomyiidae) in the field to host-produced oviposition stimulants and their non-host analogues. *Can. J. Zool.*, **59**, 872–81.

Visser, J. H. (1979) Olfaction in the Colorado Beetle at the Onset of Host Plant Selection. *Druk*, Pudoc, Wageningen.

Visser, J. H. and Ave, D. A. (1978) General green leaf volatiles in the olfactory orientation of the Colorado beetle, *Leptinotarsus decemlineata*. *Ent. exp. appl.*, **24**, 738–49.

Visser, J. H., van Straten, S. and Maarse, H. (1979) Isolation and identification of volatiles in the foliage of potato, *Solanum tuberosum*, a host plant of the Colorado beetle, *Leptinotarsa decemlineata*. *J. Chem. Ecol.*, **5**, 11–23.

Verschaffelt, E. (1910) The cause determining the selection of food in some herbivorous insect. *Proc. R. Acad. Am.*, **13**, 536–42.

Waage, J. K. (1979) Foraging for patchily-distributed hosts by the parasitoid, *Nemeritis canescens*. *J. Anim. Ecol.*, **48**, 353–71.

Waldbauer, G. P. (1962) The growth and reproduction of maxillectomized tobacco hornworms feeding on normally rejected non-solanaceous plants. *Ent. exp. appl.*, **5**, 147–58.

Wanta, R. C. and Lowry, W. P. (1976) The meteorological setting for dispersal of air pollutants. In: *Air Pollution, Third Edition*, Vol. I (Stern, A. C., ed.) pp. 327–400. *Air Pollutants: Their Transformation and Transport*. Academic Press, New York.

Whittaker, R. H. (1976) Development of flavor, odor and pungency in onion and garlic. *Adv. Food Res.*, **22**, 73–133.

Whittaker, R. H. and Feeny, P. O. (1971) Allelochemics: Chemical interactions between species. *Science*, **171**, 757–70.

Wiklund, C. (1981) Generalist vs. specialist oviposition behavior in *Papilio machaon* (Lepidoptera) and functional aspects of the hierarchy of oviposition preferences. *Oikos*, **36**, 163–70.

Wilde, J. de. (1958) Host plant selection in the Colorado beetle larva. *Ent. exp. appl.*, **1**, 14–22.

Wilde, J. de, Hille Ris Lambers-Suverkropp, K. and van Tol, A. (1969) Responses to air flow and airborne plant odours in the Colorado potato beetle. *Neth. J. Pl. Path.*, **75**, 53–7.

Williams, L. H. (1954) The feeding habits and food preferences of Acrididae and the factors which determine them. *Trans. R. Ent. Soc. Lond.*, **105**, 423–54.

Zohren, E. (1968) Laboruntersuchungen zur Massenzucht, Lebensweise, Eiablage und Eiablageverhalten der Kohlfliege *Chortophila brassicae* Bouche (Diptera: Anthomyidae) *Z. ang. Ent.*, **62**, 139–88.

7

Host-Plant Suitability

J. Mark Scriber

7.1 INTRODUCTION

The previous chapter has dealt with host plant finding and assessing. Here I will concentrate on the suitability of plants for the growth of phytophagous insects. The various plant characteristics, insect characteristics, and environmental factors influencing post-ingestive suitability for the growth of immature insects has been reviewed recently (Scriber and Slansky, 1981). The nutritional ecology of insects involves basic nutrition (requirements, concentrations and proportions; see Gordon, 1972), the dietetics (allelochemics, food consumption, behavior, and regulatory physiology; see Beck and Reese, 1976; Slansky and Scriber, 1984), and especially the ecological and evolutionary ramifications of these interactions (see Townsend and Calow, 1981; Slansky, 1982). Insect adaptations to plant chemical quality (i.e., spatial, temporal, or taxonomic differences) may be behavioral, physiological, or ecological. These adaptations include heritable and inducible detoxication mechanisms, homeostatic trade-offs in rates and efficiencies, and must be interpreted in the light of the local plant resources available and the insect's guild and life cycle. Feeding preferences and the degree of feeding specialization actually realized for an insect herbivore will be determined by its evolutionary history, its present ecological opportunities and restrictions, and its behavioral and physiological flexibility (Fox and Morrow, 1981; Scriber, 1983).

Feeny (1976) and Rhoades and Cates (1976) proposed the concept of 'plant apparency' to explain the complex mechanisms which govern insect–plant interactions (see also Gilbert, 1979). Although the ecological and evolutionary suitability of a plant will depend upon its apparency to insect herbivores, the

Chemical Ecology of Insects. Edited by William J. Bell and Ring T. Cardé
© 1984 Chapman and Hall Ltd.

biochemical and physiological suitability of plant tissues consumed by the immature insect may be largely or entirely independent of the apparency and/or choice of that tissue to and by the ovipositing adult (e.g., Straatman, 1962; Wiklund, 1975; Chew, 1977; Smiley, 1978).

The objective of this chapter is to describe larval digestive efficiency and growth performance of phytophagous insects in relation to the chemical composition of their foodplants, and to discuss various insect adaptations (primarily behavioral and biochemical) which may improve the suitability of plant tissues for larval survival and growth. A physiological efficiency model is presented in relation to ecological and evolutionary interpretations of insect–plant interactions, with particular emphasis on a feeding specialization hypothesis.

7.2 HOST-PLANT SUITABILITY

If host-finding is successful and host-assessment results in host acceptance, the consumption and post-ingestive aspects of host-suitability are to be considered next (i.e., the adequacy of the selected food to sustain growth, survival, and reproduction; see Miller and Strickler Chapter 6 and Kogan, 1975). While thorough host suitability analyses must also consider the long-term ecological and evolutionary effects of foodplants upon the population biology and life styles (i.e., guilds) of the insects (see Maxwell and Jennings, 1980; Slansky and Scriber, 1984), I will concentrate here primarily upon the larval survival and growth aspects, with little discussion of the influence of plant chemistry upon fertility and fecundity of the adults. This latter aspect has been reviewed in detail by Townsend and Calow (1981) and Slansky (1982). Subsequent community interactions between other trophic levels influenced indirectly by plant chemicals has been reviewed by Price (1981), and will not be discussed here.

7.2.1 Nitrogen

The large differences in N content of insect tissue and plant tissue may be the major reason that less than one third of the insect orders and higher taxa of terrestrial arthropods feed on seed plants (Southwood, 1972). Animals (including insects and mites) consist mainly of protein (ranging from 7–14% N), whereas plants or plant parts rarely reach concentrations of 7% N and are generally much lower ($\bar{x} = 2.1\%$ N; 894 analyses of nearly 400 species of woody plants, Russell (1947); see also Mattson (1980). Growth and reproductive success depends to a large extent on the insect's ability to ingest, digest and convert plant nitrogen efficiently and rapidly. The particular importance of nitrogen in insect–plant interactions is summarized by McNeill and Southwood (1978) and Mattson (1980). The more specific relationships between N fertilizer regimes and plant resistance to pests has been most recently reviewed by Jones (1976) and Tingey and Singh (1980).

Although it is generally assumed that increased plant nitrogen increases insect damage or populations, there is also considerable inconclusive or contradictory evidence. A survey of literature over the last five decades shows at least 115 different studies in which insect damage, growth, fecundity, or numbers increased with increased plant nitrogen (Painter, 1951; Lipke and Fraenkel, 1956; Foster, 1968; Singh, 1970; Southwood, 1972; Nickel, 1973; Wiseman, *et al.*, 1973; Leath and Radcliffe, 1974; Jones, 1976; Fox and Macauley, 1977; Onuf, *et al.*, 1977; Slansky and Feeny, 1977; White, 1978; Wolfson, 1978; Newberry, 1980; McClure, 1980; Tingey and Singh, 1980; Auerback and Strong, 1981; and references cited therein). Similar increases with increased N are observed for at least 20 studies with mites (see Huffaker *et al.*, 1969; Rodriguez, 1972; Nickel, 1973; Jones, 1976; and references therein). On the other hand, at least 44 studies with insects and a few with mites indicate a decrease in herbivore populations or damage to plants with high N concentrations or an increase in insect populations with low nitrogen (see Painter, 1951; Lipke and Fraenkel, 1956; Foster, 1968; Rodriguez, 1951; Huffaker *et al.*, 1969; Goyer and Benjamin, 1972; Carrow and Betts, 1973; Smirnoff and Bernier, 1973; Leath and Radcliffe, 1974; Baule, 1976; Jones, 1976; Pimentel, 1977; Wolfson, 1978, 1980; Tingey and Singh, 1980; and references therein).

Fertilization of plants often directly affects the preferences, survival, growth rates and reproduction of insect herbivores, but it may also alter the plant microclimate (e.g., larger internodes or more leaf surface), relationships to other plants (i.e. weeds) or plant pathogens (Hauck, 1984). These changes alter the value of the plant as a home for the herbivore and its natural enemies (Price *et al.*, 1980). The effects of such physiological and morphological alterations to plants upon the various consumers will depend upon the 'feeding guild' to which an insect might belong (e.g., phloem or xylem 'sappers'; wood, root, shoot, or stalk 'borers'; leaf or needle 'miners'; root, needle or leaf 'chewers'; fruit and seed feeders; fungus-feeders; see Martin *et al.*, 1981 and Slansky and Scriber, 1984). Variable effects on the desired invasion of plants by pollinators are also observed under different fertilization regimes (see Scriber, 1984).

Although a considerable body of information on insect responses to plant N-fertilization has accumulated since 1930 (at least 200 studies), the results are generally not comparable because of variable experimental conditions (e.g. different plant cultivars, soils, insects, temperature). Furthermore, effects of seasonal variation in quantity and quality of plant N content and biochemically associated allelochemics upon post-ingestive growth performance indices of insects were generally unknown at that time, as were the effects upon herbivores of fertilizer-induced variation in plant chemistry (Jones, 1976; Scriber, 1984).

The close relationships of larval growth rates and efficiencies (see Fig. 7.1; Waldbauer, 1968) with leaf nitrogen content is illustrated in Fig. 7.2. The symbols represent the mean performance (±SE) of the black swallowtail butterfly, *Papilio polyxenes*, larvae for 11 different species of the Umbelliferae family at particular times of the growing season. The poor correspondence of plant N

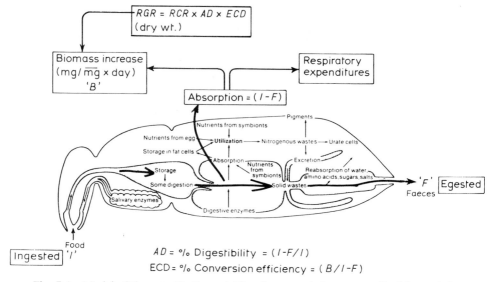

$$RGR = RCR \times AD \times ECD$$
(dry wt.)

Biomass increase
(mg/ $\overline{mg}$ × day)
'*B*'

Respiratory expenditures

Absorption = (*I* - *F*)

Ingested '*I*'

Food

'*F*' Egested
Faeces

$AD = \%$ Digestibility $= (I - F / I)$
$ECD = \%$ Conversion efficiency $= (B / I - F)$

Fig. 7.1   Model of the quantitative nutritional approach for a generalized insect (after Daly *et al.*, 1978). The relative growth rate (*RGR*) of the immature insect is a product of the relative consumption rate (*RCR*), the approximate digestibility (*AD*), the efficiency of conversion of digested biomass (*ECD*). The overall efficiency of conversion (ECI) is a product of *AD* and *ECD* (see Waldbauer, 1968; Scriber and Slansky, 1981 for further discussion).

content with calendar date is a consequence of the different phenologies and habitats of these plant species (e.g., meadow, marsh, woodland − Finke, 1977). A similar relationship between the overall efficiency of food processing (ECI) and the nitrogen content of foodplant species is illustrated by Mattson (1980).

7.2.2 Leaf water

Frequently overlooked as a nutrient, water can have a major influence on the growth of immature insects (Scriber and Slansky, 1981). Water content alterations in artificial diets (Reese and Beck, 1978) and in excised cherry leaves (Scriber, 1977) affect the metabolic costs, efficiencies and growth rates of Lepidoptera. In spite of a variety of behavioral, physiological and ecological adaptations to acquire adequate water and avoid desiccation, low leaf water content has remained a major evolutionary 'hurdle' for most phytophagous insects (Southwood, 1972).

Leaf water content provides a surprisingly useful index of larval growth performance of a variety of phytophagous (leaf-chewing) Lepidoptera (see Scriber, 1978a). Scriber and Feeny (1979) point out the correlation of leaf water content with plant growth form; forbs (which are non-graminoid, non-fern herbs) are generally higher in leaf water than leaves of shrubs or trees (see also Soo Hoo and Fraenkel, 1966a, b). Although tree-feeding insects may exhibit a

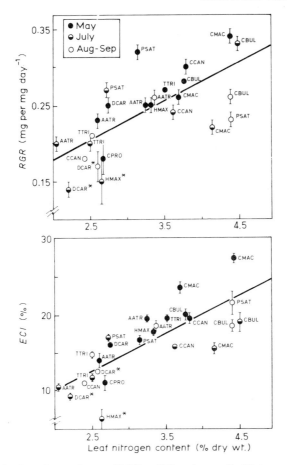

Fig. 7.2 Relative larval growth rate (*RGR* ± SE) and overall efficiency (*ECI* ± SE) of *Papilio polyxenes* in relation to total organic N content of leaves of various species of Umbelliferae at various dates (data from Finke, 1977). AATR = *Angelica atropurpurea*; DCAR = *Daucus carota*; CCAN = *Cryptotaenia canadensis*; HMAC = *Heracleum maximum*; PSAT = *Pastinaca sativa*; CMAC = *Conium maculatum*; CBUL = *Cicuta bulbifera*; CPRO = *Chaerophylum procumbens*; and TTRI = *Thaspium trifoliatum* ($n = 25$; $r = 0.685$ for RGR; $n = 25$; $r = 0.796$ for ECI).

stronger correlation of growth rate or efficiency with leaf water content than is the case with most forb-feeders, and forb-feeders may sometimes exhibit a stronger correlation of growth rate or efficiency with plant nitrogen than do tree feeders (Scriber and Feeny, 1979), these trends may be significantly altered by other nutrients or allelochemics (see Scriber and Slansky, 1981). For example, the growth response of a foodplant generalist, the southern armyworm, is correlated with both N and H_2O. However, various allelochemics may also be involved in differential performance on the 15 legume species or varieties (see Figs 7.3 and 7.4).

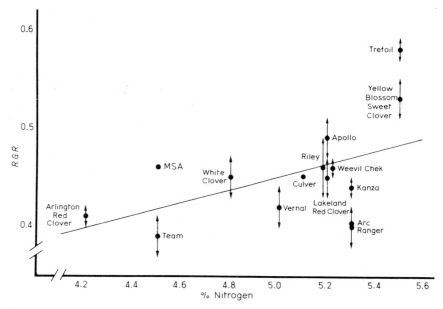

Fig. 7.3 Relative larval growth rate ($RGR \pm$ SE) of penultimate instar *Spodoptera eridania* on various legumes and alfalfa cultivars as a function of leaf nitrogen content ($r = 0.40$; $n = 15$).

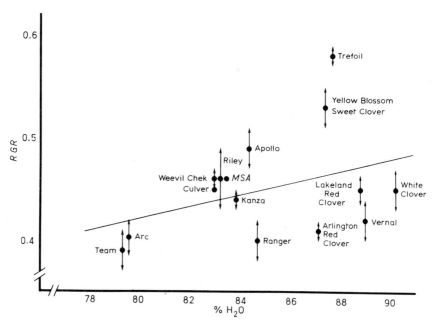

Fig. 7.4 Relative larval growth rate ($RGR \pm$ SE) of penultimate instar southern armyworm, *Spodoptera eridania*, on various legumes and alfalfa cultivars as a function of leaf water content ($r = 0.51$, $n = 15$).

7.2.3 Allelochemics

Defined as non-nutrients produced by one organism that affect the behavior, health or ecological welfare of another, 'allelochemics' includes a large array of chemicals. The allelochemic effect is both concentration-dependent and situation-dependent (see Reese, 1979). A chemical classified as a nutrient for one species may act as an allelochemic in some situations, and a chemical such as L-canavanine which is highly toxic to many herbivores may be utilized as a source of dietary nitrogen by adapted specialists (Rosenthal *et al.*, 1977; 1982). Thus, the distinction between a nutrient and an allelochemic is not always clear-cut (see also Bernays and Woodhead, 1982).

The range of potential allelochemics is diverse. Included are major classes of secondary plant metabolites: toxic amino acids, cynanogenic glycosides, alkaloids, toxic lipids, glucosinolates, sesquiterpene lactones and other ter-penoids, saponins, phytohemaglutinins, proteinase inhibitors, flavonoids, phenols, tannins and lignins, and insect hormones and antihormones (Rosenthal and Janzen, 1979; see Fenical, 1982 for a comparison in the marine environment). The following examples illustrate the manner and relative degree to which certain selected plant metabolites can affect the post-ingestive utiliz-ation of foodplant tissue for insect growth.

Nitrogen content appears to be a key nutrient for larvae of *P. polyxenes*, regardless of the season or plant species (Fig. 7.2). However, the possession and type (i.e., linear, angular) of furanocoumarins of a given umbellifer may have a major effect on patterns of distribution and abundance of insect herbivores of the Umbelliferae (Berenbaum, 1981a, b). Linear furanocoumarins are capable of binding with DNA in the presence of ultraviolet light and have been shown to be toxic to generalist herbivores (Berenbaum, 1978). Umbellifer specialists, such as *P. polyxenes*, on the other hand survive well on leaves of *Angelica atropurpurea*, *Conium maculatum* and *Pastinaca sativa*, (see Fig 7.2; Finke, 1977; Scriber and Feeny, 1979), all of which contain linear furanocoumarins as well as additional allelochemics such as angular furanocoumarins or alkaloids (Berenbaum, 1981a). The angular furanocoumarins do however reduce growth rates and adult fecundity and may represent 'an escalation of the coevolutionary arms race' between the Umbelliferae and their herbivores (see Berenbaum and Feeny, 1981). Woodland umbellifers on the other hand, characteristically are not exposed to high levels of ultraviolet light, lack furano-coumarins, and are characterized by more generalized insect feeders (Beren-baum, 1981a).

Although adult choices and larval survival or processing abilities are not necessarily correlated, even for the specialists on the Umbelliferae (Wiklund, 1975, 1981), allelochemics such as furanocoumarins have a major influence on the location, selection, and suitability (both physiological and evolutionary) of Umbelliferae to insect herbivores (Hegnauer, 1971; Berenbaum, 1981a; Berenbaum and Feeny, 1981). The same can be said for the Cruciferae with their allylglucosinolates (see Root, 1973; Kjaer, 1974; Chew, 1975; Slansky and Feeny, 1977). These two plant families and their associated insects provide valuable experimental organisms for analyses of coevolutionary processes (see Ehrlich and Raven, 1964; Feeny, 1975; Jermy, 1976; Futuyma, 1979; Connell, 1980; Janzen, 1980).

Differential interactions of umbellifer and crucifer allelochemics upon specialized and generalized herbivores has been observed. For example, the umbellifer-specialist *P. polyxenes*, which refuses to consume cruciferous plants, dies when its normal hostplants are perfused with rather low concentrations of allylglucosinolate (AG) (Erickson and Feeny, 1974; Blau *et al.*, 1978). The performance of *Pieris rapae* (the crucifer-specialized cabbage butterfly) is unaffected even by abnormally high AG concentrations in its food plants (Blau *et al.*, 1978). The polyphagous southern armyworm, *Spodoptera eridania*, successfully consumes and grows on several species of Umbelliferae and Cruciferae (Soo Hoo and Fraenkel, 1966b; Scriber 1984). Survival and growth performance of *S. eridania* larvae depend upon the particular plant species involved (e.g., they die on parsnip, *Pastinaca sativa* – Soo Hoo and Fraenkel, 1966a) and also upon the dose of chemical (e.g., they suffer a severely reduced efficiency with artificially increased AG concentrations in bean leaves; Blau *et al.*, 1978). It is therefore likely that the growth performance of

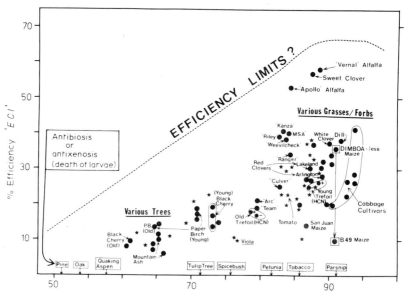

Fig. 7.5   Efficiency of biomass conversion (ECI) of penultimate-instar (5th) southern armyworm larvae on various plant leaves presented as a function of leaf water content. All experiments were performed under controlled environment conditions. Each symbol represents a single feeding experimental mean value of replicate larvae. Circles represent data of Manuwoto and Scriber (1981) and Scriber (1978b, 1979a, 1981, 1984 and unpublished); stars represent data of Soo Hoo and Fraenkel (1966b).

S. eridania larvae on various cabbage cultivars (Fig. 7.5; Scriber, 1981) may have been less than the maximum permitted by leaf water and nitrogen because of the AG in these plants.

Because of their polyphagous nature and willingness to accept many marginal host plants in no-choice situations, *S. eridania* larvae are useful bioassay organisms. Fig 7.5 represents *S. eridania* larval performance on a spectrum of plant growth forms (see Section 7.2.4(a)). A variety of plants were antibiotic or antixenotic, either of which eventually result in death of the larvae if no alternatives are provided. It appears that biomass conversion efficiencies are correlated with the plant growth form (or related factors) of acceptable hosts as indexed by leaf water content. Although specialized herbivores are generally believed to be well-adapted to the secondary plant metabolites in their respective hosts, efficiencies and growth rates of the generalized *S. eridania* are not statistically different from those of phytophagous specialists on any particular host plant (Scriber, 1978a, 1979a).

Alfalfa varieties have been bred for resistance to several alfalfa insect pest species: 'Team' alfalfa has been bred for resistance to pea aphids, spotted alfalfa aphid, potato leafhopper, and the alfalfa weevil (Barnes *et al.*, 1970); 'Arc' and 'Weevilcheck' for resistance to aphids, leafhopper and alfalfa weevils; 'Kanza' for aphid resistance; 'Culver' for leafhopper and spittlebug resistance (Wilson and Davis, 1960); and MSA CW3AN for leafhopper resistance. 'Ranger', 'Vernal', and 'Apollo' are relatively susceptible to these alfalfa pests, having been bred primarily for yield and disease resistance. To date there has been no resistance identified against either specialized or generalized Lepidoptera larvae in alfalfa cultivars (Sorensen *et al.*, 1972; Nielson and Lehman, 1980). Differential metabolic costs are observed however, for *S. eridania* larvae in the penultimate instar (Table 7.1) on these cultivars. Differential allelochemics (quality or quantity) may be responsible in part for the very high metabolic costs (affected as low efficiencies, particularly ECD's) on Team, Arc, Culver and the red clovers (Table 7.1; see also Fig. 7.5), but it is clear that leaf water (Fig. 7.4) and leaf nitrogen content (Fig. 7.3) are also involved. It seems that L-canavanine in alfalfa, coumarins in sweet clover (Manglitz *et al.*, 1976), and cyanogenic leaves of trefoil (Scriber, 1978a) pose no major problems for *S. eridania* larval growth (Fig. 7.5).

Perhaps best known among host plant resistance cases is the European corn borer, *Ostrinia nubilalis*, and Maize (Loomis *et al.*, 1957; Beck, 1965; Klun *et al.*, 1970; Guthrie, 1974; Ortega *et al.*, 1980). The mechanisms of resistance of US maize to *O. nubilalis* are believed to be mediated partly by a mixture of cyclic hydroxamates (Klun and Robinson, 1969). The aglucone, 2, 4 dihydroxy-7-methoxy-1, 4 benzoxazine-3-one (called DIMBOA) is the major component of this mixture, representing 80–90% of the hydroxamate concentration (see Fig. 7.6; Klun and Robinson, 1969; Corcuera, 1974). As with many biologically active secondary plant metabolites, the glucoside concentrations (and 'aglucone' hydrolysis products) decline rapidly with plant age. In addition,

Table 7.1 Performance of penultimate instar *Spodoptera eridania* larvae on various legumes. Data are presented as a mean (± SE) Madison, Wisconsin

Legume variety	(*n*)	Assimilation efficiency (A.D.)	Net conversion efficiency (E.C.D.)	
Vernal	5	77.2 ± 3.2	76.6 ± 7.9	a
Yellow Blossom Sweet Clover	4	80.3 ± 2.2	71.3 ± 6.8	ab
Apollo	5	79.0 ± 3.5	68.3 ± 7.5	abc
Riley	5	77.4 ± 4.8	54.4 ± 8.9	abcd
MSA CW3AN	5	76.0 ± 5.3	53.8 ± 11.1	abcd
Kanza	5	81.6 ± 2.3	50.6 ± 5.6	abcd
Weevilcheck	5	84.9 ± 4.2	48.1 ± 10.1	abcd
Empire Bird's foot trefoil	5	78.6 ± 2.2	47.8 ± 2.2	abcd
White Clover	5	86.6 ± 1.7	42.2 ± 5.7	abcd
Ranger	5	82.1 ± 1.6	41.8 ± 7.1	abcd
Lakeland Red Clover	5	80.2 ± 2.7	38.3 ± 5.5	bcd
Arlington Red Clover	5	86.5 ± 1.9	31.8 ± 1.5	cd
Culver*	1	83.3	29.8	
Arc	5	90.4 ± 2.7	24.1 ± 5.0	d
Team	4	94.0 ± 0.4	18.9 ± 2.5	d
(LSD)		(15.5)	(35.7)	

* Significant differences are indicated (*p* = 0.05) via Tukey's test for unequal sample sizes. Culver is not included in the ANOVA

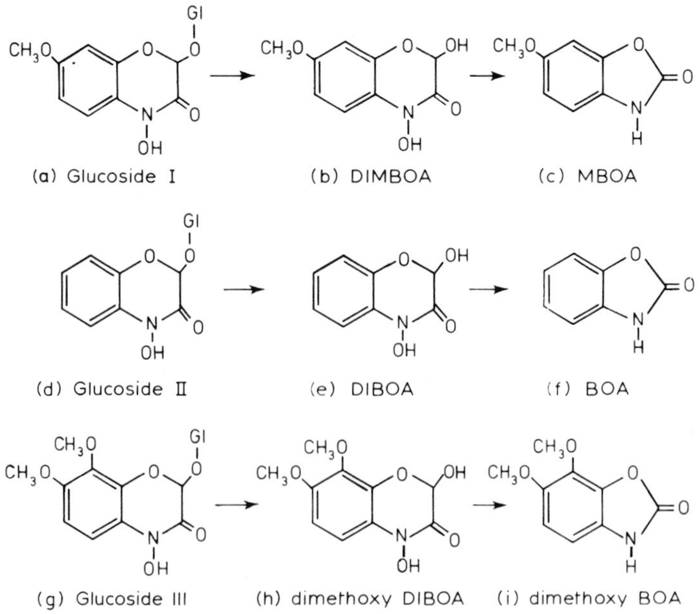

(a) Glucoside I   (b) DIMBOA   (c) MBOA

(d) Glucoside II   (e) DIBOA   (f) BOA

(g) Glucoside III   (h) dimethoxy DIBOA   (i) dimethoxy BOA

Fig. 7.6 Primary glucosides in maize, *Zea mays* and their breakdown products (from Manuwoto, 1980).

different genotypes of maize contain different DIMBOA concentrations at any particular growth stage or size, which may in part account for differential susceptibility to *O. nubilalis* feeding. The high-DIMBOA US inbreds B49 and C131A have been selected as *O. nubilalis*-resistant lines, whereas WF9 is low in DIMBOA and consistently rates as susceptible.

Recently, some Latin American germplasm has been identified and selected for resistance to stalk-boring Lepidoptera (Ortega *et al.*, 1980). In these varieties however, DIMBOA is not the key factor mediating leaf-feeding resistance, as DIMBOA concentrations are as low as in the WF9 susceptible (Sullivan *et al.*, 1974; Scriber *et al.*, 1975b). In fact, it is likely that a combination of nutritional and allelochemic factors mediate leaf-feeding and stalk-boring resistance to *O. nubilalis* both in field and laboratory situations (Beck, 1957; Manuwoto, 1980; Scriber *et al.*, unpublished Annual Report to the Rockefeller Foundation; 1975, 1976). Physical factors are also involved in antixenosis, antibiosis and tolerance, varying seasonally within the plant and with respect to environmental conditions, other plants in the community and insect ecotypes (Chiang, 1978).

Perhaps the most critical factor involved in resistance to *O. nubilalis* (and that of most Lepidoptera) is the first contact that freshly eclosed larvae have with their plant (Beck, 1957). Leaf consumption, larval survival and 72-hour larval growth rates are signficantly suppressed both on the high-DIMBOA (B49) genotype and the genotype San Juan compared to the low-DIMBOA check (WF9) and the DIMBOA-less mutant (bxbx) genotype (Fig. 7.7). Three-day survival was 99% on WF9, 86% on DIMBOA-less, 20% on B49; and 19% on San Juan (Manuwoto, 1980). These were no-choice situations, and we should realize that larval and adult host choices are important, but as yet uninvestigated aspects of this insect-plant interaction, as are the fecundity and fertility of resultant *O. nubilalis* survivors.

It is important that plant pathogens and insect pests in addition to the European corn borer be considered in host plant resistance (HPR) breeding programs. The polyphagous southern army worm, *S. eridania*, was used by Manuwoto and Scriber (1982) to assess the metabolic costs involved in consuming and processing young B49, San Juan, and DIMBOA-less maize plants. Although the digestibilities were all relatively high (68–81%), the larvae that were fed B49 and San Juan metabolized (respired) more than 87% and 82% of the assimilated biomass of B49 and San Juan, respectively, compared to only 46% on the DIMBOA-less mutant maize tissue. These extremely high metabolic costs for *S. eridania* on the *O. nubilalis*-resistant genotypes resulted in significantly suppressed overall feeding efficiencies (ECI; see also Fig. 7.5) and reduced larval growth rates. These results are similar to those discussed earlier with insect-resistant alfalfa cultivars (Scriber, 1979a).

The low efficiencies of *S. eridania* on all tree leaves tested (see Fig. 7.5) may be the result of a variety of factors (e.g., low leaf water, low nitrogen, high fiber, high tannin concentrations). The growth rates on trees generally remain

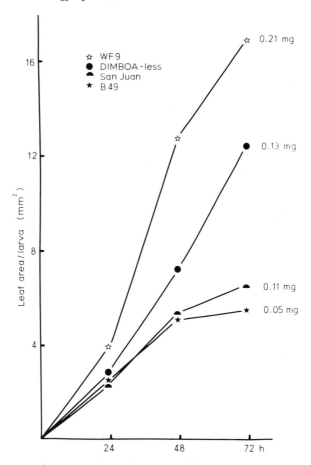

Fig. 7.7   Consumption rates of newly eclosed first-instar larvae ($n = 70-80$) of the European corn borer, *Ostrinia nubilalis*, on four genotypes of maize expressed as leaf area consumed per surviving larva (from Manuwoto, 1980). Mean weight of surviving larvae at 72-h post-eclosion are presented at the right (see text for further details).

even lower than those observed for the most resistant of the legumes, maize, and cabbage previously discussed (see also Scriber, 1981), even though insect compensation with increased consumption rates is observed (Section 7.3.2 and Scriber and Slansky, 1981).

In addition to deterrent or toxic effects in early instars or measurable metabolic costs in later instars, allelochemics may be responsible for subtle effects not manifested until pupation. Previous instar feeding experience can influence the biomass and nitrogen metabolism and dietary changes cause sufficient physiological stress (perhaps hormonal) to kill *S. eridania* during pupation (Scriber, 1982).

7.2.4 'Physiological efficiency' model of plant suitability based upon a leaf water/nitrogen 'index'

The general covariance of natural leaf water and leaf nitrogen concentration (Scriber and Feeny, 1979) and a variety of other leaf chemicals (such as tannins, fiber, lignins, oils) confounds interpretations of insect–plant interactions. Diurnal variations in plant chemistry sometimes may be comparable to seasonal changes (Scriber, 1977; Pate, 1980). Although artificial diets permit investigation of chemical mechanisms and dose-response relationships between herbivores and specific chemicals, such information cannot be readily used for prediction of larval growth on plant tissues.

With these considerations in mind and with data from several hundred feeding experiments on the nutritional physiology of Lepidoptera, I offer a 'Physiological Efficiency' model of larval growth performance. Using leaf water and leaf nitrogen content as an index of plant quality, I believe we have the basis for predictions of the optimal or maximum potential post-ingestive physiological growth performances of herbivorous insects. With such a model, it is possible to relate larval growth performance maxima to differences in plant growth form (Section 7.2.4(a)), seasonal trends in leaf chemistry (Section 7.2.4(b)), and to so-called 'qualitative' as well as 'quantitative' chemical defenses which are an integral part of the 'plant apparency' concepts (Feeny, 1975, 1976; Rhoades and Cates, 1976; see also Section 7.2.4(c)). It may also be possible to use this model to predict the effects of plant fertilization on larval growth performance, even though many other chemical factors in addition to water and nitrogen will be altered (Slansky and Feeny, 1977; Wolfson, 1981; Scriber, 1984). A variety of intrinsic (insect-related adaptations; e.g., Section 7.2.4(d)) and extrinsic (hostplant and environmental) factors will determine the observed physiological efficiency and growth rate.

The 'physiological efficiency' model may predict upper limits to postingestive growth performance of leaf-chewing insects. The extent to which the actual performance is reduced below maxima indexed by leaf water and nitrogen depends on a variety of plant chemicals and concentrations as well as the degree to which the insect is physiologically adapted to that overall chemical 'gestalt'. Thus, while the model may have predictive capabilities at the upper limits of performance, it is not useful for predicting the lower limits (cf. Fig. 7.5).

(a) Plant growth form

Classification of plants into different types of growth forms (i.e., trees, shrubs, forbs, grasses, etc.) has facilitated certain discussions of the structural organization of plant communities (Whittaker, 1975). Growth form classification also impacts on discussions of insect–plant interactions (Feeny, 1975, 1976; Rhoades and Cates, 1976; Futuyma, 1976; Scriber and Feeny, 1979). Difficulty arises when a species is intermediate in classification (e.g., prickly ash,

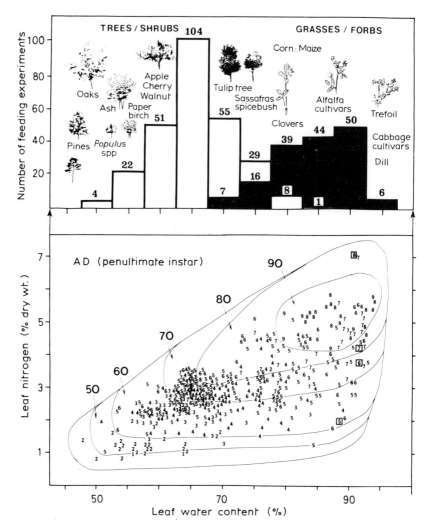

Fig. 7.8(a)   Number of penultimate-instar insect feeding experiments ($n = 436$), with each represented by numbered data points in (b) for various plant species along sections of the axis of leaf water content. For example, there are four experiments in which mean leaf water content was between 47.5 and 52.4%. The non-shaded portion of the histogram represents experiments using leaves of trees or shrubs ($n = 274$ total) and the superimposed shaded portion represents experiments with grasses or forbs ($n = 162$ total).

(b)   Approximate digestibilities (AD) of penultimate-instar insect larvae as a function of plant quality as defined by leaf water (48–94% fresh weight) and a total organic N (0.8–7.0% dry wt.). Lines enclose all digestibilities equal to or greater than those indicated for the zone, but do not exclude lower values. Each number represents the mean digestibility (AD) of replicate larvae for particular host plant species (0 = 0–4%, 1 = 5–14%, 2 = 15–24%, 3 = 25–34% . . . etc. to 9 = 85–94%). This figure includes data from 30 different insects species fed mature leaves from various of their food plants under the same controlled environmental conditions (see text; Fogal, 1974; Scriber and Slansky, 1981; and Scriber, unpublished). Performance values for *Spodoptera eridania* in relation to differential N-fertilization of B49 maize are indicated by four boxes (see Scriber, 1984).

Xanthoxylum americanum, can be a small shrub in Wisconsin and New York, but a rather large tree in the southern USA).

For phytophagous insects, leaf chemistry rather than structural diversity of plants may be the key factor. Thus, growth form is used henceforth to facilitate discussion of insect–plant interaction literature and to integrate current synecological concepts, rather than as an integral part of the physiological efficiency model of insect responses as indexed by leaf water and nitrogen.

Observations on the digestibility (AD) of 30 phytophagous insect species (Lepidoptera and Hymenoptera), in 436 experiments on various foodplants under the same controlled environmental conditions are presented in Fig. 7.8. The top of the figure represents the number of experiments using plants of different growth forms as indexed by leaf water content at the time of the experiments. Although leaf nitrogen does not correlate well with plant growth form, it does provide an important second axis of plant quality for illustration of potential larval performance (see Fig. 7.8b).

Some tree or shrub species (e.g., tulip tree, sassafras, or spicebush) have consistently higher average leaf water content than others (e.g., ash, oak, or pines), whereas the forbs and grasses were for the most part even higher in leaf water content, with a zone of separation near 75% water content (see Fig. 7.8a). Most of these experiments used mature (fully expanded) leaves. Seasonal trends superimposed upon individual plant variability (Scriber, 1977) will certainly blur the classification of growth form based on leaf water. The point is that larval performance does appear to be delineated by leaf water and leaf nitrogen content (Fig. 7.8b), regardless of plant growth form.

(b) Seasonal trends

As leaves expand, mature, and senesce, there is a general decline in daily average leaf water and leaf nitrogen across most, if not all, growth forms (Fig. 7.9a). Concurrently, there is a general increase in fiber, lignins, tannins, and leaf toughness (McHargue and Roy, 1932; Feeny, 1970; Hough and Pimentel, 1978) and decrease in the concentration of many low molecular weight secondary chemicals (Fig. 7.10 and Section 7.2.4(c)). These trends in leaf chemistry all correlate with a general reduction of digestibility for leaf chewing insects. I hypothesize that the 'physiological efficiency' maxima as indicated by the 'contour lines' of Fig 7.8b, and as indexed by water and nitrogen, will apply to most chewing insects, most food plant species, and to a large extent throughout the growing season. It seems reasonable to use this simple H_2O/N index as a model to predict digestibility maxima and even subsequent growth rate potential under a given set of environmental conditions throughout the season (see Scriber and Slansky, 1981). If valid such a model would, to a large extent, obviate the need for discussion of plant growth form and seasonal variation *per se* as intrinsic factors affecting post-ingestive suitability of leaf tissue for insect herbivores.

It is important to point out that even though a strong correlation between digestibility and N-concentration has also been observed for phloem-sucking Hemiptera (McNeill and Southwood, 1978), feeding guilds of insects other than leaf-chewers may respond in diverse manners, depending on the chemical quality and specific limiting chemicals in their food (see Scriber and Slansky, 1981;

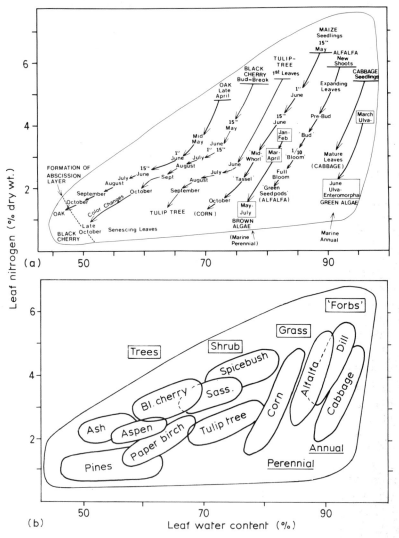

Fig. 7.9 Seasonal changes in leaf water and nitrogen contents for leaves of selected plant growth forms ((a) after Scriber Scriber and Slansky, 1981). These are generalized trends based on actual data (Smith, 1964; Feeny, 1970; Geiselman, 1980; and Scriber, unpublished). Variation in individual plant phenology, soil quality, microhabitat (sun versus shade) will alter the actual 'track' on the figure. Such variation is still rather predictable and the 'typical' range in pattern for certain mature leaves from north temperate latitudes is presented in (b) (Scriber, unpublished data).

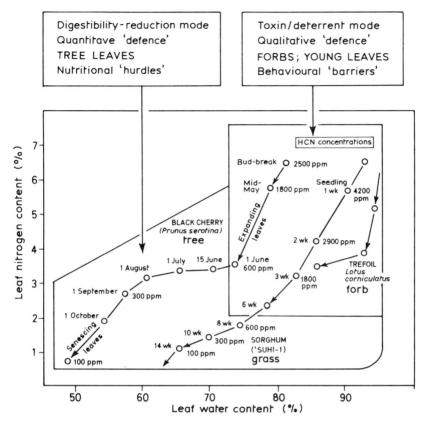

Fig. 7.10 Correlates of the plant apparency concept as illustrated with selected plant growth forms containing cyanogenic glycosides. Seasonal trends in cyanide, nitrogen and water concentrations illustrated. Although so-called 'predictable' or 'apparent' plants (or plant parts) tend to be at the left and bottom of the figure with 'unapparent' or 'unpredictable' plants at the upper right, many exceptions may exist depending upon the insects concerned (see text for further discussion).

Slansky and Scriber, 1984). Considerable variability in digestibility is introduced by differential chemical quality of the plant, by different environmental conditions, and by a variety of intrinsic factors related to age, sex, or size of the insects themselves. Nonetheless, tree-feeders on the whole have lower digestibilities (mean $AD = 39\%$, $n = 633$ treatments with 99 insect species) than do forb-feeders (mean $AD = 53\%$, $n = 716$ treatments with 38 insect species; supplementary data to Scriber and Slansky, 1981). Digestibilities of grain and seed chewers averaged 72% ($n = 48$ treatments with 9 insect species), whereas insect-feeders averaged 80% ($n = 119$ treatments with 32 species of insect parasites and predators). The water/nitrogen 'index' used for leaf-chewers does not apply to these latter two categories of resource types.

(c) Correlates of the 'plant apparency' concept

If the location, selection, acceptance, and suitability of a plant for an insect are determined largely by various chemicals, then testable hypotheses can be proposed to advance our understanding of the chemical aspects of insect-plant interactions (see Platt, 1964; Gilbert, 1979; Willson, 1981). The concept of 'Plant Apparency' as proposed by Feeny (1976) and Rhoades and Cates (1976) has spawned considerable research and debate in the last several years. Apparency, originally defined by Feeny (1976, p. 5) as 'the vulnerability of an individual plant to discovery by its enemies', depends upon many factors. The susceptibility of a plant (or plant part) to discovery is influenced not only by its size, growth form, and persistence, but also by its distribution and relative species abundance within the community (e.g., Atsatt and O'Dowd, 1976). Further, it depends on the numbers and host-locating abilities of all relevant herbivores, pathogens, and plant enemies in the community. As such, the 'apparency' concept is not realistically testable as a whole. It is nonetheless a powerful concept and very useful theoretical framework.

Since it would be extremely complicated (if not impossible) to quantify 'apparency' or 'predictability' for even a single insect on a particular plant at a given time in just the chemical dimension, it is understandably difficult to compare degrees of apparency for different plants in anything but generalized terminology. However, Feeny's (1975) concepts of 'quantitative' chemical defenses (i.e., chemicals such as tannins that reduce digestibility, and act according to their concentration in plants) and 'qualitative' defenses (i.e., chemicals such as glucosinolates whose activities are specific, that are effective in low concentration, and are relatively inexpensive for the plant to produce) fit not only his data from oaks and Cruciferae, but many other terrestrial plant species. However, it is also intriguing that many early successional plant colonizers that 'escape in space and in time', possess toxin or deterrent types of secondary metabolites.

Geiselman (1980) has drawn an interesting parallel between marine intertidal plant chemistry — herbivore interactions and terrestrial plant — herbivore interactions. As is the case with the terrestrial trees, grasses and forbs (perennial and annual), marine algae also decline in nitrogen and water content over the growing season (Fig. 7.9a). More significant perhaps is the corresponding seasonal increase in phenolics, especially in the brown (perennial) algae and the associated negative effects upon feeding of the herbivore, *Littorina littorea*, compared to feeding on the preferred brown (annual) algae with its low polyphenol concentrations. It is also intriguing that certain marine annuals (more empheral and perhaps less 'predictable' to herbivores) have some unique low molecular weight halomethane compounds (e.g., CH_2I_2, $CHBr_3$), which appear to be analogous to many of the so-called 'qualitative' deterrent or toxic secondary metabolites of herbaceous plants (see Setion 7.2.3; Allelochemics).

Although different plant growth forms do seem to some extent to possess different major classes of secondary plant metabolites, it is not possible to apply the 'qualitative' or 'quantitative' chemical categories (Feeny, 1975) consistently to any particular growth form. It is even difficult to determine in which category a given secondary plant metabolite should be placed. For example, L-canavanine, a structural analogue of arginine, inhibits the survival, growth, metamorphosis, and reproduction of many 'unadapted' insects (Rosenthal and Bell, 1979). This chemical could be easily considered a toxin, deterrent or behavioral barrier in the 'qualitative defense' category, but it occurs in concentrations as high as 13% of the tissue dry weight and might on this basis be considered as a 'quantitative defense'. On the other hand, if dose dependency is inherent in the definition of a quantitative defense, irrespective of the concentration in plant tissues, then perhaps even 'low' concentrations of toxins (Feeny, 1975, 1976) might be considered 'quantitative'. The distinction between allelochemic and nutrient is obscured in the adapted bruchid beetle, *Caryedes brasiliensis*, a specialist which not only detoxifies L-canavanine, but may in fact utilize its nitrogen as a nutrient (Rosenthal *et al.*, 1977, 1982).

Although tannins, lignins, and fiber (generally considered 'quantitative' defenses) occur at high concentrations in tree leaves (up to 10.0%) and heart-wood (up to 30–40%), these defenses are not confined to this plant growth form (Swain, 1979). For example, crown vetch is a forb which may contain up to 16% tannin and 23% fiber or lignin (Burns and Cope, 1974). Protein-binding tannins reduce the availability of nitrogenous nutrients and may also suppress the activities of digestive enzymes and symbionts in the herbivore gut (Swain, 1979). Even with certain chemical adaptations such as a high gut pH, which might inhibit complexing of tannins with protein (Berenbaum, 1980), the general increase in tannins with leaf age may be a major factor affecting digestibility and growth in most Lepidoptera. Tannin-tolerance in certain Orthoptera (Bernays, 1978; Bernays *et al.*, 1980) and Coleoptera (Fox and Macauley, 1977) contrasts to 'anti-digestive' properties supposed with Lepidoptera. Bernays (1978) suggests that these differences may be related to the evolutionary radiation of these older insect groups during the Carboniferous and Permian periods, an environment rich in tannins. The Lepidoptera radiated at a later time in an environment rich in higher plant groups (e.g., Angiosperms) and plant species devoid of tannins, and thus may be relatively more sensitive or less well adapted to tannins where they are encountered.

Along with tannins mature cell walls of all vascular plants contain lignins which are phenolic heteropolymers with a molecular weight greater than 5000 (Swain, 1979). Lignins may occur in concentrations up to 40% as in woody tissues of gymnosperms and angiosperms. In that they increase rigidity of cell walls, they are probably involved in the observed digestibility-reduction for herbivores on many mature leaves (including forage grasses) in contrast to young, expanding leaves (cf. Figs 7.8 and 7.9). Lignins, which are second only to cellulose as the most abundant of natural polymers, are probably very important factors (as are tannins) in determining the evolutionary success of land plants, especially trees and shrubs. Perhaps the most important function of tannins and lignins in plant tissues is the protection they afford against potential fungal and bacterial pathogens; their significance in the ecology of herbivores may be more 'diffuse' (Fox, 1982).

The correlation of seasonal trends in low molecular weight allelochemics as defined for representative plant growth forms by the water–nitrogen index is illustrated in Fig. 7.10. Three plants containing cyanogenic glycosides are used to illustrate the dynamics of leaf chemistry for a 'tree' (Smeathers *et al.*, 1973), a 'grass' (Loyd and Gray, 1970; Woodhead and Bernays, 1978), and a 'forb' (Smith, 1964; Scriber, 1978b). Similar trends of cyanide decreasing seasonally exist for a chaparral 'shrub', *Heteromeles arbutifolia* (Dement and Mooney, 1974) and bracken 'fern', *Pteridium aquilinum* (Lawton, 1976; Cooper-Driver *et al.*, 1977). Correlated with the decreasing cyanide concentrations is an increase in fiber, lignin, tannin.

Two sections of this two-dimensional response surface have been outlined to illustrate the manner in which certain of the components of the plant apparency concept may be superimposed upon a natural continuum of changing leaf chemistry. If we focus our attention on this general descriptive model of plant chemistry, we can observe how many insect responses align with the concept of plant apparency. For example, there are many instances of cyanide toxicity or deterrency in forbs and young tree leaves such as black cherry (Jones, 1972; Conn, 1979), and there is evidence that phenolic acids, tannins, or additional digestibility-reducing compounds may predominate as the plant matures (Feeny, 1970; Woodhead and Bernays, 1978; Fisk, 1980; and references in the previous paragraph).

On the other hand, *S. eridania* is a generalist species which survives and grows well on diets composed of cyanogenic forbs (Scriber, 1978b), shrubs (Rehr *et al.*, 1973), and trees (Scriber, 1982). In other cases, cyanide can act in a dose-dependent manner to attract certain insects (Nayar and Fraenkel, 1963).

Whereas immature plant cells are succulent and nitrogen-rich, with relatively high concentrations of various 'qualitative' allelochemics (McKey, 1979; Mattson, 1980), the 'quantitative' allelochemics (such as tannins, lignins, and fiber) can also reach concentrations of 20% in certain forbs such as crown vetch (Burns and Cope, 1974). Thus, even if discussion of 'plant apparency' correlates is restricted to post-ingestive chemical factors, it is not always possible to distinguish between toxin (qualitative) and digestibility-reduction (quantitative) modes of action (Rhoades, 1979; see also Zucker, 1983). Furthermore, the responses of different individuals of a given herbivore species will depend upon a variety of behavioral, physiological, and ecological adaptations of the insect as described in the next section.

(d) Insect adaptations

At the behavioral level, several 'choices' are available to the individual insect herbivore which could improve its post-ingestive growth performance, not all of which need to be facilitated by the ovipositing adult (see Chapter 6). For example, a switch from old to new leaves or from tree leaves to forbs could result in improved efficiency and growth rate, provided the behavioral barriers to feeding can be scaled (Figs 7.8 and 7.9). Such a shift might be visually depicted on the chemical response surface (Figs 7.8, 7.9 and 7.10) as a movement to the right (higher leaf water) or upward (higher nitrogen). A similar improvement in growth performance would result from feeding specialization on tender palisade cells in a leaf, as by leaf-miners.

In addition to pre-ingestive effects upon behavioral host choice (Jermy, *et al.*, 1968), previous feeding experience may also significantly affect post-ingestive nutritional physiology (Schoonhoven and Meerman, 1978; Scriber, 1979b, 1981, 1982). A change in consumption rate, such as that caused by behavioral induction, affects physiological efficiencies and growth of larvae (Grabstein and Scriber, 1982). Thus, the distinction between pre-ingestive and post-ingestive is not as clear as it might first seem. The importance and complexity of the interaction between behavior and physiology cannot be overemphasized.

Once consumed, allelochemics have assorted deleterious effects. Otherwise, they can pass through the insect with no effect, be sequestered, or detoxified. Many enzyme systems are involved in biochemical defense against plant allelochemics, but the most familiar and perhaps most important are the mixed function oxidases. Brattsten (1979) describes the three major characteristics that contribute to their importance in biochemical waste disposal: (i) they catalyze numerous oxidative reactions that produce more polar and hence more excretable compounds, (ii) they are non-specific in that a wide range of chemicals are acceptable substrates, (iii) they can adjust rapidly (within minutes) to the presence of allelochemics or synthetic insecticides via induction. The MFO detoxification system is not confined to insects; MFO activity has been identified in a wide range of vertebrates and invertebrates. A variety of flies, mosquitoes, cockroaches, crickets and a minimum of 40 species of Lepidoptera (as well as representatives from other insect orders) possess MFO activity (review: Brattsten, 1979).

Induction of MFO activity in insects to synthetic insecticides was discovered nearly two decades ago. This phenomenon as mediated by natural plant products has received increased attention in the last few years. Induction refers to the temporarily accelerated production of enzyme protein. A large number of natural plant chemicals are known to induce the MFO system (Brattsten *et al.*, 1977) and host plant allelochemic induction of MFO activity may mediate resistance to synthetic insecticides as well (Yu *et al.*, 1979). There can also be genetic bases to cross-resistance between pesticides and plant defenses (Gould *et al.*, 1982), thus posing another significant consideration in breeding host plant resistance.

On the other hand, a biologically active chemical (whether a natural plant product or a synthetic compound) in the presence of certain other synergistic chemicals may have increased toxicity. Such synergistic chemicals may inhibit the MFO enzymes and therefore are of considerable importance in insecticidal toxicity and drug effectiveness. The role of naturally occurring synergists on the choice, consumption and utilization of plants of phytophagous insects, however, is unknown (Scriber, 1981).

Although certain 'adapted specialist' insects have detoxification mechanisms for dealing with potentially deleterious plant metabolites (e.g., Cruciferae-feeders, Umbelliferae-feeders; see also Whittaker and Feeny, 1971), there are certain generalist species that consume a variety of secondary plant metabolites without detectable ill-effects. Prominent among these generalists are the polyphagous noctuid moths such as the black cutworm, *Agrotis ipsilon* (Reese and Beck, 1976a, b, c) and the southern armyworm, *S. eridania*. In addition to their high levels of general MFO activity relative to other Lepidoptera (Krieger *et al.*, 1971), southern armyworms also have the capacity of rapid induction (within hours) of these enzyme systems (Brattsten *et al.*, 1977).

S. eridania has another capability contributing to rapid larval growth even under severe metabolic expenditures (as reflected in low conversion efficiencies of assimilated biomass and energy; ECD), the ability to maintain extremely high consumption rates on digestively poor or allelochemically adverse host plants. For example, metabolic costs for southern armyworms to process B49 (high DIMBOA) and San Juan (low-DIMBOA, but corn borer-resistant maize tissues) are two to three times greater than observed for the DIMBOA-less mutant maize (Manuwoto and Scriber, 1981). The southern armyworm larvae appear to 'compensate' for these costs (reflected in low efficiencies) with two to three times higher consumption rates (Fig. 7.11). Similarly, increased consumption rates are observed for armyworms on certain alfalfa varieties bred for insect resistance (e.g., Team, Arc, Culver) compared to those bred primarily for forage quality (e.g., 'Vernal' and 'Apollo'; see Scriber, 1979a and Fig. 7.11). Armyworms on cabbage, however, suffer low consumption rates as well as increased metabolic costs resulting in significantly suppressed growth rates (Scriber, 1982c and Fig. 7.11).

The interactions between rates and efficiencies are complex, involving homeostatic mechanisms that influence an insect's behavior, physiology, growth and reproduction. These interactions are influenced by starvation, induction of preference, nutrients, allelochemics and various environmental (climatic) factors. Classification of a chemical as a repellent, deterrent, feeding suppressant, toxin or digestibility-reducer may be situation-dependent, dose-dependent, and involve chemicals, behavioral and physiological feedback systems (Blau *et al.*, 1978; Duffey, 1977, 1980; Reese, 1979; Grabstein and Scriber, 1981).

Feeding specialization is another type of adaptation which has widespread ecological implications, although the advantages to insect herbivores of specializing or generalizing in their choice of diet is poorly understood (see Levins and MacArthur, 1969; Otte and Joern, 1977; Fox and Morrow, 1981 for discussion). Foraging strategies cannot simply be analyzed in terms of the maximization or optimization of net energy or nutrient gain from feeding, but must

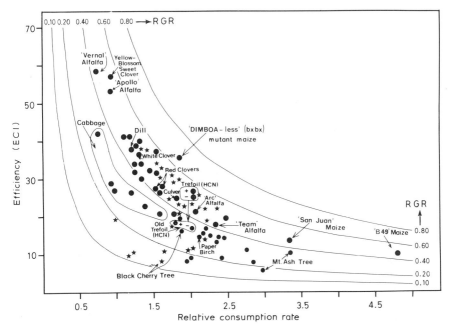

Fig. 7.11 Relationship between three performance indices (*RGR*, *ECI* and *RCR*) for penultimate-instar *Spodoptera eridania* larvae fed various plant species (after Scriber and Slansky, 1981). Each symbol represents a separate feeding experiment (see Fig. 7.5). Recall that $RGR = RCR \times ECI$.

also be considered in terms of the degree to which host plants provide shelter or otherwise minimize risks to survival (Southwood, 1972; Hassell and South-wood, 1978; Smith, 1978; Lewontin, 1979). Similarly, the evolutionary success of any insect adaptation should be evaluated in terms of reproduction as well as survival and growth. In turn, these depend upon microclimate, natural enemies, resource predictability, and competitive interactions. Nonetheless, it is plant chemistry (nutrients and allelochemics) which ultimately shapes the insect community structure by its influence on populations, energy flow, and nutrient cycling (Southwood, 1977; Gilbert, 1979; Chapin, 1980; McClure, 1980; Price *et al.*, 1980).

The degree of diet specialization is an important factor affecting interpre-tations of insect–plant interactions, as elaborated on in the following section. Although host plant suitability depends on several synecological (i.e., com-munity) factors mentioned earlier (see also Fox and Morrow, 1981), this discussion is restricted mainly to post-ingestive physiological and comparative autecological aspects of feeding specialization.

7.3 USE OF THE PHYSIOLOGICAL EFFICIENCY MODEL: INTERPRETATIONS OF THE FEEDING SPECIALIZATION HYPOTHESIS

Diet specialization has been presumed to mediate an ecological or metabolically more efficient utilization of food resources than polyphagy (Brues, 1924; House, 1962; Emlen, 1973). Dethier (1954) suggested that polyphagous and monophagous insect herbivores would reflect quantitative differences in the efficiencies of utilizing available materials, but as yet little empirical support for this hypothesis exists. Scriber and Feeny (1979) bioassayed 20 species of Lepidoptera on a variety of plant species, and found no consistent trends in efficiency of biomass or nitrogen use between those species feeding on one plant family (arbitrarily called 'specialists') versus those feeding on 2–10 ('intermediates') or more than 10 families ('generalists'). They concluded that plant growth form (or correlated chemical factors; see Section 7.2.4) accounted for the major portion of the variation in larval efficiency and growth performance. More information is required, however, on local specialization by species called generalists before the feeding specialization hypothesis can be rejected.

The importance of understanding feeding ecology at each of three levels – individual, population, and species – has been stressed by Dethier (1954). It is critical to our understanding of the evolution of feeding preferences in phytophagous insects (and therefore also to understanding insect–plant coevolution) that we have 'careful studies of feeding habit variability among populations of monophagous, oligophagous and polyphagous species of insects. Accurate information regarding individual variation in host preferences among insects and variability within polyphagous populations is completely lacking' (Dethier, 1954; p. 43).

The role which feeding specialization has played in the formation of geographic races or incipient species is largely unknown (but see Dethier, 1954; Bush, 1975; M. D. White, 1978 for current syntheses). More than a quarter of a century ago, Dethier (1954) implored researchers to 'undertake crucial experiments which would identify genetic and phenotypic races' of phytophagous insects. During the 4th International Symposium of insect–hostplant interactions he pointed out the continuing lack of genetic considerations (pp. 759–66 in Chapman and Bernays, 1978). Our understanding of the mechanisms broadening (generalization) or contraction (specialization) may depend largely upon our ability to distinguish between individual and population heterogeneity (McNaughton and Wolf, 1970; Dethier, 1978; Smiley, 1978; Fox and Morrow, 1981; Scriber, 1983). Phenotypic variability (behavioral and physiological induction and flexibility) also must be understood at the level of the individual and population before the adaptive significance of species differences in feeding ecology can be explained. Gilbert (1979, p. 145) suggests that tests of ecological generalities should be made 'within guilds defined by host–plant preferences, between local habitats within which actual resource patterns are known and with the overall life cycle of the insect fully considered.'

With the preceding considerations in mind, I will describe attempts to discriminate between the influence of host plant chemistry, environmental factors and insect adaptations (particularly feeding specialization) as factors influencing the post-ingestive growth performance of selected Lepidoptera. One question of particular concern is whether local feeding specialization may result in improved physiological efficiency and/or growth rates. Conversely, it is also of interest whether there is a loss of ability to efficiently or rapidly process ancestral hosts (i.e., those which are peripheral or non-used) after specialization upon a particular host. If the biochemical ability to effectively process marginally or infrequently used hosts is lost, various degrees of obligate monophagy might result. Smiley (1978) has suggested that this mechanism of biochemical specialization may have a major influence upon the process of insect–plant coevolution.

7.3.1 *Papilio polyxenes* foodplants

Studies with the Umbellifer-feeding *P. polyxenes* suggest that loss of digestive abilities of larvae to utilize ancestral host plants occurs at a much slower rate than changes in adult ovipositional preferences (Scriber and Feeny, 1979; see also Wiklund, 1975). Not only do the umbellifer-feeding New York populations feed upon the Rutaceae (they chew mature *Citrus* leaves and feed and pupate on *Ruta*, *Dictamnus*, *Xanthoxylum*) but they also can feed upon leaves of the Cucumber magnolia tree, *Magnolia acuminata*, which is believed to be one of the original hosts of more primitive Papilionidae (Munroe, 1960). For *P. polyxenes*, the trend in utilization efficiency and growth rate parallels the scheme of evolution of the umbellifer-feeding Papilionidae from their Rutaceae-feeding ancestors through a series of 'chemical common denominators' (Dethier, 1941, 1954; Munroe, 1960). There is a six-fold decline in relative growth rates and nearly a four-fold decline in conversion efficiency from certain Umbelliferae back through the proposed ancestral Rutaceous hosts (Table 7.2). This trend might appear to provide some enticing support for the physiological part of Smiley's (1978) hypothesis, yet examination of the plant growth forms involved illustrates the relationship between nutritional and allelochemic factors and the larval digestive efficiency and growth performance. The decline in larval growth performance of *P. polyxenes* from current umbelliferous favorites such as dill back through the presumed ancestral Rutaceous hosts such as *Xanthoxylum* and *Citrus* also parallels the trend from annual, biennial and perennial herbs to shrubs and trees. As described in Section 7.2.4, the plant growth form generally offers clues to the underlying nutritional suitability of the leaf tissues for chewing insects. The feeding transition (Table 7.2) from trees and shrubs to the annual (dill) represents an increase in leaf water content from 66% to 89% and an increase in leaf nitrogen

Table 7.2   Larval growth performance of penultimate instar *Papilio polyxenes* larvae. Ithaca, N.Y

Family	Species	Growth form	Conversion efficiency (E.C.I.)	Relative growth rate (mg gain mg^{-1} day^{-1})
Umbelliferae	Dill	(herb)	37.2 ± 2.8	$0.60 \pm .02$
	Parsnip	(herb)	28.1 ± 1.7	$0.52 \pm .05$
Rutaceae				
Gas plant	*Dictamnus fraxinella*	(perennial herb)	23.2 ± 1.3	$0.53 \pm .08$
Rue	*Ruta graveolens*	(herb)	21.6 ± 1.2	$0.23 \pm .01$
Prickly ash	*Xanthoxylum americanum*	(shrub)	10.1 ± 1.4	$0.11 \pm .02$
Lemon	*Citrus limon*	(tree)	(died)	(died)

from 2.1% (dry wt.) up to 3.6% (Scriber and Feeny, 1979). Based on these co-ordinates in the physiological efficiency model as described in Section 7.2.4, this reflects a digestibility potential (AD) of over 90% on dill, compared to a maximum of 60% on prickly ash (cf. Fig. 7.8b). Both AD values were significantly less than their theoretical maxima (supplement in Scriber and Feeny, 1979), yet the overall efficiency (ECI = AD × ECD; Table 7.2) reflected a four-fold increase, which is further amplified by consumption rate differences such that differences in growth rates are five- or six-fold (Table 7.2).

Such a correlation between plant growth forms, plant chemistry, and the physiological growth performance of *P. polyxenes* illustrates the potential usefulness of the physiological efficiency model to coevolutionary (or non-coevolutionary) interactions between phytophagous insects and their host plants. Physiological performance of a larva must be expressed relative to the underlying nutritional composition of its food, before attempting higher level autecological conclusions.

7.3.2 *Papilio glaucus* and tulip trees

Although the adaptive value for maintenance of generalized feeding abilities are difficult to assess, it is possible to determine experimentally if post-ingestive suitability of a plant for growth of such herbivores improves with individual or population experience (i.e., local specialization). Studies are needed using plants or diets of known chemical quality under controlled environmental conditions. In addition, it is important to use individuals from different geographical locations, preferably those known to have different preferred natural hosts.

The Eastern tiger swallowtail butterfly, *Papilio glaucus*, is the most polyphagous of the 540 species of Papilionidae in the world (Munroe, 1960; Scriber, 1973). It occupies a geographical area which encompasses nearly 4 million square miles (including most of Canada and the Eastern half of the US). Two

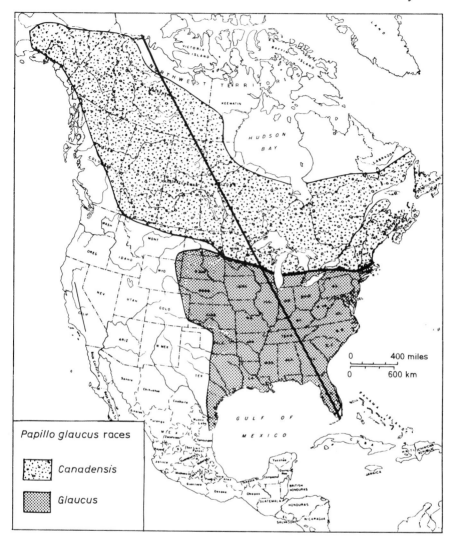

Fig. 7.12   The total geographic distribution of the two subspecies of the eastern tiger swallowtail, *Papilio glaucus*. (Data from Brower, 1957; Freeman, 1951; McGugan, 1958; Scriber *et al.*, 1981.)

subspecies of *P. glaucus* are recognized with their zone of separation at approximately 42° to 45° North latitude (Fig. 7.12; Remington, 1968; Ebner, 1970; Shapiro, 1974; Tyler, 1975; Scriber *et al.*, 1982). No single foodplant exists across the entirety of the *P. glaucus* range. A latitudinal analysis of the 20 most frequently reported food plants of *P. glaucus* along a bi-section of its range (Fig. 7.13) suggests that in general there are different hostplants on opposite sides of the plant transition zone (Fig. 7.13 and Curtis, 1959) which

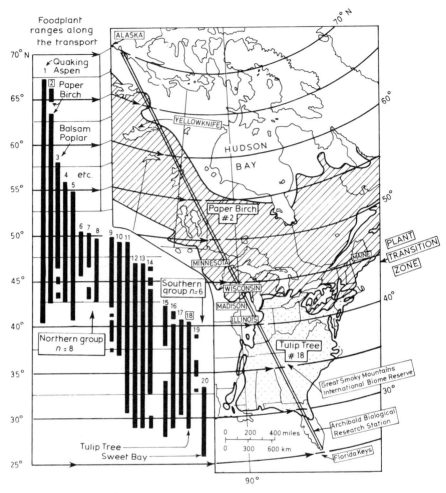

Fig. 7.13   Potential foodplants along a 3600-mile latitudinal transect of the range of *Papilio glaucus* (at the approximate bisection of the ranges of the two races; see Fig. 7.12). The most frequently reported foodplant species are numbered (1–20) for identification here and listed in Fig. 7.14. The ranges of paper birch, *Betula papyrifera* (#2), and tulip tree, *Liriodendron tulipifera* (#18) illustrate the closeness of the hostplant transition zone (i.e. where the northern meets the southern) with that for the two *P. glaucus* races (cf. Fig. 7.12).

also corresponds to the zone of separation of subspecies *P. g. glaucus* from *P. g. canadensis* (cf. Fig. 7.12). This 3600-mile latitudinal range transect (Fig. 7.12) provides an opportunity to investigate the significance of differential hostplant utilization within a phytophagous insect species existing in a variety of plant communities.

The Magnoliaceae is of tropical origin and is believed to be the ancestral host

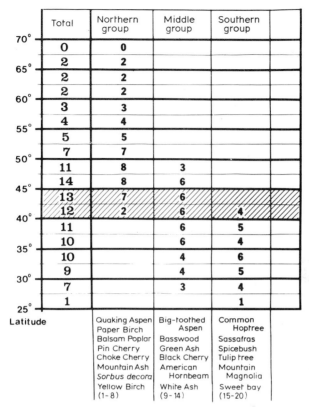

Latitude	Total	Northern group	Middle group	Southern group
70°	0	0		
	2	2		
65°	2	2		
	2	2		
60°	3	3		
	4	4		
55°	5	5		
	7	7		
50°	11	8	3	
	14	8	6	
45°	13	7	6	
	12	2	6	4
40°	11		6	5
	10		6	4
35°	10		4	6
	9		4	5
30°	7		3	4
25°	1			1
Latitude		Quaking Aspen Paper Birch Balsam Poplar Pin Cherry Choke Cherry Mountain Ash *Sorbus decora* Yellow Birch (1-8)	Big-toothed Aspen Basswood Green Ash Black Cherry American Hornbeam White Ash (9-14)	Common Hoptree Sassafras Spicebush Tulip tree Mountain Magnolia Sweet bay (15-20)

Fig. 7.14 The numbers of different potential foodplants for *Papilio glaucus* (of the 20 favorites) which occur at different latitudes along the transect presented in Figs 7.12 and 7.13. The foodplant species are numbers from top to bottom (i.e., Quaking aspen, *Populus tremuloides*, is #1, etc.). The plant transition zone is represented by shading.

family of the entire *P. glaucus* species group (Brower, 1958; Munroe, 1960). Tulip tree, *Liriodendron tulipifera* (Magnoliaceae), is perhaps the favorite foodplant wherever *P. glaucus* occurs (plant number 18 in Fig. 7.14). Tulip tree does not occur in the southern half of Florida, where sweet bay, *Magnolia virginiana*, is the primary, if not the only, host for *P. glaucus* (cf. Figs 7.13, 7.14 and 7.15). At latitudes above 43°N, plants of the Magnoliaceae, the Lauracee and the Rutaceae (numbers 15–20 in Figs 7.13 and 7.14) do not occur. The northern race, *P. canadensis*, thus cannot utilize the presumed ancestral host plants. Similarly, the southern race, *P. glaucus*, is for the most part unable to utilize the favored foodplants of *P. canadensis* (i.e., *Populus tremuloides*, *Betula papyrifera*, *Populus balsamifera*; cf. Figs 7.12–14).

The non-migratory behavior of *P. glaucus* and its patterns of feeding preference and host discrimination provide a basis for which to test the feeding specialization hypothesis (see also, Smiley, 1978). The absence of a northern

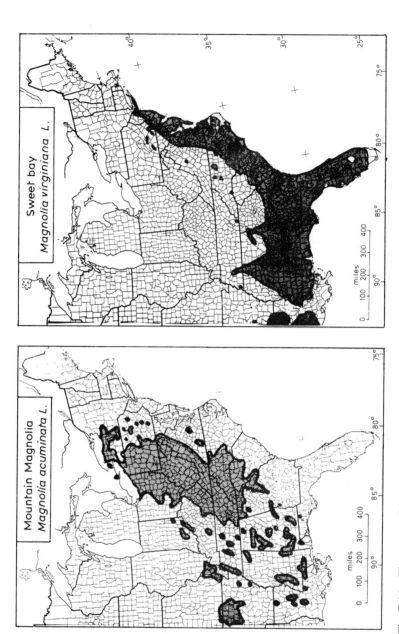

Fig. 7.15 The geographic ranges of *Magnolia acuminata* and *Magnolia virginiana* (after Fowells, 1965; Little, 1971).

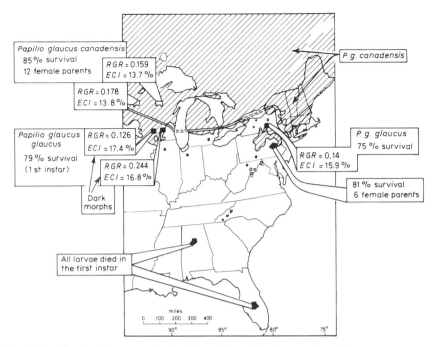

Fig. 7.16 The first-instar survival and penultimate-instar growth performance on paper birch, *Betula papyrifera*, of different populations of *Papilio glaucus canadensis* and *P. g. glaucus* as a function of geographic distance from the plant range (*RGR* = relative growth rate, *ECI* = overall efficiency of processing plant biomass; Scriber, unpublished data).

plant such as paper birch, *Betula papyrifera*, (from the *P. glaucus* populations in the southern US) might be reflected by poorer larval growth performances of southern subspecies populations relative to northern subspecies populations on this host. Larvae of Wisconsin populations of the northern subspecies (*P. canadensis*) and also the southern subspecies (*P. glaucus*) from Wisconsin, Pennsylvania, and New York, survive and grow rather efficiently and rapidly in contrast to Alabama and Florida larvae, which appear totally unadapted for successful utilization of paper birch as a host plant; all larvae died after small amounts of feeding in the first instar (Fig. 7.16). Geographical proximity of host-plants simply indicates an opportunity for biochemical adaptation *P. glaucus* populations.

With two other host plants, quaking aspen and tulip tree, the physiological adaptations appear to be linked more closely with the particular insect subspecies than with geographical proximity to the host plants (Figs 7.17 and 7.18). In Wisconsin, for example, all larvae of dark morph female *P. glaucus* died when offered quaking aspen, *Populus tremuloides*, in no-choice situations, whereas the larvae of *P. canadensis*, only slightly to the north in distribution, survive and grow well on this, one of their favorite foodplants (Fig. 7.17).

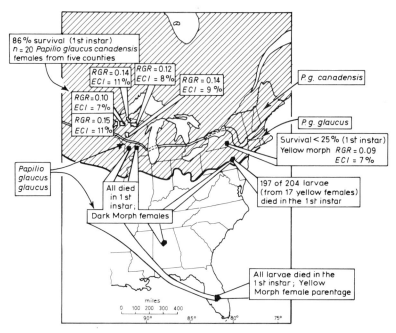

Fig. 7.17   First-instar survival and penultimate-instar growth performance (RGR = relative growth rate; ECI = overall efficiency of conversion of ingested plant biomass) on quaking aspen, *Populus tremuloides* of various populations of *P. glaucus canadensis* and *P. g. glaucus* as a function of geographical proximity to the range of the host plant.

Similar difficulties in survival of the first instar are observed for larvae of dark morph *P. glaucus* females from Pennsylvania, Alabama, and Florida.

Conversely, larvae of the northern race *P. canadensis* in northern Wisconsin did not survive the first instar on tulip tree (all larvae died after several, repeated short feedings), whereas *P. glaucus* survives and grows throughout the eastern US in an efficient and rapid manner on this host (Fig. 7.18). Even the southern Wisconsin and middle Florida populations of *P. glaucus* possess the biochemical capacity to digest the allopatric tulip tree, although they are unlikely to have encountered it in many generations, if ever.

Biochemical adaptations for more efficient post-ingestive food processing or more rapid growth rates in later instars largely represent 'fine-tuning' of the co-evolutionary patterns that are primarily determined by behavioral and toxicological barriers effective during the first instar. Furthermore, this fine-tuning operates within the limits imposed by plant chemical quality, as indexed by leaf water and nitrogen content. The growth rate (in this case, larval duration) of *Callosamia promethea* (Saturniidae) is presented in relation to these plant quality indices (Fig. 7.19). A three-year survey of survival and growth of more than 12 000 larvae from various geographic locations (including New York, Pennsylvania, Maine, Massachusetts, Ohio, Illinois, Wisconsin, Tennessee

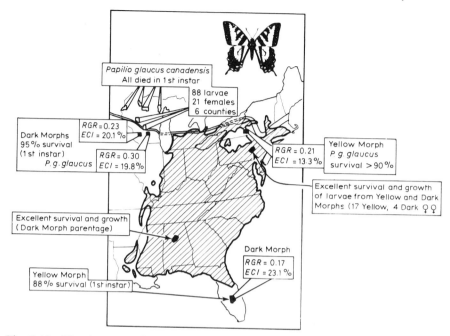

Fig. 7.18 First-instar survival and penultimate instar growth performance (*RGR* and *ECI*) on tulip tree, *Liriodendron tulipifera*, of various populations of *P. glaucus canadensis* and *P. g. glaucus* as a function of geographic proximity to the range of the host plant.

and Alabama), suggests that post-ingestive foodplant utilization is more closely determined by chemical quality than by geographic proximity and local preferences. For example, Wisconsin larvae grow as rapidly and efficiently on sassafras, spicebush, and tuliptree as do larvae from populations which can and do actually utilize these species naturally (Scriber, 1983, and unpublished results). Nonetheless, differential survival of first instar larvae based on local adaptation (or potential feeding opportunities) of *C. promethea* populations on a given foodplant does occur (Potter *et al.*, 1981; Scriber *et al.*, unpublished), further supporting the suggestion that the first instar is the critical one for Lepidoptera.

It is believed that tuliptree is the ancestral host for both the *P. glaucus* and the *C. promethea* species groups (Brower, 1958; Ferguson, 1972; Peigler, 1976; Scriber, 1983). This tree is, however, an unsuitable host for the notoriously polyphagous gypsy moth, *Lymantria dispar*, southern armyworm, *Spodoptera eridania*, and cecropia moth, *Hyalophora cecropia*. In the *P. glaucus* species group (Brower, 1959a, 1959b), females of the more stenophagous western species *Papilio rutulus* and *Papilio eurymedon* will oviposit on tulip tree leaves but, as with *P. g. canadensis*, leaves are toxic to first instar larvae. Neither the chemical basis of this toxicity nor the mechanism of detoxication by *P. g. glaucus*

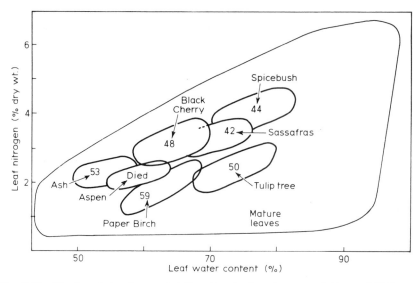

Fig. 7.19   The larval developmental time (days) of various populations of *Callosamia promethea* on different plants, presented as a function of leaf water and leaf nitrogen contents. Seasonal changes in leaf chemical composition during the rearing periods are encompassed by enclosure lines for each species. Differential survival and growth rates were observed for an initial 12 000 larvae of different *C. promethea* populations (Scriber, 1983). However, in this figure the mean duration for larvae on each plant is presented regardless of geographic source.

are known at this time. The genetics involved in tulip tree suitability are intriguing and perhaps relatively uncomplicated as *P. g. glaucus* can be hand-mated with *P. eurymedon* (Clarke and Sheppard, 1957), *P. rutulus* (Clarke and Sheppard, 1955; Scriber, unpublished), and *P. glaucus canadensis* (Scriber, unpublished data) to give F_1 larvae which are able to consume and grow on tulip tree leaves.

7.4 SUMMARY

Host plant suitability involves various plant characteristics and insect herbivore adaptations as well as modification of both the plant and the insect by abiotic and biotic factors in their community. The post-ingestive growth performance of Lepidoptera (and perhaps most other leaf-chewing insects as well) is determined to a large extent by plant chemical quality (i.e., nutrients and allelochemics) and can be indexed by leaf water–nitrogen composition.

A 'physiological efficiency' model of insect response to plant chemistry is proposed using the water–nitrogen index. This larval performance model incorporates seasonal trends in various plant growth forms and provides an empirical framework for interpreting the 'plant apparency' concept. In addition, the model is a necessary starting point for assessment of any physiological advantage (in rate or efficiency of growth) of insect feeding specialization.

A test of the feeding specialization hypothesis (i.e., for subsequently improved larval growth performance) is described using various geographical populations with different local host plant favorites. The major behavioral or toxicological barrier to favorable host suitability (of the lepidopterans tested) is determined by the first instar larvae, and post-ingestive efficiency and growth rates of later instars are dictated predominantly by plant leaf chemistry as indexed in the physiological efficiency model. The effects of host suitability upon reproductive biology and population dynamics of herbivores also play an important role in shaping the coevolutionary interactions in a community. Calow (1977) summarizes these complex interactions as follows: 'the essential function of metabolism is that it catalyses the transmission of genetic information and it is only in this sense that it can be judged efficient.'

ACKNOWLEDGMENTS

I would like to thank Drs May Berenbaum, E. A. Bernays, Michael Collins, Laurel Fox, Fred Gould, Jim Miller, Karen Strickler, and Jane Wolfson for valuable comments on the original manuscript. Research reported here was supported by NSF Grant DEB 7921749, Hatch Project 5134, NC-105 North Central States regional research project, an Indonesia-Wisconsin MUCIA (AID) Project (S. Manuwoto), EPA-USDA Grant CR-806277-030, the University of Wisconsin Graduate School, and is a contribution of the College of Agricultural Sciences of the University of Wisconsin, Madison.

REFERENCES

Atsatt, P. R. and O'Dowd, D. J. (1976) Plant defense guilds. *Science*, **193**, 24–9.

Auerbach, M. J. and Strong, D. R. (1981) Nutritional ecology of *Heliconia* herbivores: Experiments with plant fertilization and alternative hosts. *Ecol. Monogr.*, **51**, 63–83.

Barnes, D. K., Hanson, C. H., Ratcliffe, R. H., Busbice, T. H., Schillinger, J. A., Buss, G. R., Campbell, V. W., Hemken, R. W. and Blickenstaff, C. C. (1970) The development and performance of 'Team' alfalfa: A multiple pest-resistant alfalfa with moderate resistance to the alfalfa weevil. USDA (ARS) publication 34–115.

Baule, H. (1976) A note on the papers by Mssrs. Dimitri and Bogenschutz. In: *Fertilizer Use and Plant Health*. Proc. XII Colloq. Int. Potash Inst. Der Bund, Bern, Switzerland, pp. 329–32.

Beck, S. D. (1957) The European corn borer, *Pyrausta nubilalis* (Hubn.) and its principal host plant. VI. Host plant resistance to larval establishment. *J. Insect Physiol.*, **1**, 158–77.

Beck, S. D. (1965) Resistance in plants to insects. *A. Rev. Ent.*, **10**, 207–32.

Beck, S. D. and Reese, J. C. (1976) Insect–plant interactions: nutrition and metabolism. In: *Biochemical Interactions Between Plant and Insects* (Wallace, J. and Mansell, R., eds) *Rec. Adv. Phytochem.*, **10**, 41–92.

Berenbaum, M. (1978) Toxicity of a furanocoumarin to armyworms: A case of biosynthetic escape from insect herbivores. *Science*, **201**, 532–4.

Berenbaum, M. (1980) Adaptive significance of midgut pH in larval Lepidoptera. *Am. Nat.*, **115**, 138–46.

Berenbaum, M. (1981a) Patterns of furanocoumarin distribution and insect herbivory in the Umbelliferae: Plant chemistry and community structure. *Ecology*, **62**, 1254–66.

Berenbaum, M. (1981b) Patterns of furanocoumarin production and insect herbivory in a population of wild parsnip (*Pastinaca sativa*). *Oecologia*, **49**, 236–44.

Berenbaum, M. and Feeny, P. P. (1981) Toxicity of angular furanocoumarins to swallowtail butterflies: Escalation in a coevolutionary arms race? *Science*, **212**, 927–9.

Bernays, E. A. (1978) Tannins: an alternative viewpoint. *Ent. exp. appl.*, **24**, 244–53.

Bernays, E. A., Chamberlain, D. and McCarthy, P. (1980) The differential effects of ingested tannic acid on different species of Acridoidea. *Ent. exp. appl.*, **28**, 158–66.

Bernays, E. A. and Woodhead, S. (1982) Plant phenols used as nutrients by a phytophagous insect. *Science*, **216**, 201–3.

Blau, P. A., Feeny, P. P., Contardo, L. and Robson, D. (1978) Alylglucosinolate and herbivorous caterpillars: A contrast in toxicity and tolerance. *Science*, **200**, 1296–8.

Brattsten, L. (1979) Biochemical defense mechanisms in herbivores against plant allelochemicals. In: *Herbivores: Their Interation with Secondary Plant Metabolites* (Rosenthal, G. A. and Janzen, D. H., eds) pp. 199–270. Academic Press, New York.

Brattsten, L. B., Wilkinson, C. F. and Eisner, T. (1977) Herbivore-plant interactions: Mixed function oxidases and secondary plant substances. *Science*, **196**, 1349–52.

Brower, L. P. (1957) Speciation in the *Papilio glaucus* group. PhD Dissertation. Yale University, New Haven, Conn.

Brower, L. P. (1958) Larval foodplant specificity in butterflies of the *P. glaucus* group. *Lepid. News*, **12**, 103–14.

Brower, L. P. (1959a) Speciation in butterflies of the *Papilio glaucus* group. I. Morphological relationships and hybridization. *Evolution*, **13**, 40–63.

Brower, L. P. (1959b) Speciation in butterflies of the *Papilio glaucus* group. II. Ecological relationships and interspecific sexual behavior. *Evolution*, **13**, 212–28.

Brues, C. T. (1924) The specificity of foodplants in the evolution of phytophagous insects. *Am. Nat.*, **58**, 127–44.

Burns, J. C. and Cope, W. A. (1974) Nutritive value of crownvetch as influenced by structural constituents and phenolic and tannin compounds. *Agronomy J.*, **66**, 195–200.

Bush, G. L. (1975) Modes of animal speciation. *A. Rev. Ecol. Syst.*, **6**, 339–64.

Calow, P. (1977) Ecology, evolution and energetics: A study in metabolic adaptation. *Adv. Ecol. Res.*, **10**, 1–62.

Carrow, J. R. and Betts, R. E. (1973) Effects of different foliar-applied nitrogen fertilizers on balsam wooly aphid. *Can. J. For. Res.*, **3**, 12–39.

Chapin, F. S. (1980) The mineral nutrition of wild plants. *A. Rev. Ecol. Syst.*, **11**, 233–60.

Chew, F. S. (1975) Coevolution of pierid butterflies and their cruciferous foodplants. I. The relative quality of available resources. *Oecologia*, **20**, 117–27.

Chew, F. S. (1977) Coevolution of pierid butterflies and their cruciferous foodplants. II. The distribution of eggs on potential foodplants. *Evolution*, **31**, 568–79.

Chiang, H. C. (1978) Pest Management in Corn. *A. Rev. Ent.*, **23**, 101–23.

Clarke, C. A. and Sheppard, P. M. (1955) The breeding in captivity of the hybrid *Papilio rutulus* female X *P. glaucus* male. *Leppid. News*, **9**, 46–8.

Clarke, C. A. and Sheppard, P. M. (1957) The breeding of the hybrid *Papilio glaucus* female X *P. eurymedon* male. *Lepid. News*, **11**, 201–5.

Conn. E. E. (1979) Cyanide and cyanogenic glycosides. In: *Herbivores: Their Interaction with Secondary Plant Metabolites* (Rosenthal, G. A. and Janzen, D. H., eds) pp. 387–412. Academic Press, New York.

Connell, J. H. (1980) Diversity and the coevolution of competitors, or the ghost of competition past. *Oikos*, **35**, 131–8.

Cooper-Driver, G. A., Finch, S., Swain, T. and Bernays, E. (1977) Seasonal variation in secondary plant compounds in relation to the palatability of *Pteridium aquilinum*. *Biochem. Syst. Ecol.*, **5**, 177–83.

Corcuera, L. J. (1974) Identification of the major active component present in corn extracts inhibitory to soft rot, *Erwinia* species. PhD thesis. Univ. Wisconsin, Madison.

Curtis, J. T. (1959) *The Vegetation in Wisconsin*. University of Wisconsin Press, Madison.

Daly, H. V., Doyen, J. T. and Ehrlich, P. R. (1978) *Introduction to Insect Biology and Diversity*. McGraw-Hill, New York.

Dement, W. A. and Mooney, H. A. (1974) Seasonal variation in the production of tannins and cyanogenic glucosides in the chaparral shrub, *Heteromeles arbutifolia*. *Oecologia*, **15**, 65–76.

Dethier, V. G. (1941) Chemical factors determining the choice of foodplants by *Papilio* larvae. *Am. Nat.*, **75**, 61–73.

Dethier, V. G. (1954) Evolution of feeding preferences in phytophagous insects. *Evolution*, **8**, 33–54.

Dethier, V. G. (1978) Studies on insect/plant relations – past and future. *Ent. expt. appl.*, **24**, 759–66.

Duffey, S. S. (1977) Arthropod allomones: Chemical effronteries and antagonists. *Proc. XV Int. Congr. Ent.* Washington, DC, pp. 323–94.

Duffey, S. S. (1980) Sequestration of natural products by insects. *A. Rev. Ent.*, **25**, 447–8.

Ebner, J. A. (1970) *Butterflies of Wisconsin*. Milwaukee Public Mus., Pop. Sci. Handbook No. 12.

Ehrlich, P. R. and Raven, P. H. (1964) Butterflies and plants: a study in coevolution. *Evolution*, **18**, 586–608.

Emlen, J. M. (1973) Feeding Ecology. In: *Ecology: An Evolutionary Approach*. pp. 157–85. Addison Wesley, Reading, Mass.

Erickson, J. M. and Feeny, P. P. (1974) Sinigrin: A chemical barrier to the black swallow-tail butterfly, *Papilio polyxenes. Ecology*, **55**, 103–11.

Feeny, P. P. (1970) Seasonal changes in oak leaf tannins and nutrients as a cause of spring feeding by winter moth caterpillars. *Ecology*, **51**, 565–81.

Feeny, P. P. (1975) Biochemical coevolution between plants and their insect herbivores. In: *Coevolution of Animals and Plants* (Gilbert, L. E. and Raven, P. R., eds) pp. 3–19. University of Texas Press, Austin.

Feeny, P. P. (1976) Plant apparency and chemical defense. In: *Biochemical Interactions Between Plants and Insects* (Wallace, J. and Mansell, R., eds). *Rec. Adv. Phytochem.*, **10**, 1–40.

Fenical, W. (1982) Natural products chemistry in the marine environment. *Science*, **215**, 923–8.

Ferguson, D. C. (1972) *The Moths of America North of Mexico*. Fascide 20.2B: Bombycoidea, Sturniidae (Part) E. W. Classey, London.

Fisk, J. (1980) Effects of HCN, phenolic acids, and related compounds in *Sorghum bicolor* on the feeding behavior of the planthopper *Peregrinus maidis. Ent. exp. appl.*, **27**, 211–22.

Finke, M. D. (1977) Factors controlling the seasonal foodplant utilization by larvae of the specialized herbivore, *Papilio polyxenes* (Lepidoptera). MS Thesis, Wright State University, Dayton, Ohio.

Fogal, W. H. (1974) Nutritive value of pine foliage for some diprionid sawflies. *Proc. ent. Soc. Ontario*, **105**, 101–18.

Foster, A. A. (1968) Damage to forests by fungi and insects as affected by fertilizers. In: *Forest Fertilization; Theory and Practice*, pp. 42–6. TVA Nat. Fert. Dev. Center, Muscle Shoals, Alabama.

Fowells, H. A. (1965) Silvics of forest trees of the United States. *USDA Handbook*, No. 271.

Fox, L. R. (1981) Defense and dynamics in plant–herbivore systems. *Am. Zool.*, **24**, 853–64.

Fox, L. R. and Macauley, B. J. (1977) Insect grazing on Eucalyptus in response to variation in leaf tannins and nitrogen. *Oecologia*, **29**, 145–62.

Fox, L. R. and Morrow, P. A. (1981) Specialization: Species property or local phenomenon? *Science*, **211**, 887–3.

Freeman, T. N. (1951) Northern Canada and some northern butterflies. *Lepid. News*, **5**, 41–2.

Futuyma, D. J. (1976) Foodplant specialization and environmental predictability in Lepidoptera. *Am. Nat.*, **110**, 285–92.

Futuyma, D. J. (1979) *Evolutionary Biology*. Sinauer, Sunderland, Mass.

Geiselman, J. A. (1980) Ecology of chemical defenses of algae against the herbivorous snail, *Littorina littorea*, in the new England Rocky intertidal community. PhD Thesis, MIT, Cambridge, Mass/Woods Hole Oceanographic Institute, Woods Hole, Mass.

Gilbert, L. E. (1979) Development of theory in the analysis of insect-plant interactions. In: *Analysis of Ecological Systems* (Horn *et al.*, eds) pp. 117–54. Ohio State University Press, Columbus, Ohio.

Gordon, H. T. (1972) Interpretations of insect quantitative nutrition. In: *Insect and Mite Nutrition: Significance and Implications In Ecology and Pest Management* (Rodriguez, J. G., ed.). Elsevier-North Holland, Amsterdam.

Gould, F., Carroll, C. R. and Futuyma, D. J. (1982) Cross-resistance to pesticides and plant defenses: a study of the two-spotted spider mite. *Ent. exp. appl.*, **31**, 175–80.

Goyer, R. A. and Benjamin, D. M. (1972) Influence of soil fertility on infestation of jackpine plantations by the pine root weevil. *Forest Sci.*, **18**, 139–47.

Grabstein, E. M. and Scriber, J. M. (1981) The relationship between restriction of host plant consumption, and post-ingestive utilization of biomass and nitrogen in *Hyalophora cecropia*. *Ent. exp. appl.* (in press).

Guthrie, W. D. (1974) Techniques, accomplishments and future potential of breeding for resistance to European corn borers in corn. In: *Proceedings of the Summer Institute on Biological Control of Plant Insects and Diseases* (Maxwell, F. G. and Harris, F. A., eds). University Mississippi Press, Jackson.

Hassell, M. P. and Southwood, T. R. E. (1978) Foraging strategies of insects. *A. Rev. Ecol. Syst.*, **9**, 75–98.

Hauck, R. D. (ed.) (1984) *Nitrogen in Crop Production. Am. Soc. Agron.*, Madison, Wisconsin (in press).

Hegnauer, R. (1971) Chemical patterns and relationships of Umbelliferae. In: *The Biology and Chemistry of the Umbelliferae* (Heywood, V. H., ed.) pp. 267–78. Academic Press, New York.

Hough, J. A. and Pimentel, D. (1978) Influence of host foliage on development, survival, and fecundity of the gypsy moth. *Env. ent.*, **7**, 97–102.

House, H. L. (1962) Insect nutrition. *A. Rev. Biochem.*, **31**, 653–72.

Huffaker, C. B., van de Vrie, M. and McMurty, J. A. (1969) The ecology of tetranychid mites and their natural control. *A. Rev. Ent.*, **14**, 125–74.

Janzen, D. H. (1980) When is it coevolution? *Evolution,* **34**, 611–12.

Jermy, T. (1976) Insect/hostplant relationship – Co-evolution or sequential evolution? In: *The Host-plant In Relation to Insect Behavior and Reproduction* (Jermy, T., ed.) pp. 109–13. Plenum Press, New York.

Jermy, T., Hanson, F. E. and Dethier, V. G. (1968) Induction of specific food preference in Lepidopterous larvae. *Ent. exp. appl.,* **11**, 211–30.

Jones, D. A. (1972) Cyanogenic glycosides and their function. In: *Phytochemical Ecology* (Harborne, J. B., ed.) pp. 103–24. Academic Press, London.

Jones, F. G. W. (1976) Pests, resistance and fertilizers. In: *Fertilizer Use and Plant Health,* Proc. XII Int. Potash, Inst. Worflaufen, Bern, Switzerland. pp. 233–58.

Kjaer, A. (1974) Glucosinolates in the Cruciferae. In: *The Biology and Chemistry of the Umbelliferae* (Vaughan, J. G., MacLeod, A. J. and Jones, G. M. G., eds) pp. 207–20. Academic Press, London.

Klun, J. A. and Robinson, J. F. (1969) Concentrations of two 1, 4-benzoxazinones in dent corn at various stages of development of the plant and its relation to resistance of the host plant to the European corn borer. *J. Econ. Ent.,* **62**, 214–20.

Klun, J. A., Guthrie, W. D., Hallauer, A. R. and Russell, W. A. (1970) Genetic nature of 2, 4-dihydroxy-7-methoxy-2H-1, 4-benzoxazine-3-one and resistance to the European corn borer in a diallel set of eleven maize inbreds. *Crop Sci.,* **10**, 87–90.

Kogan, M. (1975) Plant resistance in pest management. In: *Introduction to Insect Pest Management* (Metcalf, R. L. and Luckman, W. H., eds) pp. 103–46. Wiley, New York.

Krieger, R. I., Feeny, P. P. and Wilkinson, C. F. (1971) Detoxication enzymes in the guts of caterpillars: An evolutionary answer to plant defenses? *Science,* **172**, 579–81.

Lawton, J. H. (1976) The structure of the arthropod community on bracken. *Bot. J. Linn. Soc.,* **73**, 187–216.

Levins, R. and MacArthur, R. (1969) An hypothesis to explain the incidence of monophagy. *Ecology,* **50**, 910–11.

Leath, K. T. and Radcliffe, R. K. (1974) The effect of fertilization on disease and insect resistance. In: *Forage Fertilization* (Mays, D. A., ed.) pp. 481–503. *Amer. Soc. Agron.,* Madison.

Lewontin, R. C. (1979) Fitness, survival and optimality. In: *Analysis of Ecological Systems* (Horn, D. H., Mitchell, R. and Stairs, G. R., eds). Ohio State University Press, Columbus, Ohio.

Lipke, H. and Fraenkel, G. (1956) Insect nutrition. *A. Rev. Ent.,* **1**, 17–44.

Little, E. L., Jr. (1971) *Atlas of United States Trees, Volume 1, Conifers and Important Hardwoods.* USDA Forest Service Misc. Publ. 1146 Washington, DC.

Loomis, R. S., Beck, S. D. and Stauffer, J. R. (1957) The European corn borer, *Pyrausta nubilalis* (Hubn.) and its principal host plant. V. A. chemical study of host plant resistance. *Plant Physiol.,* **32**, 379–85.

Loyd, R. C. and Gray, E. (1970) Amount and distribution of hydrocyanic acid potential during the life cycle of plants of three sorghum cultivars. *Agron. J.,* **62**, 394–7.

McClure, M. S. (1980) Foliar nitrogen: A basis for host suitability for elongate hemlock scale, *Fiorinia externa* (Homoptera: Diaspididae). *Ecology,* **61**, 72–9.

McGugan, B. M. (1958) *Forest Lepidoptera of Canada recorded by the Forest Insect Survey. Vol. 1. Papilionidae to Arctiidae.* Pub. 1034 Canad. Dept. Agr. Forest Biology Division.

McHargue, J. S. and Roy, W.. R. (1932) Mineral and nitrogen content of leaves of some forest trees at different times in the growing season. *Bot. Gazette,* **94**, 381–93.

McKey, D. M. (1979) The distribution of secondary compounds within plants. In: *Herbivores: Their Interaction with Secondary Plant Metabolites* (Rosenthal, G. A. and Janzen, D. H., eds) pp. 55–133. Academic Press, New York.

McNaughton, S. J. and Wolf, L. L. (1970) Dominance and the niche in ecological systems. *Science,* **267**, 131–9.

McNeill, S. and Southwood, T. R. E. (1978) The role of nitrogen in the development of insect/plant relationships. In: *Biochemical Aspects of Plant and Animal Coevolution* (Harborne, J., ed.) pp. 77–9. Academic Press, London.

Manglitz, G. R., Gorz, H. J., Haskins, F. A., Akeson, W. R. and Beland, G. L. (1976). Interactions between insects and chemical components of sweetclover. *J. Env. Qual.*, **5**, 347–52.

Manuwoto, S. (1980) Feeding and development of the European corn borer and southern armyworm on US inbred and caribbean maize genotypes. M. S. Thesis. University of Wisconsin, Madison.

Manuwoto, S. and Scriber, J. M. (1981) Consumption and utilization of three maize genotypes by the southern armyworm, *Spodoptera eridania*. *J. Econ. Ent.*, **75**, 163–7.

Martin, M. M., Kukor, J. J., Martin, J. S., O'Toole, T. E. and Johnson, M. W. (1981) Digestive enzymes of fungus-feeding beetles. *Physiol. Zool.*, **54**, 137–45.

Mattson, W. J. (1980) Herbivory in relation to plant nitrogen content. *A. Rev. Ecol. Syst.*, **11**, 119–61.

Maxwell, F. G. and Jennings, P. R. (eds) (1980) *Breeding Plants Resistant to Insects*. Wiley, New York.

Munroe, E. (1960) *The generic classification of the Papilionidae. Can. Ent.* (Suppl. 17) pp. 1–51.

Nayar, J. K. and Fraenkel, G. (1963) The chemical basis of host selection in the Mexican bean beetle, *Epilachna varivestis* (Coleoptera: Coccinellidae). *Ann. ent. Soc. Am.*, **56**, 174–8.

Newberry, D. (1980) Interactions between the coccid, *Icerya seychellarum* (Westw.), and its host tree species on Aldabra atoll. *Oecologia*, **46**, 171–9.

Nickel, J. L. (1973) Pest situation in changing agricultural systems – a review. *Bull. ent. Soc. Am.*, **19**, 136–42.

Nielson, M. W. and Lehman, W. F. (1980) Breeding approaches in alfalfa. In: *Breeding Plants Resistant to Insects* (Maxwell, F. G. and Jennings, P. R., eds) pp. 277–311. Wiley, New York.

Onuf, C. P., Teal, J. M. and Valiela, S. I. (1977) Interactions of nutrients, plant growth and herbivory in a mangrove ecosystem. *Ecology*, **58**, 514–26.

Ortega, A., Vasal, S. K., Mihm, S. K. and Hershey, C. (1980) Breeding for insect resistance in maize. In: Breeding Plants for Insect Resistance (Maxwell, F. G. and Jennings, P. R., eds) pp. 370–419. Wiley, New York.

Otte, D. and Joern, A. (1977) On feeding patterns in desert grasshoppers and the evolution of specialized diets. *Proc. Acad. Nat. Sci. Philadelphia*, **128**, 89–126.

Painter, R. H. (1951) *Insect Resistance in Crop Plants*. MacMillan, New York.

Pate, J. S. (1980) Transport and partitioning of nitrogenous solutes. *A. Rev. Plant Physiol.*, **31**, 313–40.

Peigler, R. S. (1976) Observations on host plant relationships ad larval nutrition in *Callosamia* (Saturniidae). *J. Lepid. Soc.*, **30**, 184–7.

Pimentel, D. (1977) The ecological basis of insect pest, pathogen and weed problems. In: *Origins of Pest, Parasite, Disease, and Weed Problems* (Cherrett, J. M. and Sagan, G. R., eds) pp. 3–31. Blackwell Scientific, Oxford.

Platt, J. R. (1964) Strong inference. *Science*, **146**, 347–53.

Potter, J., Evans, M. H. and Scriber, J. M. (1981) Geographic populations and nutritional ecology of the silkmoth, *Callosamia promethea. Bull. Ecol. Soc. Am.*, **62**, 86.

Price, P. W. (1981) Semiochemicals in evolutionary time. In: *Semiochemicals: Their Role in Pest Control* (Nordland, D. A., Jones, R. L. and Lewis, W. J., eds) pp. 251–79. Wiley, New York.

Price, P. W., Bauton, C. E., Gross, P., McPherson, B. A., Thompson, J. N. Weis, A. E. (1980). Interactions among three trophic levels: Influence of plants on the interactions between insect herbivores and natural enemies. *A. Rev. Ecol. Syst.*, **11**, 41–65.

Reese, J. C. (1979) Interactions of allelochemicals with nutrients in herbivore food. In: *Herbivores: Their Interaction with Secondary Plant Metabolites* (Rosenthal, G. A. and Janzen, D. H., eds) pp. 309–30. Academic Press, New York.

Reese, J. C. and Beck, S. D. (1976a) Effects of allelochemics on the black cutworm, *Agrotis ipsilon*: Effects of p-benzoquinone, hydroquinone, and duroquinone on larval growth, development, and utilization of food. *Ann. ent. Soc. Am.*, **69**, 59–67.

Reese, J. C. and Beck, S. D. (1976b) Effects of allelochemics on the black cutworm, *Agrotis ipsilon*: effects of catechol, L-dopa, dopamine, and chlorogenic acid on larval growth, development and utilization of food. *Ann. ent. Soc. Am.*, **69**, 68–72.

Reese, J. C. and Beck, S. D. (1976c) Effects of allelochemics on the black cutworm, *Agrotis ipsilon*: effects of resorcinol, phloroglucinol, and gallic acid on larval growth, development and utilization of food. *Ann. ent. Soc. Am.*, **69**, 999–1003.

Reese, J. C. and Beck, S. D. (1978) Inter-relationships of nutritional indices and dietary moisture in the black cutworm (*Agrotis ipsilon*) digestive efficiency. *J. Insect Physiol.*, **24**, 473–9.

Rehr, S. S., Feeny, P. P. and Janzen, D. H. (1973) Chemical defence in Central American non-ant acacias. *J. Anim. Ecol.*, **42**, 405–16.

Remington, C. L. (1968) Suture zones of hybrid interaction between recently joined biotas. *Evol. Biol.*, **2**, 321–428.

Rhoades, D. F. (1979) Evolution of chemical defense against herbivores. In: *Herbivores: Their Interaction with Secondary Plant Metabolites* (Rosenthal, G. A. and Janzen, D. H., eds) pp. 1–55. Academic Press, New York.

Rhoades, D. F. and Cates, R. G. (1976) Toward a general theory of plant antiherbivore chemistry. In: *Biochemical interactions between plants and insects. Rec. Adv. Phytochem* (Wallace, J. and Mansell, R., eds), **10**, 168–213.

Rodriguez, J. G. (1951) Mineral nutrition of the two-spotted spider mite, *Tetranychus bimaculatus* Harvey. *Ann. ent. Soc. Am.*, **44**, 511–26.

Rodriguez, J. G. (ed.) (1972) *Insect and Mite Nutrition*. North Holland, Amsterdam.

Root, R. B. (1973) Organization of a plant-arthropod association in simple and diverse habitats: the fauna of collards, *Brassica oleraceae*. *Ecol. Monogr.*, **43**, 95–124.

Rosenthal, G. A. and Bell, E. A. (1979) Naturally occurring toxic nonprotein amino acids. In: *Herbivores: Their Interaction with Secondary Plant Metabolites* (Rosenthal, G. A. and Janzen, D. H., eds) pp. 353–85. Academic Press, New York.

Rosenthall, G. A. and Janzen, D. H. (1979) Herbivores: Their Interaction with Secondary Plant Metabolites. Academic Press, NY.

Rosenthal, G. A., Janzen, D. H. and Dahlman, D. L. (1977) Degradation and detoxification of canavanine by specialized seed predator. *Science*, **196**, 658–60.

Rosenthal, G. A., Hughes, C. G., Janzen, D. H. (1982) L-Canavanine, a dietary nitrogen source for the seed predator, *Carydes brasiliensis* (Bruchidae). *Science*, **217**, 353–5.

Russell, F. C. (1947) The chemical composition and digestibility of fodder shrubs and trees. *St. Publs. Commonw. Agr. Bur.*, **10**, 185–231.

Schoonhoven, L. M. and Meerman, J. (1978) Metabolic cost of changes in diet and neutralization of allelochemics. *Ent. exp. appl.*, **24**, 689–93.

Scriber, J. M. (1973) Latitudinal gradients in larval feeding specialization of the World Papilionidae (Lepidoptera) *Psyche*, **80**, 355–73.

Scriber, J. M. (1977) Limiting effects of low leaf water content on the nitrogen utilization, energy budget and larval growth of *Hyalophora cecropia* (Lepidoptera: Saturniidae). *Oecologia*, **28**, 269–87.

Scriber, J. M. (1978a) The effects of larval feeding specialization and plant growth form upon the consumption and utilization of plant biomass and nitrogen: an ecological consideration. *Ent. exp. appl.*, **24**, 694–710.

Scriber, J. M. (1978b) Cyanogenic glycosides in *Lotus corniculatus*: Their effect upon growth, energy budget, and nitrogen utilization of the southern armyworm, *Spodoptera eridania*. *Oecologia*, **28**, 269–87.

Scriber, J. M. (1979a) Post-ingestive utilization of plant biomass and nitrogen by Lepidoptera: Legume feeding by the southern armyworm. *J. N.Y. Ent. Soc.*, **87**, 141–53.

Scriber, J. M. (1979b) The effects of sequentially switching foodplants upon the biomass and nitrogen utilization by polyphagous and stenophagous *Papilio* larvae. *Ent. exp. appl.*, **25**, 203–15.

Scriber, J. M. (1981) Sequential diets, metabolic costs, and growth of *Spodoptera eridania* feeding upon dill, lima bean, and cabbage. *Oecologia*, **51**, 175–80.

Scriber, J. M. (1982) The behavior and nutritional physiology of southern armyworm larvae as a function of plant species consumed in earlier instars. *Ent. expt. appl.*, **31**, 359–69.

Scriber, J. M. (1983) The evolution of feeding specialization, physiological efficiency and host plant races in selected Papilionidae and Saturniidae. In: Variable Plants and Herbivores in Natural and Managed Systems (Denno, R. F. and McClure, M. S., eds) pp. 373–412. Academic Press, New York (in press).

Scriber, J. M. (1984) Nitrogen nutrition of plants and insect invasion. In: *Nitrogen in Crop Production* (Hauck, D., ed.) Am. Soc. Agron. Madison, Wisconsin (in press).

Scriber, J. M., Lederhouse, R. C. and Contardo, L. (1975a) Spicebush, *Lindera benzoin*, a little known foodplant of *Papilio glaucus* (Papilionidae). *J. Lepid. Soc.*, **29**, 10–14.

Scriber, J. M., Tingey, W. M., Gracen, V. E. and Sullivan, S. L. (1975b) Leaf-feeding resistance to the European corn borer in genotypes of tropical (low-DIMBOA) and US Inbred (high-DIMBOA) Maize. *J. Econ. Ent.*, **68**, 823–26.

Scriber, J. M. and Feeny, P. P. (1979) The growth of herbivorous caterpillars in relation to degree of feeding specialization and to growth form of foodplant. *Ecology*, **60**, 829–50.

Scriber, J. M. and Slansky, F., Jr. (1981) The nutritional ecology of immature insects. *A. Rev. Ent.*, **26**, 183–211.

Scriber, J. M., Lintereur, G. L. and Evans, M. H. (1982) Foodplant utilization and a new oviposition record for *Papilio glaucus canadensis* R & J (Papilionidae Lepidoptera) in northern Wisconsin and Michigan. *Great Lakes Ent.*, **15**, 39–46.

Shapiro, A. M. (1974) The butterflies and skippers of New York State. *Search*, **4**, 1–59.

Singh, P. (1970) Host-plant nutrition and composition: effects on agricultural pests. *Can. Dept. Agric. Res. Inst. Inf. Bull.*, **6**, 1–102.

Slansky, F. (1982) Insect Nutrition: An adaptionist's perspective. *Florida Ent.*, **65**, 45–71.

Slansky, F. Jr. and Feeny, P. P. (1977) Stabilization of the rate of nitrogen accumulation by larvae of the cabbage butterfly on wild and cultivated food plants. *Ecol. Monogr.*, **47**, 209–28.

Slansky, F. and Scriber, J. M. (1984) Food Consumption. In: *Comprehensive Insect Physiology, Biochemistry and Pharmacology*, Vol. 4 (Kerkut, G. A. and Gilbert, L. I., eds) Pergamon Press, Oxford (in press).

Smeathers, D. M., Gray, E. and James, J. H. (1973) Hydrocyanic acid potential of black cherry leaves as influenced by aging and drying. *Agron. J.*, **65**, 775–7.

Smiley, J. (1978) Plant chemistry and the evolution of host specificity: New evidence from *Heliconius* and *Passiflora*. *Science*, **201**, 745–7.

Smirnoff, W. A. and Bernier, B. (1973) Increased mortality of the Swain jackpine sawfly, and foliar nitrogen concentrations after urea fertilization. *Can. J. For. Res.*, **3**, 112–15.

Smith, D. (1964) Chemical composition of herbage with advance in maturity of alfalfa, medium red clover, ladino clover and birdsfoot trefoil. *Res. Report 16*, University of Wisconsin, Madison.

Smith, J. M. (1978) Optimization theory in evolution. *A. Rev. Ecol. Syst.*, **9**, 31–56.

Soo Hoo, C. F. and Fraenkel, G. (1966a) The selection of foodplants in a polyphagous insect, *Prodenia eridania* (Cramer). *J. Insect Physiol.*, **12**, 711–30.

Soo Hoo, C. F. and Fraenkel, G. (1966b) The consumption, digestion, and utilization of food plants by a polyphagous insect, *Prodenia eridania* (Cramer). *J. Insect Physiol.*, **12**, 711–30.

Sorensen, E. L., Wilson, M. C. and Manglitz, G. R. (1972) Breeding for insect resistance. In: *Alfalfa Science and Technology* (Hanson, C. H., ed.) pp. 371–90. *Am. Soc. Agron.*, Madison, WI.

Southwood, T. R. E. (1972) The insect/plant relationship – an evolutionary perspective. In: *Insect/plant Relationships* (Van Emden, H. F., ed.). *Symp. R. ent. Soc. Lond. No. 6*, 3–30.

Southwood, T. R. E. (1977) The relevance of population dynamic theory to pest status. In: *Origins of Pest, Parasite, Disease, and Weed Problems* (Cherrett, J. M. and Sagan, G. R., eds) pp. 35–54. Blackwell Scientific, Oxford.

Staatman, R. (1962) Notes on Lepidoptera ovipositing on plants which are toxic to their larvae. *J. Lepid. Soc.*, **16**, 99–103.

Sullivan, S. L., Gracen, V. E. and Ortega, A. (1974) Resistance of exotic maize varieties to the European corn borer, *Ostrinia nubilalis* (Hubner). *Env. Ent.*, **3**, 718–20.

Swain, T. (1979) Tannins and lignins. In: *Herbivores: Their Interactions with Secondary Plant Metabolites* (Rosenthal, G. A. and Janzen, D. H., eds) pp. 657–82. Academic Press, New York.

Tingey, W. M. and Singh, S. R. (1980) Environmental factors influencing the magnitude and expression of resistance. In: *Breeding Plant Resistant to Insects* (Maxwell, F. G. and Jennings, P. R., eds) pp. 89–113. Wiley, New York.

Townsend, C. R. and Calow, P. (1981) *Physiological Ecology: An Evolutionary Approach to Resource Use*. Sinauer, Sunderland, Massachussetts.

Tyler, H. (1975) *The Swallowtail Butterflies of North America*. Naturegraph, Healdsburg, CA.

Waldbauer, G. P. (1968) The consumption and utilization of food by insects. *Adv. Insect Physiol.*, **5**, 229–89.

White, M. D. (1978) *Modes of Speciation*. W. H. Freeman, San Francisco.

White, T. C. R. (1978) The importance of a relative shortage of food in animal ecology. *Oecologia*, **33**, 71–86.

Whittaker, R. H. (1975) *Communities and Ecosystems*. 2nd edn. MacMillan, New York.

Whittaker, R. H. and Feeny, P. P. (1971) Allelochemics: Chemical interactions between species. *Science*, **171**, 757–70.

Wiklund, C. (1975) The Evolutionary relationship between adult oviposition preferences and larval host plant range in *Papilio machaon* L. *Oecologia*, **18**, 185–97.

Wiklund, C. (1981) Generalist vs. specialist oviposition behavior in *Papilio machaon* (Lepidoptera) and functional aspects on the hierarchy of oviposition preferences. *Oikos*, **36**, 163–70.

Willson, M. F. (1981) Ecology and Science. *Bull. Ecol. Soc. Am.*, **62**, 4–12.

Wilson, M. C. and Davis, R. L. (1960) Culver alfalfa, a new Indiana variety developed with insect resistance. *Proc. N. Cent. Branch Ent. Soc. Am.*, **15**, 30.

Wiseman, B. R., Leuck, D. B. and McMillan, W. W. (1973) Effects of fertilizers on resistance of Antigua corn to fall armyworm and corn earworm. *Florida Ent.*, **56**, 1–7.

Woodhead, S. and Bernays, E. A. (1978) The chemical basis for resistance of *Sorghum bicolor* to attack by *Locusta migratoria*. *Ent. exp. appl.* **24**, 123–44.

Wolfson, J. L. (1978) Environmentally induced variation in *Brassica nigra* (Koch): Effects of generalist and specialist herbivores. PhD Thesis, SUNY, Stony Brook.

Wolfson, J. L. (1980) Oviposition response of *Pieris rapae* to environmentally induced variation in *Brassica nigra. Ent. exp. appl.*, **27**, 223–32.

Wolfson, J. L. (1981) Developmental responses of *Pieris rapae* and *Spodoptera eridania* to environmentally induced variation in *Brassica nigra. Env. Ent.* (in press).

Yu, S. J., Berry, R. E. and Terriere, L. C. (1979) Host plant stimulation of detoxifying enzymes in a phytophagous insect. *Pest. Biochem. Physiol.*, **12**, 280–4.

Zucker, W. V. (1983) Tannins: Does structure determine function? An ecological perspective. *Am. Nat.*, **121**, 335–65.

Predators, Parasites and Prey

8

Parasitoid–Host Relationship

S. Bradleigh Vinson

8.1 INTRODUCTION

The chemical ecology of interspecific interactions is extremely diverse and complex and is exemplified by the insect parasitoid–host relationship. Because insect parasitoids represent a third trophic level (Price *et al.*, 1980), they are influenced not only by the chemicals emanating from hosts, but by chemicals emanating from the host's food or other associated organisms. The complexity becomes apparent when we consider that a host insect may support several different parasitoid species, each of which is unique in its response to the host, and these may also interact with each other. Askew (1971) estimates that there are over 100 000 species of parasitic Hymenoptera worldwide, of which over 16 000 occur in North America (Krombein *et al.*, 1979). It is estimated that only 10–35% of the Ichneumonidae, one of the better known groups of Hymenoptera, have been described (Townes, 1969). Add to this the parasitic Diptera and other parasitic groups and one can readily see the difficulties in developing overall concepts on the chemical ecology of the parasitoid–host relationship.

The term parasitoid was first used by Reuter (1913) to distinguish from the more typical parasites those species of organisms which are parasitic only during their immature stages. Doutt (1959) described parasitoids as having the following characteristics: (i) the developing individual destroys its host; (ii) only the immatures are parasitic whereas the adults are free-living; (iii) they do not live as parasites first on one host and then another (i.e., they are not heteroecious); (iv) they are relatively large in size in comparison with their host; (v) the host is usually of the same taxonomic class; and (vi) their actions resemble predators more than true parasites. In contrast to the parasite–host relationship, the

Chemical Ecology of Insects. Edited by William J. Bell and Ring T. Cardé

evolution of the parasitoid—host relationship once the host is successfully attacked, is toward physiological control of the host. Further, the effect of parasitoidism extends to the behavior of the host and the host's ecological inter-relationships. In the case of parasites, transmission between hosts and survival of each host stage is important to the dynamics of the system (Hairston, 1965). As pointed out by Holmes and Bethel (1972), the alteration of host behavior through reduced stamina and altered responses which result in increased predation would increase the transmission rate of parasites between hosts and should be favored by selection. In the case of parasitoids, the opposite situation should exist, i.e., reduced predation of parasitoided hosts should result which in turn increases parasitoid survival (see Fritz, 1982 for an in-depth discussion).

Although the evolutionary strategies of parasitoids, parasites, and phytopha-gous species may differ, the problems they encounter in the location of ovi-position sites often have many similarities (Vinson, 1981). Upon emergence, a female parasitoid is often in an alien habitat or a habitat devoid of hosts due to the presence of unsuitable host stages or because the host population may have shifted spatially. The problem is more critical if one considers the suggestion of Price (1980) that clumped host populations may be exploited to extinction.

Although Thompson and Parker (1927) believed that parasitoids selected suitable hosts, they felt it was impossible to determine whether a host was suitable in terms of its morphological or physiochemical properties, and that host selection was an instinctive process. Others (DeBach, 1964; Rogers, 1972) have suggested that host location and selection is a random process. It is difficult not to invoke the concept of random search at some point in the host selection process. Indeed, both Salt (1934) and Ullyett (1947) suggested that the searching behavior of most parasitoids is somewhere between random and systematic. Vinson (1977a) suggested that for many parasitoids searching con-sists of a series of internally controlled locomotory patterns, each serving to contract the searching space in which the parasitoid must randomly search for the next cue. Thus each random search bout is progressively more restricted in space, thereby increasing the chance of finding the next cue.

Based on early investigations of Salt (1935), Laing (1937), Doutt (1959) and recent work by Vinson and Iwantsch (1980b) and Vinson (1975a), the host selection process may be divided into five steps: (i) habitat location, (ii) host finding, (iii) host acceptance, (iv) host suitability and (v) host regulation. These steps are arbitrarily assigned, and may actually flow from one into another, overlap, some steps may require subdivisions (e.g., Schmidt, 1974 on host acceptance), or some steps may be omitted (see Vinson, 1977a for complete discussion).

A number of reviews concern the behavior (Matthews, 1974; Vinson, 1975a, 1976, 1977a; Jones *et al.*, 1976; Hirose, 1979; Lewis *et al.*, 1976; Nordlund *et al.*, 1981), physiology and pathology (Doutt, 1963; Fisher, 1971; Mellini, 1975; Stoltz and Vinson, 1979; Vinson and Iwantsch, 1980a, b), immunity (Salt, 1968, 1970a, b; Nappi, 1975; Whitcomb *et al.*, 1974; Vinson, 1977b), and the ecological impact (Price, 1980; Price *et al.*, 1980) of parasitoids. In view of the extensive literature available, I will not review completely the chemical and behavioral ecology of parasitoidism, but instead will summarize the available information, redefine some of the categories of parasitoid behavior, and examine the role of chemicals in the relationship using reviews and the more recent literature.

8.2 HOST SELECTION PROCESS

The strategies used by insects in the host selection process, whether for food or for oviposition sites, may appear superficially to differ, but actually they have a great deal in common. While host selection for most parasitoids involves a single stage (i.e., the adult female) this is not always true. For example, the female tachinid *Lixophaga diatraea*, deposits eggs near host frass. The larvae must then locate and accept the host (Roth *et al.*, 1978).

There are two major factors that result in placing a female at a point removed from the host habitat: emergence in host-poor communities and dispersal of females (Vinson, 1981). A female removed from a host population must be able to locate a number of hosts for her progeny. The first problem is to reduce the area to be searched and consequently the time required to locate hosts.

8.2.1 Habitat preference

Habitat preference is perhaps a major factor in determining the type of habitat searched and the hosts which are located or selected. Orphanides and Gonzales (1970) suggested that hosts are attacked not because they are preffered but because they are accessible in a particular habitat. General habitat preference may be influenced by such environmental factors as temperature, humidity, light intensity, and wind (Vinson, 1975a). In addition, the flying and crawling habits of the parasitoid (Weseloh, 1972), as well as adult food sources (Simmonds *et al.*, 1975) play a role in habitat preference. The role that habitat or environmental odors play in the preference for a habitat is unknown. However, it is interesting to note that Alterie *et al.* (1981) found increased parasitism in fields sprayed with water extracts of certain weeds. These results have not been explained, but they suggest that a complex of odors making up a habitat may either attract or arrest parasitoids. Read *et al.* (1970) found that aphids on sugar beets were more readily parasitized by *Diaeritiella rapae* when collard plants were interspersed among sugar beets. The collards appeared to attract the parasitoids which also searched the sugar beets.

8.2.2 Potential host community location

A particular habitat can be organized into a number of animal and plant communities. Although the orientation of a parasitoid to the plants and local environment harboring hosts has been referred to as 'habitat location', I feel that this might be better termed 'potential host-community location'. My reason for suggesting this term is two-fold. A habitat is usually divided into a number of communities consisting of the host and associated organisms. Once the parasitoid locates the preferred habitat, it often responds to cues from those organisms in the community associated with the host.

Once in the proper habitat a female parasitoid depends upon physical, chemical and biological factors to seek out the potential host community. A female's disinterest in host seeking has been attributed to a lack of oviposition appetite or to an improper physiological condition. Some species may not show a response or be repelled by host selection cues until a pre-oviposition period has passed. Other species can be attracted to and will attack hosts upon emergence. Further, females may not respond to host selection stimuli at certain times of the day or under various environmental conditions (see Vinson 1981, for details). Although indications of biological and physical factors affecting location of the host community are present in the literature, little attempt has been made to organize and validate the information.

A female must search at random within a habitat until one of the stimuli important in host selection is encountered. As with insects responding to sex pheromones, some parasitoids respond to air movements and fly upwind (anemotaxis) (Edwards, 1954). Odors stimulate a klinokinetic response in *Mormoniella* (Edwards, 1954) whereas odors may stimulate chemotaxis in other species (Read *et al.*, 1970). However, there has been relatively little data on the 'long range' orientation of parasitoids. The cues that allow a parasitoid to orient to a potential host community presumably act from a distance. The types of stimuli that meet this criterion include electromagnetic radiation, sound, or odors.

The potential host community can be perceived by parasitoids through any of the above stimuli from the host, the food or shelter of the host, organisms associated with the host, or interactions between these factors. Plants appear to play an important role in potential host community location, partly because plants are the source of food for most hosts. It has been contended that the evolution of the parasitoid habitat in Hymenoptera may have stemmed from a previous plant–parasite relationship (Malyshev, 1968). If true, such a relationship would provide insight into the role of plants in the host selection process. The importance of plants in host community location is further supported by the observation that there is less tendency for parasitoids to select phylogenetically related hosts than unrelated hosts found on the same plant (Cross and Chesnut, 1971).

The literature on potential host-community location by parasitoids (Vinson, 1981), like that for phytophagous species (Beroza and Jacobson, 1963), is full of anecdotal evidence of plant odors attracting insects. However, plant odors consist of complex mixtures (Schoonhoven, 1968) of which few have been isolated and identified as releasers of parasitoid behavior.

Not only are odors complex, but the insect response may vary with environmental conditions. Nettles and Burks (1975) found that the tachinid, *Eucelatoria sp.* responded to a greater extent to the odor of okra at low rather than high humidity. In the field, *Cardiochiles nigriceps* searches plants only in high light intensity (Vinson, 1975a). There are examples of parasitoids locating potential host communities by host-produced sound (Mangold, 1978) and by various light wave lengths (Hollingsworth *et al.*, 1970); however, the role of colors in host selection has not been firmly established.

The role of host-associated organisms in the orientation of parasitoids to potential host communities is of particular interest. One of the best examples is provided by Greany *et al.* (1977) who showed that, in rotting fuit, a fungus which is often associated with tephritid fruit fly larvae produces acetaldehyde, an attractant of the parasitoid, *Biosteres* (*Opius*) longicaudatus.

8.2.3 Host location

Once a parasitoid has reached a potential host community, a variety of strategies is used to locate hosts. Some species search randomly until they encounter cues which indicate the presence of a host. For example, *Cardiochiles nigriceps* searches plants such as tobacco or cotton by flying 2–3 cm from the plants. When damaged leaves are encountered, the parasitoid lands and briefly antennates the damaged area. If the damage is due to a non-host, the female resumes aerial searching. The host-damaged plant tissue presumably releases a factor that stimulates landing and examination. When feeding damage is due to *Heliothis virescens*, a host of *C. nigriceps*, the parasitoid is triggered to search the plant (Vinson, 1977a).

Host location cues generally act either at very close range or on contact. Weseloh (1977) reported that *Apanteles melanoscelus* is stimulated to search on contact with a chemical associated with the host's webbing, and Lewis and Jones (1971) found that frass stimulates searching in *Microplitis croceipes*. Short-range cues often stimulate searching behavior in the female and result in a search of the surrounding area rather than discovery of a host.

8.2.4 Host examination

Parasitoids exhibit several behaviors between host contact and oviposition that are part of the host selection process. These include host examination, ovipositor probing, and ovipositor drilling. The degree of development of such behaviors among various parasitoid species is probably due to different strategies. For more active hosts such as leaf-feeding Lepidoptera, sawflies, and Coleoptera, the parasitoid searches an area contaminated by host exudates. After a host is located, the process of host examination, ovipositor probing, ovipositor drilling, and oviposition are relatively fast. In the case of more sessile hosts such as eggs or pupae, the host is extensively examined and ovipositor probing, drilling, and oviposition may be prolonged. A third situation includes those species of parasitoids that attack hosts living within containers such as galls, tunnels, or rolled up leaves. The parasitoid examines the container, then probes the container with the ovipositor in search of the host inside. After contact, the host is pierced and a decision for oviposition made. These three behaviors appear to be distinct and developed to varying degrees in different parasitoid species.

Host examination as a distinct behavior has been mentioned (Weseloh, 1980; Rotheray, 1979; Tucker and Leonard, 1977), and stimuli such as shape, texture, and size (Bragg, 1974) as well as chemicals (Moran *et al.*, 1969; Henson *et al.*, 1977) may elicit host examination. Host examination includes antennal drumming which is common in the Chalcidoidea.

8.2.5 Ovipositor probing

Many authors mentioned ovipositor probing (Spradberry, 1970; Prince, 1976; Schmidt, 1974) but few have examined the factors that elicit the response. Lawrence (1981) found vibrations from the feeding host elicited ovipositor probing in *Biosteres longicaudatus*. Vinson (1975b) isolated a probing stimulant for *Chelonus insularis* from the ovary of the tobacco budworm.

8.2.6 Ovipositor drilling

The drilling or penetration of a host or container of a host has often been described (Weseloh, 1980; Tucker and Leonard, 1977; Caldwell and Wilson, 1975). The factors that stimulate ovipositor drilling have been examined for few species. In *Testrasticus haganowaii* (G. L. Piper and S. B. Vinson, unpublished data), calcium oxalate stimulated ovipositor probing (tapping) and drilling.

8.2.7 Host oviposition

Once the ovipositor penetrates the host, the decision to accept the host as an oviposition site is made. Although this has been referred to as host acceptance, the term 'acceptance' is ambiguous. The host's heart beat has been suggested as a stimulus for host oviposition (Edwards, 1954) but its importance has not been confirmed. Chemicals appear to play a role in host oviposition. Arthur *et al.* (1972) reported that certain amino acids and MgCl are important in inducing oviposition in *Itoplectis conquisitor*.

8.3 SOURCES OF PARASITOID BEHAVIORAL CHEMICALS

The source of behavioral chemicals depends on the nature and level of host selection under study. The term semiochemical has been used to designate chemicals that mediate interactions between organisms (Law and Regnier, 1971). Semiochemicals have further been divided depending on whether the interaction is intraspecific (pheromones) or interspecific (allelochemics) (see chapters in Ritter, 1979). The allelochemics proposed by Whittaker (1970) have been further subdivided into allomones, kairomones, synomones, and apneumones (Nordlund, 1981).

Long-range cues important in potential host community location and host location may emanate from the host, its food, shelter, or associated organisms. Short-range cues important in host location, examination, ovipositor probing, drilling and acceptance usually come from the host. Further, the same source and chemical may act as a pheromone in one context and a kairomone or synomone in others.

8.3.1 Kairomones

Larval parasitoids respond to silk, frass, volatiles, cuticular components, and glandular secretions (Weseloh, 1981). For example, *Cardiochiles nigriceps* responds to hydrocarbons in the mandibular glands of *H. virescens* (Vinson, 1968), *Apanteles melanoscelus* responds to the silk produced in labial glands of gypsy moth (Weseloh, 1977), and several species respond to host feces.

Parasitoids that attack eggs often respond to kairomones associated with reproduction. *Chelonus insularis* responds to a searching stimulant from female moth scales (J. R. Ables, T. Shaver and S. B. Vinson, unpublished data) and also responds to a proteinaceous material in the egg chorion (Vinson, 1975b). *Trichogramma evanescens* is stimulated to search by tricosane associated with host moth scales (Jones *et al.*, 1973). Strand and Vinson (1982b) have isolated a protein from the accessory gland of *H. virescens* females that stimulates host examination and probing in *Telenomus heliothidis*.

Host sex pheromones can serve as cues to host location for several parasitoid species (Kennedy, 1979; Sternlicht, 1973). Prokopy and Webster (1978) found that the marking pheromone of *Rhagoletis pomonella* stimulates oviposition probing of *Opius lectus*.

8.3.2 Synomones

Few synonomes have been identified, but there are a number of anecdotal reports that synomones play a role in potential host community location (Weseloh, 1981). Camors and Payne (1973) reported α-pinene released from pine trees after bark beetle attack attracts *Heydenia unica*. Allyl isothiocyanate, a volatile constituent of crucifers, attracts *Diaretiella rapae*, a parasitoid of aphids on crucifers (Read *et al.*, 1970). The tachinid, *Drino bohemica*, is attracted to host plant material of *Diprion hercyniae* in an olfactometer (Monteith, 1958), but like many other such reports (see Vinson, 1981) the synomones attracting parasitoids have not been identified.

8.3.3 Allelochemics

These materials, which are released by organisms associated with the host, pose some interesting questions concerning co-evolution. They are difficult to classify as kairomones or synomones because the benefits derived by the producing organisms are unknown. Allelochemic compounds that act at long and short range have been described. Greany *et al.* (1977) reported that a fungus in rotting peaches produced ethanol and acetaldehyde, which attracted flying *Biosteres*, a parasitoid of tephritid fruit larvae. Madden (1968) reported that several species of Ichneumonidae locate host galleries in trees at close range by chemicals released by fungi associated with the host, *Sirex* sp.

8.3.4 Apneumones

Some of the early work on parasitoid host location involved non-living host food. Thorpe and Jones (1937) found that *Venturia canescens* is attracted to the odor of oatmeal. Meat odor was found to be attractive to *Alysia manducator* and *Nasonia* (*Mormoniella*) *vitripennis* (Laing, 1937). Since the source and nature of the responsible chemicals was not determined, it may be that microorganisms were involved.

8.4 CHEMICAL-PHYSICAL INTERACTION

Although chemicals appear to play a major role in the sequence of events leading to host acceptance, physical factors may also be involved (Vinson, 1976). In host examination, ovipositor probing, drilling and possibly acceptance, both host shape and the chemical may have to be combined to elicit the appropriate behavioral response. Protein from accessory glands of *Heliothis virescens* stimulates *T. heliothidis* to examine and probe glass beads coated with the kairomone only if the beads are the approximate size of *Heliothis* eggs (Strand and Vinson, 1982b). A similar phenomenon is important in eliciting host examination and ovipositional probing in *Tetrasticus haganowaii*, a parasitoid of cockroach oötheca. Calcium oxalate elicited the appropriate behavior only when applied to rounded objects of similar size to the oötheca (Piper and Vinson, in preparation).

8.5 HOST PREFERENCE

The role that chemicals play in host preference has received little attention. Contamination of an unacceptable host by the odor of a preferred host may result in the attack of the unacceptable host (Spradberry, 1968). *Cardrochiles nigriceps*, for example, readily attacks and accepts *Galleria mellonella* and

Spodoptera frugiperda if the searching stimulant from *Heliothis virescens* is applied to these unacceptable hosts (Vinson, 1975a).

Naryanan and Rao (1955) reported that *Microbracon gelechiae* reared in the laboratory for 7 years on *Corcyra cephalonica* respond to the odor of *C. cephalonica* over its natural host, *Gnorimoschema operculella*. Whether this change in preference was due to a genetic alteration, conditioning, or an aspect of Hopkins host selection principle is unknown. Smith and Cornell (1979) indicated that Hopkins host selection principle is operational in *Nasonia*, although Salt (1935) found no evidence with *T. pretosium*. The manipulation of host preference has far-reaching implications for biological control that have not been seriously considered.

The diet of the host may play a significant role in host preference. Hendry *et al.* (1976) reported that corn contains tricosane, the kairomone for *Trichogramma evanescens* which was first isolated from *Heliothis zea*, a corn feeder. Sauls *et al.* (1979) reported that the frass of *H. zea* reared on different diets differed significantly in its kairomonal activity for *Microplitis croceipes*. Thus, the preference for a given host may be influenced by the food consumed by the host, possibly, a particular host might be able to escape parasitoidism by feeding on certain plants or food resources.

8.6 HOST DISCRIMINATION (MARKING AND EPIDEICTIC PHEROMONES)

The partitioning of food or oviposition resources has received increased attention in recent years. Peters and Barbosa (1977) developed a concept of an optimal density range for a particular resource. One of the examples of optimal density involves the Douglas fir beetle (*Dendroctonus pseudotsugae*). McMullen and Atkins (1961) found that the density of 8 adults per 1000 cm^2 of bark surface resulted in less larval mortality and a greater number of adult progeny per female than higher adult densities. With fewer adults, 4–6 per 1000 cm^2 of bark surface, the broods did not survive (Hedden and Pitman, 1978). Overcrowding may result in reduced fitness of individuals in terms of increased mortality, altered behavior, reduced size, developmental rate and fecundity (see Peters and Barbosa, 1977 for discussion). The optimal density of a particular food resource is often effectively, one host, as is the case for many solitary parastioids (DeBach, 1964).

Insects use varied mechanisms to achieve optimal density levels, including physical combat (Hubbell and Johnson, 1977), visual cues (Rothschild and Schoonhoven, 1977), accoustical cues (Russ, 1969), and chemical cues. The chemicals involved in regulating density or in preventing overcrowding are documented in detail by Prokopy *et al.* in Chapter 11.

Salt (1934) was the first to show that parasitoids could discriminate between healthy hosts and parasitized hosts. Discrimination and marking of hosts has been controversial since authors investigating the same parasitoid reported either host discrimination or no evidence of host marking. However, as van Lenteren (1981) pointed out, many studies were not designed to demonstrate host discrimination or their data are open to other interpretations. Probably most parasitoids are able to discriminate through some form of host marking (review: van Lenteren, 1981). An analogous phenomenon appears in the early stages of host selection. For example, Price (1970) found several pupal parasitoids that avoid areas previously searched.

The process of host marking also differs among the various species of parasitoids (see Prokopy, 1981; Prokopy *et al.*, Chapter 11). It appears that there are both external and internal marking substances. Salt (1937) found that *Trichogramma* rejected parasitized hosts on contact, but if the host eggs were washed the parasitoid would reject these eggs only after ovipositor insertion. Similar results were found by Guillot and Vinson (1972) for *Campoletis sonorensis* which marks its host with a pheromone perceived prior to ovipositor insertion. After 6 hours hosts were acceptable again to attack, but no new eggs were found upon dissection (Guillot and Vinson, 1972).

Cases of host discrimination occurring only in later stages of parasitoidism and after ovipositor insertion have been reported (King and Rafai, 1970). Fisher and Ganesalingam (1970) reported changes in hemolymph proteins of hosts several days after parasitism and suggested that these changes may be responsible for internal discrimination. Guillot and Vinson (1972) provided evidence that *C. sonorensis* injects an ovarial factor which is responsible for internal host discrimination but how this factor acts is unknown. The few studies on the specificity of internal marking factors (Wylie, 1971) suggest that they are not species-specific. In contrast, many studies on externally perceived marking pheromones suggest they are species-specific (McLeod, 1972; Ables *et al.*, 1981), with the notable exception of substrate-marking pheromones (Price, 1972).

The Dufours gland appears to be the source of the external marking pheromone for *Microplites croceipes*, *Cardiochiles nigriceps* and *Campoletis sonorensis* (Vinson and Guillot, 1972; Guillot and Vinson, 1972). The hydrocarbon fraction was found to be responsible for marking in *C. nigriceps* (Guillot *et al.*, 1974), a particularly interesting situation in as much as the host kairomones for *C. nigriceps* also consist of hydrocarbons (Vinson *et al.*, 1975).

Parasitoid marking pheromones, according to van Lenteren (1981) (i) prevent multiple ovipositions into the same host and consequent egg wastage, (ii) prevent host wastage because superparasitized hosts often die, resulting in death of the parasitoids, (iii) save time, particularly if the process of attack and oviposition are lengthy; and (iv) initiate migration to more productive habitats when several marked hosts are encountered.

8.7 CONDITIONING OR ASSOCIATIVE LEARNING

Both learning, which is a relatively permanent change in the behavior as a result of re-inforced practice, and conditioning a type of learning in which the organism acquires the capacity to respond to a stimulus with a reaction associated with another stimulus, have been implicated in different levels of the parasitoid–host relationship. Thorpe and Jones (1937) reported that *Venturia canescens* responded to odor of its host, *Ephestia*, even if reared apart from *Ephestia*. However, if reared from *Meliphora*, fewer parasitoids responded to the odor of *Ephestia*. Several reports suggest that a female parasitoid with a wide host range often prefers the host species from which she was reared (Eijackers and van Lenteren, 1970). Arthur (1971) reported that *V. canescens* associates the odor of geraniol with the presence of hosts after this parasitoid was exposed to the odor of geraniol and hosts simultaneously. Similar results were found with *Bracon mellitor*, a parasitoid of the boll weevil, *Anthonomus grandis* (Henson *et al.*, 1977; Vinson *et al.*, 1977).

The ability to discriminate also appears to be influenced by experience (van Lenteren, 1976). As shown by van Lenteren (1976), inexperienced females may superparasitize many hosts until they encounter unparasitized hosts and learn to discriminate. It also appears that location of the host may be a matter of experience. Female *Cardiochiles nigriceps* are more likely to search for and attack hosts if they have previously encountered a host than inexperienced females (Strand and Vinson, 1982a).

It appears that experience and associative learning are very important in the biology of insect parasitoids, particularly the Hymenoptera. Males have been reported to learn to associate certain conditions of the physical environment with the presence of mates (Robacker *et al.*, 1975). It should not be surprising that a female's preference for a particular host or host-plant complex is influenced by prior exposure and success. Many potential host populations fluctuate with respect to both their choice of food plant and habitats during different times of the year, and parasitoids with a wide host range may encounter different species of hosts at different times of the year. Conditioning allows flexibility to meet these challenges by concentrating on those hosts and habitats in which success has been achieved (Vinson, 1976).

8.8 FORAGING STRATEGIES

Up to now, the behavior and foraging strategies of only a few parasitoids have been examined and even these have only been studied within a limited environmental concept. Further, most of the species that have received some examination are ichneumonids that attack lepidopterous larvae. Yet, even within this limited group substantial differences occur. Waage (1979), examining the foraging behavior of *Venturia* (= *Nemeritis*) *canescens*, found that the parasitoid remains on a small area of grain (a patch) that had been infested with its host *Ephestia*, longer than on a similar un-infested patch. Kairomone-treated patches evoked similar foraging changes in *Apanteles melanoscelus* (Weseloh, 1980) and *Cardiochiles nigriceps* (Strand and Vinson, 1982a). Arrestment of parasitoids in kairomone-treated areas has been used by Lewis *et al.* (1975a) to explain in part the increased parasitism by *Trichogramma* in fields treated with an extract of moth scale kairomone.

Kairomones that act over short distances and increase a parasitoid's searching activity have been referred to as searching stimulants (Vinson, 1968). Searching stimulants can arrest parasitoids for a period of time in a contaminated patch or stimulate the parasitoids to increase their movement, thereby increasing the chance of host discovery (Lewis *et al.*, 1975b; Strand and Vinson, 1983a). A parasitoid on a kairomone-treated patch will leave if hosts are not contacted after a period of time, the 'giving up time' (GUT) of Waage (1979). If a female leaves a patch and re-encounters it a short time later, the GUT is decreased, as demonstrated with *A. melanoscelus* and *C. nigriceps* (Weseloh, 1980; Strand and Vinson, 1983a). These results cannot be explained by contamination of searched patches by a marking pheromone, as might be expected, because a different female or a female held for a few hours before re-exposure to the patch will have a GUT similar to the first females initial encounter with the patch. Possibly the parastitoids are habituated to the kairomone at a given concentration.

If hosts are added to the patch, the situation is different. Waage (1979) reported that when a female attacks a host, the GUT is increased a small amount. If hosts are continually added to the patch the GUT lengthened. This behavior of continuing to search a patch after a host encounter has been termed success-motivated searching (Vinson, 1977a).

The phenomenon of success-motivated searching brings another dimension to host marking because, unlike predators (Marks, 1977), the parasitoided host remains in the environment. As a parasitoid parasitizes more hosts within a patch, the frequency of encounters with parasitized hosts increases, making it advantageous for the female to leave. The presence of marking pheromones would be expected to decrease the GUT and reduce success-motivated searching.

In contrast to the foraging strategy of *Venturia* and *Apanteles*, *C. nigriceps* leaves the patch immediately after ovipositing (Strand and Vinson, 1982a), suggesting that searching is not success-motivated and that marking pheromones are unimportant. However, the apparent difference in searching strategy might be explained by the distribution of hosts. Host densities may be: (i) high density over a wide area, (ii) high density in a clumped distribution, or (iii) low density over a wide area (Southwood, 1966). In the first two cases an intensive search of the area after host encounter would be expected to result in encounters with additional hosts. In the latter case, success-motivated searching would not be expected to lead to additional hosts, as *H. virescens* is a non-aggregating species (Pieters and Sterling, 1974). Although foraging behavior of *C. nigriceps* appears to represent a 'reasonable' strategy for a low density and dispersed host distribution, the situation may be explained by patch size. Since *C. nigriceps* prefers dispersed, later-instar larvae for parasitoidism (Vinson, 1975a), searching the area immediately surrounding a parasitized larvae would be less likely to result in the location of another host than leaving the immediate area to search nearby. Thus, the 5 cm patch size used in the study of *C. nigriceps* (Strand and Vinson, 1982a) was possibly too small to detect success-motivated searching. An additional behavior may be important: after oviposition they may be dispersed.

Different strategies are exhibited by species of parasitoids attacking sessile hosts such as prepupae or eggs as well as hosts in galls, stems or containers. Although the behavior of these parasitoids has not been examined as extensively, some observations suggest a different strategy is involved. Searching stimulants (i.e., chemicals) that increase movement of parasitoids within the vicinity of a host either do not exist or the response of the parasitoid to such chemicals has not been examined. Although searching stimulants elevate the rate of host location in *Trichogramma* (Lewis *et al.*, 1972) and *Chelonus insularis* (Ables *et al.*, in preparation), both of which attack eggs; a searching stimulant for species such as *Bracon mellitor*, *Testrasticus haganowaii* and *Telenomus heliothidis* does not appear to increase their movement and search of the vicinity surrounding a sessile host. However, once the host or host container is located, then examination of the host or container is prolonged. As discussed earlier, the behaviors involved in host examination, ovipositor probing, and drilling are more elaborate in these species.

8.9 HOST EVASION AND DEFENSE

The great diversity of parasitoids and the complexity of parasitoid-host relationships is probably influenced by attempts of potential host insects to escape their predators, parasites and parasitoids. There is considerable speculation on the role that parasites (parasitoids) play in herbivore evolution. There are many examples where a host on different plants is attacked by different parasitoid species (see Vinson, 1981). As discussed by Zwolfer and Kraus (1957), and Vinson (1981), plants play an important role in the host selection process, probably by providing cues to the location of a potential host community. Theoretically, a host could escape a particular parasitoid by attacking a plant lacking those stimuli used by the parasitoid to locate the potential host community. This idea is supported by the observation that there is less tendency for parasitoids to select phylogenetically related hosts than to favor a range of hosts on a particular plant (Askew and Shaw, 1978; Cross and Chesnut, 1971).

One outstanding question that comes to mind when one describes the role kairomones play in host selection is why the host continues to produce the kairomone. No doubt elimination of the kairomone has occurred many times during the evolution of a parasitoid—host relationship, but proving such an alteration is very difficult. Most, if not all, kairomones probably serve the host in some essential way. For example, the mandibular gland secretion that acts as a kairomone for *Venturia canescens* also acts as a spacing pheromone for its host, *Ephestia kuehniella* (Corbet, 1971), and the 'glue' for eggs of *Heliothis virescens* also acts as a kairomone for *Telenomus heliothidis* (Strand and Vinson, 1983a).

An interesting possibility, that to my knowledge has not been investigated, is the development of repellents by hosts or the secretion of compounds mimicking the parasitoid's marking pheromone. Vinson (1975a) reported that when kairomones from *H. virescens* were placed on several hosts, these hosts were accepted by *C. nigriceps*, with the exception of *Spodoptera ornithogalli*, the yellow-striped armyworm. Whether this is due to a repellent is unknown. Monteith (1960) speculated that reduced parasitoidism by tachinids of the larch sawfly, *Pristiphora erichsonii*, feeding on *Larix laricina* growing in association with understory trees and shrubs, was due to a repellent or masking agent produced by associated plants. If such repellent plants exist, hosts shifting to such a plant could escape parasitoidism.

Another interesting behavior is frass throwing by *Pieris rapae* which has been found by B. Usher (personal communication). She suggests that the removal of frass, which often serves as a parasitoid cue, provides an escape from parasitoidism. It is expected that other examples of parasitoid evasion will be found in the future.

8.10 PARASITOID DEVELOPMENT PROCESS AND HOST SUITABILITY

Once a host has been accepted for oviposition, the chemical interactions between parasitoid and host continue with respect to the suitability of the host and the ability of the parasitoid to regulate the host's development. Even though a host has been accepted and an egg deposited, the host may not be suitable for development of parasitoids for the following reasons: (i) insecticide treatment, (ii) exposure to juvenile hormone mimics or other insect growth regulators, (iii) pathogenic infections, (iv) location of the host in an environment unsuitable for the parasitoid, (v) host toxicity due to food choice, (vi) competition with other parasitoids that occur within the host, (vii) nutritional suitability, (viii) hormonal balance and (ix) the ability of the host to react immunologically to the parasitoid (Vinson and Iwantsch, 1980a). In considering the chemical ecology of the parasitoid–host relationship, the last five reasons are of particular interest.

8.10.1 Competition

Most endophagous parasitoids require the entire host for their development. Although parasitoids can generally discriminate between non-parasitized and parasitized hosts (see van Lenteren 1981, for discussion) superparasitoidism and multiple parasitoidism (see Smith 1916 for definition of terms) often occur. These supernumerary progeny are eliminated either by physical attack or physiological suppression. Physical attack appears to be most common (Vinson and Iwantsch, 1980a), and it is usually the first-instar parasitoid larvae that seek and, using their mandibles, destroy competitors (Vinson and Iwantsch, 1980a). How the competing larvae or eggs are located is unknown, but chemical attraction is a possibility. Our understanding of physiological suppression is even less precise. Several authors (Jackson, 1959; Tremblay, 1966) have suggested that death of competitors is due to a toxin but as yet no such toxin is known. Spencer (1926) and Thompson and Parker (1930) suggested that, as larvae hatch a cytolytic enzyme is secreted that frees the larva from the serosa, causing degeneration of competing younger larvae, and dissolution of host tissues. However, such enzymes have not been isolated.

Salt (1934), and Fisher (1963) provided evidence that low O_2 content of host hemolymph was the cause of a competitor's death. The available evidence suggests that only eggs and young larvae of some species may be susceptible to anoxia, as older larvae of *Venturia* and *Microplitis* are able to survive in an anaerobic environment (Fisher, 1963; Edson and Vinson, 1976).

Selective starvation has also been suggested as a means by which the first larva to hatch eliminates competitors. Competing younger eggs and larvae lack the proper nutrients and die (Pschorn-Walcher, 1971; Chacko, 1969). Changes in the physiology of the host brought about by venoms, viruses or virus-like particles injected by the female during oviposition (Fisher and Ganesalingam, 1970; Stoltz and Vinson, 1979) may result in an internal environment unsuitable to the developing competitors.

From these cases, it might appear that the first larva to hatch has an advantage, but the competition for possession of a host is considerably more complicated. In this regard, the egg-larval parasitoids are of interest. Tremblay (1966) suggested that the advantage of evolving towards an egg-larval parasitoid is that the larvae gain an advantage in possessing the host before the competing larval parasitoids. However, *Chelonus insularis* an egg-larval parasitoid fails to compete with *Campoletis sonorensis* or *Microplitis croceipes*, both larval parasitoids, even though *Chelonus* successfully competes with *Cardiochiles nigriceps*, another larval parasitoid (Vinson and Ables, 1980). Even in the case where physical attack is involved, age does not necessarily provide superiority. Superparasitoidism by *C. nigriceps* revealed that first-instar larvae 1–3 days younger than their competitors have a higher survival rate (Vinson, 1972), possibly due to the greater mobility of the younger larvae. It has been suggested that ectophagous parasitoids have an intrinsic superiority over endophagous species (Flanders, 1971) through venomization, although Sullivan (1971) demonstrated that such is not always the case.

8.10.2 Nutritional suitability

The nutritional suitability of a host was proposed by Flanders (1937) and Salt (1938) to be important to the survival of a parasitoid. House (1977) concluded that the nutritional needs of most parasitoids are probably similar to most other organisms. The problem of nutritional suitability may not be lack of specific nutrients or accessory growth factors in the host, but the quality of and the ability to obtain certain nutrients present in host tissues at the proper time and to compete with host tissues for available nutrients. As discussed below under host regulation, the parasitoid may actively affect these last two conditions.

Another important factor is the amount of nutrients available to a parasitoid. A host is a finite food resource which because of its size, age, and nutritional history may affect parasitoid development (see Vinson and Iwantsch, 1980a). Smaller hosts yield smaller parasitoids with reduced emergence, fecundity and longevity, or more males (Vinson and Iwantsch, 1980a). Females may have the ability to assess host size and deposit fertilized or unfertilized (male) eggs accordingly (Nozato, 1969), or the female may oviposit eggs of both sexes with preferential survival of one sex in larger hosts (Holdaway and Smith, 1932). Flanders (1959) reported that certain Aphelinidae deposit fertile eggs in coccid hosts and unfertile eggs in lepidopterous hosts.

Often, in the case of gregarious parasitoids, the larger the number of parasitoids per host, the smaller, less fecund, and shorter-lived are the adults (Vinson and Iwantsch, 1980a). Beckage and Riddiford (1978) and Capinera and Lilly (1975) have shown that the greater the parasitoid load, the more rapidly the adults emerge; however, the opposite situation, i.e., a slower development rate in hosts with a greater parasitoid load, has also been reported (Thurston and Fox, 1972). With solitary endoparasitoids, Miles and King (1975) reported decreases in the parasitoid's development time with increasing host age at the time of parasitoidism. This may be due to nutrients in older hosts being more abundant.

8.10.3 Host food choice

The suitability of a host may be influenced by the host's food choice. Smith (1941) found that *Habrolepis rouxi* developed on the California red scale, *Aonidiella aurantii*, when the scale was reared on sago palm, but did not develop when the scale was reared on citrus. Campbell and Duffey (1979) reported that α-tomatine, an alkaloid in tomato plants, was toxic to *Hyposoter exiguae* developing in hosts fed the alkoloid.

8.10.4 Host immunity

The suitability of a host depends on whether the parasitoid can evade or suppress the host's internal defense or immune mechanism(s). It has been suggested (Doutt, 1963; Salt, 1963) that by the careful placement of their progeny on the host (ectoparasitoids), within certain tissues (i.e., fat body, ganglion), or in certain stages (eggs) many species of parasitoids evade the host's defense. However, this assumption needs critical investigation.

Many species place their progeny into the hemocoel of their host where the progeny's survival depends on whether the parasitoid can evade or suppress the host's immune mechanism or avoid being recognized as foreign. There are several ways in which parasitoids are eliminated. The host defense most often described is encapsulation (Salt, 1970a). How the foreign object is recognized and the mechanisms that lead to the formulation of a hemocytic capsule are largely unknown, despite extensive study (e.g. Whitcomb *et al.*, 1974).

In some cases a non-cellular capsule is formed, believed to consist of melanin (Vey and Gotz, 1975). Melanin may also be involved in cellular encapsulation (Nappi, 1975; Salt, 1963). In both cases it is believed that these capsules kill the parasitoid egg or larvae by blocking the oxygen supply and source of nutrition (Fisher, 1971; Salt, 1970a). In some parasitoid Diptera the encapsulation response is often manipulated to the parasitoid's benefit (Salt, 1968) by forming around the parasitoid a sheath, attached at one end to a trachea supplying O_2 and at the other end to the hemolymph through which the parasitoid feeds.

A third means of parasitoid elimination, discovered by Arthur and Ewen (1975), is referred to as cuticular encystment. After parasitism, the host forms a small cyst of the posterior dorsal cuticle which consists of a separation of the cuticle from the epidermis with a space containing hemolymph. The parasitoid egg or larva migrates into the cyst which is shed at the next molt.

There is little doubt that chemicals play an important role in the internal defense and counterdefenses that have evolved during the evolution of the various parasitoid-host relationships. Currently too little is known about the chemicals involved in internal recognition and defense to develop any overall concepts.

8.10.5 Host endocrine balance

Suggestions of interactions between the endocrine systems of a parasitoid and its host are not new (Schneider, 1950). The host's endocrine balance may play an important role in determining the suitability of a particular host or host stage for a parasitoid. One of the difficulties in determining the role of the endocrine system in host suitability is that parasitoids may not emerge, develop, or become encapsulated due to an improper hormonal balance, but the effects are attributed to some other factor(s). As the role of the endocrine system is examined, it becomes increasingly clear that the endocrine system can have profound effects on the parasitoid. The synchronization of the parasitoid's development with that of the host has been demonstrated many times (Fisher, 1971). Rouband (1924) presumed the synchronization of parasitoid development with that of its host was due to an inhibiting substance produced by the host until a certain developmental period was reached. Alternatively, Schneider (1951) suggested that the parasitoid rested until activated by specific (hormonal) changes that occurred in the host. As Mellini (1975) pointed out, it is difficult to distinguish between direct hormone action on the parasitoid and indirect hormone action on the host physiology.

Juvenile hormone stimulates the emergence of *Microtonus aethiops* from its diapausing alfalfa weevil host by breaking the parasitoid's diapause (Neal *et al.*, 1970). Baronio and Sehnal (1980) found that the first-instar tachinid, *Gonia cinerascens* is triggered to begin development by the direct action of host ecdysteroid in the absence of juvenile hormone. Lawrence (1982) found that the development of *Biosteres longicaudatus* is triggered by host hormones and that late-instar hosts are not suitable because the proper hormonal balance period has passed.

It thus appears that the suitability of a particular host stage and possibly different hosts may depend on the host's endocrine activity. However, the parasitoid also can have a profound effect on the host's endocrine system.

Johnson (1965) reported that the aphid parasitoid, *Aphidius plantensis*, had a juvenilizing effect on its host which he later attributed to the premature break-down of the prothoracic glands. Iwantsch and Similowitz (1975) observed a reduction in RNA synthesis of the prothoracic glands of *Trichoplusia ni* larvae parasitized by *Hyposoter exiquae*. Vinson (1970) found that the injection of teratocytes (cells of the braconid sorosal egg membrane released into the host upon hatching) into hosts had a juvenilizing effect. Beckage (1980) noted that the juvenile hormone titer of *Manduca sexta* larvae was higher in larvae parasitized by *Apanteles congregatus* than non-parasitized larvae; juvenile hormone esterase activity was decreased in parasitized hosts, accounting for the build-up of juvenile hormone. There is little doubt that the endocrine system of the parasitoid and its host have co-evolved in many different directions in different parasitoid-host relationships.

8.10.6 Host regulation

As previously discussed, the immature parasitoid is influenced by the physiological state of the host; however, the parasitoid is not totally passive, responding only to the host's condition. The host, once it is sucessfully attacked, is of little consequence to the evolution of the relationship other than providing food and shelter to the developing parasitoid. After successful oviposition, evolution of the parasitoid-host relationship is toward the control of the host, i.e. host regulation, for the benefit of the progeny (Vinson, 1975a). Host regulation does not imply that each physiological system of the host is controlled by the parasitoid, but that certain host systems important to the developing parasitoid are altered or modified. There is a fine line between regulation and host suitability. A particular host may not be suitable because a particular parasitoid is unable to effect those changes in the host necessary for its development.

The effects of the parasitoid on the development, morphology, behavior, physiology, and biochemistry of the host have been reviewed by Vinson and Iwantsch (1980b). These changes were attributed in much of the earlier literature to the feeding activity of the developing larva but it is increasingly clear that these host changes in response to parasitoid Hymenoptera are due to chemicals and viruses injected by female parasitoids during oviposition and to secretions released by developing eggs or larvae. Little is known concerning the parasitoid Diptera, although it may be expected that any regulatory changes are probably due to the developing eggs or larvae.

8.10.7 Ovipositing adults

There are three types of glands associated with the female reproductive system, two of which are important in parasitoid–host regulation. They are the poison gland and the epithelium of the calyx or lateral oviduct of the ovary. The poison gland is the source of substances causing various degrees of host paralysis (Beard, 1978), commonly in hosts attacked by ectoparasitoids (Muesebeck and Krombein, 1951). Unfortunately, the nature of poison gland venoms has been analyzed in very few species, with the exception of the genus *Bracon* (Beard, 1978), and the function of the poison gland of many parasitoids is unknown. Paralysis may decrease mortality of ectoparasitoids by reducing the ability of the host to remove the egg and developing larva, or it may reduce competition since some parasitoids only attack moving or active hosts (Beard, 1972). The data of Gurjanova *et al.* (1977) suggest that venom may increase the nutritional load of the hemolymph through histolysis. The poison gland of *Banchus flaverscens* induces cuticular cysts in looper larvae (Arthur and Ewen, 1975), but this appears to be a specific response of the host to the poison gland rather than a function of the poison gland. In some species, paralysis is delayed or temporary (DeLeon, 1935), while in many species there is no paralysis.

The calyx of *Venteria canescens* contains a particulate secretion (Rotheram, 1973) consisting of a glycoprotein (Bedwin, 1979). Salt (1970b) reported that this secretion coated the parasitoid eggs and inhibited the host's immune response. The calyx of *Pimpla turionellae* also contains a mucopolysaccharide (Führer, 1973) which Osman and Führer (1979) implicated in the suppression of the host's immune response.

Vinson and Scott (1975) found DNA-containing particles in the calyx of *Cardiochiles nigriceps*. Similar particles discovered in several species of braconids and ichneumonids are viral, invade host tissues, and may have profound effects on the host (see Stoltz and Vinson, 1979, for a discussion). These viruses have been implicated in affecting growth (Vinson *et al.*, 1979), hemolymph trehalose levels (Dahlman and Vinson, 1977), and suppression of the host's immune response toward the parasitoid's progeny (Edson *et al.*, 1981). It is anticipated that additional host regulatory functions of these viruses will be found.

8.10.8 Eggs and larvae

There are many species of parasitoids which do not appear to have viruses associated with their reproductive system (Vinson, unpublished); yet, these species can also alter the physiology and behavior of their host. The source and nature of the responsible factors is as yet unknown. Schneider (1950) proposed that the ability of *Diplazon pectoratorius* to induce premature diapause of its host, *Epistrophe bifasciata*, was due to a substance secreted by the larvae (probably from the salivary glands). There have been numerous, but unsubstantiated, suggestions that secretions from the egg and larva are involved in suppressing the host's immune response (Salt, 1960) or its behavior. Perhaps the development of artificial rearing of parasitoids *in vitro* will provide new impetus to determine the effects of eggs and larvae.

8.11 CONCLUSIONS

Since parasitoids are free-living as adults, and in most instances their progeny grow and develop solitarily in or on a host, the female must locate a number of hosts during her lifetime. Further, hosts may be widely distributed and a female must be able to locate these hosts. Thus, females have had to evolve effective mechanisms for host location and selection. This has included the long-range orientation of females to a potential host community, followed by a steady narrowing of the theater of search, guided by short-range and contact cues. The long-range cues often involve odors, sound or electromagnetic-radiation. The short-range cues may also involve these and, in addition, vibration, movement, and shape. Lastly, parasitoids are influenced by contact cues, these involve chemicals, shape, texture, and vibration.

To meet these needs, parasitoids have evolved to respond to cues not only produced by the host, but also cues provided by the food and shelter of the host and associated organisms. While progress has been made in isolating and identifying the cues involved in host selection, most of the effort has centered on the short-range chemical cues of egg and larval parasitoids. Only a few studies have been conducted on the long-range cues.

Once a host is found and attacked, many parasitoids mark the host or its container, reducing the parasitized host's acceptibility to other females. The identification of these marking pheromones has received little attention. After oviposition and marking of a host, many parasitoids may exhibit success-motivated dispersal, thus removing themselves from the hosts immediate vicinity. At some point these same females switch to success-motivated searching, resulting in their retention for a period of time in the general area where hosts have been located. However, success-motivated dispersal and searching have received only limited study, even though success-motivated searching has been implicated as an important behavioral management technique used to increase parasitism in field situations.

The co-evolution implied in the parasitoid–host relationship does not stop once a host has been attacked, but continues as the progeny develop within or on the host. The growth, development, physiology, and behavior of the host is altered by parasitoidism. These changes, referred to as host regulation, appear to be brought about by venoms or viruses injected into the host during oviposition, as well as by factors released by the developing parasitoid egg or larvae. The nature of factors released by eggs or developing parasitoid larvae and the effects they have on the host have received little study. The venoms injected have only been examined in a few species. The role viruses may play in the co-evolution of the parasitoid–host relationship has opened up an area of study which is only now beginning to yield new insights into parasitoid biology. The co-evolution of the relationship is very complex, involving a great diversity of interacting factors and organisms, which we are just beginning to understand. We do know that chemicals play an important role in almost every step in the co-evolutionary relationship.

ACKNOWLEDGEMENT

I would like to express my appreciation to Ms Kathryn Edson, Mr Michael Strand, Richard Thomas and Gary Elzen for their suggestions and editorial help in the preparation of this manuscript.

REFERENCES

Ables, J. R., Vinson, S. B. and Ellis, J. S. (1981) Host discrimination by *Chelonus insularis* (Hym: Braconidae), *Telenomus heliothidis* (Hym: Scelionidae), and *Trichogramma pretiosum* (Hym: Trichogrammatidae). *Entomophaga*, **26**, 149–56.

Alterie, M. A., Lewis, W. J., Nordlund, D. A., Guldner, R. C. and Todd, J. W. (1981) Chemical interactions between plants and *Tricnogramma* sp. wasps in Georgia soybean fields. *Protection Ent.* (in press).

Arthur, A. P. (1971) Associative learning by *Nemeritis canescens*. *Can. Ent.*, **103**, 1137–41.

Arthur, A. P., Hegdekar, B. M. and Batsch, W. W. (1972) A chemically defined synthetic medium that induces oviposition in the parasite *Itoplectis conquisitor*. *Can. Ent.*, **104**, 1251–8.

Arthur, A. P. and Ewen, A. B. (1975) Cuticular encystment: A unique and effective defense reaction by cabbage looper larvae against parasitism by *Banchus flaverscens*. *Ann. ent. Soc. Am.*, **68**, 1091–4.

Askew, R. R. (1971) *Parasitic Insects*. American Elsevier, New York.

Askew, R. R. and Shaw, M. R. (1978) Account of Chalcidoidea parasitizing leaf-mining insects of deciduous trees in Britain. *Biol. J. Linn.*, **6**, 289–335.

Baronio, P. and Sehnal, F. (1980) Dependence of the parasitoid *Gonia cinerascens* on the hormones of its lepidopterous hosts. *J. Insect Physiol.*, **26**, 619–26.

Beard, R. L. (1972) Effectiveness of paralyzing venom and its relation to host discrimination by Braconid wasps. *Ann. ent. Soc. Am.*, **65**, 90–3.

Beard, R. L. (1978) Venoms of Braconidae. In: *Handbuch der Experimentellen Pharmakologie*, Vol. 48 (Bettini S., ed.) pp. 773–800. Springer-Verlag, Berlin.

Beckage, N. E. (1980) Physiology of developmental interaction between the tobacco hornworm, *Manduca sexta*, and its endoparasite, *Apanteles congregatus*. Dissertation, University of Washington.

Beckage, N. E. and Riddiford, L. M. (1978) Developmental interactions between the tobacco hornworm, *Manduca sexta*, and its braconid parasite *Apanteles congregatus*. *Ent. exp. appl.*, **23**, 139–51.

Bedwin, O. (1979) An insect glycoprotein: a study of the particles responsible for the resistance of a parasitoid's egg to the defense reactions of its insect host. *Proc. R. Soc. Lond. Ser. B*, **205**, 271–86.

Beroza, M. and Jacobson, M. J. (1963) Chemical insect attractants. *World Rev. Pest Control*, **2**, 36–48.

Bragg, D. (1974) Ecological and behavioral studies of *Phaeogenes cynarae*: Ecology; host specificity; search and oviposition; and avoidance of superparasitism. *Ann. ent. Soc. Am.*, **67**, 931–36.

Caldwell, D. L. and Wilson, M. C. (1975) Studies on *Gelis* sp., a hyperparasite attacking *Bathyplectes curculionis* cocoons. *Env. Ent.*, **4**, 333–6.

Camors, F. B. Jr. and Payne, T. L. (1973) Sequence of arrival of entomophagous insects to trees infected with the southern pine beetle. *Env. Ent.*, **2**, 207–70.

Campbell, B. C. and Duffey, S. S. (1979) Tomatine and parasitic wasps – potential incompatibility of plant antibiosis with biological control. *Science*, **205**, 700–2.

Capinera, J. L. and Lilly, J. H. (1975) *Teterastichus asparagi*, a parasitoid of the asparagus beetle: Some aspects of host–parasitoid interaction. *Ann. ent. Soc. Am.*, **68**, 595–6.

Chacko, M. J. (1969) The phenomenon of superparasitism in *Trichogramma evanescens minutum* Riley. *Beitr. Ent.*, **99**, 617–35.

Corbet, S. A. (1971) Mandibular gland secretion of larvae of the flour moth. *Anagasta kuehniella*, contains an epideictic pheromone and elicits oviposition movement in a hymenopteran parasite. *Nature*, **232**, 481.

Cross, W. W. and Chesnut, T. L. (1971) Arthropod parasites of the boll weevil, *Anthonomus grandis*: 1. An annotated list. *Ann. ent. Soc. Am.*, **64**, 516–27.

Dahlman, D. L. and Vinson, S. B. (1977) Effect of calyx fluid from an insect parasitoid on host hemolymph dry weight and trehalose content. *J. Invert. Path.*, **29**, 227–9.

DeBach, P. (1964) *Biological Control of Insect Pests and Weeds*. Reinhold, New York.

DeLeon, D. (1935) The biology of *Coeloides dendroctoni* Cushman, an important parasite of the mountain pine beetle (*Dendroctonus monticolace*). *Ann. ent. Soc. Am.*, **28**, 411–24.

Doutt, R. L. (1959) The biology of parasitic Hymenoptera. *A. Rev. Ent.*, **4**, 161–82.

Doutt, R. L. (1963) Pathology caused by insect parasites. In: *Insect Pathology, an Advanced Treatise*, Vol. 2 (Steinhaus, E. A., ed.) pp. 393–422. Academic Press, New York.

Edson, K. M. and Vinson, S. B. (1976) The function of the anal vesicle in respiration and excretion in the braconid wasp, *Microplitis croceips*. *J. Insect Physiol.*, **2**, 1037–43.

Edson, K. M., Vinson, S. B. Stoltz, D. B. and Summers, M. D. (1981) Virus in a parasitoid wasp: Suppression of the cellular immune response in the parasitoid's host. *Science*, **211**, 582–3.

Edwards, R. L. (1954) The host-finding and oviposition behavior of *Mormoniella vitripennis* (Walker), a parasite of muscoid flies. *Behaviour*, **7**, 88–112.

Eijackers, H. J. P. and van Lenteren, J. C. (1970) Host choice and host discrimination in *Pseudeucoila bochei*. *Neth. J. Zool.*, **20**, 414.

Fisher, R. C. (1963) Oxygen requirements and the physiological suppression of supernumerary insect parasitoids. *J. exp. Biol.*, **40**, 531–40.

Fisher, R. C. (1971) Aspects of the physiology of endoparasitic Hymenoptera. *Biol. Rev. Camb. Phil. Soc.*, **46**, 243–78.

Fisher, R. C. and Ganesalingam, V. K. (1970) Changes in the composition of host haemolymph after attack by an insect parasitoid. *Nature*, **227**, 191–2.

Flanders, S. E. (1937) Habitat selection by *Trichogramma*. *Ann. ent. Soc. Am.*, **30**, 208–10.

Flanders, S. E. (1959) Differential host relations of the sexes in parasitic Hymenoptera. *Ent. exp. appl.*, **2**, 125–42.

Flanders, S. E. (1971) Multiple parasitism of armored coccids by host-regulative aphenlinids; ectoparasites versus endoparasites. *Can. Ent.*, **103**, 857–72.

Fritz, R. S. (1982) Selection for host behavior modification by parasitoids. *Evolution*, **36**, 283–8.

Führer, E. (1973) Sekretion von Mucopolysacchariden im weiblichen Geschlechtsapparat von *Pimpla turionellae* L. *Z. Parasitenk*, **41**, 207–13.

Greany, P. D., Tumlinson, J. H., Chambers, D. L. and Boush, G. M. (1977) Chemically-mediated host finding by *Biosteres* (*Opius*) *Longicaudatus*, a parasitoid of tephritid fruit fly larvae. *J. Chem. Ecol.*, **3**, 189–95.

Guillot, F. S. and Vinson, S. B. (1972) Sources of substances which elicit a behavioral response from the insect parasitoid, *Campoletis perdistinctus*. *Nature*, **235**, 169–70.

Guillot, F. S., Joiner, R. S. and Vinson, S. B. (1974) Host discrimination of hydro-carbons from the Dufour's gland of a braconid parasitoid. *Ann. ent. Soc. Am.*, **67**, 720–1.

Gurjanova, T. M., Molchanov, M. I. and Kochetova, N. I. (1977) Changes in the hemolymph protein composition under the infestation of the fox-colored sawfly larvae moth with the ectoparasites. *Zool. Zh.*, **56**, 648–50.

Hairston, N. G. (1965) On the mathematical analysis of schistosome populations. *Bull. W.H.O.*, **33**, 45–62.

Hedden, R. L. and Pitman, G. B. (1978) Attack density regulation: A new approach to the use of pheromones in Douglas fir beetle population management. *J. Econ. Ent.*, **71**, 633–7.

Hendry, L. B., Wichmann, J. K., Hindenlang, D. M., Weaver, K. M. and Korzeniawski, S. H. (1976) Plants: Origin of kairomones utilized by parasitoids of phytophagus insects? *J. Chem. Ecol.*, **2**, 271–83.

Henson, R. D., Vinson, S. B. and Barfield, C. S. (1977) Ovipositional behavior of *Bracon mellitor* Say, a parasitoid of the boll weevil. III. Isolation and identification of natural releasers of ovipositor probing. *J. Chem. Ecol.*, **3**, 151–8.

Hirose, Y. (1979) Behavioral ecology of parasitic Hymenoptera. In: *Ecological Aspects on Insect Behavior* Vol. 4 (Ishii, S., Oashima, C., Tateda, H. and Hidaka, T., eds) pp. 105–49. Barfuukan, Tokyo (in Japanese).

Holdaway, F. G. and Smith, H. F. (1932) A relation between size of host puparia and sex ratio of *Alysia manducator* Panzer. *Aust. J. exp. Biol. Med. Sci.*, **10**, 247–59.

Hollingsworth, J. P., Hartstack, A. W. T. and Lingren, P. D. (1970) The spectral response of *Campoletis perdistinctus*. *J. Econ. Ent.*, **63**, 1758–61.

Holmes, J. C. and Bethel, W. M. (1972) Modification of intermediate host behavior by parasites. In: *Behavioral Aspects of Parasite Transmission* (Canning, E. V. and Wright, C. A., eds) pp. 123–49. *Zool. J. Linn. Soc.*, **51**, Suppl. 1.

House, H. L. (1977) Nutrition of natural enemies. In: *Biological Control by Augmentation of Natural Enemies* (Ridgeway, R. L. and Vinson, N. B., eds) pp. 150–82. Plenum Press, New York.

Hubbell, S. P. and Johnson, L. K. (1977) Competition and nest spacing in a tropical stingless bee community. *Ecology*, **58**, 949–63.

Iwantsch, G. F. and Smilowitz, Z. (1975) Relationships between the parasitoid, *Hyposoter exiguae* and *Trichoplusia ni*: prevention of host pupation at the endocrine level. *J. Insect Physiol.*, **21**, 1151–7.

Jackson, D. L. (1959) Observations on the female reproductive organs and the poison apparatus of *Caraphractus cinctus* (Walker) (Hymenoptera: Mymaridae). *Zool. J. Linn. Soc.*, **48**, 59–81.

Johnson, B. (1965) Premature breakdown of the prothoracic glands in parasitized aphids. *Nature*, **206**, 958–9.

Jones, R. L., Lewis, W. J., Beroza, M., Bierl, B. A. and Sparks, A. N. (1973) Host-seeking stimulants (kairomones) for the egg parasite *Trichogramma evanescens*. *Env. Ent.*, **2**, 593–6.

Jones, R. L., Lewis, W. J., Gross, Jr., H. R. and Nordlund, D. A. (1976) Use of kairomones to promote action by beneficial insect parasites. In: *Pest Management With Insect Sex Attractants* (Beroza, M., ed.) pp. 119–34. *Am. Chem. Soc. Symp. Ser.*, **23**.

Kennedy, B. H. (1979) The effect of multilure on parasites of the European elm bark beetle, *Scolytus multistriatus*. *Bull. Ent. Soc. Am.*, **25**, 116–8.

King, P. E. and Rafai, J. (1970) Host-discrimination in a gregarious parasitoid, *Nasonia vitripennis* (Walker). *J. exp. Biol.*, **43**, 245–54.

Krombein, K. V., Hurd, Jr., P. D. and Smith, D. R. (1979) *Catalog of Hymenoptera in America North of Mexico*, Vol. 3. Smithsonian Institute Press, Washington, DC.

Laing, J. (1937) Host-finding by insect parasites. I. Observation on the finding of hosts by *Alysia manducator, Mormoniella vitripennis* and *Trichogramma evanescens*. *J. Anim. Ecol.*, **6**, 298–317.

Law, J. H. and Regnier, F. E. (1971) Pheromones. *A. Rev. Biochem.*, **40**, 533–48.

Lawrence, P. O. (1981) Host vibration: A cue to host location by the parasite, *Biosteres longicaudatus*. *Oecologia*, **48**, 249–51.

Lawrence, P. O. (1982) *Biosteres longicaudatus*: Developmental dependence on host physiology. *Exp. Parasitol.*, **53**, 396–405.

Lewis, W. J. and Jones, R. L. (1971) Substance that stimulates host-seeking by *Microplitis croceipes*, a parasite of *Heliothis* species. *Ann. ent. Soc. Am.*, **64**, 71–3.

Lewis, W. J., Jones, R. L. and Sparks, A. N. (1972) A host-seeking stimulant for the egg parasite *Trichogramma evanescens*; its source and demonstration of its laboratory and field activity. *Ann. ent. Soc. Am.*, **65**, 1087–9.

Lewis, W. J., Jones, R. L., Nordlund, D. A. and Sparks, A. N. (1975a) Kairomones and their use for management of entomophagous insects: I. Evaluation for increasing rate of parasitization by *Trichogramma* spp. in the field. *J. Chem. Ecol.*, **1**, 343–7.

Lewis, W. J., Jones, R. L., Nordlund, D. A. and Gross, Jr., H. R. (1975b) Kairomones and their use for management of entomophagous insects: II. Mechanisms causing increases in rate of parasitization by *Trichogramma* spp. *J. Chem. Ecol.*, **1**, 343–7.

Lewis, W. J., Jones, R. L., Gross, Jr., H. R. and Nordlund, D. A. (1976) The role of kairomones and other behavioral chemicals in host finding by parasitic insects. *Behav. Biol.*, **16**, 267–89.

McLeod, J. M. (1972) A comparison of discrimination of density responses during oviposition by *Exenterus amictorius* and *Exenterus diprionis*, parasites of *Neodiprion swainei*. *Can. Ent.*, **104**, 1–157.

McMullen, L. H. and Atkins, M. D. (1961) Intraspecific competition as a factor in the natural control of the Douglas fir beetle. *Forest Sci.*, **7**, 197–203.

Madden, J. (1968) Behavioral responses of parasites to symbiotic fungus associated with *Sirex noctilio* F. *Nature*, **218**, 189–90.

Malyshev, S. I. (1968) *Genesis of the Hymenoptera and the Phases of Their Evolution* (Richards, O. W. and Uvalrov, B., eds) Translated from Russian by the National Lending Library for Science and Technology. Methuen, London.

Mangold, J. R. (1978) Attraction of *Euphasiopteryx ochracea. Corethrella* sp. and gryllids to broadcast songs of the southern mole cricket. *Florida Ent.*, **61**, 57–61.

Marks, R. J. (1977) Laboratory studies of plant-searching behavior by *Coccinella septempunctata* L. larvae. *Bull. Ent. Res.*, **67**, 235–41.

Matthews, R. W. (1974) Biology of Braconidae. *A. Rev. Ent.*, **19**, 15–32.

Mellini, E. (1975) Studi sui ditteri larvevoridi XXV. Sul determinismo ormonale delle influenze esecitate dagli ospiti sui loro parassiti. *Boll. Int. Ent. vin. Bologna*, **31**, 165–203.

Miles, L. R. and King, E. G. (1975) Development of the tachinid parasite, *Lixophaga* diatraeae, on various developmental stages of the sugarcane borer in the laboratory. *Env. Ent.*, **4**, 811–14.

Monteith, L. G. (1958) Influence of the host and its food plant on host-finding by *Drino bohemica* Msn. (Diptera: Tachinidae) and interaction of other factors. *Proc. X. Int. Cong. Ent.*, **2**, 603–6.

Monteith, L. G. (1960) Influence of plants other than the food plants of their host on host-finding by tachinid parasites. *Can. Ent.*, **92**, 641–52.

Moran, V. C., Brothers, D. J. and Case, J. J. (1969) Observations on the biology of *Tetrastichus flavigaster* Brothers and Moran parasitic on psyllid nymphs. *Tran. R. Ent. Soc. Lond.*, **121**, 41–58.

Muesebeck, C. F. W. and Krombein, K. U. (1951) Hymenoptera of America north of Mexico, Synoptic Catalog. USDA Agr. Monogr. 2.

Nappi, A. J. (1975) In: *Invertebrate Immunity* (Maramorosch, K. and Shope, R. E., eds) pp. 293–326. Academic Press, New York.

Naryanan, E. S. and Subba Rao, B. R. (1955) Studies in insect parasitism I–III: The effect of different hosts on the physiology, on the development, behavior and on the sex ratio of *Microbracon gelechiae* Ashmead. *Beitre. Ent.*, **5**, 36–60.

Neal, Jr., J. W., Bickley, W. E. and Blickenstaff, C. C. (1970) Recovery of the braconid parasite *Microtonus aethiops* from the alfalfa weevil after hormonal treatment. *J. Econ. Ent.*, **63**, 681–2.

Nettles, W. C. and Burks, M. L. (1975) A substance from *Heliothis virescens* larvae stimulating larviposition by females of the tachinid, *Archytas marmoratus. J. Insect Physiol.*, **21**, 965–78.

Nordlund, D. A. (1981) Semiochemicals: a review of the terminology. In: *Semiochemicals Their Role in Pest Control* (Nordlund, D. A., Jones, R. L. and Lewis, W. J., eds) pp. 13–28. John Wiley, New York.

Nordlund, D. A., Jones, R. L. and Lewis, W. J. (1981) *Semiochemicals: Their Role in Pest Control*. John Wiley, New York.

Nozato, K. (1969) The effect of host size on the sex ratio of *Itoplectis cristitae* Momoi, a pupil parasite of the Japanese pine shoot moth, *Petrova* (= *Evetria*) *cristata* (Walsingham). *Kontyu*, **37**, 134–46.

Orphanides, G. M. and Gonzales, D. (1970) Effects of adhesive materials and host location on parasitization by uniparental race of *Trichogramma pretiosum. J. Econ. Ent.*, **63**, 1891–8.

Osman, S. E. and Führer, E. (1979) Histochemical analysis of accessory genital gland secretions in female *Pimpla turionellae* L. (Hymenoptera: Ichneumonidae) *Int. J. Invert. Repro.*, **1**, 323–32.

Peters, T. M. and Barbosa, P. (1977) Influence of population density on size, fecundity, and developmental rate of insects in culture. *A. Rev. Ent.*, **22**, 431–50.

Pieters, E. P. and Sterling, W. L. (1974) Aggregation indices of cotton arthropods in Texas. *Env. Ent.*, **4**, 598–600.

Price, P. W. (1970) Trial odors: recognition by insects parasitic on cocoons. *Science*, **170**, 546–7.

Price, P. W. (1972) Behavior of the parasitoid *Pleolophus basizonus* in response to changes in host and parasitoid density. *Can. Ent.*, **104**, 129–40.

Price, P. W. (1980) *Evolutionary Biology of Parasites*. Princeton University Press, Princeton N.J.

Price, P. W., Bouton, C. E., Gross, P., McPheron, B. A., Thompson, J. N. and Weis, A. E. (1980) Interactions among three trophic levels: influence of plants on interactions between insect herbivores and natural enemies. *A. Rev. Ecol. Syst.*, **11**, 41–65.

Prince, G. J. (1976) Laboratory biology of *Phaenocarpa persimilis* Papp (Braconidae: Alysiinea), a parasitoid of *Drosophila. Aust. J. Zool.*, **24**, 240–64.

Prokopy, R. J. (1981) Epideictic pheromones that influence spacing patterns of phytophagous insects. In: *Semiochemicals Their Role in Pest Control* (Nordlund, D. A., Jones, R. L. and Lewis, W. J., eds) pp. 181–213. John Wiley, New York.

Prokopy, R. J. and Webster, R. P. (1978) Oviposition deterring pheromone of *Rhagoletis pomonella*: A kairomone for its parasitoid *Opius lectus*. *J. Chem. Ecol.*, **4**, 481–94.

Pschorn-Walcher, H. (1971) Experiments on inter-specific competition between three species of tachinids attacking the sugar cane moth borer, *Diatraea saccharalis* (F.). *Entomophaga*, **16**, 125–31.

Read, D. P., Feeny, P. P. and Root, R. B. (1970) Habitat selection by the aphid parasite *Diaretiella rapae*. *Can. Ent.*, **102**, 1567–78.

Reuter, O. M. (1913) *Lebensgewohnheiten und Instinkte der Insekten*. Friedlander, Berlin.

Ritter, F. J. (ed.) (1979) *Chemical Ecology: Odour Communications in Animals*. Elsevier/North Holland Biomedical Press, Amsterdam.

Robacker, D. C., Weaver, K. M. and Hendry, L. B. (1975) Sexual communication and associative learning in the parasitic wasp *Itoplectis conquisitor* (Say). *J. Chem. Ecol.*, **2**, 39–48.

Rogers, D. (1972) Random search and insect population models. *J. Anim. Ecol.*, **41**, 369–83.

Roth, J. P., King, E. G. and Thompson, A. C. (1978) Host location behavior by the tachinid, *Lixophaga diatraeae*. *Env. Ent.*, **7**, 794–8.

Rotheram, S. (1973) The surface of the egg of a parasite insect. II. The ultrastructure of the particulate coat on the egg of *Nemeritis*. *Proc. R. Soc. Lond. Ser. B*, **183**, 195–204.

Rotheray, G. E. (1979) Biology and host-searching behavior of a cynipoid parasite of *Aphidophagous* syrphid larvae. *Ecol. Ent.*, **4**, 75–82.

Rothschild, M. and Schoonhoven, L. M. (1977) Assessment of egg load by *Pieris brassicae*. *Nature*, **266**, 352–55.

Roubaud, E. (1924) Histoire des Anacamptomyies, mouches parasites des g quepes socialis d'Afrique. *Ann. Soc. Nat. Zool.*, **7**, 197–248.

Russ, K. (1969) Beitrage zum Territorialverthalten der Rauped des Springwurmwicklers. *Sparganothis pilleriana* (Lepidoptera: Tortricidae). *Pflanzenschutzberichte*, **40**, 1–9.

Salt, G. (1934) Experimental studies in insect parasitism. II. Superparasitism. *Proc. R. Soc. Lond. Ser. B*, **114**, 455–76.

Salt, G. (1935) Experimental studies in insect parasitism. III. Host selection. *Proc. R. Soc. Lond. Ser. B*, **117**, 413–35.

Salt, G. (1937) Experimental studies in insect parasitism. V. The sense used by *Trichogramma* to distinguish between parasitized and unparasitized hosts. *Proc. R. Soc. Lond. Ser. V*, **122**, 57–75.

Salt, G. (1938) Experimental studies in insect parasitism. VI. Host suitability. *Bull. Ent. Res.*, **29**, 223–46.

Salt, G. (1960) Experimental studies in insect parasitism. XI. The haemocytic reaction of a caterpillar under varied conditions. *Proc. R. Soc. Lond. Ser. B*, **151**, 446–67.

Salt, G. (1963) The defense reactions of insects to metazoan parasites. *Parasitology*, **53**, 527–642.

Salt, G. (1968) The resistance of insect parasitoids to the defense reactions of their hosts. *Biol. Rev. Camb. Phil.*, **43**, 200–32.

Salt, G. (1970a) *The Cellular Defense Reactions of Insects*. Cambridge University Press, London.

Salt, G. (1970b) Experimental studies in insect parasitism. XV. The means of resistance of a parasitoid larvae. *Proc. R. Soc. Lond. Sen. B*, **176**, 105–14.

Sauls, C. E., Nordlund, D. A. and Lewis, W. J. (1979) Kairomones and their use for management of entomophagous insects. VIII. Effect of diet on the kairomone activity of frass from *Heliothis zea* (Boddie) larvae for *Microplitis croceipes* (Cresson). *J. Chem. Ecol.*, **5**, 363–9.

Schmidt, G. T. (1974) Host-acceptance behavior of *Campoletis sonorensis* toward *Heliothis zea*. *Ann. ent. Soc. Am.*, **67**, 835–44.

Schneider, F. (1950) Die Entwicklund des Syrephidenparasiten *Diplazon Fissorius* Grav. in uni-, oligo- und polyvoltinen Wirten und sein Verhalten bei parasitarer Aktivierung der Diapauselarven durch *Diplazon pectoratorius* Grav. *Mitt. Schweiz. Ent. Gesell.*, **23**, 155–94.

Schneider, F. (1951) Einige physiologische Beziehungen zwischen Syrphidenlarven und ihren Parasiten. *Z. Angew. Ent.*, **33**, 150–62.

Schoonhoven, L. M. (1968) Chemosensory basis of host plant selection. *A. Rev. Ent.*, **13**, 115–36.

Simmons, G. A., Leonard, D. E. and Chen, C. W. (1975) Influence of tree species density and composition of parasitism of the spruce budworm, *Choristoneura fumiferana* (Clem.) *Env. Ent.*, **4**, 832–6.

Smith, H. A. (1916) An attempt to redefine the host relationships exhibited by entomophagous insects. *J. Econ. Ent.*, **9**, 477–86.

Smith, H. S. (1941) Status of biological control of scale insects. *Calif. Citrogr.*, **26**, 75–7.

Smith, M. A. and Cornell, H. V. (1979) Hopkins host-selection in *Nasonia vitripennis* and its implications for sympatric speciation. *Anim. Behav.*, **27**, 365–70.

Southwood, T. R. E. (1966) *Ecological Methods*. Methuen, London.

Spradbery, J. P. (1968) A technique for artificially culturing ichneumonid parasites of wood wasps. *Ent. exp. appl.*, **11**, 257–60.

Spradbery, J. P. (1970) Host-finding by *Rhyssa persuasora* (L.) an ichneumonid parasite of sircid woodwasps. *Anim. Behav.*, **18**, 103–14.

Spencer, H. (1926) Biology of parasites and hyperparasites of aphids. *Ann. ent. Soc. Am.*, **19**, 119–57.

Sternlicht, M. (1973) Parasitic wasps attracted by the sex pheromone of their coccid host. *Entomophaga*, **18**, 339–42.

Stoltz, D. B. and Vinson, S. B. (1979) Viruses and parasitism in insects. *Adv. Virus Res.*, **24**, 125–71.

Strand, M. R. and Vinson, S. B. (1982a) Behavioral response of the parasitoid, *Cardiochiles nigriceps* to a kairomone. *Ent. expt. appl.*, **31**, 305–15.

Strand, M. R. and Vinson, S. B. Source (1982b) and characterization of a host-recognition kairomone of *Thelonus heliothidis*: an egg parasitoid of *Heliothis virescens*. *Physiol. Ent.*, **7**, 83–90.

Sullivan, D. J. (1971) Comparative behavior and competition between two aphid hyperparasites: *Alloxysta victrix* and *Asaphes californicus*. *Env. Ent.*, **1**, 234–44.

Thompson, W. R. and Parker, H. L. (1927) The problem of host relations with special reference to entomophagous parasites. *Parasitology*, **19**, 1–34.

Thompson, W. R. and Parker, H. L. (1930) The morphology and biology of *Enlimneria crossifernus*, an important parasite of the European corn borer. *J. Agric. Res.*, **40**, 321–45.

Thorpe, W. H. and Jones, F. G. W. (1937) Olfactory conditioning in a parasite insect and its relation to the problem of host selection. *Proc. R. Ent. Soc. Lond. Ser. B*, **124**, 56–82.

Thurston, R. and Fox, P. M. (1972) Inhibition by nicotine of emergence of *Apanteles congregatus* from its host, the tobacco hornworm. *Ann. ent. Soc. Am.*, **65**, 547–50.

Townes, H. (1969) The genera of Ichneumonidae, Part I. *Mem. Ann. Ent. Inst.*, No. 11.

Tremblay, E. (1966) Ricerche sugli imenotteri parassiti. III. Osservazioni sulla competizione intra-specifica negli Aphidiinae. *Boll. Lab. Ent. Agrar. Portici*, **24**, 209–25.

Tucker, J. E. and Leonard, D. E. (1977) The role of kairomones in host recognition and host acceptance behavior of the parasite *Brachymeria intermedia*. *Env. Ent.*, **6**, 527–31.

Ullyett, G. C. (1947) Mortality factors in populations of *Plutella maculipennis* Curtis, and their relation to the problems of control. *Union S. Africa, Dept. Agric. Ent. Mem.*, **2**, 77–202.

Van Lenteren, J. C. (1976) The development of host discrimination and the prevention of superparasitism in the parasite *Pseudeucoila biochei* (Hym., Cynipidae). *Neth. J. Zool.*, **26**, 1–83.

Van Lenteren, J. C. (1981) Host discrimination by parasitoids. In: *Semiochemicals: Their Role in Pest Control* (Nordlund, D. A., Jones, R. L. and Lewis, W. J., eds) pp. 153–79. John Wiley, New York.

Vey, A. and Gotz, P. (1975) Humoral encapsulation in Diptera (Insecta) comparative studies *in vitro. Parasitology*, **70**, 77–86.

Vinson, S. B. (1968) Source of a substance in *Heliothis virescens* that elicits a searching response in its habitual parasite, *Cardiochiles nigriceps. Ann. ent. Soc. Am.*, **61**, 8–10.

Vinson, S. B. (1970) Development and possible function of teratocytes in the host-parasite association. *J. Invert. Path.*, **16**, 93–101.

Vinson, S. B. (1972) Competition and host discrimination between two species of tobacco budworm parasitoids. *Ann. ent. Soc. Am.*, **65**, 229–36.

Vinson, S. B. (1975a) Biochemical coevolution between parasitoids and their hosts. In: *Evolutionary Strategies of Parasitic Insects and Mites* (Price, P. W., ed.) pp. 14–48. Plenum Press, New York.

Vinson, S. B. (1975b) Source of material in the tobacco budworm involved in host recognition by the egg-larval parasitoid, *Chelonus texanus. Ann. ent. Soc. Am.*, **68**, 381–4.

Vinson, S. B. (1976) Host selection by insect parasitoids. *A. Rev. Ent.*, **21**, 109–33.

Vinson, S. B. (1977a) Behavioral chemicals in the augmentation of natural enemies. In: *Biological Control by Augmentation of Natural Enemies* (Ridgeway, R. L. and Vinson, S. B., eds) pp. 237–79. Plenum Press, New York.

Vinson, S. B. (1977b) Insect host responses against parasitoids and the parasitoid's resistance with emphasis on the Lepidoptera–Hymenoptera association. *Comp. Pathobiol.*, **3**, 103–25.

Vinson, S. B. (1981) Habitat location. In: *Semiochemicals: their Role in Pest Control* (Nordlund, D. A., Jones, R. L. and Lewis, W. J., eds) pp. 51–77. John Wiley, New York.

Vinson, S. B. and Guillot, F. S. (1972) Host-marking: source of a substance that results in host discrimination in insect parasitoids. *Entomophaga*, **17**, 241–5.

Vinson, S. B., Jones, R. L., Sonnet, P., Beirl, B. A. and Beroza, M. (1975) Isolation, identification and synthesis of host-seeking stimulants for *Cardiochiles nigriceps*, a parasitoid of the tobacco budworm. *Ent. exp. appl.*, **18**, 443–50.

Vinson, S. B. and Scott, J. R. (1975) Particles containing DNA associated with the oocyte of an insect parasitoid. *J. Invert. Path.*, **25**, 375–8.

Vinson, S. B., Barfield, C. S. and Henson, R. D. (1977) Ovipositional behavior of *Bracon mellitor* Say, a parasitoid of the boll weevil (*Anthonomus grandis* Boh.). II. Associative learning. *Physiol. Ent.*, **2**, 157–64.

Vinson, S. B., Edson, K. M. and Stoltz, D. B. (1979) Effects of a virus associated with the reproductive system of the parasitoid wasp, *Campoletis sonorensis*, in host weight gain. *J. Invert. Path.*, **34**, 133–7.

Vinson, S. B. and Ables, J. R. (1980) Interspecific competition among endoparasitoids of tobacco budworm larva. *Entomophaga*, **25**, 357–62.

Vinson, S. B. and Iwantsch, G. F. (1980a) Host suitability for insect parasitoids. *A. Rev. Ent.*, **25**, 397–419.

Vinson, S. B. and Iwantsch, G. F. (1980b) Host regulation by insect parasitoids. *Q. Rev. Biol.*, **55**, 143–65.

Waage, J. K. (1979) Foraging for patchily distributed hosts by the parasitoid *Nermeritis canescens* (Grav.). *J. Anim. Ecol.*, **48**, 353–71.

Weseloh, R. M. (1972) Spatial distribution of the Gypsy moth and some of its parasitoids within a forest environment. *Entomophaga*, **17**, 339–51.

Weseloh, R. M. (1977) Behavioral responses of the parasite, *Apanteles melanoscelus*, to Gypsy moth silk. *Env. Ent.*, **5**, 1128–32.

Weseloh, R. M. (1980) Behavioral changes in *Apanteles melanoscelus* females exposed to Gypsy moth silk. *Env. Ent.*, **9**, 345–9.

Weseloh, R. M. (1981) Host location by parasitoids. In: *Semiochemicals: their Role in Pest Control* (Nordlund, D. A., Jones, R. L. and Lewis, W. J., eds) pp. 79–95. John Wiley, New York.

Whitcomb, R. F., Shapiro, M. and Granados, R. R. (1974). In: *The Physiology of Insects*, Vol. 5 (Rockstein, M., ed.) pp. 447–536. Academic Press, New York.

Whittaker, R. N. (1970) *Communities and Ecosystems*. Macmillan, New York.

Wylie, H. G. (1971) Oviposition restraint of *Muscidifurax zaraptor* (Hymenoptera: Pteromalidae) on parasitized housefly pupae. *Can. Ent.*, **103**, 1537–44.

Zwolfer, H. and Kraus, M. (1957) Biocoenotic studies on the parasites of two fir and two oak-tortricids. *Entomophaga*, **2**, 173–96.

Chemical Protection

9

Alarm Pheromones and Sociality in Pre-Social Insects

L. R. Nault and P. L. Phelan

9.1 INTRODUCTION

Chemical alarm signals operate when interacting individuals are aggregated or in close proximity. It is not unexpected, then, that such signals are well-developed in the social insects and relatively uncommon in pre-social species. Outside the social insects, chemical alarm systems are best developed in two homopteran groups: aphids and treehoppers. In this chapter, we discuss alarm pheromones and their social context in these and other pre-social insects and suggest modes of their evolution.

9.2 APHIDS (APHIDIDAE)

9.2.1 Aphid aggregation

Aphids are seldom distributed evenly over their host plants, but are usually found clustered together on leaves and stems. Such clustering is generally a result of parthenogenetically borne young not dispersing from their viviparous mothers. Aggregations of genetically identical individuals may remain clustered for a complete generation as in *Chaitophorus populicola* (Fig. 9.1) or *Brevicoryne brassicae*, whereas in other species such as *Acyrthosiphon pisum* (Fig. 9.2), third-instar nymphs begin to disperse away from the cluster.

Aphid clusters result not only from females and their offspring but certain species aggregate following dispersal flights or disruption from clusters. Aestivating nymphs of *Periphyllus acericola* will reform in compact groups if

Chemical Ecology of Insects. Edited by William J. Bell and Ring T. Cardé
© 1984 Chapman and Hall Ltd.

Fig. 9.1   The poplar aphid, *Chaitophorus populicola*, tightly clustered on its host leaf.

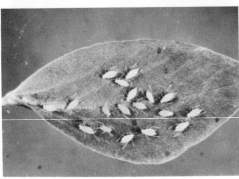

Fig. 9.2   A dispersed colony of mid-instar pea aphids, *Acyrthosiphon pisum*.

dispersed (Hille Ris Lambers, 1947) and alate viviparous *Drepanosiphum platanoides* will alight and aggregate on leaves of its sycamore host (Kennedy and Crawley, 1967). Ibbotson and Kennedy (1951) have shown that crawling *Aphis fabae* will aggregate around aphids or aphid models. Aphid species differ in their propensity to aggregate following mechanical or chemical disruption. This propensity appears to be correlated roughly with the degree of myrmecophily (ant attendance) (Nault *et al.*, 1976). Tight clusterings of myrmecophilous aphids minimize the 'tending territory' of attendant ants, facilitating honeydew collection and protection of aphids from predators (Figs 9.3 and 9.4).

Fig. 9.3   The formicine ant, *Formica subsericea*, soliciting honeydew from *Chaitophorus populicola*.

Fig. 9.4   Ants converge on a coccinellid predator in response to alarm pheromone from *Chaitophorus populicola*.

Maintenance of aggregations is generally regarded as a function of the location of stems and leaf veins on which aphids prefer to settle and feed. Nevertheless, experiments with leaf cages have demonstrated that aggregation in *A. fabae* can take place independently of variations in the leaf surface (Ibbotson and Kennedy, 1951). Even more convincing is the evidence that *A. fabae* will aggregate on a synthetic diet where all nutritional and physical factors are uniform (Klingauf and Sengonca, 1970). Visual, mechanical and olfactory stimuli have all been suggested as being important cues in the aggregation of aphids (Kennedy and Crawley, 1967). *C. populicola* dispersed by alarm pheromone will re-aggregate in the dark, eliminating visual stimuli as an essential factor (Nault and Montgomery, 1977). Critical evidence which separates tactile from chemical factors has also been found. Kay (1976) has shown that *A. fabae* will probe and turn in response to odors from settled, non-feeding aphids hidden behind gauze screens.

It is well known that most aphid species prefer to feed on young succulent growth or senescing leaves, both of which presumably offer aphids optimal nutrients for growth and development. Additionally, group feeding of aphids may improve the nutritional status of the leaf. *A. fabae* or *B. brassicae* reared singly on leaves are smaller and initially less fecund than aphids reared in small clusters (Way and Cammell, 1970; Dixon and Wratten, 1971). In both studies, experimental design eliminated mutual tactile stimuli as a factor, suggesting that altered nutritional status of the leaf was involved. Aphids may alter host tissues and therefore nutrient availability in the immediate vicinity of the cluster. *B. brassicae* can also alter translocation patterns of the whole plant. This aphid characteristically colonizes mature leaves that are normally deficient in phloem nutrients as compared to young or senescent leaves. Way (1973) has shown that ^{14}C-labeled sucrose is diverted from non-infested leaves to leaves infested by *B. brassicae*. These aphid-colonized leaves apparently divert photosynthate and act as powerful 'sinks' for nutrients.

9.2.2 Aphid defensive mechanisms

The gregarious habit of aphids, coupled with their sedentary nature, make them particularly vulnerable to predators and parasites. However, despite their small size and soft bodies, aphids are not entirely defenseless. Certain species, such as *Microlophium evansi*, can see approaching predators and are active enough to avoid capture (Dixon, 1958). If an appendage of this species or of *Drepanosiphum platanoides* is seized by a predator, the aphid will attempt to kick the predator away, or failing this, pull free from the appendage (Dixon, 1958; Dixon and Stewart, 1975). Another conspicuous reaction to predators is 'jerking', a behavior well documented in *Schizaphis graminum* (Boyle and Barrows, 1978). Jerking can prevent attacks from parasitoids if the behavior is initiated as or soon after wasps approach.

For most aphid species, defense is centered around paired, tubular structures arising from the fifth or sixth tergites. These cornicles or 'siphunculi' are found only in the aphids, although not all aphids possess them. Among the Aphidoidea, the primitive root-dwelling or gall-forming Adelgidae and Phylloxeridae lack cornicles. Within the Aphididae various types of cornicles are found. In the Hormaphidinae and Eriosomatinae, cornicles, when present, are no more than rimmed pores. In the Lachninae, the cornicle is a ring affixed to a broad, cone-shaped base. The cornicle is varied in the Drepanosiphinae, ranging from a pore to elongated cylindrical structures. The Chaitophorinae have reticulated, truncated cornicles. In the Aphidinae, the cornicle is usually elongated and always articulated at the base.

The cornicle is a extension of the wall of the tergite with a semilunar pore at the tip. It consists of the cuticle, a cellular hypodermis, and a thin inner, acellular, basal lamina (Chen and Edwards, 1972) or a single layer of flat cells with compressed nuclei (Wynn and Boudreaux, 1972). The cornicle pore is closed from the inside by a valve-like flap and opened inward by a slender retractor muscle (Edwards, 1966; Strong, 1967; Lindsay, 1969; Wynn and Boudreaux, 1972). Expulsion of the cornicle contents is evidently by abdominal turgor pressure, for there is no contractile structure surrounding the base of the cornicle (Edwards, 1966). In species with an articulated cornicle, a large muscle is attached at one end of the cornicle base or just anterior to the base and at the other end to the dorsal body wall. This muscle serves to elevate and rotate the cornicle (Lindsay 1969; Wynn and Boudreaux, 1972).

Contained within the cornicle and cornicle sac are numerous round globules (Strong, 1967). These are specialized, $30-60\,\mu m$ diam. secretory cells, thought to be modified oenocytes (Edwards, 1966; Chen and Edwards, 1972). The cells contain a single large lipid globule which crowds the cytoplasm and nucleus to the cell periphery (Chen and Edwards, 1972). The cornicle secretory cells are bathed in a fluid which differs from the hemolymph and is probably derived from cornicle secretory cells which have ruptured. Additional cells rupture when the cornicle droplet is secreted. This fluid is the cornicle secretion.

The principal non-volatile constituents of the cornicle droplets of *A. pisum* and *Myzus persicae* were reported by Strong (1967) to be triglycerides. In *M. persicae*, the major fatty acid is myristic acid. Callow *et al.* (1973) later examined the cornicle droplets of 28 aphid species and confirmed triglycerides as the major components. Fatty acid composition was found to be independent of the aphid's host plant, stage of development, and taxonomic position. Among the species studied, hexanoic, sorbic, palmitic as well as myristic acid predominated.

Cornicle secretions have a defensive function. When attacked by a predator (Dixon, 1958; Nault *et al.*, 1973; Dixon and Stewart, 1975) or parasitoid (Goff and Nault, 1974; Boyle and Barrows, 1978), an aphid secretes sticky droplets of triglycerides from its cornicles (Figs 9.5 and 9.6). The release of droplets is reflexive (Edwards, 1966) and under nervous control (Strong, 1967). Treatment of aphids with a nerve anesthetic, such as ether, inhibits droplet secretion whereas an anesthetic, such as CO_2, which does not affect the nervous system does not block droplet secretion (Strong, 1967). Early instars of aphids secrete droplets more readily than adults (Callow *et al.*, 1973; Nault and Montgomery, 1977) and certain species secrete droplets more readily than others (Nault *et al.*, 1973). Aphids do not secrete droplets readily when their appendages are subjected

Fig. 9.5 Pea aphid, *Acyrthosiphon pisum*, responds to attack by predatory nabid by secreting cornicle droplets. Note that in an attempt to smear a droplet on the nabid, part of the secretion has been applied to the aphid's antenna.

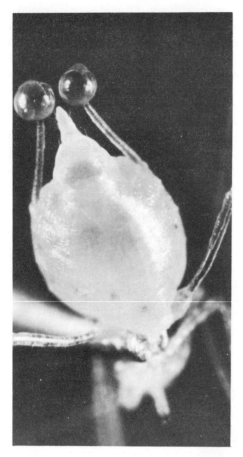

Fig. 9.6 Cornicle droplets
produced by the foxglove aphid,
Aulacorthum solani in response to
mechanical stimulus.

to mechanical stimuli simulating predator attack, i.e., pinched with forceps
(Nault *et al.*, 1973; Dixon and Stewart, 1975), but are very likely to do so when
their head or thorax is pinched with forceps or their abdomen is pierced with a
minuten pin. If daubed on a predator's head, the cornicle secretions can gum the
predator's mouthparts and irritate its sensory organs, resulting in release of the
aphid prey (Dixon, 1958; Dixon and Stewart, 1975). Articulation and elongation of the cornicle may have evolved to assist the aphid in applying droplets
to an attacking predator or parasitoid. Several authors have shown the cornicle
secretions to be ineffective in deterring attack by predators and parasitoids
(Strong, 1967; Lindsay, 1969; Wynn and Boudreaux, 1972; Nault *et al.*, 1973;
Goff and Nault, 1974; Boyle and Barrows, 1978). Dixon and Stewart (1975),
however, stressed the importance of relative predator–prey size in effectiveness
of the system, a point not considered by others. Nevertheless, 'waxing' of the
predator is of uncertain importance, for less than 10% of *Microlophium evansi*
that escaped a coccinellid predator utilized this method (Dixon, 1958). More

studies are needed to assess the defensive importance of waxing. Although triglycerides could serve to thwart predator attack, their most important function may be in dispensing the smaller hydrocarbon, alarm pheromone fraction of the cornicle droplet.

9.2.3 Aphid alarm pheromone

Discovery that crushed aphids cause siblings to react in alarm was first reported by Dahl (1971). Independently, Kislow and Edwards (1972) noted the same behavior and observed that the response was elicited by odors from the aphid's cornicle droplets. Nault *et al.* (1973) called these odors 'alarm pheromones' because aphids were observed to secrete cornicle droplets in response to attack by predators.

Shortly after discovery of aphid alarm pheromones, several groups of investigators (Bowers *et al.*, 1972; Edwards *et al.*, 1973; Wientjens *et al.*, 1973) independently discovered that (E)-β-farnesene (EBF) is the principal alarm pheromone utilized by several aphid species. The pheromone is now known to occur in species from 19 genera in 3 aphid subfamilies (Aphidinae, Chaitophorinae, Callaphidinae), making it one of the most broadly interspecific pheromones known (Nault and Montgomery, 1979; Pickett and Griffiths, 1980). Aphids in the genus *Thereoaphis* (Callaphidinae) utilize the cyclic sesquiterpene, $(-)$-germacrene A, as a pheromone (Bowers *et al.*, 1977b; Nishino *et al.*, 1977). Partial identification of a pheromone from two other callaphidines in the genera *Eucallipterus* and *Calaphis* have revealed a third hydrocarbon that may be β-selinene (Nault and Montgomery, 1979), a breakdown product of germacrene A (Nishino *et al.*, 1977). These compounds, especially EBF and germacrene A, appeared to be well-suited to their role as alarm pheromones as their relatively high volatility results in a more rapid transmission of the alarm message through the air. The high lability of these hydrocarbons provides a rapid decomposition of the molecule so that the alarm message does not persist and the site of release may be re-infested by aphids at a later time.

There is evidence that some aphids use more than one component in their alarm pheromone. Wohlers (1980) noted a higher response by *A. pisum* to pheromone from cornicle droplets than to synthetic EBF, regardless of concentration. This suggests that another component may be involved, although an alternative explanation might be the differences in dispensing the chemicals (cornicle droplets vs. filter paper). In *Megoura viciae*, a mixture of $(-)$-α-pinene, $(-)$-β-pinene, and EBF produces a greater alarm response than any of these compounds alone (Pickett and Griffiths, 1980). *Lipaphis* (= *Hyadaphis*) *erysimi* and *Hyadaphis foeniculi* both secrete a second unidentified component in addition to EBF (Nault and Montgomery, 1979). The second compound is evidently essential for alarm response, since these aphids respond only marginally or not at all to EBF (Nault and Bowers, 1974). The Calaphidinae is a polymorphic group varying greatly in morphology, including the shape and size of their cornicles. These aphids utilize at least three different hydrocarbons as alarm pheromones (Nault and Montgomery, 1979). Further characterization of alarm pheromones from aphids in this and other groups should be useful to systematists in understanding aphid phylogeny.

The receptors for aphid alarm pheromones are located on the antennae (Nault *et al.*, 1973). Shambaugh *et al.* (1978) detailed the ultrastructure of potential alarm pheromone receptors in 17 aphid species. Excision of the terminal (6th) antennal segments of *M. persicae*, that bear the primary sensoria eliminated its response to alarm pheromone. Such excisions reduced but did not eliminate the response of *A. pisum* and *Aulacarthum solani*, both of which bear secondary sensoria on their 3rd antennal segments. Only removal of this segment eliminated response in these species.

Aphid response to alarm pheromone may vary with temperature and humidity. Using a synthetic analogue of EBF (Nishino *et al.*, 1976), Wiener and Capinera (1979) found that aphid response is diminished at both high and low temperatures and humidities. Moreover, Wiener and Capinera (1979) and Wohlers (1981) have shown that the response to alarm pheromone habituates after prolonged exposure or exposure to subthreshold dosages.

Apids respond to alarm pheromones in a variety of ways. The most marked response is by falling (Nault *et al.*, 1973; Phelan *et al.*, 1976; Montgomery and Nault, 1977a, b, 1978; Roitberg and Myers, 1978) or jumping from the host (Bowers *et al.*, 1977a; Nishino *et al.*, 1977). In contrast to falling from the host, walking away from the feeding site is a conservative or perhaps restrained response. Certain aphids will either 'back up,' 'waggle,' or jerk, but not remove their mouthparts from feeding sites (Nault *et al.*, 1973, 1976, Roitberg and Myers, 1978). The function of this jerking behavior, as noted earlier, may be to discourage attacking predators and parasites. In ant-attended species, this behavior may alert ants to the presence of predators, thereby augmenting ant response to aphid alarm pheromone (see later discussion).

Pheromone from cornicle droplets is effective over a range of about 1–3 cm (Nault *et al*, 1973). The proportion of aphids that fall rather than walk decreased with distance of the pheromone source (Montgomery and Nault, 1977a). Aphids, such as *S. graminum*, usually fall in response to pheromone; others, as *M. persicae*, both fall and walk, whereas some, such as *Chaitophorus populicola* only walk or waggle (Nault *et al.*, 1976; Montgomery and Nault, 1977a). Falling aphids walk more rapidly and do not orient as readily to vertical images following exposure to alarm pheromone as compared to aphids dislodged by tactile stimuli (Phelan *et al.*, 1976; Wohlers, 1981). Such behavior causes aphids to disperse to other potentially predator-free hosts. Aphids that walk in response to alarm pheromone will move to other leaves but most remain or return to leaves on which they were originally feeding (Montgomery and Nault, 1977b; Calabrese and Sorensen, 1978). These aphids are not necessarily more vulnerable to attack from predators. Following pheromone exposure they are more sensitive to tactile stimuli (Roitberg and Myers, 1978; Montgomery and Nault, 1978) and better able to avoid capture by walking away from an approaching predator (Montgomery and Nault, 1978). The differential strategies of response to alarm pheromone in which some individuals disperse to other hosts whereas others remain may ensure that some aphids will survive under different situations. When other hosts are nearby and predator effectiveness is high, the optimal strategy is to disperse to other hosts; when hosts are scarce or the ground environment is hostile, the optimal strategy is to remain on the host. The latter point is well illustrated by a biotype of *A. pisum* from a hot, arid region; this aphid rarely responds to alarm pheromone by falling (Roitberg and Myers, 1978).

Variation in response to pheromone not only occurs between aphid species but is present among various morphs and ages of the same species. Young of *A. pisum* (Roitberg and Myers, 1978) and *M. persicae* (Montgomery and Nault, 1978) are less responsive than adults. Dispersing adults, particularly those that fall from their hosts, are more likely to survive and locate their host than are their younger siblings (Montgomery and Nault, 1978). The smaller, less mobile *A. pisum* nymphs, when knocked off a plant, take longer than adults to find a host (Frazer and Gilbert, 1976) and are more susceptible to death by desiccation and heat (Broadbent and Hollings, 1951).

Aphid species forming dense aggregations usually respond only to high dosages of pheromone and their response is to walk (Montgomery and Nault, 1977a) or waggle (Nault *et al.*, 1976) rather than to fall. As noted earlier, these aphids are often myrmecophiles. Previous exposure of myrmecophilous aphids to ants further reduces their innate weak response to alarm pheromone. Maintenance of an intact aphid aggregation may be crucial to success of the mutualistic association between ants and aphids. A weak alarm pheromone response prevents disruption of the aggregation. It would also appear that in these aphids a 'risky' strong response has been replaced with a dependence on ants for protection. The most intriguing aspect of this association is the response of the attendant ant to aphid alarm pheromone: rapid searching and aggression toward intruding aphid predators (Nault *et al.*, 1976). The aphid pheromone is acting as a synomone (Nordlung and Lewis, 1976), evoking a behavioral response in a second species that is adaptively favorable to both the emitter and the receiver.

The evolutionary success of the aphid appears closely linked to the development of chemical alarm signals and the cornicle. The primitive Aphidoidea that lack or have poorly developed cornicles have adopted a strategy of predator avoidance; most live within protective plant galls or are soil dwellers. There is a strong correlation between presence of well-developed cornicles in aphids and their exposed habitat. One of us (L.R.N.) discovered what may represent an evolutionary primitive form of aphid alarm communication. *Hamamelistes spinosus*, a Hormaphidinae that lacks cornicles, releases low levels of alarm chemicals from its punctured body wall. The aphid lives within cupped birch leaves that afford only partial protection from predators. From such an aphid, the cornicle and its related secretory glands could have evolved as a more efficient method of releasing alarm pheromone. Further evolution resulting in elongation and articulation of the cornicle would enhance ability of the aphid to smear the sticky pheromone-bearing droplet on the attacking predator. Since cornicle droplets retain alarm pheromone for several minutes (Nault *et al.*, 1973), a marked predator continues to advertise its presence as it moves about the aphid colony. We favor this explanation over Dixon's (1958) defensive 'waxing' as the driving force in the evolution of the aphid cornicle. Possibly both mechanisms are correct and together they shaped the cornicle of modern aphids.

9.3 TREEHOPPERS (MEMBRACIDAE)

Alarm pheromones represent only one aspect of social organization in the Membracidae. At least 35 species of treehoppers have maternal care of eggs and/or young (Hinton, 1977), most are gregarious as nymphs, and a great many species are myrmocophilous. Unfortunately, descriptions of membracid biology are frequently anecdotal and based on limited observations. In spite of this, it is possible to derive some general patterns of treehopper behavior. Wood (1979) lists four levels of sociality in the Membracidae on the basis of nymphal gregariousness and female parental involvement: (i) solitary nymphs with no maternal care, (ii) gregarious nymphs with no maternal care but ant attendance, (iii) gregarious nymphs with both maternal involvement and ant attendance, and (iv) continuous maternal care until muturation of young. Treehopper alarm pheromones are found in all categories but the first. None of the tree-hopper pheromones have been chemically identified.

As an example of the first category of treehopper sociality, *Sphongophorus* sp. nymphs are found alone or in groups of only two or three (Wood, 1979). Ant attendance of this treehopper is very sporadic, probably due to the small amounts of honeydew produced and little protection afforded by the tree-hoppers. *Hyphinoe camelus* is solitary as an adult and for at least the last three instars (Hinton, 1977). Adults are aposematically colored, whereas nymphs are cryptic. No parental care has been observed in either of these species. Although further study of this group may yet reveal alarm pheromones, their presence would seem to have limited benefit in species with little or no aggregation.

Enchenopa binotata females insert eggs in the woody tissue of their host in late summer and early fall (Wood, 1980). These eggs do not hatch until the following spring and the female parent does not overwinter. The egg mass is covered with a frothy secretion after deposition and contains an ovipositional stimulant that causes other females to lay their eggs in the same location. Clusters containing 100–150 egg masses have been reported on a single branch (Wood, 1979). In as much as the honeydew from a large number of nymphs would provide a greater food supply for ants, the benefit of these aggregations of egg masses most likely involves the 'attraction' and 'maintenance' of attendant ants. With no parental care, discovery of the resulting nymphal aggregations by ants shortly after hatching is critical to the survival of the colony (Wood, 1979), and dispersal of the aggregation occurs in the absence of ants.

Eggs and nymphs of the treehopper *Entylia bactriana* are dependent both on parent females and/or attendant ants for their survival (Wood, 1977) (Fig. 9.7). Parental females remain with their eggs during development, sitting on or near the egg mass and protection continues for at least the first two instars. Brooding females are reluctant to leave the mass even when disturbed. Such maternal care appears to be important for the reduction of predation; egg masses with females were observed to have more than five times as many eggs as those without females present (Wood, 1977). In more than half the egg masses without females, all eggs had been eaten by predators. A female responds to visual contact with predators by decoy behavior and aggressive displays. Positioning herself between the predator and her offspring, she twists laterally, moves back and forward, tilts, and wing-fans. An alarm pheromone has been demonstrated for *E. bactriana* (Nault *et al.*, 1974) and discs treated with crushed nymphs elicit these same aggressive behaviors by the female (Wood, 1977). *E. bactriana* nymphs respond to crushed individuals by moving away from the source of the pheromone; the intensity of this response is reduced by the presence of a brooding female.

A reduction in alarm response is also observed when ants are attending the colony. Ants respond to the treehopper alarm pheromone with directed movements toward the source much as *Formica subsericea* attending aphids responds to aphid alarm pheromone (Nault *et al.*, 1976). When treehopper nymphs respond to the pheromone, ants increase their antennation of nymphs. Dorsal stroking slows movement of nymphs and its effectiveness is dependent on the amount of time spent stroking. Wood (1977) investigated the relative benefit of attendance by parent females vs. ants on survival of nymphs. Survival was always higher in colonies attended by either females or ants than in non-attended colonies. Survival of colonies attended by ants was not different than that of colonies attended by females until about 15 days after egg hatch.

Fig. 9.7   Female treehopper, *Entylia bactriana*, tending newly hatched nymphs.

The drop in survival of maternal-attended nymphs at that time was attributable to desertion by the female at about 10 days. The presence of attending ants caused parental females to remain with the colony significantly longer.

Similar studies of the relative effectiveness of ants and parental females on nymphal survival have been made for *Publilia concava* (McEvoy, 1979) (Fig. 9.8). In this treehopper survival of nymphs is significantly reduced if ants are excluded from the treehopper colony. Although maternal care is provided during part of the nymphal development, when ants are removed, the presence of the female has surprisingly little effect on the survival of offspring. The benefit from female brooding may be indirect. Since early-instar treehoppers produce relatively small quantities of honeydew and eggs produce none, the presence of a female adult appears to be sufficient to attract and maintain attendant ants during the early period of low honeydew production in the colony. Although the presence of an alarm pheromone has been demonstrated in *P. concava* (Nault *et al.*, 1974), no behavioral analysis has been made of response by adults, nymphs, or attending ants.

Parental investment in *Umbonia crassicornis* is at a level higher than any of the treehoppers discussed thus far. The female parent remains with her offspring until maturity without ant mutualism. Nymphs depend on the mother for: (i) brooding eggs, (ii) preparation of feeding sites for early instars by making spiral slits in the bark of the host, (iii) active maintenance of the aggregation, and (iv) providing protection from predators (Wood, 1976). Several factors appear to contribute to egg mortality: parasitoidism, predation, mold, and desiccation. Wood (1976) found about three times as many nymphs hatched from egg masses attended by females as from non-attended egg masses, thus demonstrating the importance of the mother in reducing the impact of these factors. Once the eggs hatch and the nymphs move to the spiral feeding slits, the female maintains the aggregation. She remains below the aggregation and intercepts young that walk down the branch. By tapping their dorsum, she causes her young to stop and return to the aggregation. Response of the female to predators is similar to that described earlier for *E. bactriana*. *U. crassicornis*, however, appears to be more aggressive, making directed movements toward

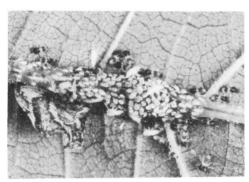

Fig. 9.8 Aggregation of immature *Publia concava* treehoppers.

the source of disturbance while wing-fanning and sometimes making an audible buzzing. These behaviors are sometimes intense enough to knock predators from the plant (Wood, 1976). Aggressive behavior by the parental female is also directed towards filter paper discs treated with crushed nymphs. Unlike most other aggregating treehoppers, *U. crassicornis* nymphs do not respond to alarm pheromone from their siblings.

Parental investment is also high in the Central American treehopper, *Guayaguila compressa*. Females remain with their offspring throughout nymphal development and aggregations are maintained through the teneral adult stage (Wood, 1978). Although the mother broods the eggs, she is very sensitive to disturbances and readily takes flight. After time away from the egg mass, the female returns. Following egg hatch, the female maintains the colony by 'herding'. The female and offspring may relocate several times during a five to six day period (Wood, 1979). Females presented with crushed nymphs respond by aggressive displays followed by escape flight. Nymphs respond by immediate dispersal. It has been suggested that female displays distract potential predators and allow the young time to escape (Wood, 1978). Nymphs re-aggregate after dispersal and the parent female also returns to the colony. Teneral adult aggregations of this species become progressively more sensitive to disturbance; Wood (1978) describes the 'explosive' dispersal of these adult aggregations when physically disturbed. This is presumed to elicit a startle response in potential predators. Re-aggregation of individuals occurs at a later time. Whether this type of dispersal is elicited by an alarm pheromone has not been tested.

An interesting aspect of membracid biology is that in spite of considerable maternal investment, female parents demonstrate a surprising inability to recognize their own eggs or nymphs. In the species studied, *E. bactriana* (Wood, 1977), *Bilimekia broomfieldi* (Hinton, 1977), and *Umbonia ataliba* (Ekkens, 1972), none distinguish egg masses of another individual from their own. In some species this might be explained by the nature of egg mass distribution. If egg masses are scattered sporadically and the probability of contacting the eggs from another female is low, selective pressure for recognizing one's own eggs may not be great. Another explanation may be the reluctance of some parental females to leave their eggs even after extensive disturbances. Hinton (1977) attempted to disrupt a female *B. broomfieldi* brooding her eggs. In one experiment, the pronotal projection of the female was depressed to the substrate 53 times in 6 days without causing the female to leave her eggs. In any case, the mistaken brooding of unrelated eggs would not be expected to be common.

9.4 HEMIPTERA

Levinson *et al.* (1974) report the presence of an alarm pheromone in the bedbug, *Cimex lectularius* (Cimicidae). Although bedbugs are solitary feeders, they feed for only a small part of the day, needing only a single blood meal per molt. Most time is spent in hiding. During this period bedbugs form aggregations from a combination of negative phototaxis, thigmotaxis (Rivnay, 1932), and an 'assembling scent' (Levinson and Bar Ilan, 1971). Bedbug alarm pheromone is produced by both larvae and adults, males and females, and is emitted by the scent glands which are located ventrally on the metathorax of the adult and dorsally on the abdomen of the larva. The glandular contents are 73–92% (*E*)-2-octenal and (*E*)-2-hexenal and either of these compounds is sufficient to bring about the alarm behavioral sequence (Levinson *et al.*, 1974). (*E*)-2-octenal is about 5 times as effective as (*E*)-2-hexenal in eliciting an alarm response and the ratio of the two compounds appears not to be important because it differs between adults and larvae. In addition to acting as an alarm pheromone, these compounds also serve a defensive role, for they can effectively repel Pharoah's ant, *Monomorium pharaonis*, a natural enemy of bedbugs (Levinson *et al.*, 1974). These glandular secretions also make bedbugs distasteful to bat species which are host to *C. lectularus*. This defensive role may have been the primary function of the secretion and the alarm response of conspecifics evolved secondarily.

A parallel situation is reported for another hemipteran, *Dysdercus intermedius* (Pyrrhocoridae) (Calam and Youdeowei, 1968), which forms tight feeding aggregations. Analysis of the posterior scent gland of fifth-instar nymphs revealed eight compounds which account for 99.9% of the total secretion: dodecane, tridecane, pentadecane, hexanal, (*E*)-2-hexenal, 4-keto-2-hexenal, 2-octenal, and 4-keto-2-octenal (Calam and Youdeowei, 1968). This mixture is similar to the defensive secretions of other Hemiptera, with hexanal being widespread in the Coreidae, and hexenal and octenal common in the Pentatomidae (Calam and Youdeowei, 1968). Short-chain saturated and unsaturated aldehydes are very effective chemical irritants (Remold, 1963). But as in *C. lectularius*, the scent gland secretion will cause dispersal of *D. intermedius* aggregations. Either hexanal or (*E*)-2-hexenal alone will also bring about the alarm response.

A dual role for (*E*)-2-hexenal is again found in three species of stinkbugs (Pentatomidae); *Eurydema rugosa*, *E. pulchra* and *Nezara viridula*, although secondary compounds are suspected in the alarm pheromone for these species (Ishiwatari, 1974). *E. rugosa* respond to scent gland secretions either by walking away from the feeding aggregation or falling from the plant. Although all instars respond with about the same frequency (ca. 80%), there is a slight tendency for early instars to fall more often than late instars. Similarly, scent gland secretions of three Alydidae elicit an interspecific alarm response: *Megalotomus quinquespinosus*, *Alydus eurinus* and *A. pilosulus* (Oetting and Yonke, 1978). Alarm behavior in the three alydid species is characterized as either dispersal from the feeding site or for those that remain, an increased sensitivity to minor disturbances. Hexanal, found in nymphal defensive secretions, will produce the alarm response, but again this response is not as strong as that caused by the scent gland fluid.

9.5 CONCLUSIONS

A prerequisite to the evolution of alarm pheromones is the evolution of group living. Evolution of groups has occurred only in specific circumstances because the benefits of such a system were sufficient to overcome the obvious disadvantages (Alexander, 1974). The disadvantages include increased competition for resources, probability of disease transmission, parasitism, and predation. In aphids, the benefits derived from aggregating involve the altering of host physiology, enhancement of mutualism with ants, and possibly use of the group for cover. Hamilton (1971) proposed that it is advantageous for animals to aggregate as a form of cover-seeking in which an individual animal reduces the chance of predation. In this model, Hamilton assumes that predators attack the first prey encountered and that prey can reduce their domain of danger by aggregating, leaving only those individuals at the edge of the 'herd' the most vulnerable. It is noteworthy that Klingauf and Sengonca (1970) observed only aphids at the periphery of *B. brassicae* clusters were attacked by parasitoid wasps. Behaviorists might find aphids to be a model system to provide additional evidence for Hamilton's theory.

In addition to the benefits of modification of host physiology and greater ant mutualism, McEvoy (1979) suggests two additional driving forces for evolving group formation in treehoppers; increased effectiveness of maternal care and enhancement of aposematic coloration in distasteful species. The latter is probably also the cause of grouping in many of the Hemiptera discussed.

The occurrence of alarm signalling and other examples of apparent altruistic behavior has been explained by the following theories:

(i) *Group manipulation* – As an example, Charnov and Krebs (1975) hypothesize that when a member of a flock of birds sees a predator, it may convey alarm by a call that offers few directional cues to the predator (or to members of the flock). The caller then may take advantage of the group response by 'seeking cover on the far side or middle of the flock.'

(ii) *Group benefit* – Selection at the group level or the species level has been used to explain acts of altruism in animal populations (Wynne-Edwards, 1962). However, the necessary conditions of small group size and restricted gene flow are so stringent as to suggest that group selection is of questionable evolutionary consequence (Levin and Kilmer, 1974; Maynard Smith, 1976).

(iii) *Kin benefit* – Because fitness of an individual is dependent on the genetic contribution made to future generations (inclusive fitness), altruistic acts which decrease one's own survival are advantageous only if the recipient of the benefit is a relative and thus shares a high proportion of genes with the altruistic individual (Hamilton, 1964).

(iv) *Reciprocity* – Proposed by Trivers (1971), this model states that altruism may be possible if the cost to the giver has a high probability of being reciprocated in the future by the recipient. This system assumes a long-term association between giver and recipient and requires a system for discriminating against 'cheaters,' i.e., those who will not reciprocate.

The model with greatest support is kin selection (e.g., Sherman, 1977). It is widely held that sociality in Hymenoptera has evolved within the framework of kin selection. The haplo-diploid system of sex determination in this order (Hamilton, 1964, 1972; Trivers and Hare, 1976) results in a high degree of relatedness among sisters, thus making sister–sister altruism more profitable in terms of genetic contribution to future generations*. Kin selection is also the preferred explanation for occurrence of alarm pheromones in aphids. Except for the myrmecophiles, aphids do not emit their alarm pheromone until death is almost imminent, and thus, they derive no direct benefit from its production such as in manipulation or reciprocity. Because summer forms of aphids reproduce by ameiotic parthenogenesis, sisters are identical genetically. Thus, by warning sisters of a predatory danger, aphids increase their own inclusive fitness. The expenditure of energy in the synthesis of an alarm pheromone and the evolution of an elaborate system to dispense it is thus warranted.

The evolution of treehopper alarm pheromones can be explained in a similar manner. Since treehoppers develop from fertilized eggs, the probability of treehopper siblings sharing a given gene, i.e., the coefficient of relatedness (r), is 0.25. The selective advantage of warning siblings is lower, therefore, than in aphids ($r = 1.0$). However, the increase in inclusive fitness for the treehopper may be enough to offset the cost of alarm pheromone production. Since no chemical identification of treehopper pheromones have been made, it is not known if these compounds are specifically manufactured for the alarm system or if they have some other biochemical role. It may be that treehoppers are simply responding to an odor that is released when the treehopper body wall is pierced (Nault *et al.*, 1974). The cost of maintaining such an alarm system would be lower than the aphid system, but possibly not as effective.

In the evolution of hemipteran alarm pheromones, we submit that kin selection probably does not represent the principal driving mechanism. In all cases of Hemiptera discussed, the alarm pheromone also acts as a defensive secretion. Thus, by secreting this material when attacked by a predator, an individual derives direct benefit by attempting to repel the attacker. The alarm response among other members of the aggregation probably evolved secondarily; since the presence of this defensive secretion is associated with a predator, individuals that respond by dispersing would be favored. This explanation does not rule out additional benefit by alerting related individuals in the aggregation. Unfortunately, this benefit is difficult to assess since the degree of relatedness within aggregations has not been thoroughly studied.

The occurrence of alarm pheromones in insects outside the social groups is an infrequent phenomenon, so far found in only a few families of pre-social insect orders. This pattern of distribution suggests that rather unusual conditions are

Not all authors agree on the importance of haplo-diploidy in evolution of sociality in the Hymenoptera. See Evans (1977) for a contrasting view.

necessary for its evolution. The first step in alarm pheromone evolution is the development of group living, which is in itself uncommon in insects. Future investigations of alarm pheromone would be most profitable among those pre-social groups with high coefficients of relatedness due to parthenogenetic reproduction, e.g., (i) Aleyrodoidea and some Coccoidea in the Homoptera, (ii) in the Thysanoptera, and (iii) Scolytidae and Micromalthidae in the Coleoptera (Hartl and Brown, 1970).

REFERENCES

Alexander, R. D. (1974) The evolution of social behavior. *A. Rev. Ecol. Syst.*, **5**, 325–83.

Bowers, W. S., Nault, L. R., Webb, R. E. and Dutky, S. R. (1972) Aphid alarm pheromone: Isolation, identification, synthesis, *Science*, **177**, 1121–2.

Bowers, W. S., Nishino, C., Montgomery, M. E. and Nault, L. R. (1977a) Structure–activity relationships of analogs of the aphid alarm pheromone, (*E*)-β-farnesene. *J. Insect Physiol.*, **23**, 697–701.

Bowers, W. S., Nishino, C., Montgomery, M. E., Nault, L. R. and Nielsen, M. W. (1977b) Sesquiterpene progenitor, Germacrene A: an alarm pheromone in aphids. *Science*, **196**, 680–1.

Boyle, H. and Barrows, E. M. (1978) Oviposition and host feeding behavior of *Aphelinus asychis* (Hymenoptera: Chalcidoidea: Aphelinidae) on *Shizaphis graminum* (Homoptera: Aphididae) and some reactions to aphids to this parasite. *Proc. ent. Soc. Wash.*, **80**, 441–55.

Broadbent, L. and Hollings, M. (1951) The influence of heat on some aphids. *Ann. appl. Biol.*, **38**, 577–81.

Calabrese, E. J. and Sorensen, A. J. (1978) Dispersal and recolonization by *Myzus persicae* following aphid alarm pheromone exposure. *Ann. ent. Soc. Am.*, **71**, 181–2.

Calam, D. H. and Youdeowei, A. (1968) Identification and functions of secretion from the posterior scent gland of fifth-instar larva of the bug, *Dysdercus intermedius*. *J. Insect Physiol.*, **14**, 1147–58.

Callow, R. K., Greenway, A. R. and Griffiths, D. C. (1973) Chemistry of the secretions from the cornicles of various species of aphids. *J. Insect Physiol.*, **19**, 737–48.

Charnov, E. R. and Krebs, J. R. (1975) The evolution of alarm calls; altruism or manipulation? *Am. Nat.*, **109**, 107–12.

Chen, S. and Edwards, J. S. (1972) Observations on the structure of the secretory cells associated with aphid cornicles. *Z. Zellforsch.*, **130**, 312–17.

Dahl, M. L. (1971) Uber einen Schreckstoff bei Aphiden. *Dtsch. Ent. Z*, **18**, 121–8.

Dixon, A. F. G. (1958) Escape responses shown by certain aphids to the presence of the coccinellid *Adalia decempunctata* (L.). *Trans. R. Ent. Soc. Lond.*, **110**, 319–34.

Dixon, A. F. G. and Wratten, S. D. (1971) Laboratory studies on aggregation size and fecundity in the black bean aphid, *Aphis fabae* Scop. *Bull. Ent. Res.*, **61**, 97–111.

Dixon, A. F. G. and Stewart, W. A. (1975) Function of the siphunculi in aphids with particular reference to the sycamore aphid, *Drepanosiphum platanoides*. *J. Zool. Lond.*, **175**, 279–89.

Edwards, J. S. (1966) Defense by smear: Supercooling in the cornicle wax of aphids. *Nature*, **211**, 73–4.

Edwards, L. J., Siddall, J. B., Dunham, L. L., Uden, P. and Kislow, C. J. (1973). Trans-β-farnesene, alarm pheromone of the green peach aphid, *Myzus persicae* (Sulzer). *Nature*, **241**, 126–7.

Ekkens, D. (1972) Peruvian treehopper behavior (Homoptera: Membracidae). *Ent. News*, **83**, 257–71.

Evans, H. E. (1977) Extrinsic versus intrinsic factors in the evolution of insect sociality. *Bioscience*, **27**, 613–17.

Frazer, B. D. and Gilbert, N. (1976) Coccinellids and aphids: A quantitative study of the impact of adult ladybirds (Coleoptera: Coccinellidae) preying on field populations of pea aphids (Homoptera: Aphididae). *J. ent. Soc. Brit. Columbia*, **73**, 33–56.

Goff, A. M. and Nault, L. R. (1974) Aphid cornicle secretions ineffective against attack by parasitoid wasps. *Env. Ent.*, **3**, 565–6.

Hamilton, W. D. (1964) The genetical evolution of social behaviour. *J. Theoret. Biol.*, **7**, 1–52.

Hamilton, W. D. (1971) Geometry for the selfish herd. *J. Theoret. Biol.*, **31**, 295–311.

Hamilton, W. D. (1972) Altruism and related phenomena, mainly in social insects. *An. Rev. Ecol. Syst.*, **3**, 193–232.

Hartl, D. L. and Brown, S. W. (1970) The origin of male haploid genetic systems and their expected sex ratio. *Theoret. Pop. Biol.*, **1**, 165–90.

Hille Ris Lambers, D. (1947) Notes on the genus *Periphyllus* v. d. Hoeven (Hom. Aph.) *Tijdschr. v. Ent.*, **88**, 225–42.

Hinton, H. E. (1977) Subsocial behaviour and biology of some Mexican membracid bugs. *Ecol. Ent.*, **2**, 61–79.

Ibbotson, A. and Kennedy, J. S. (1951) Aggregation in *Aphis fabae* Scop. I. Aggregation on plants. *Ann. appl. Biol.*, **38**, 65–78.

Ishiwatari, T. (1974) Studies on the scent of stink bugs (Hemiptera: Pentatomidae). I. Alarm pheromone activity. *Appl. ent. Zool.*, **9**, 153–8.

Kay, R. H. (1976) Behavioral components of pheromonal aggregation in *Aphis fabae* Scopoli. *Physiol. Ent.*, **1**, 249–54.

Kennedy, J. S. and Crawley, L. (1967) Spaced-out gregariousness in sycamore aphids, *Drepanosiphum platanoides* (Schrank) (Hemiptera: Callaphididae). *J. Anim. Ecol.*, **36**, 147–70.

Kislow, C. J. and Edwards, L. J. (1972) Repellent odour in aphids. *Nature*, **235**, 108–9.

Klingauf, F. and Sengonca, C. (1970) Koloniebildung von Rohrenblattlausen (Aphididae) unter Feindeinwirkung. *Entomophaga*, **15**, 359–77.

Levin, B. R. and Kilmer, W. L. (1974) Interdemic selection and the evolution of altruism: a computer simulation study. *Evolution*, **28**, 527–45.

Levinson, H. Z. and Bar Ilan, A. R. (1971) Assembling and alerting scents produced by the bedbug, *Cimex lectularius* L. *Experientia*, **27**, 102–3.

Levinson, H. Z., Levinson, A. R. and Maschwitz, U. (1974) Action and composition of the alarm pheromone of the bedbug, *Cimex lectularius* L. *Naturwiss*, **12**, 684–5.

Lindsay, K. L. (1969) Cornicles of the pea aphid, *Acyrthosiphon pisum*, their structure and function. A light and electron microscopic study. *Ann. ent. Soc. Am.*, **62**, 1015–21.

McEvoy, P. B. (1979) Advantages and disadvantages to group living in treehoppers (Homoptera: Membracidae). *Misc. Publ. ent. Soc. Am.*, **11**, 1–13.

Maynard Smith, J. (1976) Group selection. *Q. Rev. Biol.*, **51**, 277–83.

Montgomery, M. E. and Nault, L. R. (1977a) Comparative response of aphids to the alarm pheromone, (E)-β-farnesene. *Ent. exp. appl.*, **22**, 236–42.

Montgomery, M. E. and Nault, L. R. (1977b) Aphid alarm pheromones: dispersion of *Hyadaphis erysimi* and *Myzus persicae*. *Ann. ent. Soc. Am.*, **70**, 669–72.

Montgomery, M. E. and Nault, L. R. (1978) Effects of age and wing polymorphism on the sensitivity of *Myzus persicae* to alarm pheromone. *Ann. ent. Soc. Am.*, **71**, 788–90.

Nault, L. R., Edwards, L. J. and Styer, W. E. (1973) Aphid alarm pheromones: Secretion and reception. *Env. Ent.*, **2**, 101–5.

Nault, L. R., Wood, T. K. and Goff, A. M. (1974) Treehopper (Membracidae) alarm pheromones. *Nature*, **149**, 387–8.

Nault, L. R., Montgomery, M. E. and Bowers, W. S. (1976) Ant-aphid association: role of aphid alarm pheromone. *Science*, **192**, 1349–51.

Nault, L. R. and Bowers, W. S. (1974) Multiple alarm pheromones in aphids. *Ent. exp. appl.*, **17**, 455–7.

Nault, L. R. and Montgomery, M. E. (1977) Aphid pheromones. In: *Aphids as Virus Vectors* (Harris, K. and Maramorosch, K., eds) Academic Press, New York.

Nault, L. R. and Montgomery, M. E. (1979) Aphid alarm pheromones. *Misc. Pub. ent. Soc. Am.*, **11**, 23–31.

Nishino, C., Bowers, W. S., Montgomery, M. E. and Nault, L. R. (1976) Aphid alarm pheromone mimics: Sesquiterpene hydrocarbons. *Agric. Biol. Chem.*, **40**, 2303–4.

Nishino, C., Bowers, W. S., Montgomery, M. E., Nault, L. R. and Nielson, M. W. (1977) Alarm pheromone of the spotted alfalfa aphid, *Therioaphis maculata* Buckton (Homoptera: Aphididae). *J. Chem. Ecol.*, **3**, 349–57.

Nordlund, D. A. and Lewis, W. J. (1976) Terminology of chemical releasing stimuli in intraspecific and interspecific interactions. *J. Chem. Ecol.*, **2**, 211–20.

Oetting, R. D. and Yonke, T. R. (1978) Morphology of the scent-gland apparatus of three Alydidae (Hemiptera). *J. Kansas ent. Soc.*, **51**, 294–306.

Phelan, P. L., Montgomery, M. E. and Nault, L. R. (1976) Orientation and locomotion of apterous aphids dislodged from their hosts by alarm pheromone. *Ann. ent. Soc. Am.*, **69**, 1153–6.

Pickett, J. A. and Griffiths, D. C. (1980) Composition of aphid alarm pheromones. *J. Chem. Ecol.*, **6**, 349–60.

Remold, H. (1963) Scent-glands of land-bugs, their physiology and biological function. *Nature*, **198**, 764–8.

Rivnay, E. (1932) Studies in tropisms of the bed bug, *Cimex lectularius* L. *Parasitol.*, **24**, 121–36.

Roitberg, B. D. and Myers, J. H. (1978) Adaptations of alarm pheromone responses to the pea aphid, *Acyrthosiphon pisum*. *Can. J. Zool.*, **56**, 103–8.

Shambaugh, G. F., Frazier, J. L., Castell, A. E. M. and Coons, L. B. (1978) Antennal sensilla of seventeen aphid species (Homptera: Aphidinae). *Int. J. Insect Morphol. Ebryol.*, **7**, 389–404.

Sherman, P. W. (1977) Nepotism and the evolution of alarm calls. *Science*, **197**, 1246–53.

Strong, F. E. (1967) Observations on aphid cornicle secretions. *Ann. ent. Soc. Am.*, **60**, 668–73.

Trivers, R. L. (1971) The evolution of reciprocal altruism. *Q. Rev. Biol.*, **46**, 35–57.

Trivers, R. L. and Hare, H. (1976) Haplo-diploidy and the evolution of the social insects. *Science*, **191**, 249–63.

Way, M. J. (1973) Population structure in aphid colonies. In: *Perspectives in Aphid Biology* (Lowe, A. D., ed.) Bull. No. 2 ent. Soc., New Zealand, Aukland.

Way, M. and Cammell, M. (1970) Aggregation behavior in relation to food utilization by aphids. In: *Animal Populations in Relation to Their Food Resources* (Watson, A., ed.) pp. 229–47. Sym. No. 18, British Ecological Society, London.

Wiener, L. F. and Capinera, J. L. (1979) Greenbug response to an alarm pheromone analog: temperature and humidity effects, disruptive potential and analog releaser efficacy. *Ann. ent. Soc. Am.*, **72**, 369–71.

Wientjens, W. H. J. M., Lakwijk, A. C. and Van Der Marel, T. (1973) Alarm pheromone of grain aphids. *Experientia*, **29**, 658–60.

Wohlers, P. (1980) Die Fluchtreaktionen der Erbenlaus *Acyrthosiphon pisum* Ausgelost durch Alarmpheromon und zusatzliche Reize. *Ent. exp. appl.*, **27**, 156–68.

Wohlers, P. (1981) Effects of the alarm pheromone (E)-β-farnesene on dispersal behavior of the pea aphid *Acyrthosiphon pisum*. *Ent. exp. appl.*, **29**, 117–24.

Wood, T. K. (1976) Alarm behavior of brooding female *Umbonia crassicornis* (Homoptera: Membracidae). *Ann. ent. Soc. Am.*, **69**, 340–44.

Wood, T. K. (1977) Role of parental females and attendant ants in the maturation of the treehopper, *Entylia bactriana* (Homoptera: Membracidae). *Sociobiol*, **2**, 257–72.

Wood, T. K. (1978) Parental care in *Guayaquila compressa* Walker (Homoptera: Membracidae). *Psyche*, **85**, 134–45.

Wood, T. K. (1979) Sociality in the Membracidae (Homoptera). *Misc. Pub. ent. Soc. Am.*, **11**, 15–22.

Wood, T. K. (1980) Divergence in the *Enchenopa binotata* Say complex (Homoptera: Membracidae) affected by host plant adaptation. *Evolution*, **34**, 147–60.

Wynne-Edwards, V. C. (1962) *Animal Dispersion in Relation to Social Behaviour*. Oliver and Boyd, Edinburgh and London.

Wynn, G. G. and Boudreaux, H. B. (1972) Structure and function of the aphid cornicles. *Ann. ent. Soc. Am.*, **65**, 157–66.

10

Warning Coloration and Mimicry*

James E. Huheey

* Dedicated to the memory of Philip M. Sheppard, one of those few zoologists who turned the study of mimicry from 'stamp collecting' into a true science (an idea derived from the philosophy of Lord Kelvin)

10.1 INTRODUCTION

The purpose of this review is to provide a current analysis of warning coloration and mimicry to accompany the other discussions in this volume of the chemical ecology of insects. Recent literature in this area includes a chapter devoted principally to the chemical aspects of mimetic and aposematic associations (Eisner, 1970) and two excellent chapters on mimicry provided by Ford (1975) in his book on ecological genetics. The only recent book on mimicry available in English (Wickler, 1968) provides an introduction to many aspects of the subject but sometimes goes rather far afield. Pasteur (1972) has written a useful little book, but it is not readily available in this country. Rettenmeyer (1970) and Turner (1977) have written thorough reviews of insect mimicry and butterfly mimicry, respectively, and the reader is referred to those articles for the earlier discussions. The present review attempts to place mimicry in the ecological and evolutionary context of the chemical ecology of insects. Most of the examples discussed in this chapter will be from that group, but since the primary purpose is to develop the principles, selective forces, limitations, and results of mimicry, examples from other animals will also be included where they provide relevant insights. The extremely interesting topic of plant mimicry has been completely omitted (Wiens, 1978). The literature is reviewed through the first half of 1983, but some references have reluctantly been omitted because of space considerations.

10.2 WARNING COLORATION AND PREDATOR LEARNING

Warning coloration and mimicry are so straightforward that they make an

Chemical Ecology of Insects. Edited by William J. Bell and Ring T. Cardé
© 1984 Chapman and Hall Ltd.

obviously simple, indeed, too often simplistically presented, example of natural selection and adaptation. However, the *mechanisms* involved in the evolution of mimicry are still a matter of some disagreement and debate. Therefore the reader will be presented with opposing points of view and the conflicting arguments. Such disagreements notwithstanding, the crux of the ecology and evolution of mimicry is *warning coloration* or *aposematism*. Once the principle of warning coloration is accepted, the concept of mimicry follows naturally and inevitably. (Obviously there are other signals that noxious prey can send besides *visual* ones. Other types of warning signals will be discussed in Section 10.6.)

10.2.1 Color vision and pigmentation

The extent of color vision in the various groups of animals continues to be a matter of some controversy. Color vision is well-developed in insects, lizards, turtles, birds, and primates, but apparently in no other mammals. Reports of limited color vision are known for other vertebrate groups, but most workers in the field believe birds to be the primary predators responsible for mimicry with perhaps some importance played by lizards (Sexton *et al.*, 1966; Boyden, 1976), primates (Cott, 1940), and amphibians (L. Brower *et al.*, 1960).

Since the warning colors must be distinguished from background colors that principally derive from the green of chlorophyll during the growing season and the browns typical in the dry season (tropics) and winter (temperate regions), typical aposematic colors are reds, oranges, and yellows (pteridines and carotenoids) with contrasting black (melanin) and occasionally blue. This list is meant to be representative, not exhaustive (see Fox, 1976, Ford, 1975). The pattern is often composed of red and black, yellow and black, etc. in a way that emphasizes the 'signal', exactly opposite from the way that pattern may be used disruptively in crypsis (Cott, 1940). The yellow and black stripes on loading docks as well as most highway signs are good anthropogenic examples of warning coloration. Aposematism is thus self-advertisement, and its success depends upon experienced predators recognizing the signal and avoiding the prey. The conspicuous advertisement often includes behavior that has been described as bold and exhibitionist (Cott, 1940; Fisher, 1958; Sheppard, 1958). The correlation between toxic deterrents, warning coloration, and behavior has often been noted. For example, most recently Bakus (1981) observed that 73% of all exposed coral reef invertebrates were toxic to fish. Only 25% of the cryptic species were toxic.

10.2.2 The chemical nature of the deterrents

Chemical deterrents in insects belong to so many different chemical classes that there is little in the way of chemical similarity. Among the toxins identified have been cardiac glycosides (**1**), histamine (**2**), acetylcholine (**3**), hydrogen cyanide (**4**), aristolochic acid (**5**), alkaloids (**6**) (Rothschild, 1972b), and benzoquinones (**7**) (Eisner *et al.*, 1980).

1

2

3

4

The source of the toxins appears varied. In perhaps the best-studied case, that of the cardiac glycosides in the monarch butterfly, *Danaus plexippus*, the toxin has been well documented as coming from the food chain and this is true for many other species feeding on milkweeds (for further discussion, see Sections 10.4.1–10.4.2). Cochineal insects, *Dactylopius* spp., a pyralid moth, *Laetilia coccidivora*, and the ant *Monomorium destructor* present a fascinating variant on the above. The cochineals protect themselves from ant predation with the well-known red dye, carminic acid (7), which has been shown to be efficacious in deterring ants. The *Laetilia coccidivora* caterpillar feeds exclusively on scale insects and seems unaffected by the carminic acid, but the latter retains its potency in the gut. The caterpillar responds to tactile stimuli with a rapid about-face, bringing the mouth containing a drop of the contents of the crop (including carminic acid) to bear on the point of stimulus. Ants wetted with the fluid immediately break off the attack. *Laetilia* thus provides another example of an organism that has broken through a defensive chemical barrier of a food-stuff and turned it to its own defenses (Eisner *et al.*, 1980).

5

6

7

Little evidence is available for the *de novo* synthesis of toxins in insects. Pasteels and Daloze (1977) have presented evidence for the production of cardiac glycosides in the defensive secretion in chrysomelid beetles. The situation is very difficult to define and involves the detection limits of our chemistry. If cardiac glycosides were to be detected in a butterfly but not in larval food plant, one could not eliminate the possibility that concentration of the toxin by the larva raised the formerly indetectable levels in plants above the experimental detection limits. However, in the one test of cardiac glycosides occurring in milkweeds at low concentrations, they did *not* appear to be concentrated by the host (L. Brower *et al.*, 1982).

10.2.3 Toxicity versus distastefulness

Traditionally, the viewpoint has been that warningly colored prey are *distasteful* to their predators, and that upon eating a distasteful prey the predator remembers the encounter and avoids that prey in the future. In the past two decades there has been an increasing number of workers, led by Rothschild (1961), who have looked askance at simple 'distastefulness' as a sufficient basis upon which warning coloration and mimicry might evolve. However, it must be admitted that there is as yet no unanimity on this question. Some workers claim that an aposematic species *must be toxic* in order to possess an effective deterrent; others state that it *cannot* be toxic if warning coloration is to work. The most widely debated case of this sort is that of the coral snake, *Micrurus fulvius*, with potentially lethal neurotoxic venom and its less toxic (rear-fanged colubrids) and non-toxic mimics (see Grobman, 1978; Greene and McDiarmid, 1981; and references therein), but the same arguments apply equally well to the burnet moth, *Zygaena ephialtes*, which releases the potent toxin hydrogen cyanide (Jones *et al.*, 1962).

The chief argument given by those who doubt the importance of dangerous toxins in warning coloration and mimicry is that, if an encounter with the prey proves fatal to the predator, no learning can take place. However, this is an all-or-nothing argument. *Only if* all *of the interactions are lethal does this argument carry any weight.* Otherwise *those predators that die and do not learn are irrelevant to a learning situation: They might as well never have existed.* This is simply a war of attrition of doubtful value (see also Eisner, 1970; p. 196). Indeed, if the lethality is sufficiently high, there should be selection for a different kind of warning coloration and mimicry, one involving genetic fixation of response to the warning coloration rather than learning. S. Smith (1975, 1977) has suggested that such hereditary information is present with respect to the coral snake in two species of predatory birds. The question is certainly intriguing, but the phenomenon seems to be rare at best since Smith's results appear to be unique, though it must be admitted that little effort has been expended in this direction. This is an area for further work.

To return to the learning model, *it is the presence of learning survivors, few or many, that will contribute to learning and mimicry.* Of course if the prey wins a war of attrition, mimicry will be impossible, but modest losses of predators will have no effect on the survivors learning and avoiding. Another attempt to circumvent the perceived problem of lethality has been the suggestion of empathic learning, that is, learning by one predator observing the distress, trauma, and possibly warning call of a lethally poisoned comrade encountering a warningly colored prey (Gans, 1965; see also Rothschild and Ford (1968) and references therein).

Many workers have explored various aspects of predator learning when presented with mimetic complexes (see Section 10.4.4 for list of experiments and references). Many facets of Batesian mimicry such as degree of resemblance, degree of punishment and frequency of mimics have been investigated successfully by these methods.

Many years ago Mühlmann (1934) showed that experimental birds soon became accustomed to such strong-tasting substances as oil of cloves, mustard, quinine, and magnesium sulfate. Only tartar emetic acting physiologically on the crop and stomach, not just through taste, was an effective deterrent. There are a number of studies wherein administering severely unpleasant stimuli such as X-radiation (Revusky and Bedarf, 1967), injection of lithium chloride solutions (Burghardt *et al.*, 1973; Chen and Amsel, 1980) and ingestion of drugs (Bernstein, 1978) produced strong, induced aversions to co-administered flavors. In many of these cases the discomfort and sickness did not take place until well after the initial sampling yet the two were associated by the predator. In addition, it is likely that the bitter and otherwise unpleasant flavors observed in warningly colored prey are, like the color, merely another signal reminding the predator of the more potent effects of a toxic deterrent.

One can envisage a multiple-line defense employed in aposematism: (i) warning coloration (and behavior, see below), acting at long range to avoid contact; (ii) aversive taste or odors or both, acting upon contact; (iii) toxinosis (usually) acting only upon ingestion (and, often, death). Each of the preliminary signals reduces the likelihood of the ultimate confrontation between predator and prey, detrimental to both. Indeed aposematic prey are often noted for their 'toughness,' the ability to be pecked, grasped, 'tasted,' and identified by a potential predator without suffering serious harm (Rothschild, 1971; Ford, 1975). If the bad taste of a toxic butterfly serves to remind the predator of the untoward effects of the toxin, the taste serves a useful purpose even though it may not be sufficiently noxious to be a deterrent *per se* (L. Brower and Glazier, 1975). Furthermore, Rothschild (1961) has argued cogently for odors as an important cue in protecting aposematic insects at an 'early warning line'. Evidence for the effectiveness of such a multiple-line defense comes from the common observation that experienced predators attack prey more cautiously and slowly (J. Brower, 1958; 1960; Brandon *et al.*, 1979). They would thus be more apt to perceive and respond to the multiple defenses. In fact the presence of ill-tasting toxin preferentially in sites most likely to be tasted, e.g., butterfly wings, would cause the predator to encounter it before consuming the body. This situation has been found in the monarch (L. Brower and Glazier, 1975) and, on the basis of beak-mark data, it has been hypothesized for other butterflies (D. Smith, 1979).

Detailed chemical knowledge of the structure and action of deterrents is often absent, although this is a situation that is rapidly changing. One pattern that is

Fig. 10.1   (a) Naive blue jay eating a monarch butterfly. If the monarch contains cardiac glycosides, emesis (b) soon follows. After drinking water it vomits again (c). The bird recovers, but rejects (d) subsequent monarchs on sight (from Brower, 1969, with permission).

(c)

(d)

emerging as widespread, though far from universal, is the emetic action of many deterrents (recall tartar emetic, above). Thus, cardiac glycosides (Parsons, 1965), iridoid glycosides (Bowers, 1980), aristolochic acid (von Euw *et al.*, 1968), acetylcholine (Rothschild *et al.*, 1972), and *Photinus* firefly deterrents (Sydow and Lloyd, 1975) are strong emetics. Although the toxins are quite capable of being fatal, usually emesis prevents death by halting the absorption of lethal amounts of toxin. Monarch-induced emesis has been studied in blue jays and has been shown to be effective in predator learning (see Fig. 10.1). The advantage of emetics over mere distastefulness was also shown with experiments on artificial models and mimics (Alcock, 1970). Upon emesis the predator is reminded of the taste of the prey and may more readily associate it with the trauma of indigestibility (Eisner, 1970, p. 196; L. Brower, 1969).

10.2.4 Other attributes of aposematic insects

There is a tendency towards longevity on the part of aposematic insects. This may in part be related to a certain 'robustness' akin to the 'toughness' mentioned above – a 'Long-term investment in plant', as it were. This tendency is especially notable for species that are not completely aposematic in all life history phases: The most aposematic phases are extended. Thus the monarch butterfly migrates as far as from the Eastern USA to Mexico and lives as long as 9 months (L. Brower, 1977). In tropical butterflies belonging to the genus *Heliconius*, the extremely long adult life span compared to a short larval stage, where the predation risk is greater, is particularly advantageous because of the aposematism of the adults (Turner, 1971; Gilbert, 1972). To support this long adult life span and reproduction, the adults eat pollen, unique among butterflies (Gilbert, 1972), and the female receives a nutrient contribution from the male on mating (probably in the spermatophore), further strengthening her energy reserves (Boggs and Gilbert, 1979). Furthermore, as Turner (1971) has said, '(this) must result in a butterfly flying in the same population as its parents and even grandparents, which will again favor group selection (actually kin selection) for the development of altruism' (see Section 10.7).

10.2.5 Predation on aposematic butterflies

The presence of a toxic deterrent (or any other defense) does not confer absolute protection on its possessor. The learning process of the predator will incur a certain loss among the prey (see Section 10.5.1). However, occasionally predators seem to specialize in preying heavily on 'protected' butterflies, for example tanagers (*Pipraeidea melanota*) upon ithomiine butterflies (Brown and Neto, 1976) and orioles (*Icterus* spp.) and black-headed grosbeaks (*Pheuticus melanocephalus*) upon monarch butterflies (Calvert *et al.*, 1979; Fink and Brower, 1981). It should not be surprising that specialized predators can breech the toxic defense, just as the butterflies themselves have done.

10.2.6 Autotoxicity and its prevention

To the extent that chemical deterrents in insects are definitely toxic, not just 'distasteful,' an additional stress may be placed upon their bearers in the form of autotoxinosis. The solution to this problem may involve simple isolation – having glands lined with cuticular membrane that prevents the toxicants from seeping into the body cavity (see Eisner, 1970). In addition, the remaining integument is often impermeable to the exudate or spray (these insects usually attacking actively compared to the passive ingestion of butterflies). Alternatively, the poison may be mde *in situ*. For example, the millipede *Apheloria corrugata* manufactures mandelonitrile which is the addition compound of hydrogen cyanide and benzaldehyde. Presumably the biosynthesis of mandelonitrile is made by a route which avoids *free* hydrogen cyanide entirely or requires it only in small, viably tolerable concentrations. However, the enzymatic decomposition of the mandelonitrile:

$$\text{(10.1)}$$

readily releases large quantities of HCN during the short time of the discharge.

Finally, resistance to the toxin may make it possible for the organism to tolerate relatively high concentrations of the toxin. For example, many cyanogenic moths and butterflies are resistant to HCN. Rothschild (1971) discusses at some length the resistance of *Zygaena* to HCN, including those in otherwise overpowering 'killing bottles'.

A grasshopper, *Poekilocerus bufonius*, that feeds on milkweeds is 300 times less sensitive to the cardenolide ouabain than are two other species that do not feed on milkweeds (von Euw *et al.*, 1967). Likewise, the monarch butterfly, *D. plexippus*, has become highly tolerant of the cardiac glycosides found in its foodplants, the various milkweeds, *Asclepias* spp. Similar species feeding on milkweeds include lygaeid bugs and cerambycid beetles (Isman *et al.*, 1977). If a species has overcome a chemical defensive barrier of its host plant, it may then turn loss to gain by using these chemical deterrents against its own predators. However, for those species utilizing a toxic deterrent, *autotoxicity* may yet present a serious and often unappreciated cost. Even though an insect may have developed some tolerance to a toxin, the large amounts sequestered when used as a secondary deterrent may put it at risk in several ways. At the present time there are few data, and most discussions of the problem are conjectural but reasonable.

There is some evidence that the monarch butterfly suffers physiological cost as a result of sequestering cardiac glycosides. For example melanic larvae, in addition to other aberrations have been observed in the monarch (L. Brower *et al.*, 1972; L. Brower and Glazier, 1975; Seiber *et al.*, 1980). Dixon *et al.* (1978)

have suggested that *Danaus* may overcome the inherent adverse physiological cost of cardiac glycosides by their sequestration in non-metabolic storage sites and the use of altered enzyme systems. As is the case with most stresses placed on an organism, adaptation is expected to minimize the cost. In fact, evidence has recently been presented that *Danaus chrysippus* grows faster and attains significantly greater size when fed as a larva on several species of asclepiads containing cardiac glycosides than on one which does not (D. Smith, 1978b), indicating that not only is there no physiological stress from the glycosides, but the reverse. This is puzzling and contrary to what we expect and observe in other species.

There may be an unexpected side benefit arising from possession of toxins in the body. D. Smith (1978a) has found that larvae of *Danaus chrysippus*, an African relative of the monarch, are less susceptible to attack by endoparasitic Diptera if they have fed on cardiac glycoside-containing plants.

Indirect evidence for a toxic load on possessors of sequestered toxins comes from species that do not sequester them. Isman *et al.* (1977) determined the cardiac glycoside content of eight species of insects eating leaves and stems of *Asclepias* spp. and assumed to be aposematic. The cardiac glycoside content varied with several species having little or none. They thus appear to be *Batesian mimics* rather than being truly aposematic. This may involve selection to avoid autotoxicity and remain Batesian mimics, or the evolution of Müllerian mimics into Batesian mimics (see Section 10.7). Another example of evolutionary attempts to escape autotoxic effects is automimicry (see Section 10.4). In any event the impression remains in all of these examples that, though the prey may be more at risk as Batesian mimics than as Müllerian mimics, their overall fitness is improved by no longer being susceptible to autotoxinosis.

The ultimate in avoiding autotoxicity is for the warningly colored prey to forego a toxin altogether. This seems to contradict what was said earlier about *toxins vs. distastefulness*. Indeed, we should probably not find it common, but Gibson (1980) has shown that birds can be taught to avoid models when the only stimulus is complete frustration, i.e., the models 'escape'. This concept may be applicable to certain brightly colored leafhoppers, but it has received little attention.

10.3 MÜLLERIAN MIMICRY

Müllerian mimicry is normally viewed as the convergence arising between two or more warningly colored species, first explained by the German zoologist Fritz Müller (1879), though of course two closely related species could be Müllerian mimics simply through parallel evolution (L. Brower and J. Brower, 1972). Since it is another example of warning coloration, most of the interesting questions have already been asked in Section 10.2. What remains is to explore how two warningly colored species interact ecologically and evolutionarily.

In the traditional view, such convergence or sharing of the warning 'signal' reduces the predation load on each of the Müllerian mimics. This has usually been considered in terms of naive fledglings entering the predator population (but see Section 10.5 for predators forgetting). In either case, all of the mimics should benefit from such mutual 'advertising'. In any event, the fact is that a mutant which looks more like a similarly warningly colored species will be more apt to survive than one that changes in the opposite direction. Because predators of warningly colored prey are encouraged to generalize, flexibility in color-pattern mutations is allowed.

The use of the words 'mimic' and 'mimicry' in this instance is unfortunate since it implies deception. Note that in Müllerian mimicry as defined here the predator is not 'deceived' since it is to its advantage to be able to recognize all deterrent and trauma-inducing species. The 'mimicry' does not have to be close enough to cause misidentification on the part of predator, merely close enough to remind it of a past experience with *similar* prey. The phenomenon might better be termed 'mutual warning coloration'. Thus although there is selection for convergence of pattern onto presumably the most common, 'average' pattern, it need not be rapid nor precise. Indeed, the resemblance among Müllerian mimics is often not close. Unlike Batesian mimicry (see Section 10.4), Müllerian mimicry does not create the selective forces operating to encourage close mimicry with subsequent intraspecific divergence and polymorphisms (see Section 10.4). The existence of a community of sympatric *polymorphic* Müllerian mimics makes no sense since it defeats the original advantage of Müllerian mimicry by increasing the numbers of the signals to be learned, and sympatric polymorphic Müllerian mimics appear to be unknown.

Likewise, parapatric polymorphism (subspecies) should not be expected in Müllerian mimicry. This is not so much because of forces opposing it as from a lack of forces creating it. The existence therefore, of several geographically variable Müllerian rings is puzzling in terms of classical mimicry theory (see Section 10.5).

Much less experimental work has been done on Müllerian mimicry than on Batesian mimicry. Thus feeding experiments have been few. Benson (1972) was able to establish natural selection operating to stabilize the mimetic pattern of *Heliconius erato* in Costa Rica by comparing survival times and wing damage in altered 'non-mimics' versus controls.

10.4 BATESIAN MIMICRY

Batesian mimicry is the familiar phenomenon known simply as 'mimicry' to most people. It is named after the English naturalist Henry W. Bates who proposed it as an explanation of certain resemblances among Amazonian butterflies (Bates, 1862). It is, unlike the situations we have examined thus far, exemplified by the *deception of the predator by prey*. As Rothschild (1972a) so succinctly puts it, a Batesian mimic is a 'sheep in wolf's clothing'. In the traditional view the Batesian mimic benefits at the expense of both its predator and its model (the aposematic and deterrent-bearing species which it resembles). The predator's cost is clear – the loss of palatable food that it would otherwise have eaten. The model's loss is indirect. By confusing the predator, the Batesian mimic jumbles up the clear-cut, mutually beneficial system of warning coloration. By encountering and eating 'aposematic' (actually *pseud*aposematic) mimics, the learning of the predator is impeded, causing a greater predation load on the model than otherwise would be the case. This has been demonstrated both theoretically (Huheey, 1964) and experimentally (Lea and Turner, 1972). Thus, unlike the situation of pure aposematism (including Müllerian mimicry) in which all agents benefit mutually, Batesian mimicry involves three different species, each acting under different selective forces, often at cross-purposes.

Because the true Batesian mimic has no deterrent, not even a weak one, its sole means of surviving predation is to prevent attack through its resemblance to its model. Hence Batesian mimicry is *close mimicry*, so much so that it is the *exceptions* that attract attention. One exception that has received considerable discussion (Ford, 1975; Nur, 1970; Rothschild, 1963) is the frequent tendency for the mimic to be somwhat smaller than its model. It has been suggested that in this way the unprotected Batesian mimic avoids attack because of the tendency of birds to attack the larger of two prey items. D. Smith (1978b) has even suggested that the apparent dwarfing of *Danaus chrysippus* when fed on plants not containing cardiac glycosides is selectively advantageous by making the mimics smaller than their models. These specimens would thus simultaneously be innocuous automimics of their cardiac glycoside-containing neighbors while, as Batesian mimics, they exhibit a smaller size. Although it is obvious that a Batesian mimic gains by such deflection of predation onto its larger model, it is not apparent that the larger model gains any advantage in terms of predation (see Section 10.7). J. E. Lloyd (personal communication) has suggested that there are differing strategies for a species feeding on mixed foodplants: (i) if nurtured on a poisonous foodplant, then maximize growth, energy storage, egg production, and reproductive potential; maximize protection from toxin; or (ii) if nurtured on a non-poisonous foodplant, then reduce maximum size, energy, eggs, and reproduction, but increase fitness by being less apt to be noticed, tasted, and eaten.

The situation differs somewhat in amphibians and reptiles where the size of the adult is not fixed but continues to increase as long as the individual lives. This results in the unusual (in insect mimicry) situation that the mimic may outgrow its model. There have been a number of discussions suggesting the idea of a 'super-mimic', a Batesian mimic that would give a 'super visual stimulus' thus deterring the predator more than the model would. However, it has been concluded that a 'super-mimic' is a poor mimic, and that several mechanisms have developed to circumvent the problem (Huheey and Brandon, 1974). As an example of a 'super-mimic' in insects, one could consider drones as Batesian mimics (automimics) of aposematic worker bees. However, Beal (1918) found drones, but no workers, in the stomachs of swallows and kingbirds. The birds were clearly distinguishing the two on the basis of size, but not in favor of the drones ('super-mimics').

An interesting example of size effects in mimicry is given by the ant *Camponotus planatus* and one of its mimics in Central America (Jackson and Drummond, 1974). The mantid *Mantoida maya* mimics *C. planatus* in its early instars with individuals between 3 and 9 mm resembling ants. Although the largest of these are larger than *C. planatus*, they may be generally protected by the common ponerine ants. The last nymphal instars of *M. maya* (larger than 9 mm) are strikingly different and resemble vespid wasps instead of the ants. The latter are now too small to be effective models.

One further example that not only illustrates the importance of size matching, but also illustrates the semantic problems that beset one trying to make neat watertight compartments into which to place examples of mimicry is illustrated by an excellent color plate by Rothschild (1971, p. 214, facing). The larva of the Alder Moth, *Apatele alni*, clearly resembles shiny, wet, fresh bird droppings. After the third molt it assumes an obviously warningly colored pattern of black and yellow bars – it is now too big to be mistaken for a bird dropping. The resemblance during the early moults is usually considered cryptic coloration, though by resembling something inedible, it would be considered Batesian. Finally, since the distastefulness presumably re-inforcing the warning coloration probably comes from its diet, one could even make a tenuous case for Müllerian mimicry between the feces and the caterpillar, both of limited palatability!

10.4.1 The monarch butterfly, *Danaus plexippus*, and the viceroy butterfly, *Limenitis archippus archippus*

These two butterflies together with closely related species have probably been studied in more different ways and in more detail than any other Batesian systems. Although the viceroy has been considered a Batesian mimic of the monarch for at least a century, there have been repeated denials and re-affirmations of the presence of a deterrent in the monarch. These seemingly contradictory results have been rationalized by the phenomena of a *palatability spectrum* and *automimicry* (see below). The first strong experimental evidence for Batesian mimicry was presented by Jane V. Z. Brower (1958) who, in a series of careful experiments, showed that Florida scrub jays (*Aphelocoma c. coerulescens*) not only learned to avoid monarchs after encountering them, but also transferred this avoidance to specimens of the viceroy butterfly which otherwise proved comparatively palatable to naive birds. The avoidance of the monarch and the protection of the viceroy was confirmed in similar feeding experiments with blue jays (*Cyanocitta c. cristata*) (Platt *et al.*, 1971).

The chemical agents involved in the monarch deterrent are well-studied. Parsons (1965) partially purified the active ingredient and showed that it was digitalis-like. Thereafter, von Euw *et al.* (1967), Reichstein *et al.* (1968), and Rothschild *et al.* (1970) isolated and characterized the cardiac glycosides from the monarch and other insects feeding on milkweeds. Three of the glycosides are calotropin **(1)**, calotoxin, and calactin. These three glycosides, among others, had previously been characterized from asclepiads. These milkweeds are avoided by most herbivores such as cattle, insects, etc., because of the presence of the cardiac glycosides with pharmacological actions similar to digitalis. Thus it appears that the cardiac glycosides originated in the asclepiads as a defense against herbivores, one that *Danaus* (and some other insects) was able to breach by developing increased physiological tolerance. Then it was possible for the monarch butterfly, in turn, to utilize those same toxins derived from its diet as a deterrent coupled with warning coloration (orange and black), later to be mimicked by the viceroy butterfly, *Limenitis a. archippus* (J. Brower, 1958) to form a case of Batesian mimicry.

Further evidence on the relationship between the plant, the toxin, and the butterfly was provided by L. Brower, J. Brower, and Corvino (1967) who raised Monarchs on cabbage and found that this foodplant renders larvae, prepupae, and adults palatable to blue jays. However, monarchs raised on *Asclepias curassavica* caused the birds to vomit after ingestion of less than one adult butterfly. Surprisingly, butterflies fed on a related asclepiad, *Gonolobus rostratus*, proved to be completely palatable, and subsequent investigation showed that the plant did not contain the heart poisons. This result immediately obviated the assumption that all asclepiads and all butterflies feeding on them would have the same cardiac glycosides and in the same amounts. It simultaneously answered long-puzzling questions, but raised new ones in the ecology and evolution of mimicry.

10.4.2 A palatability spectrum

One of the more obvious answers provided was a simple explanation of the seemingly disparate observations of edibility vs. inedibility of the monarch. If certain monarchs used non-toxic asclepiads (cf. *Gonolobus*, above) as food-plants, they could not sequester cardiac glycosides. A survey of five common species of *Asclepias* used as foodstuffs in the eastern USA, showed that some contain substantial amounts of cardiac glycosides. Others contain little or none.* A butterfly raised on one of the former pair will be highly toxic, but if raised on one of the non-cardenolide-containing plants, it will be harmless. We thus find a very interesting circumstance: Two monarch butterflies can have grown to maturity within a meter of one another, one on a cardenolide-rich

---

* The situation is considerably complicated by the fact that the different species, *or individuals or races* within a species, may contain different amounts of, or even chemically different, toxins (Miriam Rothschild, personal communication). These convolutions on the basic model should not be allowed to obscure the fundamental idea – they merely give it exciting added dimensions.

milkweed and the other on a non-cardenolide-containing milkweed, with one being completely potent with regard to cardiac glycoside toxicity, the second not. The first will fit all of the characteristics for warning coloration, the second not. In fact, *the second butterfly is a harmless Batesian mimic of the first, even though both belong to same species.* L. Brower, J. Brower, and Corvino (1967) have termed this phenomenon *automimicry*, though others have suggested that *Browerian mimicry* would be a better term (Pasteur, 1972; Bees, 1977; Rothschild, 1979). Note that all of the antagonisms raised by Batesian mimicry will arise, but now the model and the mimic are conspecific.

If we carry this thought one step further and consider several possible food plants, we might expect great diversity in the concentration and variation of cardiac glycosides in various foodplants of the monarch. Such studies have been done using both blue jays (ED_{50} = mean dose to cause emesis in 50% of the animals tested) and spectrophotometry as estimators of the cardenolides present (L. Brower and Moffitt, 1974; L. Brower and Glazier, 1975). Thus a butterfly may range in noxiousness from the maximum possible down through many levels to being completely innocuous. There will thus be a complete *palatability spectrum* of monarchs if they have a sufficient variety of foodplants from which to feed (L. Brower *et al.*, 1972). A similar palatability spectrum has been demonstrated in the closely related African butterfly, *Danaus chrysippus* (Owen, 1970; L. Brower *et al.*, 1975; 1978). Palatability spectra have also been suggested for some Müllerian rings (Huheey, 1961; 1976).

10.4.3 Frequency of Batesian mimics

The presence of innocuous monarch butterflies flying with toxic ones raises interesting questions for mimicry. From the point of view of the predator automimics are 'perfect' Batesian mimics if we assume that they differ in no way from their 'models' except as a result of toxins in the diet of one and not the other. They therefore make a useful model for 'perfect' Batesian mimicry, that is, where the predator cannot distinguish visually between model and mimic.

If a predator samples a flock of monarch butterflies having a mixed feeding history of the larvae, the chances that it will get a noxious specimen rather than an innocuous one will go as the frequencies, p and q, of those forms. It is obvious that if the innocuous mimic becomes too numerous, the predator will eat an increased number of both models and mimics, with the mimetic situation tending to break down. This is not unique to perfect automimicry. Naturalists have long realized that a Batesian mimic places itself and its model at risk by becoming too numerous. If the predator encounters innocuous mimics too frequently compared to traumata from the models, the learning will be impaired. Unfortunately, this has often been simplistically stated either as: 'If the mimic outnumbers its model the system breaks down,' somehow implying a magic quality to the 50–50% ratio, or 'The mimic must be rare with respect to its model,' which has a certain truth to it – mimics are often rare – and a vague indecisiveness about it that defies testing the hypothesis even semi-quantitatively. More realistically, there will be some function of *p* and *q* that will tend to maximize the number of mimics that can exist while allowing the models to achieve some maximum of their own, perhaps limited by some non-predatory factor such as food supply.

Owen (1971, pp. 131–135) has presented data on several African associations of butterfly models (*Bematistes* spp.) and mimics (*Pseudacraea eurytus*). In most examples the mimic is considerably less frequent than its model, but it outnumbers it in one sample. As might be expected, the relative frequencies vary considerably depending on the nature of the model/mimic pair (relative palatability, closeness of resemblance, sex-limited or not) and on the presence or absence of other species in the mimicry ring.

Sheppard (1959) has also discussed this correlation of mimics to models, as well as the related phenomenon of pattern breakdown in Batesian mimics. If for reasons unrelated to mimicry there is a sudden increase in the number of mimics or a sudden decrease in the number of models, the selective advantage of close mimicry declines, perhaps even becomes disadvantageous and the variability of the mimics increases.

The limitations of numbers have been vexing to some who see a paradox in: (i) the mimetic pattern must confer an advantage upon its possessor, otherwise it wouldn't exist; (ii) the mimic must be 'rare', and a gene with a selective advantage should be common, shouldn't it? The subtleties of 'advantage', 'selection', 'rare/common', etc., will be taken up in Section 10.7, but the following quote from E. B. Ford (1975) may help:

'We here touch upon a subject of difficulty and violent controversy: the significance of selective advantage. Selection will favour the increasing perfection of Batesian mimicry since, other things being equal, those individuals which best resemble their model will contribute most to posterity; but it will also ensure that the quality selected for will cease to be an asset. For any increase in the relative numbers of a mimic compared with its model, whether due to more perfect deception or other cause, can only be carried up to the point of equipoise, that at which the value of the resemblance is balanced by making a species conspicuous when it is in fact, relatively palatable and defenseless.'

Jane Brower (1960) performed an interesting experiment using mealworms (*Tenebrio*) as prey, dipped either in a solution of quinine hydrochloride (models) or distilled water (mimics). The test predators consisted of starlings (*Sturnus vulgaris*). She showed that the number eaten was a function, though nonlinear, of the frequency of models (see Section 10.5.1). Furthermore, with a sufficiently strong deterrent, the models could support even more than their numbers of mimics, though in all probability that is a rare situation.

If the frequency of mimic is thus severely limited, polymorphism in the mimic species is strongly encouraged. Thus, while there are many species like the viceroy that consist of a single Batesian morph, many others consist of a cryptic morph and one or more Batesian morphs. The latter is a well-known phenomenon in Batesian mimicry. Just as all populations tend to expand within the limits imposed by the constraints of predators, disease, and food and habitat availability, Batesian populations expand until checked by the frequency-dependent selection of learning predators (see Section 10.5.1). Individuals in a Batesian population at equilibrium that 'switch' models through an appropriate mutation will be less heavily selected and will have increased fitness *until their population increases to equilibrium numbers*. This has happened repeatedly in some species (see frontispiece, Owen, 1971).

10.4.4 Experimental tests of Batesian mimicry

Many workers have explored various aspects of learning when caged predators are presented with mimetic complexes. Among them are: butterflies (J. Brower, 1958; Boyden, 1976; Platt *et al.*, 1971), bees (L. Brower *et al.*, 1960; J. Brower and L. Brower, 1962; 1965; Huheey, 1980b), fireflies (Sexton *et al.*, 1966), meal worms made distasteful by quinine (J. Brower, 1960; Schuler, 1980) or artificial prey including pastry, colored drinking water, and flavored seeds (see Terhune, 1977, and references therein). Various aspects of Batesian mimicry such as degree of resemblance, degree of punishment and frequency of mimics have been investigated successfully by these methods.

There have been a few field studies testing the selective advantage of Batesian mimicry in the wild. Such experiments are fraught with difficulties: Too few 'mimics' and the statistics are poor; too many and the natural system is perturbed. Nevertheless valuable data on the action of Batesian mimicry have been obtained (L. Brower *et al.*, 1967; Cook *et al.*, 1969; Sternburg *et al.*, 1977).

10.4.5 Non-butterfly Batesian systems

Although the vast majority of the work on Batesian mimicry has been done on butterflies, it is by no means limited to this order. The following summary is meant to be representative, not exhaustive. Perhaps the best-known examples of Batesian mimicry outside of the Lepidoptera are those involving bees and wasps as models and other insects, especially flies, as mimics. The Browers (1960, 1962, 1965) have studied the reactions of toads (*Bufo terrestris*) to honeybees (*Apis mellifera*) and their dronefly mimics (*Eristalis* spp.) as well as to bumblebees (*Bombus americanorum*) and their robberfly mimics (*Mallophora bomboides*).

Heal (1981) has studied sexual dimorphism in *Eristalis arbustorum*. The females mimic several small, dark bees (mainly mining bees) well. The males less specifically mimic yellow and black Hymenoptera. The sexual dimorphism of pattern also extends to behavior as well. Heal raises the question of 'whether there may also be unnoticed behavioural differences between mimetic morphs . . . with loci affecting behaviour associated with supergenes for colour pattern.'

This is an interesting suggestion with regard to the evolution of mimicry (see Section 10.6.3).

In as much as bumblebee (*Bombus*) males are stingless, they are automimics of the stinging females. As such, like all Batesian mimics, one would expect the resemblance to be as close as possible. However, often it is not. Stiles (1979) has suggested that because male bees are subjected to diurnal temperature fluctuations, which the female bees are not, the sexual dimorphism may be a trade-off for the males between best mimicry and best thermoregulation.

An interesting bee/wasp mimic has been described among beetles of the family Buprestidae. The elytra of *Acmaeodera* spp. are marked with yellow and black bands. What apparently makes individuals of *Acmaeodera* unique is that when they *fly*, the elytra are held parallel to the abdomen in a distinctly un-beetle-like pose, but one that resembles the appearance of the model hymenopteran (Silberglied and Eisner, 1969).

Ants are probably involved to a considerable extent in many mimetic associations, but unfortunately they have been studied relatively little (Wickler, 1968). Reiskind (1977) has listed thirteen species of clubionid and salticid ant-mimicking spiders from Barro Colorado Island, Panama. He notes several modifications of the spiders to mimic prominent structures of ants, such as constrictions of cephalothorax and various pigmentations to resemble insect body segmentation; one of the four pairs of legs raised to resemble the antennae (plus three pairs of legs); prominent pedipalps, often flattened to resemble mandibles.

An interesting example of a mimicry complex based on ants has recently been reported (Jackson and Drummond, 1974). The ant *Camponotus planatus* is mimicked by (i) a clubionid spider, *Myrmecotypus fuliginosus*; (ii) a salticid spider, *Sarinda linda*; (iii) a mirid bug, *Barberiella* sp.; and (iv) a mantid, *Mantoida maya*. This is the largest known Batesian mimicry complex based upon an ant model, and this 'fact may be attributed to the abundance of *C. planatus*.

A final, cross-taxa example of Batesian mimicry comes from southern Africa. The model consists of small, black 'oogpister' beetles (*Anthia* spp.; Carabidae) which resemble the *Eleodes* beetles of North America. The latter have beetle mimics that are well known (Eisner, 1970). In the case of the oogpister, the mimic is a juvenile form of a small lizard, *Eremias lugubris*, of the Kalahari desert. The adult lizard is cryptically colored to match the sand, but the juvenile is jet black with some white accent marks resembling the beetle. The gait of the juvenile lizard is thought to enhance the resemblance (Huey and Pianka, 1977). The juvenile switches defense strategy as it becomes an adult since, as it outgrows its model, it cannot remain a Batesian mimic (see Section 10.4).

10.5 PROBLEMS IN BATESIAN AND MÜLLERIAN SYSTEMS

The Batesian and Müllerian forms of mimicry are not the only ones known (see Section 10.6), but they are the best defined and have more quantitative data than some other forms. It is natural, then, that more theoretical work and analysis should have been done in this context. As we have seen, the criteria and results of Batesian and Müllerian mimicry are clear-cut and the two systems often lead to diametrically opposite results. A brief review of these attributes is given in Table 10.1.

Table 10.1   Comparison of attributes of traditional Müllerian and Batesian mimicry

Characteristic or trait	Müllerian mimicry	Batesian mimicry
Close mimicry?	Not necessarily	Yes
Deception involved?	No, merely mutual warning coloration	Yes, if predator can distinguish mimic from the model, the protection of former is lost
Who benefits?	Model, 'mimic', and predator	Mimic, but *neither* model *nor* predator
Microdistributional aspects?	Clumping, to enhance the learning by the predator	Dispersive, on mimic's part, to inhibit repeated contact and learning by predator
Sympatry, geographic?	Yes, but never strict	Yes, and generally quite strict
Sympatry, temporal?	Yes, but not very strict	Yes, but not necessarily as strict as was once thought
Frequency-dependent?	No	Yes; frequency of the mimic is controlled by the frequency of its model
Behavior?	Open, slow movements, conspicuous	Mimic more like cryptic species; inconspicuous
Toughness?	Yes	Models, yes; mimics, no
Will a mild deterrent help?	Yes, though not so much as stronger one	Not for the mimic
Invest in a long life span?	Yes	Models, yes; mimics, no
Evolutionary tendency?	Models converge	Mimics converge on models but models attempt 'to escape' the mimics

10.5.1 Mathematical models of mimetic systems

Because of the advent of data and the desire to predict and to model mimetic systems, the past two decades have seen a number of attempts to construct mathematical models, often utilizing the simulative power of the computer. Thus Huheey (1964, 1976) has presented models for Batesian and Müllerian mimicry, respectively, based originally on the experimental data of J. Brower (1960) with respect to the feeding habits of starlings on quinine-dipped meal-worms, but augmented by data on the reaction of toads and treefrogs to honey-bees, *Apis mellifera* (Huheey, 1980b). Other workers who have made valuable contributions to these models have been L. Brower *et al.* (1970), Pough *et al.* (1973), Estabrook and Jespersen (1974), Bobisud and Potratz (1976), Arnold (1978), and Turner (1978). The differences of viewpoint among these papers are not great and for want of space will not be discussed here – suffice it to say the differences were more in approach, applications, or emphasis than in theory or conclusions.

The basis of the Batesian model is a simplistic assumption about predator behavior: *A predator will cease eating a particular noxious model upon experiencing a traumatic encounter with it*, and thereafter *it will avoid eating that model (and its Batesian mimics) for the next n-1 encounters.* A simple mathematical model can then be derived from this assumption to give (Huheey, 1964):

$$P = \frac{1}{p + qn} \tag{10.2}$$

where P is the percentage of prey eaten, and p and q are the frequencies of mimics and models in the population. This equation satisfactorily handles the feeding habits of starlings on mealworms (J. Brower, 1960), and those of toads (*Bufo*) and treefrogs (*Hyla*) on honeybees (*Apis mellifera*) (Huheey, 1980b). Results are given in Table 10.2. Furthermore, the model reflects the frequency-dependent selection acting on Batesian mimics. The probability of a mimic being eaten decreases as its frequency *decreases* and as the noxious quality (as measured by n) *increases*. Although predation is a nonlinear function of the frequency, p, a simple transformation yields a linear relationship with the memory parameter n (Fig. 10.2).

Table 10.2   Results of feeding experiments with different predators and prey*

Predator	Prey	Value of n
Sturnus vulgaris	*Tenebrio* (quinine)	11.0
Bufo terrestris	*Apis mellifera*	8.0
Hyla cinerea	*Apis mellifera*	28.7

* Data from J. Brower (1960) and Huheey (1980b)

The main difficulties with this mathematical model, aside from the simplistic picture of predator memory, are that (i) time does not enter into the model as a real variable, and (ii) there is no allowance for alternative prey. The latter should be particularly important since the reluctance of the predator to accept 'unpalatable prey' will be a function of its hunger. Holling (1965) has presented a mathematical treatment of mimicry which considers type and quantity of alternative food. Dill (1975) has extended these ideas and provided some experimental support (see also Schuler, 1980).

A number of workers have been intrigued by the value and meaning of n. Most have agreed that long-waiting strategies are inefficient and have opted for an 'eat nothing—eat everything' strategy (Estabrook and Jesperson, 1974) or one where the predator samples the prey population at frequent intervals to ascertain changing model/mimic frequencies (Dill, 1975; Huheey, 1980b).

Arnold (1978) has called attention to the fact that all previous workers had assumed a uniform distribution of models and mimics in space. He was able to show mathematically that if the models clump, there will be reduced predation upon them. Selection will favor values for the predator's mean movements and waiting time (n) such that it will move out of the clump of models and into a region with a higher density of mimics. Individual models should be selected for clumping including traits, such as pheromones, that promote it (Eisner and Kafatos, 1962).

In contrast, it is generally better for Batesian mimics to disperse since a predator encountering and sampling an edible clump of mimics would decimate it. Thus while models may employ pheromones to promote aggregation, this is unexpected among Batesian mimics, though it would be advantageous for them to respond to the model's pheromones, the better to infiltrate a model clump.

10.5.2 Can there be Batesian—Müllerian intermediates?

Traditionally, the concepts of Batesian and Müllerian mimicry have been considered completely distinct. Certainly, this seems true in many of the factors operating and in the predictions that one would make on the basis of Table 10.1. Therefore most students of mimicry have tended to view Batesian and Müllerian mimicry as discrete, isolated compartments. Some have gone so far as to say that 'mimicry' *is* Batesian mimicry, that Müllerian mimicry is merely warning coloration, and the two should not even be considered at the same time. As we shall see, this will not be the only time that semantics raises its head. Still, vexing questions continue to be posed. One is the very problem of the evolution of mimicry which, needs be, raises further questions concerning how species cross over the 'absolute' boundaries of crypsis/aposematism or pseudaposematism/aposematism (see Section 10.7).

Despite the apparently discrete characters of Batesian and Müllerian mimicry, many authors (Fisher, 1958; Huheey, 1961; Linsley *et al.*, 1961; Eisner *et al.*, 1962; L. Brower and J. Brower, 1964; L. Brower *et al.*, 1970; Pough *et al.*, 1973; Huheey, 1976; Brandon *et al.*, 1979; Sbordoni *et al.*, 1979) have discussed, from either an empirical or theoretical viewpoint, apparent intermediate cases between classical Batesian and Müllerian mimicry or the evolution of Batesian mimicry into Müllerian mimicry, or *vice versa*.

An interesting example is provided by certain lycid and cerambycid beetles. The lycids are unpalatable to vertebrates, warningly colored, and mimicked by palatable cerambycids. The situation would thus seem to be simply one of Batesian mimicry, but the cerambycids are carnivorous: We thus have the unusual situation of the Batesian mimic preying upon its model, thereby ingesting the model's toxin, and presumably in turn becoming toxic – a Müllerian mimic!

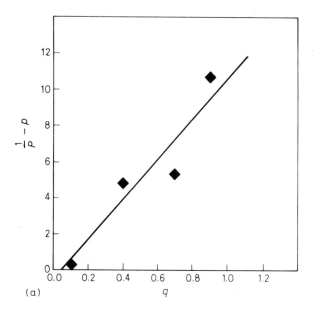

(a)

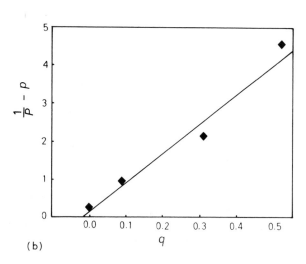

(b)

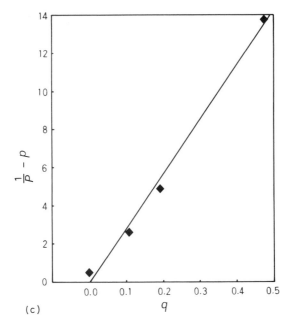

Fig. 10.2 Predation rates as a function of frequency of models (q): (a) starlings on meal worms (quinine); (b) toads on bees; (c) treefrogs on bees (from Huheey, 1964, 1980b, with permission).

Furthermore, since the cerambycids apparently gradually detoxify the material ingested from the lycid prey, whether a particular cerambycid is a Batesian mimic or Müllerian mimic could well depend on how recently it had ingested a lycid (Eisner *et al.*, 1962).

Jane Brower (1958) in her classic studies on monarchs and viceroys observed a fact that was little appreciated at the time: Although the monarchs were obviously unpalatable and the viceroys comparatively palatable (and therefore Batesian mimics), the viceroys themselves were not as palatable as control butterflies – they are weak Müllerian mimics.

Thus, as long as a quarter century ago questions were raised concerning the distinction of Batesian and Müllerian mimicry: '. . . it is not easy to distinguish between Batesian and Müllerian mimicry, because abundance and edibility are relative matters, and a range of intergrades occurs' (Remington and Remington, 1957); '(They) are not mutually exclusive, but merely limiting cases of a presumably complete spectrum of possibilities, with the mimic ranging in undesirability from being completely innocuous on the one hand to being fully as potent as the model' (Huheey, 1961); 'In practice one cannot distinguish between Batesian and Müllerian mimicry because different species will show quite a range of palatability' (Sheppard, 1961); 'from a practical viewpoint Müllerian mimicry is inseparable from Batesian mimicry because the (disagreeable) qualities of the Müllerian mimic are relative' (Pasteur, 1972; quoted by Rothschild, 1979); '. . . the results of Brower (1958) . . . suggest that there is in fact no sharp line between Batesian and Müllerian mimicry. These should be thought of as extremes of a continuum' (Ehrlich and Raven, 1965). Nevertheless, this question has not been examined in detail until recently.

A surprising set of inferences was drawn from close inspection of the Batesian model proposed above: Equation 10.2 is only a special example of a more generalized equation:

$$P = \frac{1}{pm + qn + ro \ldots} \tag{10.3}$$

which is applicable to both Müllerian *and* Batesian mimicry. The frequencies of the species in the Müllerian ring are p, q, r . . . with memory parameters (proportional to deterrents) of m, n, o . . ., etc. From this equation it follows that for all cases in which the deterrents of two species are not equal, e.g., $n \neq m$, the weaker Müllerian model/mimic will actually be a Batesian mimic of the stronger model (Huheey, 1976). It had previously been suggested (Pough *et al.*, 1973) that the weaker of a pair of Müllerian mimics benefits most, but the reciprocal idea that it does so at the expense of the stronger Müllerian mimic seems to be new, so much so that it attracted prompt criticism concerning the fundamental concepts of Batesian and Müllerian mimicry (Benson, 1977; Sheppard and Turner, 1977). However, as we have seen above, there have been numerous workers questioning the absolute distinctness of Batesian and Müllerian mimicry, preferring to view them as useful limiting extremes of a continuous spectrum. That this spectrum may be more heavily populated towards the extremes than otherwise does not detract from the usefulness of the concept of possible intermediate situations. Indeed, many of the difficulties appear to be more semantic than biological (Huheey, 1980c).

Heliconiid butterflies provide interesting examples of Batesian–Müllerian mimicry. For example, *Heliconius erato* and *H. melpomene* are sibling species with many subspecies ranging through tropical Central and South America (see Turner, 1971, Plate II.I; 1976a; 1976b; 1981, Fig. 10.2). The presence of some three dozen subspecies of these two species that are undoubtedly Müllerian mimics (both species are unpalatable, and the local subspecies parallel each other in pattern remarkably well) is unexpected since Müllerian mimicry is, as we have seen above, expected to promote *convergence*. It is *Batesian mimicry* that should result in extensive polymorphism and subspeciation.

Some clue to this problem comes from the discovery (L. Brower *et al.*, 1963; Marsh and Rothschild, 1974) that the species *H. erato* and *melpomene* are not unpalatable to the same degree. Whichever local form has the weaker deterrent will act in some ways like a Batesian mimic on the other. This will tend to cause an evolutionary chase in which the model 'flees' the mimic while the latter 'pursues' (see Section 10.7). Since local populations will differ in the random pattern changes that take place, divergence is to be expected. If, as has been argued forcefully and cogently, changing climates have isolated populations of these butterflies in forest refugia (Turner, 1971, 1976a, 1976b; Brown *et al.*, 1974), then the suggested mechanism will operate even more effectively having selection, not genetic drift alone, altering the gene pool.

A third species, *H. doris* presents further problems for the notion that Batesian and Müllerian mimicry form neat pigeonholes. Although it is unpalatable, L. Brower *et al.* (1963) found that it is the least so of the five species tested. Complementing this weak deterrent, *H. doris* is polymorphic, mimicking two aposematic patterns of other members of the genus, and has a third, apparently non-mimetic, morph. Thus, as a weak Müllerian mimic, *H. doris* has evolved in ways typical of a classical Batesian mimic (Huheey, 1976; 1980c).

Brown and Benson (1974) have described an extensive polymorphism in *H. numata*. They ascribe the unexpected polymorphism in this Müllerian mimic to the spatial and temporal fluctuations of the larger and much more abundant ithomiine model/mimics, providing a further example of 'Batesian' selection in a 'Müllerian' system.

A number of Müllerian assemblages have been described based upon a dominant lycid beetle accompanied by an arctiid moth and sometimes a pyromorphid moth. Edible longicorn beetles and sometimes geometrid moths appear to act as Batesian mimics. 'However, the division of these assemblages into Müllerian and Batesian elements according to traditional schemes should be approached with caution. In each assemblage the dominance of a single species is so strong, that the impression is given that it is the exclusive model, all other forms, Batesian and Müllerian, being subservient' (Linsley *et al.*, 1961). Or, in the terminology of this review, the Müllerian mimics are acting like Batesian mimics.

Zygaena ephialtes is an unpalatable burnet moth that ranges widely over Europe (Turner, 1971). Sbordoni *et al.* (1979) describe a complex of Italian moths wherein *Z. ephialtes*, which in Northern and Central Europe is a member of a red-and-black-patterned Müllerian ring consisting of *Z. filipendulae* and other burnets, becomes a member of the yellow-and-black *Amata phegea* complex. By mimicking the presumably more potent *A. phegea*, *Z. ephialtes* has evolved like a Batesian mimic – classical Müllerian mimicry would have *Z. ephialtes* converge on its fellow burnets in all regions. Other 'Batesian' traits which *Z. ephialtes* possesses are that it is quite rare with regard to its model, and that it emerges *after* its model (see Section 10.7). Rothschild (1979) has examined the conditions under which a Müllerian mimic can be 'captured' by a strong model, leaving a Müllerian ring to become a Batesian mimic (see Section 10.7).

Marsh *et al.* (1977) have investigated the relationship, usually considered Batesian mimicry, between the butterflies *Hypolimnas bolina* (mimic) and *Euploea core* (model). They found that both species stored cardioactive substances, depending upon the food plant, but that the activity was less than the well-known poisonous African butterfly, *Danaus chrysippus*, which in turn is less poisonous than the North American butterfly, *D. plexippus* (L. Brower *et al.*, 1978). Marsh and co-workers point to difficulties in the Batesian/Müllerian categories and suggest that the roles of 'model' and 'mimic' could change, depending upon diet.

Such Batesian-Müllerian mimicry problems are not limited to the butterflies. Dressler (1979) describes *Müllerian rings* of bees (*Eulaema* spp. and *Euplusia* spp.), yet admits, 'Though the mimetic relationships between *Euplusia* and *Eulaema* are Müllerian, in most cases I consider *Eulaema* to be the model, and *Euplusia* the mimic'. He goes on to list other similar species, and the reasons for his conclusions.

10.6 MODES OF MIMETIC RESEMBLANCE

Visual mimicry is the only form that has been discussed thus far. Indeed, the majority of work in mimicry has dealt with color patterns, especially of butterflies. In part this is probably a true reflection of the actual state of things, but it certainly also reflects the fact that man is a visually oriented animal prejudiced to observe visual mimicry more readily than other aspects.

10.6.1 Acoustic or auditory mimicry

A number of chemically protected arthropods emit sounds when disturbed (see Eisner, 1970, p. 208; Rettenmeyer, 1970). Such aposematic and pseudoaposematic auditory signals are certainly important, especially in the Hymenoptera and their mimics. Gaul (1952) found that the vespid wasp *Dolichovespula arenaria* ('yellowjacket') has a dominant frequency of 150 Hz arising from its wingbeat. A visual mimic, *Spilomyia hamifera* (Syrphidae), appears also to be an auditory mimic with a wingbeat generated sound of 147 Hz. Lane and Rothschild (1965) have described a similar example between the stridulations of a burying beetle, *Necrophurus investigator*, and the bumblebees it mimics.

10.6.2 Olfactory or odor mimicry

Rothschild (1961, 1964) lists several cases of odor mimicry, both Müllerian and Batesian. This raises an interesting question: If a Batesian mimic develops both a noxious odor and taste that furthers the resemblance to the model (Batesian mimicry), has it not built the first of multiple-lines-of-defense and thus become an incipient Müllerian mimic?

The millipede *Apheloria corrugata*, as well as other species, produces hydrogen cyanide and benzaldehyde by enzymatic decomposition of the cyanohydrin adduct, mandelonitrile (equation 10.1). One can question: 'Why benzaldehyde?' since all normal aldehydes add hydrogen cyanide reversibly across the $C = O$ double bond. However, to the human nose at least, HCN and benzaldehyde have a similar odor, that of almonds. Could the millipede be extending the threat of the HCN mixture by having it smell more strongly of cyanide than would otherwise be possible in order to remind an experienced predator without undergoing undue harassment?

10.6.3 Behavioral mimicry

This category is not exclusive of the others inasmuch as it involves visual, olfactory, or auditory perceptions by the predator. The emphasis here, however, is upon behavioral patterns that lead to deception of the predator. That a Müllerian and Batesian mimic should adapt in such a way as to resemble its model in as many ways as possible is to be expected, but mimicry has gone beyond such simple one-to-one comparisons.

Ford (1975) describes an extremely interesting case of mimicry in the African butterfly *Hypolimnas dubius*. On the east coast of Africa there are two forms of this species, *mima* and *wahlbergi*. These mimic *Amauris echeria* and *A. albimaculata* (both by *mima*) and *A. niavius* (by *wahlbergi*). Now both *wahlbergi* and its model are 'sun-loving . . . (with) . . . a floating flight rest(s) upon the uppersides of leaves, and emerges from the pupa in the morning'. In contrast, *mima* and its models 'rest(s) in shady places on the underside of leaves, takes shorter and more rapid flights, and emerges from the pupa in the afternoon'. These observations come from naturalists who wrote nearly three-quarters of a century ago, but appear to have been careful observers. If this example proves correct, it would be the first example of complicated behavior being controlled by a 'switch gene', a situation well-known in pattern genetics of polymorphic butterflies (see Ford, 1975; Sheppard, 1959).

It is a well-known fact that fireflies of the genera *Photuris*, *Photinus* and *Pyractomena* (Lampyridae) form mating pairs by means of their bioluminescent signals. Typically, a flying, flashing male, is seen by a female observing from her position on vegetation. The female responds repeatedly with her species-specific signal. The male approaches her, usually alighting on the leaf beside her. In the normal course of events, copulation ensues.

Females of most species of *Photuris*, such as *P. versicolor*, *P. jamaicensis*, and *Photuris* 'B' (an undescribed species) emit signals, in addition to their own mating flashes, that mimic the species-specific signals of females of species of *Photinus*, *Photuris* and *Pyractomena*. Males of these species respond to the signals and land near the female producing them, believing her to be a conspecific, potential mate. The *Photuris* females are carnivorous; they attack and eat the lured males. They have been dubbed *femmes fatales* (Lloyd, 1975, 1981). If the mimetic flashes are insufficient to lure the males completely, the females may simply rise to attack them in the air (Lloyd and Wing, personal communication).

This is clearly a case of *aggressive mimicry*. In a sense, it is closely related to Batesian mimicry in that deception is involved, and that deception acts to the benefit of the originator and the detriment of the receiver. However, most workers would classify it separately inasmuch as the mimic plays an *active* role unlike the *passive* role of the Batesian mimic (Lloyd, 1980; Wickler, 1968).

However the story does not end here. Males of several species of *Photuris* have evolved 'false' signals of their own. They occasionally search for a 'hunting' female *Photuris* with the signal of the hunted prey rather than their 'own' species-specific signal. The significance of this is not completely clear, but most likely the males are using the false signals to find hunting, but possibly sexually receptive (*seduction mimicry*) females (Lloyd, 1980).

Furthermore, male *Photinus macdermotti*, on close (walking) approach to a female *macdermotti* will inject an extra flash into the flash patterns of other conspecific males courting the female. The timing of these injected flashes is like that of flashes that aggressive mimics (*Photuris* females) sometimes use when attracting males as prey. Presumably the 'mistake' is read by the other *macdermotti* males as a *Photuris* female imperfectly mimicking a *Photinus* female, and they are consequently repelled. In this case, the original *macdermotti* that gave the false signal (injection) may acquire unrivaled access to the female he is approaching.

Finally, the *macdermotti* males will be selected to approach cautiously the 'mistake flash' since it may actually come from a *macdermotti* male near a *macdermotti* female, and their reproductive success will be improved. *However*, if *Photuris* again counters in the co-evolutionary process, it can in turn mimic this last signal into a mimetic advantage (Lloyd, 1981). While all the details of these complex signals have not been worked out and there may well be variants, it is obvious that the entire mimetic complex is of considerable interest.

Thus male mate-rivals mimic their predators and predators mimic mate-rivals, each to its own benefit. The current *reciprocal mimicry* situation seems to be the result of a long co-evolutionary arms race involving mate-rivals and the predators, begun by mate-rivals mimicking and making use of flash errors predators made when attracting them (see Fig. 10.3). At the moment there are no alternative explanations for the behavior shown by these species, and field experiments are now being made to test the reciprocal mimicry explanation (Lloyd, personal communication).

Robberflies (Asilidae) are often called 'bee-killers' from their heavy predation on honeybees, *Apis mellifera*. Because of a resemblance between the two, some have suggested that aggressive mimicry is involved, allowing the robber fly to approach its victim more closely. However, *A. mellifera* is an exotic species. Linsley (1960) studied native bees and wasps as possible models in the American south-west. He concluded that 'the mimicry seems to be of the Batesian type' since the models seemed to be out of the preferred size range and may appear but rarely in the mimic's prey. L. Brower *et al.* (1960) came to essentially the same conclusions with regard to their work with bumblebees (*Bombus americanorum*) and robberflies (*Mallophora bomboides*) in Florida.

Most descriptions of aggressive mimicry are somewhat anecdotal since the

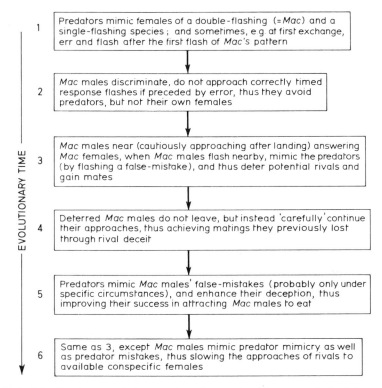

EVOLUTIONARY TIME

1 | Predators mimic females of a double-flashing (=*Mac*) and a single-flashing species; and sometimes, e.g. at first exchange, err and flash after the first flash of *Mac*'s pattern

2 | *Mac* males discriminate, do not approach correctly timed response flashes if preceded by error, thus they avoid predators, but not their own females

3 | *Mac* males near (cautiously approaching after landing) answering *Mac* females, when *Mac* males flash nearby, mimic the predators (by flashing a false-mistake), and thus deter potential rivals and gain mates

4 | Deterred *Mac* males do not leave, but instead 'carefully' continue their approaches, thus achieving matings they previously lost through rival deceit

5 | Predators mimic *Mac* males' false-mistakes (probably only under specific circumstances), and enhance their deception, thus improving their success in attracting *Mac* males to eat

6 | Same as 3, except *Mac* males mimic predator mimicry as well as predator mistakes, thus slowing the approaches of rivals to available conspecific females

Fig. 10.3 Model of an evolutionary 'arms' race (conflict co-evolution) leading to reciprocal mimicry of *Photinus macdermotti* males and their predators, *Photuris* females. A number of variants and origins are possible (from Lloyd, 1981, with permission).

phenomenon is not nearly so well studied as Batesian and Mullerian mimicry. Each one is, therefore, all the more interesting since it provides us with new facets on the phenomenon of mimicry. Consider the scorpionfly *Hylobittacus apicalis*. Males present a prey arthropod to the female during courtship and copulation. However, some males mimic the behavioral attitude and movements of a female, inducing the male possessing the nuptial gift to offer it to them. The mimicking males reduce their time hunting nuptial prey and probably incur reduced predation as well (Thornhill, 1979).

10.6.4 Aggressive chemical mimicry

Bolas spiders, *Mastophora* spp., capture prey by swinging a sticky ball on the end of a thread at passing insects. Once struck, the prey adheres to the ball whereupon the spider descends the connecting strand and feeds on the prey. The success of such a short-range means of attack, requires that a substantial number of prey passes within the range of the spider's attack. Eberhard (1977) has presented convincing evidence that the spider emits a volatile substance that mimics the female sex attractant pheromone of the Fall Armyworm, *Spodoptera frugiperda*. The evidence, while indirect, is strong. Of almost two hundred prey items either captured by the spider or netted downwind as they hovered near the spider, all were male noctuid moths, most *S. frugiperda* but some *Leucania* sp. The chemistry of this aggressive mimicry has not as yet been elucidated.

There are further examples of similar chemical mimicry such as in termitophiles (Howard *et al.*, 1980) and myrmecophilous beetles (Vander Meer and Wojcik, 1982), but at this point Müllerian chemical mimicry merges imperceptibly with a general study of interspecific pheromones and is out of place in this review.

10.7 Evolutionary aspects of mimetic systems

Before delving into the evolution of various kinds of mimicry, we should explore a facet of mimicry barely touched upon above: sympatry, or the idea that all models, mimics, and predators must co-exist (at least figuratively within the memory span of the predator) if successful learning is to be feasible. Most often attention has been focused on *geographic sympatry*. This subject has been examined extensively by Sheppard (1959), Ford (1975), and Turner (1975) and will not be discussed here beyond calling attention to two vertebrate examples of accurate sympatry: coral snakes (Greene and McDiarmid, 1981) and salamanders (Huheey and Brandon, 1974).

Less extensively studied is the subject of *temporal sympatry* or a synchony of activity on the part of models and mimics. There are a number of studies that show that a Batesian mimic is most protected if it flies under the protection of its model (Rothschild, 1963; 1971; Waldbauer and Sheldon, 1971; Marsh and Rothschild, 1974; Bobisud, 1978; Sbordoni *et al.*, 1979). Shapiro (1965) has described a case of temporal dimorphism in *Antepione thiosaria*. The spring generation appears to be cryptic and to be patterned like related members of its tribe. The summer generation is much more brightly colored and appears to mimic *Xanthotype* spp. with which it flies. Its flight time coincides with the second half of the flight time of *X. sospeta* and the first half of the second brood of *X. urticaria*. The non-mimetic spring brood of *A. thiosaria* does not overlap with any *Xanthotype* at all.

Waldbauer and Sheldon (1971) found that mimetic dipterans emerged in spring *before* their hymenopteran models. At first this might appear to contradict the other examples cited, but it was suggested that the mimics were taking advantage of the education of the previous year's adults and escaping the massive predation that would accompany the current year's fledgling cohort later.

It is easy to see that if a Batesian mimic adjusts its emergence (in whatever way) to encounter only predators previously educated by the model, its survival will be maximized, and selection will move in that direction. It is not so simple to see why the model will remain in the vanguard. By delaying its appearance until it flies with the mimic, the latter would share the load, or by emerging *after* the mimic it could allow the unprotected mimic to suffer severe predation, and it could then emerge at a time when there were fewer mimics to confuse the learning process. It should be noted that as long as one assumes some variant of a 'predator-forgetting model' (see Section 10.5), the presence of a small number of models does not appreciably alter the rate of predation. To achieve an advantage for the model by emerging early, one must assume that the predator, encountering a certain number of models without mimics, can be 'programmed' to accept neither models nor mimics. Under these conditions it behooves the model to get out early and educate the predator before the noise of a mixed model/mimic system makes it impossible (Huheey, 1980a).

The evolution of mimicry begins with the evolution of a noxious deterrent and warning coloration. Here a problem that has received recurrent attention, perhaps out of proportion to its importance, namely, 'How can warning coloration evolve, if the new mutant is destroyed (eaten) in educating the predator?' Too often discussions of this type have been of the sort attempting to prove that the evolution of warning coloration and mimicry is 'impossible', usually in the style of McAtee's invectives against protective coloration. This leads to the often discussed and highly controversial subjects of kin selection and group selection, topics far beyond the scope of this review, especially since different authors use these terms in different ways (see Johnson, 1976; Mayr, 1970; Ricklefs, 1979; Turner, 1971; Wilson, 1975). It must be noted, however, that there is no evidence that selection in mimetic situations is anything other than selection on *individuals*.

10.7.1 Müllerian evolution

The idea of the evolution of Müllerian mimicry has always appeared simple: Two or more warningly colored species converge toward a common pattern. There has been a consensus that the species with the strongest deterrent (and probably most abundant) will act as a model for the convergence of the other species (Brown *et al.*, 1974; Huheey, 1980c; Turner, 1977, 1981). Thus, if we find a race of *Heliconius erato* with a pattern deviating from the *melpomene–erato* complex (Turner, 1975) but similar to that of sympatric *H. cydno*, we may conclude that the former has converged on the latter. One could expect that *H. cydno* thus has a stronger deterrent than the local form of *H. erato*. However, there is some disagreement as to the evolutionary behavior of *H. cydno*: The conventional wisdom states that all forms in the evolution of a Müllerian mimicry complex will converge. However analysis of a mathematical model (see Section 10.5) indicates that under certain conditions the weaker Müllerian mimic will act as a Batesian mimic on the stronger Müllerian mimic (Huheey, 1976; 1980c). The consequences of Batesian mimicry (see below) would be expected to obtain, including the possibility of the stronger model evolving *away* from the weaker.

Another method for the evolution of Müllerian mimicry has been suggested. In a sense, a Batesian mimic is pre-adapted for evolution towards Müllerian mimicry. Furthermore, we have seen adjustments in the behavior of predators after being exposed to unpalatable prey: They attack more slowly and cautiously, and sometimes 'taste' but fail to maintain the attack (J. Brower, 1958, 1960). Mimetic prey is thus not exposed to as severe attack as a non-mimic (until the sham is pierced). It would be more likely to survive and continue its evolution towards Müllerian mimicry if some incipient offensive distastefulness was developed (Huheey, 1961). It has been suggested that this is the *only* means available for the initiation of Müllerian mimicry in rare species (L. Brower *et al.*, 1970). The conditions favoring the evolution of a Batesian mimic into a Müllerian mimic have been determined from the mathematical model and are not surprising: The model must not be overly abundant nor have a strong protective deterrent. In other words, if the Batesian mimicry is weak, it behooves the mimic to evolve its own deterrent (Huheey, 1976).

10.7.2 Batesian evolution

We have already discussed several aspects of the evolution of Batesian mimicry. One point that may not have been sufficiently stressed is the rate of evolution of the model and its mimic. This has been discussed by Fisher (1958) and more recently by Nur (1970). Since the mimic gains at the model's expense, it is expected that selection will operate on the mimic for convergence upon its model, but that selection will favor divergence by the model. Detection of differences between the model and mimic by the predator will be fatal for the mimic. Therefore selection upon it will be strong and highly directional, whereas selection on the model for divergence will be weaker and non-directional. Furthermore, a striking divergence by the model is apt to be non-adaptive since it may take the mutant out of the range of the ability of the predator to generalize (i.e., it will not be recognized as warningly colored at all). Therefore, it is to be expected that the mimic will evolve towards the model faster than the latter can escape. Were this not so, we should never observe Batesian mimicry.

Another interesting phenomenon is the evolution of a Müllerian mimic into a Batesian mimic. One might think this an unlikely event since a Müllerian mimic has a deterrent, a Batesian mimic none. However, if a *weaker* Müllerian model is under the protection of a *very strong model*, it may lose its deterrent at little predation cost to itself, with a possible gain in foodplant choice or reduction in autotoxicity (Huheey, 1976). In fact this is exactly what the automimicry in the monarch butterfly is all about (see Section 10.4). Some individuals are able to exploit otherwise neglected foodplants or avoid autotoxicity, or both, yet escape predation 'under the wing' of their toxic conspecifics. L. Brower and J. Brower (1964) have suggested that *Lycorea*, which shares a color pattern with many ithomiines and heliconiines, may have evolved 'regressively' from unpalatability to a greater palatability in order to reduce the adverse effects of its toxins while it retained its protection from predation by keeping a Batesian mimetic role.

Rothschild (1979) has made a strong case that various pre-adaptations play a very important, even vital, role in the evolution of Batesian mimics. These traits, such as rarity, would not predispose a species to mimicry but, if present, facilitate or 'allow' such evolution toward Batesian mimicry. Thus rarity would not arise after the occurrence of Batesian mimicry, but would be almost a pre-condition as a result of predator learning. If this suggestion is correct, it would place severe limits on the evolution of Müllerian mimics into Batesian mimics as discussed above.

10.7.3 Balanced polymorphism

There are frequently populations of Batesian mimics that are polymorphic, often with one cryptic morph. If one assumes that the mimetic pattern is superior, the existence of the cryptic appears paradoxical. Actually, in such populations at equilibrium, *all* morphs are equally fit through density-dependent selection. The common supposition (which often *is* true), that the mimetic morphs are less successful at courtship is unnecessary, and may even be mis-leading. No superiority on the part of the cryptic morph need be shown – by overabundance the mimetic morphs negate their initial advantage at low fre-quencies. See Ford (1975), Ayala and Campbell (1974), and Barrett (1976) and references therein.

Finally, all too often aposematic mimetic adaptations have been assumed to form an ecological and evolutionary panacea for the possessor. This they certainly do *NOT*; no more than any other adaptation. Thus, as we have seen, in polymorphic Batesian mimicry, a Batesian morph is not *better* protected than a cryptic morph, it is just *differently* protected. At equilibrium, all morphs will be equally protected. Once multiple species are involved, it is quite possible for a Batesian mimic to become so tied to its model that an independent exist-ence is impossible. Furthermore, even its own numbers will be governed by the abundance and deterrence of its model. In fact, as I have said elsewhere (Huheey, 1980b), 'If, indeed, a Batesian mimic courts extinction by its rarity, it would not be the first example of selective forces which have led species into evolutionary dead ends from which they could not emerge.'

10.8 THE FUTURE: PROBLEMS AND POSSIBLE SOLUTIONS IN INSECT MIMICRY

The first one hundred years of the study of mimicry consisted almost entirely of visual observations, data that were often anecdotal, seldom even semi-quantitative, with few means of pushing beyond field observations. Even with these limitations, by combining careful observation and careful reasoning, a tremendous literature on mimicry, most of it excellent and far too voluminous even to refer to here, had developed. The past two decades have seen a veritable explosion in the use of novel and sophisticated tools such as electronics, chemistry, statistics and genetics to attack the problem.

An example of a 'future trend' is already well underway. Increasingly, the traditional question, 'palatable or unpalatable?' will be supplemented or supplanted by chemical characterization of the toxin and quantitative estimates of the amount present. Many of the puzzling cases that we are presented with today may be resolved with good data of this sort. And, increasingly, the data must represent statistically sound and geographically representative samples.

Increasingly, automation and electronic equipment will replace or supplement human senses. Examples to date at Lloyd's (1980) recording of temporal data on firefly flashes, the recording of auditory signals (Gaul, 1952; Lane and Rothschild, 1965), or merely the ever increasingly accessibility of computer facilities (now handheld!) to process the statistical data.

Lastly, I hope that the future will see a continuation of the current trend in the 'philosophy' of the discipline. Too often in the past mimicry was viewed by those outside the discipline as a 'gee whiz' sort of sideshow that said that earthworms must mimic beets because both live in the soil and both are dark red! To be sure, there was some of this in the past, but it should not have been allowed to overshadow the excellent work that has been done in the field and the laboratory. If the next twenty years bring about as many advances as the last twenty, we may enter the new century without worry in this regard. I believe we will.

ACKNOWLEDGMENTS

Two events occurred in my graduate studies that greatly influenced my later zoological work – the reading of Jane van Zandt Brower's papers on her experimental work with mimicry in butterflies (1958) and that of Philip M. Sheppard's insightful little book (1958) and following review (1959) on natural selection and evolution especially as regards mimicry.

I have profited greatly from discussions of mimicry with many people over the years. Specifically, Ronald A. Brandon, Edmond D. Brodie, Jr., Lincoln P. Brower, and The Hon. Miriam Rothschild have been particularly important in shaping the ideas presented herein. Ronald A. Brandon, Lincoln P. Brower, James E. Lloyd, Martha G. Price, and The Hon. Miriam Rothschild have read and criticized the manuscript. This does not mean that they agree with all that I wrote – often disagreement is scientifically more useful! Jay E. McPherson helped with problems in entomology with which I was not conversant. My studies of mimicry have been supported by grants G-17005, GB-3889, BMS 74-14371, and DEB 78-05959 from the National Science Foundation.

REFERENCES

Alcock, J. (1970) Punishment levels and the response of white-throated sparrow (*Zonotrichia albicollis*) to three kinds of artificial models and mimics. *Anim. Behav.*, **18**, 733–9.

Arnold, S. J. (1978) The evolution of a special class of modifiable behaviors in relation to environmental pattern. *Am. Nat.*, **112**, 415–27.

Ayala, F. J. and Campbell, C. A. (1974) Frequency-dependent selection. *A. Rev. Ecol. Syst.*, **5**, 115–38.

Bakus, G. J. (1981) Chemical defense mechanisms on the Great Barrier Reef, Australia. *Science*, **211**, 497–9.

Barrett, J. A. (1976) The maintenance of non-mimetic forms in a dimorphic Batesian mimic species. *Evolution*, **30**, 82–5.

Bates, H. W. (1862) Contributions to an insect of the Amazon Valley. Lepidoptera: Heliconidae. *Trans. Linn. Soc. Zool.*, **23**, 495–566.

Beal, F. E. L. (1918) Food habits of the swallows, a family of valuable native birds. *U.S. Dept. Agric. Bull. No.* **619**, 1–28.

Bees, J. (1977) Terminal terminology. *Antenna*, **1**, 35.

Benson, W. W. (1972) Natural selection for Müllerian mimicry in *Heliconius erato* in Costa Rica. *Science*, **176**, 936–9.

Benson, W. W. (1977) On the supposed spectrum between Batesian and Müllerian mimicry. *Evolution*, **31**, 454–5.

Bernstein, I. L. (1978) Learned taste aversions in children receiving chemotherapy. *Science*, **200**, 1302–3.

Bobisud, L. E. (1978) Optimal time of appearance of mimics. *Am. Nat.*, **112**, 962–5.

Bobisud, L. E. and Potratz, C. J. (1976) One-trial versus multi-trial learning for a predator encountering a model-mimic system. *Am. Nat.*, **110**, 121–8.

Boggs, C. L. and Gilbert, L. E. (1979) Male contribution to egg production in butterflies: evidence for transfer of nutrients at mating. *Science*, **206**, 83–4.

Bowers, M. D. (1980) Unpalatability as a defense strategy of *Euphydryas phaeton* (Lepidoptera: Nymphalidae). *Evolution*, **34**, 586–600.

Boyden, T. C. (1976) Butterfly palatability and mimicry: Experiments with *Ameiva* lizards. *Evolution*, **30**, 73–81.

Brandon, R. A., Labanick, G. M. and Huheey, J. E. (1979) Relative palatability, defensive behavior, and mimetic relationships of *Pseudotriton ruber*, *P. montanus*, and red efts. *Herpetologica*, **35**, 289–303.

Brower, J. V. Z. (1958) Experimental studies of mimicry in some North American butterflies. Part I. The monarch, *Danaus plexippus*, and viceroy, *Limenitis archippus archippus*. *Evolution*, **12**, 32–47.

Brower, J. V. Z. (1960) Experimental studies of mimicry. IV. The reactions of starling to different proportions of models and mimics. *Am. Nat.*, **94**, 271–82.

Brower, J. V. Z. and Brower, L. P. (1962) Experimental studies of mimicry. 6. The reaction of toads (*Bufo terrestris*) to honeybees (*Apis mellifera*) and their dronefly mimics (*Eristalis vinetorum*). *Am. Nat.*, **96**, 297–308.

Brower, J. V. Z. and Brower, L. P. (1965) Experimental studies of mimicry. 8. Further investigations of honeybees (*Apis mellifera*) and their dronefly mimics (*Eristalis* spp.). *Am. Nat.*, **99**, 173–88.

Brower, L. P. (1969) Ecological chemistry. *Sci. Am.*, **220**, 22–29 (Feb).

Brower, L. P. (1970) Plant poisons in a terrestrial foodchain and implications for mimicry theory. In: *Biochemical Coevolution* (Chambers, K. L., ed.) pp. 69–82. Oregon State University Press, Corvallis.

Brower, L. P. (1977) Monarch migration. *Nat. Hist.*, **86**., 41–53 (June–July).

Brower, L. P., Brower, J. V. Z. and Westcott, P. W. (1960) Experimental studies of mimicry. 5. The reactions of toads (*Bufo terrestris*) to bumblebees (*Bombus americanorum*) and their robberfly mimics (*Mallophora bomboides*), with a discussion of aggressive mimicry. *Am. Nat.*, **94**, 343–55.

Brower, L. P., Brower, J. V. Z. and Collins, C. T. (1963) Experimental studies of mimicry. 7. Relative palatability and Müllerian mimicry among Neoptropical butterflies of the subfamily Heliconiinae. *Zoologica*, **48**, 65–84.

Brower, L. P. and Brower, J. V. Z. (1964) Birds, butterflies, and plant poisons: A study in ecological chemistry. *Zoologica*, **49**, 137–59.

Brower, L. P., Brower, J. V. Z. and Corvino, J. M. (1967) Plant poisons in a terrestrial food chain. *Proc. Nat. Acad. Sci. USA.*, **57**, 892–8.

Brower, L. P., Cook, L. M. and Croze, H. J. (1967) Predator responses to artificial Batesian mimics released in a neotropical environment. *Evolution*, **21**, 11–23.

Brower, L. P., Ryerson, W. N., Coppinger, L. L. and Glazier, S. C. (1968) Ecological chemistry and the palatability spectrum. *Science*, **161**, 1349–51.

Brower, L. P., Pough, F. H. and Meck, H. R. (1970) Theoretical investigations of automimicry, I. Single trial learning. *Proc. Nat. Acad. Sci. U.S.A.*, **66**, 1059–66.

Brower, L. P. and Brower, J. V. Z. (1972) Parallelism, convergence , divergence, and the new concept of advergence in the evolution of mimicry. *Trans. Connecticut Acad. Sci.*, **44**, 59–67.

Brower, L. P., McEvoy, P. B., Williamson, K. L. and Flannery, M. A. (1972) Variation in cardiac glycoside content of monarch butterflies from natural populations in eastern North America. *Science*, **177**, 426–9.

Brower, L. P. and Moffitt, C. M. (1974) Palatability dynamics of cardenolides in the monarch Butterfly. *Nature*, **249**, 280–3.

Brower, L. P., Edmunds, M. and Moffitt, C. M. (1975) Cardenolide content and palatability of a population of *Danaus chrysippus* butterflies from West Africa. *J. Ent. (A)*, **49**, 183–96.

Brower, L. P. and Glazier, S. C. (1975) Localization of heart poisons in the monarch butterfly. *Science*, **188**, 19–25.

Brower, L. P., Gibson, D. O., Moffitt, C. M. and Panchen, A. L. (1978) Cardenolide content of *Danaus chrysippus* butterflies from three areas of East Africa. *Biol. J. Linn.*, **10**, 251–73.

Brower, L. P., Seiber, J. N., Nelson, C. J., Lynch, S. P. and Tuskes, P. M. (1982) Plant-determined variation in the cardenolide content, thin-layer chromatography profiles, and emetic potency of monarch butterflies, *Danaus plexippus*, reared on the milkweed, *Asclepias eriocarpa*, in California. *J. Chem. Ecol.*, **8**, 579–633.

Brown, Jr., K. S. and Benson, W. W. (1974) Adaptive polymorphism associated with multiple Müllerian mimicry in *Heliconius numata* (Lepid. Nymph.). *Biotropica*, **6**, 205–28.

Brown, Jr., K. S., Sheppard, P. M. and Turner, J. R. G. (1974) Quaternary refugia in tropical America: Evidence from race formation in *Heliconius* butterflies. *Proc. R. Soc. Lond. Ser. B*, **187**, 369–78.

Brown, Jr., K. S. and Neto, J. V. (1976) Predation on aposematic ithomiine butterflies by Tanagers (*Pipraeidea melanota*). *Biotropica*, **8**, 136–41.

Burghardt, G. M., Wilcoxon, H. C. and Czaplicki, J. A. (1973) Conditioning in garter snakes: Aversion to palatable prey induced by delayed illness. *Anim. Learn. Behav.*, **1**, 317–20.

Calvert, W. H., Hedrick, L. E. and Brower, L. P. (1979) Mortality of the Monarch Butterfly (*Danaus plexippus* L.): Avian predation at five overwintering sites in Mexico. *Science*, **204**, 847–50.

Chen, J-S. and Amsel, A. (1980) Recall (versus recognition) of taste and immunization against aversive taste anticipations based on illness. *Science*, **209**, 831–833.

Cook, L. M., Brower, L. P. and Alcock, J. (1969) An attempt to verify mimetic advantage in a neotropical environment. *Evolution*, **23**, 339–45.

Cott, H. B. (1940) *Adaptive Coloration in Animals*. Methuen, London.

Dill, L. M. (1975) Calculated risk-taking by predators as a factor in Batesian mimicry. *Can. J. Zool.*, **53**, 1614–21.

Dixon, C. A., Erickson, J. M., Kellett, D. N. and Rothschild, M. (1978) Some adaptations between *Danaus plexippus* and its food plant, with notes on *Danaus chrysippus* and *Euploe core* (Insecta: Lepidoptera). *J. Zool., Lond.*, **185**, 437–67.

Dressler, R. L. (1979) *Eulaema bombiformis, E. meriana*, and Müllerian mimicry in related species (Hymenoptera: Apidae). *Biotropica*, **11**, 144–51.

Eberhard, W. G. (1977) Aggressive chemical mimicry by a Bolas Spider. *Science*, **198**, 1173–5.

Ehrlich, P. R. and Raven, P. H. (1965) Butterflies and plants: a study in co-evolution. *Evolution*, **18**, 586–608.

Eisner, T. (1970) Chemical defense against predation in arthropods. In: *Chemical Ecology* (Sondheimer, E. and Simeone, J. B., eds). Academic Press, New York.

Eisner, T. and Kafatos, F. C. (1962) Defense mechanisms of arthropods. X. A pheromone promoting aggregation in an aposematic distasteful insect. *Psyche*, **69**, 53–61.

Eisner, T., Kafatos, F. C. and Linsley, E. G. (1962) Lycid predation by mimetic adult Cerambycidae (Coleoptera). *Evolution*, **16**, 316–24.

Eisner, T., Nowicki, S., Goetz, M. and Meinwald, J. (1980) Red cochineal dye (carminic acid): Its role in nature. *Science*, **208**, 1039–42.

Estabrook, G. F. and Jespersen, D. C. (1974) Strategy for a predator encountering a model–mimic system. *Am. Nat.*, **108**, 443–57.

von Euw, J., Fishelson, L., Parsons, J. A., Reichstein, T. and Rothschild, M. (1967) Cardenolides (heart poisons) in a grasshopper feeding on milkweeds. *Nature*, **214**, 35–9.

von Euw, J., Reichstein, T., and Rothschild, M. (1968) Aristolochic acid-I in the swallowtail butterfly *Pachlioptera aristolochiae* Fabr. (Papilionidae), *Israel J. Chem.*, **6**, 659–70.

Fink, L. S. and Brower, L. P. (1981) Birds can overcome the cardenolide defence of monarch butterflies in Mexico. *Nature*, **291**, 67–70.

Fisher, R. A. (1958) *The Genetical Theory of Natural Selection*. Dover Press, New York.

Ford, E. B. (1975) *Ecological Genetics*, 2nd ed. Oxford University Press, Oxford.

Fox, D. L. (1976) *Animal Biochromes and Structural Colors*, 2nd ed., University of California Press, Berkeley.

Gans, C. (1965) Empathic learning and the mimicry of African snakes. *Evolution*, **18**, 705.

Gaul, A. T. (1952) Audio mimicry: An adjunct to color mimicry. *Psyche*, **59**, 82–3.

Gibson, D. O. (1980) The role of escape in mimicry and polymorphism: I. The response of captive birds to artificial prey. *Biol. J. Linn. Soc.*, **14**, 201–14.

Gilbert, L. E. (1972) Pollen feeding and reproductive biology of *Heliconius* butterflies. *Proc. Nat. Acad. Sci. USA.*, **69**, 1403–7.

Greene, H. F. and McDiarmid, R. W. (1981) Coral snake mimicry: Does it occur? *Science*, **213**, 1207–11.

Grobman, A. B. (1978) An alternative solution to the Coral Snake mimic problem (Reptilia, Serpentes, Elapidae). *J. Herp.*, **12**, 1–11.

Heal, J. R. (1979) Colour patterns of Syrphidae. II. *Eristalis intricarius*. *Heredity*, **43**, 229–38.

Heal, J. R. (1981) Colour patterns of Syrphidae. III. Sexual dimorphism in *Eristalis arbustorum*. *Ecol. Ent.*, **6**, 119–27.

Holling, C. S. (1965) The functional response of predators to prey density and its role in mimicry and population regulation. *Mem. Ent. Soc. Can.*, **45**, 1–60.

Howard, R. W., McDaniel, C. A. and Blomquist, G. J. (1980) Chemical mimicry as an integrating mechanism: cuticular hydrocarbons of a termitophile and its host. *Science*, **210**, 431–3.

Huey, R. B. and Pianka, E. R. (1977) Natural selection for juvenile lizards mimicking noxious beetles. *Science*, **195**, 201–2.

Huheey, J. E. (1961) Studies in warning coloration and mimicry. III. Evolution of Müllerian mimicry. *Evolution*, **15**, 567–8.

Huheey, J. E. (1964) Studies of warning coloration and mimicry. IV. A mathematical model of model-mimic frequencies. *Ecology*, **45**, 185–8.

Huheey, J. E. (1976) Studies in warning coloration and mimicry. VII. Evolutionary consequences of a Batesian–Müllerian spectrum: A model for Müllerian mimicry. *Evolution*, **30**, 86–93.

Huheey, J. E. (1980a) The question of synchrony or "temporal sympatry" in mimicry. *Evolution*, **34**, 614–6.

Huheey, J. E. (1980b) Studies in warning coloration and mimicry. VIII. Further evidence for a frequency-dependent model of predation. *J. Herp.*, **14**, 223–30.

Huheey, J. E. (1980c) Batesian and Müllerian mimicry: Semantic and substantive differences of opinion. *Evolution*, **34**, 1212–15.

Huheey, J. E. and Brandon, R. A. (1974) Studies in warning coloration and mimicry. VI. Comments on the warning coloration of red efts and their presumed mimicry by red salamanders. *Herpetol.*, **30**, 149–55.

Isman, M. B., Duffey, S. S. and Scudder, G. G. E. (1977) Cardenolide content of some leaf- and stem-feeding insects on temperate North American milkweeds (*Asclepias* spp.). *Can. J. Zool.*, **55**, 1024–8.

Jackson, J. F. and Drummond, B. A. III (1974) A Batesian ant-mimicry complex from the Mountain Pine Ridge of British Honduras, with an example of transformational mimicry. *Am. Midl. Nat.*, **91**, 248–51.

Jones, D. A., Parsons, J. and Rothschild, M. (1962) Release of hydrocyanic acid from crushed tissues of all stages in the life-cycle of species of the Zygaeninae. *Nature*, **193**, 52–3.

Johnson, C. (1976) *Introduction to Natural Selection*. University Park Press, Baltimore.

Lane, C. and Rothschild, M. (1965) A case of Müllerian mimicry of sound. *Proc. R. Ent. Soc. Lond. Ser. A*, **40**, 156–8.

Lea, R. G. and Turner, J. R. G. (1972) Experiments on mimicry: II. The effect of a Batesian mimic on its model. *Behaviour*, **38**, 131–51.

Linsley, E. G. (1960) Ethology of some bee- and wasp-killing robberflies of Southeastern Arizona and Western New Mexico. (Diptera: Asilidae). *Univ. Cal. Publ. Ent.*, **16**, 357–92.

Linsley, E. G., Eisner, T. and Klots, A. B. (1961) Mimetic assemblages of sibling species of lycid beetles. *Evolution*, **15**, 15–29.

Lloyd, J. E. (1975) Aggressive mimicry in *Photuris* fireflies: Signal repertoires by femmes fatales. *Science*, **187**, 452–3.

Lloyd, J. E. (1980) Male *Photuris* fireflies mimic sexual signals of their females' prey. *Science*, **210**, 669–71.

Lloyd, J. E. (1981) Firefly mate-rivals mimic their predators and vice versa. *Nature*, **290**, 498–500.

Marsh, N. and Rothschild, M. (1974) Aposematic and cryptic Lepidoptera tested on the mouse. *J. Zool., Lond.*, **1974**, 89–122.

Marsh, N. A., Clarke, C. A., Rothschild, M. and Kellett, D. N. (1977) *Hypolimnas bolina* (L.), a mimic of danaid butterflies, and its model *Euploea core* (Cram.) store cardioactive substances. *Nature*, **268**, 726–8.

Mayr, E. (1970) *Populations, Species, and Evolution*. Harvard University Press, Cambridge.

Mühlmann, H. (1934) Im Modellversuch künstlich erzeugte Mimikry unde ihre Bedeutung für den 'Nachahmer'. *Z. Morph. Ökol.*, **28**, 259–96.

Müller, F. (1879) *Ituna* and *Thyridia*; a remarkable case of mimicry in butterflies. *Proc. R. Ent. Soc. Lond.*, 20–29.

Nur, U. (1970) Evolutionary rates of models and mimics in Batesian mimicry. *Am. Nat.*, **104**, 477–86.

Owen, D. F. (1970) Mimetic polymorphism and the palatability spectrum. *Oikos*, **21**, 333–6.

Owen, D. F. (1971) *Tropical Butterflies*. Clarendon Press, Oxford.

Parsons, J. A. (1965) A digitalis-like toxin in the monarch butterfly *Danaus plexippus* L. *J. Physiol.*, **178**, 290–304.

Pasteels, J. M. and Daloze, D. (1977) Cardiac glycosides in the defensive secretion of chrysomelid beetles: evidence for their production by the insects. *Science*, **197**, 70–2.

Pasteur, G. (1972) *Le Mimétisme*. Presses Universitairres de France, Paris.

Platt, A. P., Coppinger, R. P. and Brower, L. P. (1971) Demonstration of the selective advantage of mimetic *Limenitis* butterflies presented to caged avian predators. *Evolution*, **25**, 692–701.

Pough, Jr., F. H., Brower, L. P., Meck, H. R. and Kessell, S. R. (1973) Theoretical investigations of automimicry: multiple trial learning and the palatability spectrum. *Proc. Nat. Acad. Sci. USA*, **70**, 2261–5.

Reichstein, M., von Euw, J., Parsons, J. A. and Rothschild, M. (1968) Heart poisons in the monarch butterfly. *Science*, **161**, 861–866.

Reiskind, J. (1977) Ant-mimicry in Panamanian clubionid and salticid spiders (Araneae: Clubionidae, Salticidae), *Biotropica*, **9**, 1–8.

Remington, J. E. and Remington, C. L. (1957) Mimicry, a test of evolutionary theory. *Yale Sci. Mag.*, **32**, 10–11, 13–14, 16–17, 19, 21.

Rettenmeyer, C. W. (1970) Insect mimicry. *A. Rev. Ent.*, **15**, 43–74.

Revusky, S. H. and Bedarf, E. W. (1967) Association of illness with prior ingestion of novel foods. *Science*, **155**, 219–20.

Ricklefs, R. E. (1979) *Ecology*, 2nd edn, Chiron Press, New York.

Rothschild, M. (1961) Defensive odours and Müllerian mimicry among insects. *Trans. R. ent. Soc. Lond.*, **113**, 101–21.

Rothschild, M. (1963) Is the buff ermine (*Spilosoma lutea* (Huff.)) a mimic of the White Ermine (*Spilosoma lubricipeda* (L.))? *Proc. Roy. Ent. Soc. Lond.*, **38**, 159–64.

Rothschild, M. (1964) A note on the evolution of defensive and repellent odours of insects. *Entomologist*, **97**, 276–80.

Rothschild, M. (1971) Speculations about mimicry with Henry Ford. In: *Ecological Genetics and Evolution* (Creed, R., ed.) pp. 202–23. Blackwell, Oxford.

Rothschild, M. (1972a) Colour and poisons in insect protection. *New Sci.* (11 May), 318–20.

Rothschild, M. (1972b) Secondary plant substances and warning colouration in insects. In: *Insect/Plant Relationships* (van Emden, H. F., ed.) pp. 59–83. *Symp. R. Ent. Soc. Lond. No.*, **6**.

Rothschild, M. (1979) Mimicry: Butterflies and Plants. *Proc. Symp. on Parasites as Plant Taxonomists, Symb. Bot. Upsal.*, **XXII**, 82–99.

Rothschild, M. and Ford, B. (1968) Warning signals from a starling *Sturnus vulgaris* observing a bird rejecting unpalatable prey. *Ibis*, **110**, 14–105.

Rothschild, M., von Euw, J. and Reichstein, T. (1970) Cardiac glycosides in the oleander aphid, *Aphis nerii*. *J. Insect Physiol.*, **16**, 1141–1145.

Rothschild, M., von Euw, J. and Reichstein, T. (1972) Some problems connected with warningly coloured insects and toxic defense mechanism. *Mitt. Basler Afrika Bibliogr.*, **4–6**, 135–58.

Sbordoni, V., Bullini, L., Scarpelli, G., Forestiero, S. and Rampini, M. (1979)

Mimicry in the burnet moth *Zygaena ephialtes*: Population studies and evidence of a Batesian–Müllerian stiuation. *Ecol. Ent.*, **4**, 83–94.

Schuler, W. (1980) Zum Meidenlernen ungeniessbarer Beute bei Vögelin: Der Einfluss der Faktoren Umlernen, neue Alternativebeute and Ahnlichkeit der Alternativebeute. *Z. Tierpsychol.*, **54**, 105–43.

Seiber, J. N., Tuskes, P. M., Brower, L. P. and Nelson, C. N. (1980). Pharmacodynamics of some individual milkweed cardenolides fed to larvae of the monarch butterfly (*Danaus plexippus* L.). *J. Chem. Ecol.*, **6**, 321–39.

Sexton, O. J., Hoger, C. and Ortleb, E. (1966) *Anolis carolinensis*: Effects of feeding on reaction to aposematic prey. *Science*, **153**, 1140.

Shapiro, A. M. (1965) *Antepione thiosaria* and *Xanthotype*: A case of mimicry. *J. Res. Lepid.*, **4**, 6–11.

Sheppard, P. M. (1958) *Natural Selection and Heredity*. Harpers, New York.

Sheppard, P. M. (1959) The evolution of mimicry; a problem in ecology and genetics. *Cold Spring Harbor Symp. Quant. Biol.*, **24**, 131–40.

Sheppard, P. M. (1961) Recent work on polymorphic mimetic papilios. In: *Insect Polymorphism* (Kennedy, J. S., ed.) pp. 20–9. *Symp. R. Ent. Soc. Lond.*

Sheppard, P. M. and Turner, J. R. G. (1977) The existence of Müllerian mimicry. *Evolution*, **31**, 452–3.

Silberglied, R. E. and Eisner, T. (1969) Mimicry of Hymenoptera by beetles with unconventional flight. *Science*, **163**, 486–8.

Smith, D. A. S. (1976) Phenotypic diversity, mimicry and natural selection in the African butterfly *Hypolimnas misippus* L. (Lepidoptera: Nymphalidae). *Biol. J. Linn. Soc.*, **8**, 183–204.

Smith, D. A. S. (1978a) Cardiac glycosides in *Danaus chrysippus* (L.) provide some protection against an insect parasitoid. *Experientia*, **34**, 844–5.

Smith, D. A. S. (1978b) The effect of cardiac glycoside storage on growth rate and adult size in the butterfly *Danaus chrysippus* (L.). *Experientia*, **34**, 845–6.

Smith, D. A. S. (1979) The significance of beak marks on the wings of an aposematic, distasteful and polymorphic butterfly. *Nature*, **281**, 215–16.

Smith, S. M. (1975) Innate recognition of coral snake pattern by a possible avian predator. *Science*, **187**, 759–60.

Smith, S. M. (1977) Coral-snake pattern recognition and stimulus generalisation by naive great kiskadees (Aves: Tyrannidae). *Nature*, **265**, 535–6.

Sternburg, J. G., Waldbauer, G. P. and Jeffords, M. R. (1977) Batesian mimicry: Selective advantage of color pattern. *Science*, **195**, 681–3.

Stiles, E. W. (1979) Evolution of color pattern and pubescence characteristics in male bumblebees: Automimicry vs. themoregulation. *Evolution*, **33**, 941–57.

Sydow, S. L. and Lloyd, J. E. (1975) Distasteful fireflies sometimes emetic, but not lethal. *Florida Ent.*, **58**, 312.

Terhune, E. C. (1977) Components of a visual stimulus used by scrub jays to discriminate a Batesian model. *Am. Nat.*, **111**, 435–51.

Thornhill, R. (1979) Adaptive female-mimicking behavior in a scorpionfly. *Science*, **205**, 412–14.

Turner, J. R. G. (1971) Studies of Müllerian mimicry and its evolution in Burnet Moths and heliconid butterflies. In: *Ecological Genetics and Evolution* (Creed, R., ed.) pp. 224–60. Blackwell, Oxford.

Turner, J. R. G. (1975) A tale of two butterflies. *Nat. Hist.*, **84** (Feb) 28–37.

Turner, J. R. G. (1976a) Müllerian mimicry: classical 'beanbag' evolution and the role of ecological islands in adaptive race formation. In: *Population Genetics and Ecology* (Karlin, S. and Nevo, E., eds) pp. 185–218. Academic Press, New York.

Turner, J. R. G. (1976b) Adaptive radiation and convergence in subdivisions of the butterfly genus *Heliconius* (Lepidoptera: Nymphalidae). *Zool. J. Linn. Soc.*, **58**, 297–308.

Turner, J. R. G. (1977) Butterfly mimicry: The genetical evolution of an adaptation. In: *Evolutionary Biology*, Vol. 10 (Hecht, M. K., Steere, W. C. and Wallace, B., eds) pp. 163–206. Plenum Press, New York.

Turner, J. R. G. (1978) Why male butterflies are non-mimetic: natural selection, sexual selection, group selection, modification and sieving. *Biol. J. Linn. Soc.*, **10**, 385–432.

Turner, J. R. G. (1981) Adaptation and evolution in *Heliconius*: A defense of Neo-Darwinism. *A Rev. Ecol. Syst.*, **12**, 99–121.

Vander Meer, R. K. and Wojcik, D. P. (1982) Chemical mimicry in the myrmecophilous beetle *Myrmecaphidus excavaticollis*. *Science*, **218**, 806–808.

Waldbauer, G. P. and Sheldon, J. K. (1971) Phenological relationships of some aculeate Hymenoptera, their dipteran mimics, and insectivorous birds. *Evolution*, **25**, 371–82.

Wickler, W. (1968) *Mimicry in Plants and Animals*. McGraw-Hill, New York.

Wiens, D. (1978) Mimicry in plants. In: *Evolutionary Biology*, Vol. 11 (Hecht, M. K., Steere, W. C. and Wallace, B., eds) pp. 365–403. Plenum Press, New York.

Wilson, E. O. (1975) *Sociobiology: The New Synthesis*. Harvard University Press, Cambridge.

NOTE ADDED IN PROOF: A very useful review of various types of mimicry has recently been provided by Georges Pasteur (1982) A classificatory review of mimicry systems. *Ann. Rev. Ecol. Syst.*, **13**, 169–199.

Chemical-mediated Spacing

11

Resource Partitioning

Ronald J. Prokopy, Bernard D. Roitberg and Anne L. Averill

11.1 INTRODUCTION

At some point during their lifetime, individuals of most insects are probably involved in competition for an essential resource, be it for food, mating territories, egg-laying sites, or refugia. Should the density of competing individuals exceed the carrying capacity of a unit of resource, the result may be detrimental to fitness. Therefore, selection should favor avoidance of overcrowding. In some species, however, too low a density of individuals on a unit of resource may have equally detrimental consequences to fitness. Here, selection should favor some degree of aggregation, short of overcrowding. Thus, we may consider the concept of an optimal density range of individuals per unit of resource as valid (Peters and Barbosa, 1977; Prokopy, 1981a).

To achieve density levels within an optimal range, insects may employ a variety of mechanisms, including physical combat (Hubbell and Johnson, 1977; Whitham, 1979; Mitchell, 1980), or use of visual (Benson *et al.*, 1975; Rausher, 1979; Williams and Gilbert, 1981; Shapiro, 1981; Gilbert, 1982), acoustic (Russ, 1969; Rudinsky *et al.*, 1973; Otte and Joern, 1975; Doolan, 1981), or chemical stimuli. In the following chapter, Birch focuses on aggregation in bark beetles as achieved through response to pheromones and chemicals of plant origin. In this chapter, we focus principally on chemical stimuli of insect or plant origin which mediate against overcrowding. In addition, we consider chemically mediated recognition of uncrowded but previously searched or utilized resource sites.

Three recent reviews illustrate the role which pheromones may play in deterring overcrowding of resources by insects. The first (van Lenteren, 1981)

Chemical Ecology of Insects. Edited by William J. Bell and Ring T. Cardé
© 1984 Chapman and Hall Ltd.

analyzes host discrimination by parasitoids. The second (Prokopy, 1981a) deals with a survey of phytophagous insect species in which pheromones are known or suspected of functioning in this manner. The third (Prokopy, 1981b) treats the oviposition deterring pheromone system of the apple maggot fly, *Rhagoletis pomonella*, as a model case.

Resource availability, foraging behavior, and resource exploitation will be considered during our discussion of the ways in which chemical stimuli act to deter overcrowding of resources or to signal recognition of previously searched or utilized resource sites. Types of resources include food, mating territories and mates, egg-laying sites, and refugia.

11.2 AVAILABILITY OF RESOURCES IN TIME AND SPACE

Resource availability to foraging insects depends upon resource density and distribution in space and time. Assuming that they can be identified individually and that they occupy definite spatial and/or temporal locations, resource units can generally be characterized by one of three major dispersion patterns: (i) uniform – e.g., plants in agricultural crop fields, (ii) random – e.g., greenhouse whiteflies, *Trialeurodes vaporarium*, which are egg-laying sites for the parasitoid *Encarsia formosa* (van Lenteren *et al.*, 1976), or (iii) clumped or aggregated – e.g., *Passiflora* flowers which are food resources for *Xylocopa* bees (see Section 11.5) (Frankie and Vinson, 1977). These three categories may also describe the dispersion patterns of groups of resource units. In each case, both physical and biological factors may shape the dispersion pattern.

Measurement and statistical analysis of resource dispersion patterns is readily and often done (see Pielou, 1977). However, such analysis may tell us little about the availability of particular resource units to the foraging insect. A more useful measure might be that of 'effective distribution', which we define as the spatial and/or temporal dispersion of those resource units that are available for exploitation by the foragers. For example, in Section 11.5 we discuss several insect species wherein some females are rendered unmatable through the courting activities of successful mate-foraging males. As a second example, individual olive fruits, the only egg-laying site for the monophagous tephritid fly *Dacus oleae* (see Section 11.5), mature at different rates. At any one time within a given tree, only a select number of olives may be available for exploitation by *D. oleae* flies (Louskas *et al.*, 1980). Thus, not all resources are as readily available as they might appear (see also Caughley and Lawton, 1981; Whitham, 1983).

Several factors may influence the effective distribution pattern of resources:

11.2.1 Habitat structure

Physical features of the landscape may directly or indirectly influence effective resource distribution (e.g. *Passiflora* plants grow on earthen mounds in disturbed areas). In some cases, such locations may be physically inaccessible to foragers. In a direct manner, the landscape may influence search behavior of foragers (e.g., search paths, speed of movement), thereby rendering some resource units less accessible than others (Price *et al.*, 1980). Different kinds of foragers probably respond differently to physical features of the environment. Thus, if we were to compare the influence of tree habitat structure on the effective distribution of apples as egg-laying sites for crawling foragers (e.g., *Conotrachelus* beetles; Owens *et al.*, 1982) versus flying ones (e.g., *Rhagoletis* flies; see Section 11.5), we might expect to find major differences.

11.2.2 Age and size

Natural age and size variation among resource units within a population make some resources less accessible or utilizable than others, as in the aforementioned variation in olive fruit maturity.

11.2.3 Background information

Individual resource units may be more or less easy to locate and exploit depending upon the composition of the community in which they occur. As pointed out by Miller and Strickler in Chapter 6, insects that search for resources within a multispecies community are assailed by many different sensory inputs. Different combinations of sensory information may alter effective resource distribution by affecting the host location mechanisms of foragers. Non-utilizable members of the resource community may influence host location processes by releasing sensory information that either (i) masks cues required for location of the primary resource (e.g., odors from nearby *Picea* trees interfere with odors of larch sawflies and their host tree from perception by the sawfly parasite *Bessa harveyi*; Monteith, 1960), or (ii) repels foragers from sites where resources are located (e.g., male-scent-marked mated female *Tenebrio* beetles are repellent to conspecific males − see Section 11.5). Alternatively, resources located in association with particular community members may be more readily located interactions between the two produce attractive sensory signals (e.g., tobacco plant leaf tissue, a non-utilizable resource damaged by feeding by tobacco budworms, *Heliothis virescens*, attracts *Cardiochiles nigriceps*, a parasitoid of *H. virescens* − Vinson, 1975).

11.2.4 Resource experience

Many organisms respond to attempted exploitation by foragers by altering their physiology or behavior to render themselves less vulnerable to attack or exploitation. For example, birch leaves reduce their quality as food for the autumnal moth, *Oporinia autumnata*, when adjacent leaves are damaged by *O. autumnata* larval feeding (Haukioja, 1980). As a second example, once-mated female apple maggot flies, *Rhagoletis pomonella*, are not as attracted to male-released sex pheromone as are virgin females (Prokopy, 1975).

11.2.5 Resource depression

As pointed out by Charnov *et al.* (1976), foraging animals tend to depress the availability of resources through their foraging activities. Thus, effective resource distribution may depend upon the manner in which foragers search for, exploit, and depress resource units. We separate resource depression states into four major classes: (i) the resource unit is removed from the foraging site coincident with its exploitation, (ii) resource remains within the sensory field but has no resource value (e.g., an already mated female), (iii) resource remains within sensory field and replenishes itself (e.g., flower produces new nectar after original supply has been extracted), and (iv) resource becomes less vulnerable to attack and/or exploitation as a result of previous use (e.g., previously damaged birch leaves; Haukioja, 1980).

Several other factors may also influence effective resource distribution. However, the aforementioned examples constitute those of primary importance to our discussion here. We wish to re-emphasize that the distribution of resource units is often very different and the availability of resources less abundant than first appears. We now turn our attention to the means by which foraging insects search for resources.

11.3 FORAGING FOR RESOURCES

Insects search for resources that are distributed in various spatial and/or temporal patterns. By far the most common and most studied ecological situation is the foraging of insects for clumped or patchily distributed resources. Insects that search for clumps or patches of resources must make several 'decisions': (i) how much energy to expend searching for resource-containing patches within a given habitat; (ii) how much energy to expend searching for individual resources within a given resource-containing patch (i.e., how long to remain); (iii) the particular search methodology (e.g., search path characteristics) that will be employed to locate resource units; and (iv) how much energy should be expended attempting to exploit individual resource units once they have been located.

Decisions (i) and (ii), and to some extent (iv), may be similar, but each of them considers foraging strategies at a different level of resource distribution. It is important to consider that resource patches exist at different hierarchical levels (e.g., forest, tree, limb, leaf, etc.) and that each patch level and each patch boundary can be defined by the behavior of the foraging insect. Whatever the patch level, the problem the forager faces is as follows: individuals that remain 'too long' in resource-poor areas accumulate resources at lower rates than they might elsewhere, whereas those individuals that leave resource patches 'too quickly' may spend inordinately large amounts of time moving between resource patches.

If we consider the parameters of effective resource distribution within and among patches, together with the search behavior of a particular insect and resource depression by that insect, there is a point at which emigration from a particular resource patch becomes the optimal strategy because the gains to be made by remaining are less than the search costs. Several theoretical models have been developed to predict the optimal amount of energy animals should invest in within-patch search in patches and habitats of varying resource richness (see Pyke *et al.*, 1977). Whether insects, or animals in general, do make optimal foraging decisions is a topic of current debate (Maynard Smith, 1978; Lewontin, 1979). It is useful to keep in mind that more than one factor usually shapes the evolution of a particular behavior. For example, if minimization of energy expenditure were the only factor driving the evolution of egg-laying behavior in insects, then individuals would deposit all of their eggs on the first host encountered. Many factors, however, including predation on adults, quality and quantity of food for offspring, predation or parasitism on eggs and/or larval offspring, etc., act within physiological, morphological and ecological constraints to shape the foraging process (see Stamp, 1980, for a review on egg deposition patterns in butterflies).

Whether insects forage optically is extremely difficult to prove or disprove (as pointed out above). Nonetheless, there is considerable qualitative evidence that at least some insects can be highly efficient foragers (Heinrich, 1979). There exist several means by which insects may enhance their foraging efficiency through response to chemical messages (both intra- and inter-specific): (i) location of potential resources through recognition of and attraction to resource-derived chemicals (see Miller and Strickler, Chapter 6); (ii) assessment of resource density through perception of, and differential response to, varying concentration levels of resource-derived chemicals (see Miller and Strickler, Chapter 6; Vinson, Chapter 8); (iii) avoidance of chemically marked, previously *searched* resource patches (see Section 11.5); and (iv) avoidance of chemically marked, previously *exploited* resource patches (see Section 11.5). Insects presumably do not make 'conscious' foraging decisions, but rather their response to chemical messages modifies their behavior in ways that may enhance their foraging efficiency.

Until now, we have considered how resources may be distributed in space and time, and how individuals forage for resources. A further complexity arises from influence of competitors on the exploitation of available resources.

11.4 RESOURCE EXPLOITATION AND COMPETITIVE INTERACTIONS

The degree of exploitation of a resource may depend on a variety of traits of the resource as well as those of the insect exploiter.

One factor of critical importance may be resource quality. For example, insects as resources (hosts or prey) may vary in quality as a result of their diet (Campbell and Duffey, 1979), females may be mated (see Section 11.5.2), or plants as resources may vary in quality as a result of their production of toxins or feeding deterrents (Rosenthal and Janzen, 1979). Further, resource quality may be unpredictable both in space and time. For example, as mentioned in Section 11.2, local variability occurs in birch stands where trees defoliated by *O. autumnata* exhibit a marked decrease in leaf digestibility for up to three years as compared to non-defoliated trees. Additionally, birches growing under favorable conditions are less vulnerable to moth attack than those growing under less favorable conditions (Haukioja, 1980). Highly ephemeral resources such as cattle droppings utilized by the common yellow dung fly, *Scatophaga sterocoraria*, deteriorate within hours (Parker, 1978b), as do *Melandrium* flowers, the host of the noctuid moth *Hadena bicruris* (see Section 11.5.3) (Brantjes, 1976).

Turning our attention now to the exploiter, we noted in Section 11.1 that there may exist an optimal density range of individuals per unit of resource. A low density of individuals per unit of resource may be disadvantageous because some insects may be unable to overcome host defenses (see Birch, Chapter 12), survive natural enemy attack, locate a mate, or effectively utilize host nutrients (Prokopy, 1981a). One example of benefits of aggregation is that of egg clusters of certain butterfly species being aposematically colored, thereby providing a conspicuous signal of egg toxicity and lowering predation or parasitism (Stamp, 1980). Other examples include formation of mating leks (see Section 11.5.2), the enhanced feeding efficiency of larval aggregations of the burnet moth, *Pryeria sinica* on *Euonymus* (Tsubaki and Shiotsu, 1982), and the communal feeding habit of young larvae of the tropical butterfly, *Mechanitis isthmia*, which allows the larvae to overcome host defense via collective removal of leaf trichomes (Young and Moffett, 1979).

In some cases, a tendency toward grouping or solitariness within a species may change over space or time as a result of selective advantages accrued. Thus, in the case of *O. autumnata*, aggregated larvae grow better on previously defoliated trees, whereas solitary larvae grow better on high quality, non-defoliated trees (Haukioja, 1980). Other insects such as the fall webworm, *Hyphantria cunea*, and *M. isthmia* may aggregate as younger larvae and then disperse as larger larvae, or just prior to pupation (Suzuki *et al.*, 1980, Young and Moffett, 1979). Overall, a dual strategy of some degree of aggregation together with avoidance of overcrowding may give rise to maximal use of presently occupied resources as well as untapped resources within a habitat.

Individuals that participate in large aggregations per unit of resource may lower their individual fitness owing to local resource exhaustion or deterioration. Under crowded conditions, an individual may exercise one of three options: stay put and attempt to obtain sufficient resource from the present unit, enter diapause, or attempt to locate other resource units and risk mortality during dispersal.

The portion of the population which remains in an overcrowded habitat may necessarily engage in competition, that is, the active demand by two or more individuals for a common resource (Wilson 1975). Competition has been broadly classified into two processes: interference and exploitation (Park, 1954).

Interference competition embraces any activity which directly or indirectly limits a competitor's access to a resource (Miller, 1967). 'Indirect' interference invariably involves a form of chemical communication, or signal, which is effective in the owner's absence. 'Direct' interference includes a range of inter-actions such as territoriality, dominance, physiological or physical suppression, and cannibalism. Exploitation is the joint utilization of a limited resource once access has been gained (Miller, 1967). Thus, rather than deal with other individuals, an exploiter deals exclusively with available resources and is successful only through competitive ability. Nicholson (1954) proposed a comparable, but not identical, distinction between modes of competition by designating contest and scramble competition, terms which relate to interference and exploitation competition, respectively.

Well-documented studies of interference competition exist. Elimination of supernumerary larval parasitoids by physical attack of conspecifics is a dramatic case of direct interference by a competitor (Salt, 1961). Other examples include physical defense of territories by *Pogonomyrmex* harvester ants and *Trigona* bee species, and cannibalism in *Tribolium* beetles (Young, 1970) (see Section 11.5). Examples of indirect interference competition via chemical mediators have been found increasingly over the past decade (Prokopy, 1981a; see also Section 11.5).

Fewer well-documented cases of exploitation competition are available owing to the fact that (i) many studies focus on the end result of increasing population density rather than on the underlying modes and mechanisms of competition (Case and Gilpin, 1974; Peters and Barbosa, 1977), and (ii) elements of both interference and exploitation are often interacting within crowded populations (Gilpin, 1974; Miller, 1967). Thus, it is difficult to extricate exploitation competitive effects from interference effects, such as stress or fouling of the habitat with waste products (Peters and Barbosa, 1977). Studies which provide an inferred demonstration of resource exploitation as a major competitive element include experimental crowding of *Ephestia kuehni-ella* larvae (see Section 11.5.3), which results in progressive diminution of body size, a delay in development, and increased mortality (Smith, 1969). Similarly, retarded larval development has been noted in crowded *Aedes aegypti* mosquitoes (Dye, 1982), increased mortality is manifest in crowded populations of both *Callosobruchus* and *Tribolium* beetles (Peters and Barbosa, 1977; Giga and Smith, 1981), and retarded larval development, increased mortality, and an altered sex ratio have been observed in crowded *Culex* mosquitoes (Suleman, 1982) (see Section 11.5.3). Additionally, crowding of *Plodia interpunctella* larvae (see Section 11.5.3) causes earlier pupation and emergence of males compared to females, owing to reduced sensitivity of males to increasing density (Podoler, 1974).

Many of the effects of overcrowding discussed above may directly decrease fitness of surviving individuals. For example, decreased longevity, smaller body size and weight, and poor nutrition frequently result in lowered fecundity (Peters and Barbosa, 1977). Mating frequency may diminish with increasing interference or damage from physical encounters with conspecifics (Parker, 1978b).

However, an assumed prevalence of exploitation competition among insects (Miller, 1967) may be inaccurate. Numerous species utilize strategies which lead to avoidance of over-exploitation of exhaustible resources. For example, the polyphagous parasitoid *Trichogramma embryophagum*, adjusts the number of eggs deposited (1–25) to host size (Klomp and Teerink, 1962). Also, several phytophagous insects utilize visual assessments of conspecific density to avoid oviposition on previously exploited resources (see Section 11.1). Furthermore, several interference competition mechanisms (territoriality, dominance, mutual interference, and chemical communication) may reduce the frequency of exploitative conspecific encounters in insects (Case and Gilpin, 1974).

Finally, it is important to realize that situations such as the above, which we discern and analyze within the framework of present time, may not be representative of the ecology and competitive interactions which may have shaped various competitive avoidance mechanisms over evolutionary time.

11.5 CHEMICALLY MEDIATED PARTITIONING OF RESOURCES

This section treats diverse examples of chemical stimuli which mediate against overcrowding or which signal recognition of previously searched or utilized resource sites. The examples comprise cases where: the resource is mobile vs. stationary; the forager has high mobility vs. slight mobility; the forager is a herbivore vs. a carnivore; the chemical stimulus can be perceived at a distance vs. only upon direct contact; the chemical stimulus is of insect origin (a pheromone) vs. resource site origin; and the resource constitutes principally food, mating territory or a mate, an oviposition site, or a refuge. They are organized according to resource type, while acknowledging that for some species, the same site may offer a combination of resources.

11.5.1 Food

Here, we focus upon situations where the resource is principally one of food for the larval or adult stage. Examples where the resource offers adult food as well as a mating and oviposition site will be discussed under the latter category.

The twelve-spotted asparagus beetle, *Crioceris duodecimpunctata*, oviposits on foliar or stem tissue of female asparagus plants bearing flowers or berries. Though reproductive plants may be patchy in distribution, each usually bears hundreds of berries. Van Alphen and Boer (1980) observed that if a newly-eclosed beetle larva locates a berry which is unoccupied, the larva usually starts gnawing a hole between the sepals, enters the berry, and feeds. Most larvae complete development within a single berry. If the berry is occupied, the searching larva usually leaves upon discovery of the entry hole of the occupant. Data from Fink (1913) and Van Alphen and Boer (1980) reveal that a single berry usually contains and supports no more than one larva. Van Alphen and Boer (unpublished data) demonstrated that the principal factor deterring searching larvae from entering occupied berries is chemical. They have not yet determined whether the chemical is (i) a pheromone deposited via mouthpart dabbing of the berry during pre-entry search, (ii) a pheromone released into the feces, or (iii) a plant constituent emanating from the feces or chewed tissue. Whatever the principal factor, asparagus beetle larvae exemplify deterrence in response to chemical stimuli from discrete sites occupied by conspecifics to avoid over-crowding of exhaustible food resources.

A parallel situation exists in larvae of the European apple sawfly, *Hoplocampa testudinea*, which are repelled by chemical components of larval frass from entering developing apples occupied by conspecific larvae (Roitberg and Prokopy, 1980). Similarly, carnivorous larvae of the staphylinid beetle *Aleochara curtala* searching for fly pupae as food are repelled from entering occupied pupae by pheromone in excretions of the conspecific occupant (Fuldner and Wolf, 1971). Larvae of the Khapra beetle, *Trogoderma granarium*, feeding on grains of wheat, appear to be olfactorily repelled by pheromone emitted by high densities of neighboring conspecific larvae (Tannert and Hein, 1976), though the same or another pheromone may be attractive to larvae when neighbors are at low density (Finger *et al.*, 1965).

Larvae of the lime aphid, *Eucallipterus tiliae*, feeding on plant leaves assess the degree of aphid crowding on that plant, with the ensuing adults emigrating if they were at high density as immatures (Kidd, 1977). Kidd postulates that the cue triggering dispersal might be in the form of aphid-induced changes in leaf physiology or in the form of salivary substances (of aphid or plant origin) injected into the plant by other aphids. If either or both of these explanations turn out to be true, it would represent a most intriguing manner of chemical communication of overcrowding. In the realm of pure speculation, one might also imagine that aphid alarm pheromone (see Nault and Phelan, Chapter 9), aphid-aggregating pheromone (Kay, 1976), or certain repellent fatty acids contained within aphid bodies but released in an unknown manner (Greenway *et al.*, 1978, Griffiths *et al.*, 1978; Sherwood *et al.*, 1981) might, under appropriate concentration or circumstances, function to prevent aphid overcrowding of food resources (Roitberg *et al.*, 1979).

Let us turn now from the realm of chemical mediation against overcrowding *per se* to the realm of chemical partitioning of food foraging activities.

Xylocopa virginica bees forage for nectar from *Passiflora incarnata* flowers. Because of high energy requirements, *Xylocopa* foragers are constrained to visit flowers with large nectar rewards if they are to profit energetically (Heinrich, 1975). In Texas, where Frankie and Vinson (1977) studied the food-foraging activities of these bees, *P. incarnata* vines are patchily distributed on earthen mounds in disturbed areas. These workers observed that in conjunction with nectar-foraging visits, *Xylocopa* females appeared to be depositing a chemical marker on the flowers. When an extract of the contents of female DuFour's glands was applied to several flowers, approaching females turned away just before landing and did not attempt to take any nectar. Indeed, often such bees moved to adjacent subpatches before resuming foraging. The repellent pheromone remained active for about 10 minutes, and was found by Vinson *et al.* (1978) to be in part a mixture of the methyl esters of myristic and palmitic acids. Frankie and Vinson (1977) suggest that the approximate 10-minute duration of pheromone effectiveness may be selectively adjusted to the time required for partial floral replenishment of the nectar supply after a foraging bout. Hence, the pheromone apparently acts to enhance overall foraging efficiency of females within and among *P. incarnata* flower patches.

Similar examples of pheromonal recognition of previously searched food resource sites have been demonstrated or are thought to occur in *Apis mellifera* workers searching for nectar (Ferguson and Free, 1979), in predatory *Coccinella septempunctata* larvae searching for aphid prey (Marks, 1977), and in predatory females of *Amblyseius fallacis* mites searching for phytophagous mites as prey (Hislop and Prokopy, 1981).

11.5.2 Mating territories and mates

As far as we have been able to determine, there exists only one convincing published example among insects of chemical repulsion of males by an intrasexual occupying a mating territory.

When waiting for arrival of females as potential mates, male *Pteronymia veia* butterflies in tropical cloud forests perch on sun-exposed twigs and leaves of trees. Parker (1978a) points out that such 'sunspotting' behavior may be resource-based in that an increase in body temperature may benefit both sexes by increasing the rate at which ingested food is converted into sperm or eggs. As observed by Pliske (1975), the wings of perched males are open and the hairpencils, located on the costae of the hind wings, are splayed and erect. Edgar *et al.* (1976) found the principal chemical component of the hairpencils to be a lactone, precursors of which the males obtain through imbibing pyrrolizidine alkaloids from dead shoots of *Heliotropium indicum* plants. Pliske *et al.* (1976) applied this lactone to attractive *H. indicum* plants and found it to be repellent to the males. Often, males refused to take up perches within 5–10 meters of a lactone-treated site. Other males of the same tribe of butterflies (Ithomiine) likewise were repelled by the lactone. The lactone also repelled other males

during intra- and interspecific homosexual courtship pursuits. It had no observable attractant or repellent effect on the females, though it apparently serves as an aphrodisiac during the latter phase of courtship behavior, once male pursuit of arriving females has already begun. There appear to be two sorts of selective advantage in using this volatile territorial pheromone (i) some degree of uniformity of dispersion of males among sunlit perches favorable for entry and pursuit of females, and (ii) facilitation of energy conservation though rapid termination of fruitless courtship pursuits of other males.

Males in at least seven families of bees and two families of wasps are territorial (Alcock *et al.*, 1978). In some of these species (e.g. *Centris* bees, *Philanthus* wasps), males pheromone-mark areas in the vicinity of nest sites, food resources, and vantage points favorable to scanning for females. Apparently such pheromone marking serves to attract receptive females as well as to establish and partition territories favorable for mating (Raw, 1975; Alcock, 1975; Gwynne, 1978; Evans and O'Neill, 1978; Eickwort and Ginsberg, 1980; Vinson *et al.*, 1982). Although such pheromones are described more fully by Duffield *et al.* in Chapter 14, we are unaware of any conclusive evidence that arriving males are repelled by the marking pheromone of the resident. Indeed, as Alcock *et al.* (1978) speculate, other non-territory-holding males may be attracted to pheromone deposited by a resident at a favorable site as a way of gaining free access to females, without depositing pheromone. Physical combat may then determine the occupant of the territory.

In addition to Ithomiine males, males of other insects are known to release pheromone repellent to intrasexuals during close range intra- or intersexual encounters. Thus, a male armyworm moth, *Pseudaletia unipuncta*, when courting a female, releases from abdominal scent brushes a pheromone which not only has an aphrodisiac effect upon the female, but which may inhibit the approach and courtship activity of other conspecific males (Hirai *et al.*, 1978; Hirai, 1982). Males of the beetle *Tenebrio molitor* and the butterfly *Heliconius erato* transfer anti-aphrodisiac pheromone to females during mating, thereby repelling other males (Happ, 1969; Gilbert, 1976). In all of these cases, males responding to the pheromone avoid wasting time and energy attempting to exploit unproductive (unreceptive) resources.

Several insects form leks, or communal display areas, where males congregate for the sole purpose of attracting and courting females and to which females come for mating (Emlen and Oring, 1977; Borgia, 1979). In some species, such as certain tephritid flies (Prokopy, 1980), males release pheromone which attracts other conspecific males, thus chemically facilitating male aggregation and lek formation. One may postulate that too great a number of males in a single lek may be maladaptive in terms of individual male probability of procuring a mate within a lek as well as in terms of efficient utilization of available lekking sites within a habitat harboring females. However, to our knowledge, there is no evidence as yet of individual males being repelled by pheromone emanating from a large lek.

11.5.3 Egg-laying sites

By far the majority of known cases in which insects respond to chemical cues to avoid resource overcrowding involve reaction to pheromones or plant components emanating from occupied oviposition sites.

The first convincing demonstration in insects of a chemical stimulus having this effect appears to be the work of Salt (1937) on the chalcid parasitoid *Trichogramma evanescens*, which oviposits into a variety of lepidopteran eggs. The host eggs may be clumped or random in distribution, according to species. The parasitoid larva is constrained to complete development within a single host egg (Salt, 1934a). Each egg contains enough nutrients to support only a limited number of such larvae, usually one (Salt, 1934b). After first demonstrating (Salt, 1934b) that *T. evanescens* adults were able to discriminate between parasitized and unparasitized eggs of its host *Sitotroga cerealella*, Salt (1937) went on to prove that this discriminating ability arose from the fact that parasitized hosts had been marked with pheromone. Two sorts of pheromone appeared to be involved. The first is deposited externally on the surface of the host egg during pre-oviposition searching behavior and may be perceived at a distance by the antennae. The second is deposited internally in association with egg deposition and is perceived on contact by the ovipositor.

Since this discovery by Salt, numerous other cases of external and/or internal pheromone marking of host eggs, larvae, or pupae by parasitoids, or of detection of parasitized hosts on the basis of changes in hemolymph chemistry, have been reported (see Vinson, Chapter 8). In addition to such cases of chemical recognition of occupied oviposition sites, some parasitoids, like several aforementioned herbivores or predators searching for food, are capable of discriminating between presearched and unsearched habitat areas containing potential hosts, thereby enhancing oviposition-site foraging efficiency (Price, 1970; Galis and van Alphen, 1981).

Among phytophagous insects, the first evidence suggesting the existence of a chemical deterrent to egg-laying at an occupied oviposition site seems to be the work of Yoshida (1961) on the azuki bean weevil, *Callosobruchus chinensis*, and the southern cowpea weevil, *C. maculatus*. The larvae feed and develop only within seeds of growing or stored legumes. A single egg is laid on each seed. The larva eats its way into the seed, and, unable to move to another seed, is constrained to develop to maturity therein. Most seeds support only a single larva (Mitchell, 1975). Although, in nature, reproducing host plants may be rather distant from one another (Janzen, 1970), each reproductive plant may produce a large number of seeds. Thus, those beetles which locate a host plant or a seed storage site may, within a short time, have access to abundant egg-laying sites. Together, the works of Yoshida (1961), Oshima *et al.* (1973) and Wasserman (1981) demonstrate that during oviposition, the females deposit a marking pheromone on the seed surface which deters repeated egg-laying on the seed. The pheromone is present in both sexes, but is released in a much greater amount by the females. The population consequence of pheromone deposition is uniformity in egg distribution among available seeds, and thereby a reduction in the number of occasions in which there is a deleterious effect on fitness owing to larval competition for the limited resources of a single seed (Mitchell, 1975; Giga and Smith, 1981). A similar phenomenon occurs in the bean weevils *Zabrotes subfasciatus* (Umeya, 1966) and *Acanthoscelides obtectus* (Szentesi, 1981).

The majority of known examples of responses by phytophagous insects to pheromonal cues from occupied oviposition sites occur among tephritid fruit flies. The females oviposit in the flesh of growing fruit. The larvae feed and develop there until maturity, being unable to move from one fruit to another. Several studies have shown a tendency toward uniformity in egg distribution among available oviposition sites in nature (Le Roux and Mukerji, 1963; Prokopy, 1976; Remund *et al.*, 1980; Prokopy *et al.*, 1982). The apple maggot fly, *Rhagoletis pomonella*, is the first tephritid in which such a pheromone was discovered (Prokopy, 1972). In this and all other of those tephritids studied to date which deposit pheromone, it is deposited during ovipositor dragging on the fruit surface immediately after egg-laying and, upon direct contact, deters repeated oviposition in that fruit. The larger the fruit, the greater the number of larvae which can be supported to maturity, and the greater the number of ovipositor dragging bouts which can accumulate before deterrence arises. The *R. pomonella* oviposition-deterring pheromone system is perhaps the best understood deterrent system among Tephritidae. The following elements have been reviewed by Prokopy (1981b): sites of pheromone production and reception; dynamics of pheromone release, residual activity, and partitioning of larval resources from overcrowding; female foraging activities in relation to pheromone-marked and unmarked oviposition sites in nature; and inter-population and inter-species recognition of *R. pomonella* pheromone.

In addition to *R. pomonella*, oviposition-deterring pheromone is now known to occur in many other species of frugivorous Tephritidae (Prokopy, 1981b), although there is evidence against its existence in the frugivorous melon fly, *Dacus cucurbitae* (Prokopy and Koyama, 1982). The latter species, constrained to be opportunistic and oviposit in pre-existing wounds made in fruit by other agents, refrains from ovipositing in fruit occupied by conspecifics on the basis of ability to detect directly conspecific larvae or their effects.

There are several other phytophagous insect groups in which the existence of oviposition-deterring pheromone deposited by adults or emanating from larvae has been demonstrated: the agromyzid fly, *Agromyza frontella* (Quiring, 1982); two species of anthomyiid flies, *Hylemya* spp. (Zimmerman, 1979) and *Atherigona soccata* (Raina, 1981); five species of pyralid moths infesting grains, other stored products or maize, including the Mediterranean flour moth, *E. kuehniella* (Corbet, 1973), dried currant moth, *E. cautella*, cocoa moth, *E. elutella*, and Indian meal moth, *P. interpunctella* (Mudd and Corbet,

1973), and European corn borer, *Ostrinia nubilalis* (Dittrick *et al.*, 1983); the noctuid moths *H. bicruris* (Brantjes, 1976), and corn earworm, *Heliothis zea* (Gross and Jones, 1977); cabbage butterflies *Pieris brassicae* and *P. rapae* (Rothschild and Schoonhoven, 1977; Behan and Schoonhoven, 1978; Den Otter *et al.*, 1980; Schoonhoven *et al.*, 1981); and the monarch butterfly, *Danaus plexippus* (Dixon *et al.*, 1978). Indirect evidence suggests such pheromone may also exist in the cotton boll weevil, *Anthonomus grandis* (McKibben *et al.*, 1982).

In none of the phytophagous insects releasing oviposition-deterring pheromone has the pheromone been fully chemically characterized. However, substantial progress toward this end has been made by Oshima *et al.* (1973) on *Callosobruchus*, by Mudd and Corbet (1973) on *Ephestia*, by Hurter *et al.* (1976) on *R. cerasi*, and by Dundulis, Tumlinson, and Prokopy (unpublished data) on *R. pomonella*.

The residual properties of oviposition-deterring pheromones are of particular importance to the host discrimination process. In cases where deterrent components from occupied resources may be emitted until completion of larval development, such as pheromonal release by larvae of *Ephestia*, *Plodia* and *Heliothis*, host discrimination mediated by pheromone may remain at a high level. However, in cases where deterrent components have only moderate residual activity under dry conditions or are water-soluble, both of which characterize tephritid pheromones, host discrimination mediated by pheromone may break down well before the completion of larval development. Here, one might suspect selection would favor female detection of larvae, or their effects. This does in fact seem to be the case in *R. completa* (Cirio, 1972) and *R. pomonella* (Averill and Prokopy, unpublished data), but further research is needed to explore this aspect more fully.

The dipterans *Hylemya* spp., *A. fraterculus*, and *R. pomonella* are of special interest in that females evidently regulate the amount of pheromone deposited according to quality or size of oviposition site (Zimmerman, 1980, 1982; Prokopy *et al.*, 1982; Averill and Prokopy, unpublished data). This aspect of pheromone deposition behavior deserves careful examination in other species.

There is some evidence which hints that spruce budworm females, *Choristoneura fumiferana*, as well as *Heliothis armigera* and *H. zea* females, when at high population density, are caused disperse by volatile pheromone components emitted by conspecific females (Palanaswamy and Seabrook, 1978; Saad and Scott, 1981). It has not yet been established whether the concentration of female pheromone at potential mating or oviposition sites ever reaches a level causing female dispersal under natural population conditions nor have the females' behavioral reactions to pheromone been linked conclusively to dispersal. However, such dispersal would represent a situation analogous to the aforementioned Ithomiinae butterfly males being repelled from valuable resource sites (sunlit perches) by volatile pheromone from other males. This largely neglected area of the possible response by female moths to intrasexual adult pheromone demands more attention.

In some insects, a single resource may serve not only as an oviposition and larval development site, but offer adult food and serve as a mating site as well. Two insect groups which exploit a single resource in this manner and are known to use chemical cues to avoid overcrowding are *Tribolium* flour beetles and *Dendroctonus* bark beetles. Through the combined work of Alexander and Barton (1943), Loconti and Roth (1953), Naylor (1959, 1961), and Faustini and Burkholder (1980), crowded *Tribolium castaneum* and *T. confusum* adults have been shown to produce and release quinones into flour and thereby effectively repel other conspecific females from entering that site to feed, mate, and oviposit. Similarly, adults of several *Dendroctonus* and *Ips* bark beetles, at high densities in trees or logs, emit pheromones which inhibit arrival of additional conspecific or interspecific adults at that site for feeding, mating, and oviposition (see Birch, Chapter 12).

Up to this point, all situations discussed in which chemicals have been implicated in preventing overcrowding or in recognition of previously searched or utilized resource sites have involved compounds of insect (pheromonal) or uncertain origin. A clear example of a chemical blend solely of plant origin mediating against overcrowding of resources by a phytophagous insect occurs in the tephritid *Dacus oleae*. This insect, apparently monophagous, oviposits singly into growing olive fruits. The original host biotype produces olives so small that no more than one *D. oleae* larva can complete development in a single fruit. Cirio (1971) found that immediately after withdrawal of the ovipositor from the fruit flesh, a *D. oleae* female uses her mouthparts to suck up the juice exuding from the oviposition puncture and proceeds to dab this juice onto the fruit surface while circling the fruit. Cirio showed that arriving females reject juice-marked olives for oviposition, depositing eggs uniformly among otherwise acceptable fruits. The deterrent is apparently detected via both antennal and tarsal chemoreceptors (Haniotakis and Voyadjoglou, 1978). There is debate as to the precise chemical nature of the juice components eliciting deterrence, with dihydroxyphenylethanol (Vita, 1978) as well as acetophenone and benzaldehyde (Girolami *et al.*, 1981) apparently playing active roles. The former compound is generated from glucosides through enzymatic oxidation following female application of juice to the fruit surface (Vita, 1978).

In a situation such as *D. oleae*, where plant components released to the exterior serve apparently as the sole signal of insect occupancy, one wonders whether resource occupancy by a different species may elicit a similar-avoidance response. Comparatively few herbivores feed on olive fruit. To our knowledge, the feeding effects of these other species have not been assayed for deterrency to *D. oleae* oviposition. Interestingly, however, volatile fruit components liberated from the pulp by the feeding of *D. oleae* larvae appear to deter female egg-laying (Girolami *et al.*, 1981). Some of these may be the same components present in the deterrent juice.

While host plant components in abnormally high concentration (see Finch, 1978) or inorganic or organic non-host plant components (see Jermy and Szentesi, 1978; Saxena and Basit, 1982) may repel or inhibit egg-laying in phytophagous insects, we are aware of only two cases other than *D. oleae* where host plant components at normal concentration have been shown convincingly to mediate oviposition deterrence. The first is *O. nubilalis*, where volatiles from (i) corn plants infested with corn borer larvae, (ii) mechanically injured corn plants, or (iii) a distillate of corn juice proved repellent to ovipositing corn borer females (Schurr and Holdaway, 1970). The active volatile components may or may not be the same oviposition-deterring components which, as shown by Dittrick *et al.* (1983), are present in the larval frass of *O. nubilalis*.

The second is the cabbage looper, *Trichoplusia ni*, where compounds in crucifers and other non-cruciferous hosts, released through the feeding activities of the larvae and present also in the larval frass, deter adult egg-laying (Renwick and Radke 1980, 1981). As in the case of *D. oleae*, one wonders about other sorts of biotic and abiotic agents which may cause release of oviposition deterrents from a host plant, thereby indicating the plant may be incapable of supporting larvae of *O. nubilalis* or *T. ni* to maturity.

Turning to detritivores, several species of *Culex* and *Aedes* mosquitoes are attracted to potential oviposition sites or stimulated to oviposit by pheromones emitted by conspecific eggs, larvae, or pupae (Mulla, 1979). Presumably such pheromones indicate locales of comparatively ample resources which have proved favorable to recent development of conspecifics. On the other hand, larvae of some of these same species, when overcrowded, emit chemicals which are toxic or growth-retarding to conspecifics (Mulla, 1979). How these two facets of chemical regulation of mosquito egg and larval density relate to one another and to individual fitness is uncertain and merits further study.

11.5.4 Refugia

A refuge may be considered a place offering shelter or protection from danger or stress. Among non-social insects, 'individual' space may constitute a refuge from stress of physical agitation, stress of high concentration of toxic excretory products from neighboring conspecifics, or stress of competition for nesting or feeding sites. Among social insects, the scope of refuge may be extended to include the nesting site of certain species that form colonies which are rather fixed in location. This concept stems from the interpretation of Hamilton and Watt (1970), who consider 'refuging' to be the 'rhythmical dispersal of groups of animals from and their return to a fixed point in space' (e.g., a nesting site). Resources hitherto viewed principally as being food, mating territories, or egg-laying sites, may of course additionally provide refuge to the occupants. In this section, we consider resources whose principal value appears to be that of refuge.

With respect to non-social insects, as mentioned earlier, larvae of *E. kuehni-ella*, which feed upon a variety of stored food products, emit an oviposition-deterring pheromone. As shown by Corbet (1971), the pheromone is contained within droplets of fluid emitted from larval mandibular glands when larvae meet one another head to head. However, not only does the pheromone deter oviposition, it also elicits larval dispersal away from areas of high pheromone concentration. While such dispersal may be selectively advantageous in terms of reducing larval competition for food, Cotter (1974) found that food remained in excess, irrespective of the degree of larval crowding. He went on to demonstrate that doubling the larval living space by diluting food with non-nutritive, non-toxic sawdust yielded an increase in development rate, possibly owing to a reduction in stress from repeated physical encounters with other larvae or from harmful biotic or abiotic agents associated with larval feces. Whichever, the evidence suggests that larval dispersal in response to high concentration of mandibular gland pheromone is adaptively advantageous in terms of refuge partitioning. The same sort of advantage may accrue to mosquito larvae should they be found to disperse from locales containing high concentrations of larval-emitted, growth-retarding chemicals (Mulla, 1979). Partitioning of refuge sites may also be a principal selective advantage in European house cricket, *Acheta domestica*, dispersal from locales marked with a high concentration of phero-mone emitted by conspecifics (Sexton and Hess, 1968). Finally, individual larvae of the swallowtail *Iphiclides podalirius*, which are solitary and deposit pheromone in association with silk trails laid from relatively permanent resting sites to feeding sites on host foliage, apparently respond positively only to their own pheromonal blend and reject pheromone-marked trails deposited by other conspecific individuals (Weyh and Maschwitz, 1982). This behavior results in establishment of individual resting and feeding territories.

With respect to social insects, perhaps the clearest illustration of the role of pheromones in partitioning refuge sites (*sensu* Hamilton and Watt, 1970) is that of the African weaver ant, *Oecophylla longinoda*. This ant is a dominant species in African forest canopies. The workers bind leaves into tight nest compartments. A single colony, which may consist of more than 500 000 individuals, usually builds hundreds of nests. These are distributed over several nest trees and are concentrated in the peripheral canopy of the trees (Hölldobler and Lumsden, 1980). The workers forage for insect honeydews, plant exudates, insect prey, and a variety of other foods. Hölldobler and Wilson (1978) observed that *O. longinoda* colonies are more or less uniformly distributed within a habitat, and that each colony occupies one to several trees over which it maintains exclusive possession. They showed that the workers mark newly acquired home-range territory with large numbers of pheromone droplets emitted mainly from the rectal vesicle. The pheromone remains active for at least 12 days and enables workers to distinguish their own domain from that of alien colonies. Aliens respond to the pheromone with aversion and aggressive displays.

Worker ants of *Myrmica rubra, M. scabrinodis, M. ruginodis, M. sabuleti* and *Atta cephalotes* likewise deposit territorial marking pheromones (Cammaerts *et al.*, 1977, 1978, 1981; Jaffe *et al.*, 1979). In other ants, such as the harvester ants, *Pogonomyrmex barbatus* and *P. rugosus*, territorial partitioning of resources is not accomplished directly through release of territorial pheromone, but rather indirectly through deposition of colony-specific pheromones by foraging workers along trunk trails (Hölldobler, 1976). If the trunk trails of neighbors overlap or cross, aggressive physical confrontation occurs, resulting in subsequent divergence of the trails from one another and maintenance of territorial integrity. This method of establishing territory is analogous to that of several *Trigona* bee species, which mark foraging areas and potential nest sites with pheromone (Hubbell and Johnson, 1977). The pheromone is attractive to other workers from the same colony, which then defend the marked areas against individuals of other colonies. The end result is uniformity of nest spacing among rival colonies. Chapters 14 and 15 provide further treatment of pheromone mediation of nest and colony spacing in ants and bees.

11.6 STABILITY OF RESOURCE-PARTITIONING CHEMICAL STIMULI

The stability of a chemical partitioning system over time may be the outcome of a diversity of influences on factors such as production, reception, and residual properties of the stimulus involved. Each of these factors will be considered in turn.

Changes in diet as a result of host shifts, polyphagy, or temporary insufficiency of host material may give rise to changes in production of ample quantity or quality of deterrent chemical stimuli emitted by the insect. This may be especially true for insects which release pheromone in frass and in cases where a pheromone precursor is obtained from the host (see Section 11.5). For example, the chemical structure of available host tree precursors determines which isomer of the *Ips paraconfusus* aggregation pheromone will be produced (Renwick *et al.*, 1976). Chemical stimuli of resource site origin (i.e., where a plant component has been shown to act as a resource partitioning cue, see Section 11.5) may also vary with plant changes.

Symbiotic micro-organisms associated with a pheromone-producing insect may add yet another level of complexity to a chemically-mediated system. Thus, a hindgut bacterium is capable of producing two *I. paraconfusus* aggregation pheromones. Moreover, the repellent pheromone of *Dendroctonus frontalis* can be produced by a symbiotic fungus (see Birch, Chapter 12). Thus, changes in the presence or composition of these important micro-organisms may effect attendant changes in a given insect's spacing communication system.

Another major element affecting stability of spacing communication systems may be variability of insect response to chemical stimuli. Several factors may influence this response:

11.6.1 Genetically based intraspecific variation

Within a population, thresholds for response to chemical stimuli may vary. For example, three geographic strains of the pine engraver beetle, *Ips pini*, infesting similar host material, exhibited regional variation in production and reception of an attractant pheromone (Lanier *et al.*, 1972, 1980) (see also Birch, Chapter 12).

11.6.2 Threshold modification based upon physiological state

Individuals deprived of suitable resource sites for increasing lengths of time may demonstrate proportionately greater tendency to accept unsuitable (i.e., previously exploited) resources than do undeprived individuals. For example, female *R. pomonella* will more readily accept fruits marked with oviposition-deterring pheromone following deprivation than following periods of access to suitable oviposition sites (Roitberg and Prokopy, 1983). Similarly, pheromone response may vary with environmental changes such as barometric flux associated with unstable weather (Lanier and Burns, 1978; Prokopy, 1981b), time of day, or season of year (Roberts *et al.*, 1982).

11.6.3 Threshold modification over time

Age effects may influence an individual's reaction to a chemical stimulus. Thus, in *Myrmica rubra* ants (see Section 11.5.3) callow individuals respond only to Dufour's gland secretions whereas older ants are attracted by both mandibular and Dufours' gland pheromone (Cammaerts, 1974). Additionally, sensory adaptation or habituation can influence response level to a stimulus (Waage, 1979; Weseloh, 1980).

11.6.4 Learning as a factor in host discrimination

Naive individuals may require exposure to or previous experience of a host or pheromonal stimulus before host discrimination occurs (van Lenteren, 1981; Roitberg and Prokopy, 1981).

Still another factor in spacing communication system stability is residual activity of a chemical stimulus, which may vary over time according to the insect species and the nature of the message conveyed. For example, the repellent pheromone deposited by *Xylocopa* bees following extraction of nectar from passion flowers persists for only about 10 minutes, which may coincide with the time required for replenishment of some of the nectar, whereas *Oecophylla* ants deposit territorial pheromone which persists for up to 12 days, designating areas of continuous exploitation (see Section 11.5). Residual activity may be particularly unstable in cases of water-soluble chemical stimuli. In the case of *R. pomoriella*, our assessment indicates definite loss of activity of the water-soluble oviposition-deterring pheromone, with 90% activity remaining following gentle showers and only 55% activity remaining following downpours (Averill and Prokopy, unpublished data). Other, as yet un-investigated, factors such as substrate properties, heat, or desiccation probably also affect residual activity of deterrent chemical stimuli.

To this point, we have considered only those factors which may affect spacing communication stability over the short term (e.g., within a single generation). Long-term, evolutionary changes must also be considered. For example, we speculate that changes in insect pheromone production may occur over time as a result of natural enemy pressure. Price (1981) notes that decoding of intraspecific chemical communication systems by kairomonally cueing parasitoids and predators may result in strong selection pressure for changes in the code, either in quality or quantity of the message. Though no example of such a phenomenon is available, Price (1981) projects that given present pest management practices, such shifts can be anticipated for some pest insects. Pheromone production could similarly be altered under relaxed selection pressure characteristic of agricultural situations where resources are both abundantly and uniformly available. Under such conditions, resource partitioning systems would be of decreasing selective advantage. Further, in cases where competition from interspecific individuals for a resource (e.g., Vinson and Ables, 1980) constitutes additional selection pressure, intraspecific communication systems may become obsolete.

Finally, the stability of resource-partitioning chemical stimuli over the long term, as well as the short term, may be affected by the stochastic influences of both abiotic and biotic components of an unpredictable environment.

11.7 CONCLUSIONS

This chapter focuses on chemical stimuli which mediate against insect overcrowding or which signal recognition of previously searched or utilized insect resources of food, mating territories or mates, egg-laying sites, or refugia. In addition to discussing known examples of such chemical influences on resource partitioning, we have attempted to provide an appropriate ecological framework for viewing partitioning by treating factors which affect resource distribution, insect foraging behavior, competition for resources, and stability of spacing signals over time.

It is instructive to note that the vast majority of known examples of insect response to resource-partitioning chemical signals have come to light only within the past decade. Thus, we would predict that chemical messengers of this sort are more widespread among insects than presently realized. More specifically, we anticipate in the near future a substantial increase in the number of species known to release oviposition-deterring pheromone. In addition, we suggest that research aimed at detecting chemical stimuli of host plant origin which deter repeated egg-laying (Section 11.5) and chemical stimuli which elicit dispersal of intrasexuals from overcrowded favorable mating sites (Section 11.5) could prove particularly rewarding.

If we were to speculate on which sorts of insects might be employing resource-partitioning signals, we would hypothesize that populations associated with more permanent habitats comprised of relatively predictable resource sites would be more likely to have evolved such a signalling system than populations associated with more temporary habitats comprised of relatively unpredictable resource sites. By the same reasoning, monophagous-oligophagous (specialist) insects might be more likely to have evolved resource-partitioning signals in association with plant exploitation processes than polyphagous (generalist) insects. We do not intend this to suggest that all, or even a majority, of monophagous–oligophagous insects associated with predictable resource sites utilize such signals. Indeed, our own recent study of the codling moth, *Laspeyresia pomonella*, an oligophagous insect whose larvae infest apple fruit (a limited resource), suggests that no detectable resource-partitioning chemical messengers are emitted by the females, eggs, or larvae during the resource exploitation process (Roitberg and Prokopy, 1982). In addition, over-exploitation of larval host plant resources may be common in certain herbivore–plant systems, as illustrated by the cinnabar moth, *Tyria jacobaeae*, and certain *Heliconius* and *Euptychia* butterflies (Myers and Campbell, 1976; van der Meijden, 1979; Young, 1980; Singer and Mandracchia, 1982).

While many new examples of resource-partitioning stimuli in insects will undoubtedly be uncovered through laboratory investigations, we believe that analysis of the foraging behavior patterns of individuals in space and time in diverse semi-natural or natural (non-agricultural) habitats will prove more productive. Too often, the laboratory test arena is not conducive to the full expression of a behavioral repertoire as exhibited in a more natural situation where the entire constellation of relevant resource and background stimuli may be simultaneously interacting.

Even in nature, the wide array of variables which may affect the degree to which individual insects respond to a resource-partitioning chemical may be of such a magnitude that the response may be only slightly or not at all manifest under certain circumstances. Such variables may include: (i) genetic polymorphism among or even within populations in resource utilization patterns (Wiklund, 1981), (ii) the physiological state of individuals, particularly in cases where resources are scarce or individuals have failed to locate a resource in the recent past (Section 11.6), (iii) resource availability, wherein individuals congregate at already occupied resource sites because available, unoccupied sites go undetected (Section 11.2), or where host range expansion has occurred, as in the case of *R. pomonella*, in which the new resource, apple fruit, is so much larger than the native host fruit of hawthorne (*Crataegus*) that response to the pheromone mediating oviposition deterrence on apples in nature is usually difficult to discern, and (iv) competition effects, wherein intra- or inter-specific individuals are so abundant that most or all available resources have already been utilized (Section 11.4).

Although variability in insect response to resource-partitioning signals may often frustrate the investigator, such variability may be the product of an evolutionarily stable strategy (Maynard Smith, 1974; Parker and Knowlton, 1980). Thus, the future payoff of any one foraging strategy in a habitat containing a mixture of unutilized and utilized resources may be dependent upon the frequency of other foraging strategies in the population, leading to a mixture of strategies maintained in stable equilibrium.

As a final remark, we put forth the prospect of an exciting biological adventure to those trying to 'decode' the chemical mediation of resource partitioning in insects.

ACKNOWLEDGMENTS

Part of the research on *Rhagoletis pomonella* reported here was supported by Massachusetts Agricultural Experiment Station Project No. 488 and by the Science and Education Administration of the U.S. Department of Agriculture under Grant No. 7800168 from the Competitive Research Grants Office and Cooperative Agreement No. 12-14-1001-1205. We appreciate the constructive criticisms of the following persons on an earlier draft of this manuscript: A. Renwick, M. Singer and T. Whitham.

REFERENCES

Alexander, P. and Barton, D. H. R. (1943) The excretion of ethylquinone by the flour beetle. *Biochem. J.*, **37**, 463–5.

Alcock, J. (1975) Territorial behavior by males of *Philanthus multimaculatus* (Hymenoptera: Sphecidae) with a review of territoriality in male sphecids. *Anim. Behav.*, **23**, 889–95.

Alcock, J., Barrows, E. M., Gordh, G., Hubbard, L. J., Kirkendall, L., Pyle, D. W., Ponder, T. L. and Zalom, F. G. (1978) The ecology and evolution of male reproductive behavior in the bees and wasps. *Zool. J. Linn. Soc. Lond.*, **64**, 293–326.

Behan, M. and Schoonhoven, L. M. (1978) Chemoreception of an oviposition deterrent associated with eggs in *Pieris brassicae*. *Ent. exp. appl.*, **24**, 163–79.

Benson, W. W., Brown, K. S. and Gilbert, L. E. (1975) Co-evolution of plants and herbivores: passion flower butterflies. *Evolution*, **29**, 659–80.

Borgia, G. (1979) Sexual selection and the evolution of mating systems. In: *Sexual Selection and Reproductive Competition in Insects* (Blum, M. and Blum, A., eds) pp. 19–80. Academic Press, New York.

Brantjes, N. B. M. (1976) Prevention of superparasitation of *Melandrium* flowers by *Hadena*. *Oecologia*, **24**, 1–6.

Cammaerts, M. C. (1974) Production and perception of attractive pheromones by differently aged workers of *Myrmica rubra* (Hymenoptera: Formicidae). *Insectes Soc.*, **21**, 235–48.

Cammaerts, M. C., Morgan, E. D. and Tyler, R. C. (1977) Territorial marking in the ant *Myrmica rubra* L. (Formicidae). *Biol. Behav.*, **2**, 263–72.

Cammaerts, M. C., Inwood, M. R., Morgan, E. D., Parry, K. and Tyler, R. C. (1978) Comparative study of the pheromones emitted by workers of the ants *Myrmica rubra* and *Myrmica scabrinodis*. *J. Insect Physiol.*, **24**, 207–14.

Cammaerts, M. C., Evershed, R. P. and Morgan, E. D. (1981) Comparative study of the Dufour gland secretions of workers of four species of *Myrmica* ants. *J. Insect Physiol.*, **27**, 59–65.

Campbell, B. C. and Duffey, S. S. (1979) Tomatine and parasitic wasps: potential incompatibility of plant antibiosis with biological control. *Science*, **205**, 700–2.

Case, T. J. and Gilpin, M. E. (1974) Interference competition and niche theory. *Proc. Nat. Acad. Sci. U.S.A.*, **71**, 3073–7.

Caughley, G. and Lawton, J. H. (1981) Plant–herbivore systems. In: *Theoretical Ecology: Principles and Applications* (May, R. M., ed.) pp. 132–66. Sinauer Assoc., Sunderland, MA.

Charnov, E. L., Orians, G. H. and Hyatt, K. (1976) Ecological implications of resource depression. *Am. Nat.*, **110**, 247–59.

Cirio, U. (1971) Reperti sul meccanismo stimolo-risposta nell ovideposizione del *Dacus oleae* Gmelin. *Redia*, **52**, 577–600.

Cirio, U. (1972) Osservazioni sul comportamento di ovideposizione della *Rhagoletis completa* in laboratorio. *Proc. IX Cong. Ital. Ent. Siena*, pp. 99–117.

Corbet, S. A. (1971) Mandibular gland secretion of larvae of the flour moth, *Anagasta kuehniella*, contains an epideictic pheromone and elicits oviposition movements in a hymenopteran parasite. *Nature*, **232**, 481–4.

Corbet, S. A. (1973) Oviposition pheromone in larval mandibular glands of *Ephestia kuehniella*. *Nature*, **243**, 537–8.

Cotter, W. B. (1974) Social facilitation and development in *Ephestia kuehniella*. *Science*, **183**, 747–8.

Den Otter, C. J., Behan, M. and Maes, F. W. (1980) Single cell responses in female *Pieris brassicae* to plant volatiles and conspecific egg odors. *J. Insect Physiol.*, **26**, 465–572.

Dittrick, L. E., Jones, R. L. and Chiang, H. C. (1983) An oviposition deterrent for the European corn borer, *Ostrinia nubilalis*, extracted from larval frass. *J. Insect Physiol.*, **29**, 119–21.

Dixon, C. A., Erickson, J. M., Kellert, D. N. and Rothschild, M. (1978) Some adaptations between *Danaus plexippus* and its food plant, with notes on *Danaus chrysippus* and *Euploea core*. *J. Zool. Lond.*, **185**, 437–67.

Doolan, J. M. (1981) Male spacing and the influence of female courtship behaviour in the bladder cicada, *Cystosoma saundessii* Westwood. *Beh. Ecol. Sociobiol.*, **9**, 269–76.

Dye, C. (1982) Intraspecific competition amongst larval *Aedes aegypti*: food exploitation or chemical interference? *Ecol. Ent.*, **7**, 39–46.

Edgar, J. A., Culvenor, C. C. J. and Pliske, T. E. (1976) Isolation of a lactone, structurally related to the esterifying acids of pyrrolizidine alkaloids, from the costal fringes of male Ithomiinae. *J. Chem. Ecol.*, **2**, 263–70.

Eickwort, G. C. and Ginsberg, H. S. (1980) Foraging and mating behavior in Apoidea. *A. Rev. Ent.*, **25**, 421–6.

Emlen, S. T. and Oring, L. W. (1977) Ecology, sexual selection, and the evolution of mating systems. *Science*, **197**, 215–23.

Evans, H. E. and O'Neill, K. M. (1978) Alternative mating strategies in the digger wasp *Philanthus zebratus* Cresson. *Proc. Nat. Acad. Sci. USA*, **75**, 1901–3.

Faustini, D. L. and Burkholder, W. E. (1980) Interaction of aggregation and epideictic pheromones in *Tribolium castaneum*. Paper presented at the *National Meeting of the Entomological Society of America*, Atlanta. Ga.

Ferguson, A. W. and Free, J. B. (1979) Production of a forage-marking pheromone by the honeybee. *J. Apic. Res.*, **18**, 128–35.

Finch, S. (1978) Volatile plant chemicals and their effect on host plant finding by the cabbage root fly, *Delia brassicae. Ent. exp. appl.*, **24**, 150–9.

Finger, A., Heller, D. and Shulov, A. (1965) Olfactory response of the Khapra beetle (*Trogoderma granarium*) larva to factors from larvae of the same species. *Ecology*, **46**, 542–3.

Fink, D. E. (1913) The asparagus miner and the twelve-spotted asparagus beetle. *Cornell University Agriculture Experiment Station Bulletin No.*, **331**, 422–35.

Frankie, G. W. and Vinson, S. B. (1977) Scent marking of passion flowers in Texas by females of *Xylocopa virginica texana. J. Kansas ent. Soc.*, **50**, 613–25.

Fuldner, D. and Wolf, H. (1971) Staphyliniden-larven beeinflussen chemisch das Orientierungsverhalten ihrer Konkurrenten. *Naturwissenschaft*, **58**, 418.

Galis, F. and van Alphen, J. J. M. (1981) Patch time allocation and search intensity of *Asobara tabida*, a larval parasitoid of *Drosophila. Neth. J. Zool.*, **31**, 596–611.

Giga, P. P. and Smith, R. H. (1981) Varietal resistance and intraspecific competition in the cowpea weevils *Callosobruchus maculatus* and *C. chinensis. J. Appl. Ecol.*, **18**, 755–61.

Gilbert, L. E. (1976) Post-mating female odor in *Heliconius* butterflies: a male-contributed antiaphrodisiac. *Science*, **193**, 419–20.

Gilbert, L. E. (1982) The co-evolution of a butterfly and a vine. *Sci. Am.*, **247**, 110–21.

Gilpin, M. E. (1974) Intraspecific competition between *Drosophila* larvae in serial transfer systems. *Ecology*, **55**, 1154–9.

Girolami, V., Vianello, A., Strapazzon, A., Ragazzi, E. and Veronese, G. (1981) Ovipositional deterrents in *Dacus oleae. Ent. exp. appl.*, **29**, 177–88.

Greenway, A. R., Griffiths, D. C. and Lloyd, S. L. (1978) Response of *Myzus persicae* to components of aphid extracts and to carboxylic acids. *Ent. exp. appl.*, **24**, 169–74.

Griffiths, D. C., Greenway, A. R. and Lloyd, S. L. (1978) The influence of repellent materials and aphid extracts on settling behavior and larviposition of *Myzus persicae. Bull. Ent. Res.*, **68**, 613–19.

Gross, H. R. and Jones, R. L. (1977) Evaluation of oviposition deterrents for *Heliothis zea* on corn. Paper presented at the National Meeting of the Entomological Society of America, Washington, DC.

Gwynne, D. T. (1978) Male territoriality of the bumblebee wolf, *Philanthus bicinctus*: Observations on the behavior of individual males. *Z. Tierpsychol.*, **47**, 89–103.

Hamilton, W. J. and Watt, K. E. F. (1970) Refuging. *A. Rev. Ecol. Syst.*, **1**, 263–86.

Haniotakis, G. E. and Voyadjoglou, A. (1978) Oviposition regulation in *Dacus oleae* by various olive fruit characters. *Ent. exp. appl.*, **24**, 187–92.

Happ, G. M. (1969) Multiple sex pheromones of the mealworm beetle, *Tenebrio molitor. Nature*, **222**, 180–1.

Haukioja, E. (1980) On the role of plant defenses in the fluctuation of herbivore populations. *Oikos*, **35**, 202–13.

Heinrich, B. (1975) Energetics of pollination. *A. Rev. Ecol. Syst.*, **6**, 139–70.

Heinrich, B. (1979) *Bumble-Bee Economics*. Harvard University Press, Cambridge.

Hirai, K. (1982) Directional flow of male scent released by *Pseudaletia separata* Walker and its repellent effect on adults and larvae of four noctuid and one phycitine moth. *J. Chem. Ecol.*, **8**, 1263–70.

Hirai, K., Shorey, H. H. and Gaston, L. K. (1978) Competition among courting male moths: Male-to-male inhibitory pheromones. *Science*, **202**, 644–5.

Hislop, R. G. and Prokopy, R. J. (1981) Mite predator responses to prey and predator-emitted stimuli. *J. Chem. Ecol.*, **7**, 895–904.

Hölldobler, B. K. (1976) Recruitment behavior, home range orientation, and territoriality in harvester ants, *Pogonomyrmex. Behav. Ecol. Sociobiol.*, **1**, 3–44.

Hölldobler, B. K. and Wilson, E. O. (1978) The multiple recruitment systems of the African weaver ant *Oecophylla longinoda. Behav. Ecol. Sociobiol.*, **3**, 19–60.

Hölldobler, B. and Lumsden, C. J. (1980) Territorial strategies in ants. *Science*, **210**, 732–9.

Hubbell, S. P. and Johnson, L. K. (1977) Competition and nest spacing in a tropical stingless bee community. *Ecology*, **58**, 949–63.

Hurter, J., Katsoyannos, B., Boller, E. F. and Wirz, P. (1976) Beitrag zur Anreicherung und Teilweisen Reinigung des eiablageverhindernden Pheromons der Kirschenfliege, *Rhagoletis cerasi* L. (Dipt., Trypetidae). *Z. Angew. Ent.*, **80**, 50–6.

Jaffe, K., Bazire-Benazet, M. and Howse, P. E. (1979) An integumentary pheromone-secreting gland in *Atta* sp: territorial marking with a colony-specific pheromone in *Atta cephalotes. J. Insect Physiol.*, **25**, 833–9.

Janzen, D. H. (1970) Herbivores and the number of tree species in tropical forests. *Am. Nat.*, **104**, 501–28.

Jermy, T. and Szentesi, A. (1978) The role of inhibitory stimuli in the choice of oviposition site by phytophagous insects. *Ent. exp. appl.*, **24**, 258–71.

Kay, R. (1976) Behavioral components of pheromonal aggregation in *Aphis fabae. Physiol. Ent.*, **1**, 249–54.

Kidd, N. A. C. (1977) The influence of population density on the flight behavior of the lime aphid, *Eucallipterus tiliae. Ent. exp. appl.*, **22**, 251–61.

Klomp, H. and Teerink, B. J. (1962) Host selection and number of eggs per oviposition in the egg-parasite *Trichogramma embryophagum* Htg. *Nature*, **195**, 1020–1.

Lanier, G. N., Birch, M. C., Schmitz, R. F. and Furniss, M. M. (1972) Pheromones of *Ips pini* (Coleoptera: Scolytidae): Variation in response among three populations. *Can. Ent.*, **104**, 1917–23.

Lanier, G. N. and Burns, B. W. (1978) Barometric flux: effects on the responsiveness of bark beetles to aggregation attractants. *J. Chem. Ecol.*, **4**, 319–47.

Lanier, G. N., Classon, A., Stewart, T., Piston, J. J. and Silverstein, R. M. (1980) *Ips pini*: The basis for interpopulational differences in pheromone biology. *J. Chem. Ecol.*, **6**, 677–87.

LeRoux, E. J. and Mukerji, M. K. (1963) Notes on the distribution of immature stages of the apple maggot, *Rhagoletis pomonella*, on apples in Quebec. *Ann. Ent. Soc. Quebec*, **8**, 60–70.

Lewontin, R. C. (1979) Sociobiology as an adaptationist program. *Behav. Sci.*, **24**, 5–14.

Loconti, J. D. and Roth, L. M. (1953) Composition of the odorous secretion of *Tribolium castaneum. Ann. ent. Soc. Am.*, **46**, 281–9.

Louskas, C., Lariopoulos, C., Canard, M. and Laudeho, Y. (1980) Early summer infestation of olives by *Dacus oleae* (Gmel.) (Diptera, Trypetidae) and its control by the parasite *Eupelmus urozonus* Dalm. (Hymenoptera, Eupelmidae) in a Greek olive grove. *Z. Angew. Ent.*, **90**, 473–84.

McKibben, G. H., McGovern, W. L. and Dickerson, W. A. (1982) Boll weevil oviposition behavior: a simulation analysis. *J. Econ. Ent.*, **75**, 928–31.

Marks, R. J. (1977) Laboratory studies of plant searching behavior by *Coccinella septempunctata* larvae. *Bull. Ent. Res.*, **67**, 235–41.

Maynard Smith, J. (1974) The theory of games and evolution of animal conflicts. *J. Theoret. Biol.*, **47**, 209–22.

Maynard Smith, J. (1978) Optimization theory in evolution. *A. Rev. Ecol. Syst.*, **9**, 31–56.

Miller, R. S. (1967) Pattern and process in competition. *Adv. Ecol. Res.*, **4**, 1–74.

Mitchell, R. (1975) The evolution of oviposition tactics in the bean weevil, *Callosobruchus maculatus. Ecology*, **56**, 696–702.

Mitchell, P. L. (1980) Combat and territorial defense of *Acanthocephala femorata*. *Ann. ent. Soc. Am.*, **73**, 404–8.

Monteith, L. G. (1960) Influence of plants other than the food plants of their host on host-finding by tachinid parasites. *Can. Ent.*, **92**, 641–52.

Mudd, A. and Corbet, S. A. (1973) Mandibular gland secretion of larvae of the stored products pests *Anagasta kuehniella, Ephestia cautella, Plodia interpunctella* and *Ephestia elutella*. *Ent. exp. appl.*, **16**, 291–3.

Mulla, M. (1979) Chemical ecology of mosquitoes: auto and transpecific regulating chemicals in nature. *Proc. 47th Ann. Conf. California Mosquito and Vector Control Ass.*, 65–8.

Myers, J. H. and Campbell, B. J. (1976) Distribution and dispersal in populations capable of resource depletion: a field study on cinnabar moth. *Oecologia*, **24**, 7–20.

Naylor, A. F. (1959) An experimental analysis of dispersal in the flour beetle *Tribolium confusum*. *Ecology*, **40**, 453–65.

Naylor, A. F. (1961) Dispersal in the red flour beetle *Tribolium castaneum*. *Ecology*, **42**, 231–7.

Nicholson, A. J. (1954) An outline of the dynamics of animal populations. *Aust. J. Zool.*, **2**, 9–65.

Oshima, K., Honda, H. and Yamamoto, I. (1973) Isolation of an oviposition marker from Azuki bean weevil, *Callosobruchus chinensis*. *Agric. Biol. Chem.*, **37**, 2679–80.

Otte, D. and Joern, A. (1975) Insect territoriality and its evolution: population studies of desert grasshoppers on creosote bushes. *J. Anim. Ecol.*, **44**, 29–54.

Owens, E. D. Hauschild, K. I., Hubbell, G. L. and Prokopy, R. J. (1982) Diurnal behavior of plum curculio adults within host trees in nature. *Ann. ent. Soc. Am.*, **75**, 357–62.

Palanaswamy, P. and Seabrook, W. D. (1978) Behavioral responses of the female Eastern spruce budworm, *Choristoneura fumiferana*, to the sex pheromone of her own species. *J. Chem. Ecol.*, **4**, 649–55.

Park, T. (1954) Experimental studies of interspecies competition. II. Temperature, humidity, and competition in two species of *Tribolium. Physiol. Zool.*, **27**, 177–238.

Parker, G. A. (1978a) Evolution of competitive mate searching. *A. Rev. Ent.*, **23**, 173–96.

Parker, G. A. (1978b) Searching for mates. In: *Behavioural Ecology: An Evolutionary Approach*. (Krebs, J. R. and Davies, N. B., eds) pp. 214–44. Sinauer Ass., Sunderland, MA.

Parker, G. A. and Knowlton, N. (1980) The evolution of territory size – some ESS models. *J. Theoret. Biol.*, **84**, 445–76.

Peters, T. M. and Barbosa, P. (1977) Influence of population density on size, fecundity and developmental rate of insects in culture. *A. Rev. Ent.*, **22**, 431–50.

Pielou, E. C. (1977) *An Introduction to Mathematical Ecology*. Wiley Interscience, New York.

Pliske, T. E. (1975) Courtship behavior and use of chemical communication by males of certain species of Ithomiine butterflies. *Ann. ent. Soc. Am.*, **68**, 935–42.

Pliske, T. E., Edgar, J. A. and Culvenor, C. C. J. (1976) The chemical basis of attraction of Ithomiine butterflies to plants containing pyrrolizidine alkaloids. *J. Chem. Ecol.*, **2**, 255–62.

Podoler, H. (1974) Effects of intraspecific competition in the Indian meal-moth (*Plodia interpunctella* Hübner) (Lepidoptera: Phycitidae) on populations of the moth and its parasite *Nemeritis canescens* (Gravenhorst) (Hymenoptera: Ichneumonidae). *J. Anim. Ecol.*, **43**, 641–51.

Price, P. W. (1970) Trail odors: recognition by insects parasitic on cocoons. *Science*, **170**, 546–7.

Price, P. W. (1981) Semiochemicals in evolutionary time. In: *Semiochemicals: Their Role in Pest Control* (Nordlund, D. A., Jones, R. L. and Lewis, W. J., eds) pp. 251–79. Wiley Press, New York.

Price, P. W., Bouton, C. E., Gross, P., McPheron, B. A., Thompson, J. N. and Weis, A. E. (1980) Interactions among three trophic levels: Herbivores and natural enemies. *A. Rev. Ecol. Syst.*, **11**, 41–66.

Prokopy, R. J. (1972) Evidence for a marking pheromone deterring repeated oviposition in apple maggot flies. *Env. Ent.*, **1**, 326–32.

Prokopy, R. J. (1975) Mating behavior of *Rhagoletis pomonella*. V. Virgin female attraction to male odor. *Can. Ent.*, **107**, 905–8.

Prokopy, R. J. (1976) Significance of fly marking of oviposition site (in Tephritidae). In: *Studies in Biological Control* (Deluchhi, V., ed.) pp. 23–7. Cambridge University Press, Cambridge.

Prokopy, R. J. (1980) Mating behavior of frugivorous Tephritidae in nature. *Proc. Symp. Fruit Fly Problems, Kyoto. Nat. Inst. Agric. Sci., Japan*, pp. 37–46.

Prokopy, R. J. (1981a) Epideictic pheromones that influence spacing patterns of phytophagous insects. In: *Semiochemicals: Their Role in Pest Control* (Nordlund, D. A., Jones, R. L. and Lewis, W. J., eds) pp. 181–213. Wiley Press, New York.

Prokopy, R. J. (1981b) Oviposition-deterring pheromone system of apple maggot flies. In: *Management of Insect Pests with Semiochemicals*. (Mitchell, E. R., ed.) pp. 477–94. Plenum Press, New York.

Prokopy, R. J. and Koyama, J. (1982) Oviposition site partitioning in *Dacus cucurbitae*. *Ent. exp. appl.*, **31**, 428–32.

Prokopy, R. J., Malavasi, A. and Morgante, J. S. (1982) Oviposition-deterring pheromone in *Anastrepha fraterculus*. *J. Chem. Ecol.*, **8**, 763–71.

Pyke, G. H., Pulliam, H. R. and Charnov, E. L. (1977) Optimal foraging: A Selective review of theory and tests. *Q. Rev. Biol.*, **52**, 137–54.

Quiring, D. T. W. (1982) An epideictic pheromone in the alfalfa blotch leafminer, *Agromyza frontella*. Paper presented at the National Meeting of the Entomological Society of America, Toronto, Canada.

Raina, A. K. (1981) Deterrence of repeated oviposition in sorghum shootfly, *Atherigona soccata*. *J. Chem. Ecol.*, **7**, 785–90.

Rauscher, M. D. (1979) Egg recognition: its advantage to a butterfly. *Anim. Behav.*, **27**, 1034–40.

Raw, A. (1975) Territoriality and scent marking by *Centris* males in Jamaica. *Behaviour*, **54**, 311–21.

Remund, U., Katsoyannos, B. I. and Boller, E. F. (1980) Zur Eiverteilung der Kirschenfliege, *Rhagoletis cerasi*, im Freiland. *Mitt. Schweiz. Ent. Gesell.*, **53**, 401–5.

Renwick, J. A. A. and Radke, C. D. (1980) An oviposition deterrent associated with frass from feeding larvae of the cabbage looper, *Trichoplusia ni*. *Env. Ent.*, **9**, 318–20.

Renwick, J. A. A. and Radke, C. D. (1981) Host plant constituents as oviposition deterrents for the cabbage looper, *Trichoplusia ni*. *Ent. exp. appl.*, **30**, 201–4.

Renwick, J. A. A., Hughes, P. R. and Krull, I. S. (1976) Selective production of *cis*- and *trans*- verbenol from (−)- and (+)- α-pinene by a bark beetle. *Science*, **191**, 199–201.

Roberts, E. A., Billings, P. M., Payne, T. L., Richerson, J. V., Berisford, C. W., Hedden, R. L. and Edson, L. J. (1982) Seasonal variation in laboratory response to behavioral chemicals of the southern pine beetle. *J. Chem. Ecol.*, **8**, 641–52.

Roitberg, B. D., Myers, J. H. and Frazer, B. D. (1979) Influence of predators on the movement of apterous aphids between plants. *J. Anim. Ecol.*, **48**, 111–22.

Roitberg, B. D. and Prokopy, R. J. (1980) Spacing chemicals in European apple sawfly. Paper presented at the Nat. Meeting of the Ent. Soc. Am., Atlanta, Ga.

Roitberg, B. D. and Prokopy, R. J. (1981) Experience required for pheromone recognition by the apple maggot fly. *Nature*, **292**, 540–1.

Roitberg, B. D. and Prokopy, R. J. (1982) Resource assessment by adult and larval codling moths. *J. NY Ent. Soc.*, **90**, 260–6.

Roitberg, B. D. and Prokopy, R. J. (1983) Host deprivation influence on response of *Rhagoletis pomonella* to its oviposition-deterring pheromone. *Physiol. Ent.*, **8**, 69–72.

Rosenthal, G. A. and Janzen, D. H. (eds) (1979) *Herbivores: Their Interactions with Secondary Plant Metabolites*. Academic Press, New York.

Rothschild, M. and Schoonhoven, L. M. (1977) Assessment of egg load by *Pieris brassicae*. *Nature*, **266**, 352–5.

Rudinsky, J. A., Morgan, M. Libbey, L. M. and Michael, R. R. (1973) Sound production in Scolytidae: 3-Methyl-2-cyclohexen-1-one released by the female Douglas fir beetle in response to male sonic signal. *Env. Ent.*, **2**, 505–9.

Russ, K. (1969) Beitrage zum Territorialverhalten der Raupen des Springwurmwicklers., *Sparganothis pilleriana* (Lepidoptera: Tortricidae). *Pflanzenschutzberichte*, **40**, 1–9.

Saad, A. D. and Scott, D. R. (1981) Repellency of pheromones released by females of *Heliothis armigera* and *H. zea* to females of both species. *Ent. exp. appl.*, **30**, 123–7.

Salt, G. (1934a) Experimental studies in insect parasitism. I. Introduction and technique. *Proc. R. Soc. Lond. Ser. B.*, **114**, 450–4.

Salt, G. (1934b) Experimental studies in insect parasitism. II. Superparasitism. *Proc. R. Soc. Lond. Ser. B.*, **114**, 455–76.

Salt, G. (1937) Experimental studies in insect parasitism. V. The sense used by *Trichogramma* to distinguish between parasitized and unparasitized hosts. *Proc. Roy Soc. Lond. Ser. B.*, **122**, 57–75.

Salt, G. (1961) Competition among insect parasitoids. In: *Mechanisms in Biological Competition. Symp. Soc. exp. Biol.*, **15**, 96–119.

Saxena, K. N. and Basit, A. (1982) Inhibition of oviposition by volatiles of certain plants and chemicals in the leafhopper *Amrasca devastaus* (Distant). *J. Chem. Ecol.*, **8**, 329–38.

Schoonhoven, L. M., Sparnaay, T., Van Wissen, W. and Meerman, J. (1981). Seven-week persistance of an oviposition-deterring pheromone. *J. Chem. Ecol.*, **7**, 583–8.

Schurr, K. and Holdaway, F. G. (1970) Olfactory responses of female *Ostrinia nubilalis*. *Ent. exp. appl.*, **13**, 455–61.

Sexton, O. J. and Hess, E. H. (1968) A pheromone-like dispersant affecting the local distribution of the European house cricket, *Acheta domestica*. *Biol. Bull.*, **134**, 490–502.

Shapiro, A. M. (1981) Egg mimics of *Streptanthus* (Cruciferae) deter oviposition by *Pieris sisymbrii* (Lepidoptera: Pieridae). *Oecologia*, **48**, 142–3.

Sherwood, M. H., Greenway, A. R. and Griffiths, D. C. (1981) Responses of *Myzus persicae* to plants treated with fatty acids. *Bull. Ent. Res.*, **71**, 133–6.

Singer, M. C. and Mandracchia, J. (1982) On the failure of two butterfly species to respond to the presence of conspecific eggs prior to oviposition. *Ecol. Ent.*, **7**, 327–30.

Smith, S. D. (1969) The effects of crowding on larvae of the meal moth, *Ephestia kuehniella*. *J. exp. Zool.*, **170**, 193–204.

Stamp, N. E. (1980) Egg deposition patterns in butterflies: Why do some species cluster their eggs rather than deposit them singly. *Am. Nat.*, **115**, 367–80.

Suleman, M. (1982) The effects of intraspecific competition for food and space on the larval development of *Culex quinquefaciatus*. *Mosq. News*, **42**, 347–56.

Suzuki, N., Kunimi, Y., Uematsu, S. and Kobayashi, K. (1980) Changes in spatial distribution pattern during the larval stage of the fall webworm, *Hyphantria cunea* Drury (Lepidoptera: Arctiidae). *Res. Popul. Ecol.*, **22**, 273–83.

Szentesi, A. (1981) Pheromone-like substances affecting host-related behavior of larvae and adults in the dry bean weevil, *Acanthoscelides obtectus*. *Ent. exp. appl.*, **30**, 219–26.

Tannert, W. and Hein, B. C. (1976) Nachweis und Funktion eines repellent Pheromones der Larven von *Trogoderma granarium*. *Zool. Jahrb. Physiol.*, **80**, 69–81.

Tsubaki, Y. and Shiotsu, Y. (1982) Group feeding as a strategy for exploiting food resources in the burnet moth *Pryeria sinica*. *Oecologia*, **55**, 12–20.

Umeya, K. (1966) Studies on the comparative ecology of bean weevils. I. On the egg distribution and the oviposition behaviors of three species of bean weevils infesting Azuki bean. *Res. Bull. Plant Prot. Japan.*, **3**, 1–11.

van Alphen, J. J. M. and Boer, H. (1980) Avoidance of scramble competition between larvae of the spotted asparagus beetle. *Crioceris duodecimpunctata* by discrimination between unoccupied and occupied asparagus berries. *Neth. J. Zool.*, **30**, 136–43.

van Lenteren, J. C. (1981) Host discrimination by parasitoids. In: *Semiochemicals: Their Role in Pest Control* (Nordlund, D. A., Jones, R. L. and Lewis, W. J., eds) pp. 153–80. Wiley Press, New York.

van Lenteren, J. C., Nell, H. W., Sevenster-van der Lelie, L. A. and Woets, J. (1976) The parasite-host relationship between *Encarsia formosa* and *Trialeurodes vaporarium*. I. Host finding by the parasite. *Ent. exp. appl.*, **20**, 123–30.

van der Meijden, E. (1979) Herbivore exploitation of a fugitive plant species: local survival and extinction of the Cinnabar moth and the ragwort in a heterogeneous environment. *Oecologica*, **42**, 307–23.

Vinson, S. B. (1975) Biochemical coevolution between parasitoids and their hosts. In: *Evolutionary Strategies of Parasitic Insects and Mites* (Price, P., ed.) pp. 14–48. Plenum Press, New York.

Vinson, S. B. Frankie, G. W., Blum, M. S. and Wheeler, J. W. (1978) Isolation, identification, and function of the Dufour gland secretion of *Xylocopa virginica texana*. *J. Chem. Ecol.*, **4**, 315–23.

Vinson, S. B. and Ables, J. R. (1980) Interspecific competition among endoparasitoids of tobacco budworm larvae (Lep.: Noctuidae). *Entomophaga*, **25**, 357–62.

Vinson, S. B., Williams, H. J., Frankie, G. W., Wheeler, J. W., Blum, M. S. and Coville, R. E. (1982) Mandibular glands of male *Centris adani*, their morphology, chemical constituents, and function in scent marking and territorial behavior. *J. Chem. Ecol.*, **8**, 319–27.

Vita, G. (1978) Individuazione del composto chimico che regola la deposizione delle uova del *Dacus oleae*. *Rapporto Tecnico CNEN, RT/BIO* (78) **17**, 1–8.

Waage, J. K. (1979) Foraging for patchily distributed hosts by the parasitoid, *Nemeritis canescens*. *J. Anim. Ecol.*, **48**, 353–71.

Wasserman, S. S. (1981) Host-induced oviposition preferences and oviposition markers in the cowpea weevil, *Callosobruchus maculatus*. *Ann. ent. Soc. Am.*, **74**, 242–5.

Weseloh, R. M. (1980) Behavioral changes in *Apanteles melanoscelus* females exposed to gypsy moth silk. *Env. Ent.*, **9**, 345–9.

Weyh, R. and Maschwitz, U. (1982) Individual trail marking by larvae of the scarce swallowtail *Iphiclides podalirius* L. (Lepidoptera; Papilionidae). *Oecologia*, **52**, 415–6.

Whitham, T. G. (1979) Territorial behavior of *Pemphigus* gall aphids. *Nature*, **279**, 324–5.

Whitham, T. G. (1983) Host manipulation of parasites: within-plant variation as a defense against rapidly evolving pests. In: Variable Plants and Herbivores in Natural and Managed Systems (Denno, R. F. and McClure, M. S., eds) pp. 15–41. Academic Press, New York.

Wiklund, C. (1981) Generalist vs. specialist oviposition behavior in *Papilio machaon* (Lepidoptera) and functional aspects on the hierarchy of oviposition preferences. *Oikos*, **36**, 163–70.

Williams, K. S. and Gilbert, L. E. (1981) Insects as selective agents on plant vegetative morphology: egg mimicry reduces egg laying by butterflies. *Science*, **212**, 467–9.

Wilson, E. O. (1975) *Sociobiology: The New Synthesis*. Belknap Press, Cambridge, Massachusetts.

Yoshida, T. (1961) Oviposition behavior of two species of bean weevils and interspecific competition between them. *Mem. Fac. Lib. Arts Educ., Miyazaki Univ.*, **11**, 41–65.

Young, A. M. (1970) Predation and abundance in populations of flour beetles. *Ecology*, **51**, 602–19.

Young, A. M. (1980) Over-exploitation of larval host plants by the butterflies *Heliconius cydno* and *Heliconius sapho* (Lepidoptera: Nymphalidae: Heliconiinae: Heliconiini) in Costa Rica. *J. N. Y. ent. Soc.*, **88**, 217–27.

Young, A. M. and Moffett, M. W. (1979) Studies on the population biology of the tropical butterfly *Mechanitis isthmia* in Costa Rica. *Am. Midl. Nat.*, **101**, 309–19.

Zimmerman, M. (1979) Oviposition behavior and the existence of an oviposition-deterring pheromone in *Hylemya*. *Env. Ent.*, **8**, 277–9.

Zimmerman, M. (1980) Selective deposition of an oviposition-deterring pheromone by *Hylemya*. *Env. Ent.*, **9**, 321–4.

Zimmerman, M. (1982) Facultative deposition of an oviposition-deterring pheromone by *Hylemya*. *Env. Ent.*, **11**, 519–22.

12

Aggregation in Bark Beetles

M. C. Birch

12.1 INTRODUCTION

Bark beetles belong to the family Scolytidae, many species of which aggregate on their host trees and then breed in the host tissues. Aggregation is mediated by pheromones in combination with other stimuli. Some species colonize living trees, some, weakened or dying trees, and others, recently fallen timber.

The aggregation and boring activity of large numbers of beetles over a short period of time may be enough to overcome and kill even an apparently healthy tree. Dead timber may be scarce, and what is available may desiccate rapidly or be colonized first by competing species. In either case there is an advantage in colonizing the host quickly. This rapid aggregation of large numbers of beetles on trees or timber is called 'mass attack', and it is this phenomenon, and the role of chemical communication in its initiation, continuance, and termination that are discussed here.

Several scolytid species are of great economic importance. *Dendroctonus* species primarily attack living trees, often valuable for timber or shade, aggregate on them, kill them, and breed in the phloem and bark tissues. *Ips* species can attack living trees, but usually colonize fallen trees, broken limbs or trees already partly colonized by *Dendroctonus* species. One notable exception is the spruce *Ips, I. typographus*, which kills spruce in Europe. *Scolytus* species attack a wide variety of host trees, including conifers, and two species, *S. scolytus* and *S. multistriatus* have recently become more widely known as vectors of the Dutch elm disease pathogen, *Ceratocystis ulmi*. Ambrosia beetles, primarily *Trypodendron* and *Gnathotrichus* species, cause problems by attacking felled and stored lumber.

Chemical Ecology of Insects. Edited by William J. Bell and Ring T. Cardé
© 1984 Chapman and Hall Ltd.

When a host tree is colonized, female beetles oviposit along galleries chewed out of the phloem or at the phloem—xylem or phloem—bark interfaces (or into the xylem in the case of ambrosia beetles). When the eggs hatch, the larvae feed and tunnel into the surrounding host tissues, producing systems of galleries which are distinctive for each species. The larvae pupate at the end of their galleries, and on eclosion, they tunnel out through the bark, fly and colonize new trees.

Clearly, a single tree cannot support an infinite number of beetles. The process of colonization must end before the number of attacks reaches a density above which the developing larvae do not have enough food for all to develop. Some species will lay many more eggs in lightly colonized wood than when the population density is high, but at the higher densities there are clearly mechanisms by which mass attacks can be terminated and a balance achieved between aggregation and optimal spacing in the host tree.

The interaction between one beetle and its host tree cannot be studied in isolation. Several bark beetle species co-exist in most environments and may be competing for the same resources for breeding. Aggregation pheromones released into the environment by any one species are also available to be perceived by other species and may trigger their own colonization or avoidance of the same resource. Parasites and predators of bark beetles respond to the beetle aggregation pheromones and are guided to the trees just when their bark beetle prey are arriving in greatest numbers. Besides these interactions, the presence of fungi and of other micro-organisms, which may be involved in pheromone production, mean that chemical communication in these systems is fascinatingly complex.

This sequence of events will be described in roughly the order in which they occur. Successful selection of host material leads to pheromone production, aggregation, and subsequent termination of responses, as a tree is fully colonized. Inter-specific responses may modify the behavior at all stages by influencing the location or portion of the tree that is attacked or by influencing density through competition, predation or parasitism. The complex of species colonizing ponderosa pine in California will be used to illustrate the principal concepts and other species are mentioned where they add different ideas. References are recent rather than comprehensive and give a lead into the various directions of research. Wood (1982) gives an excellent overall review of the role of pheromones, kairomones and allomones in host tree selection and colonization. Other recent reviews cover the topic from different perspectives (Borden, 1974; 1977; Vité and Francke, 1976; Birch, 1978).

12.2 SELECTION OF HOST MATERIAL

The mechanisms by which a tree or dead limb is first located by a bark beetle and then determined to be suitable or unsuitable for colonization are little known. However, two hypotheses have been extensively debated (Wood, 1972; 1982; Moeck, 1981). One hypothesis is that the beetles respond to 'primary' attractants produced by a tree which are symptomatic of disease or other stress factors that may weaken the tree. The second hypothesis is that beetles land randomly on trees. Only after landing, and possibly feeding, does a beetle perceive a tree as suitable or not, and either start to release pheromone or leave. There is some evidence to support both hypotheses.

Primary attraction has been demonstrated in several species. The clearest example is that of the ambrosia beetle *Trypodendron lineatum* which is attracted to ethanol produced under anaerobic conditions in dying trees (Moeck, 1970). Using host material cut or damaged to expose phloem and xylem tissues, primary attraction has also been demonstrated in four other ambrosia beetles, two *Scolytus*, two *Dendroctonus*, *Ips typographus* and five other species (Moeck, 1981).

Conversely, other studies have failed to demonstrate the existence of primary attraction in eleven *Ips* and *Dendroctonus* species. It is exceedingly difficult to determine whether selection occurs without primary attraction since the events in selection, aggregation and colonization are not discreet, and as soon as the first beetle that has landed and started to bore produces a 'secondary' attractant (i.e., the aggregation pheromone), other beetles respond to this rather than to any 'primary' attractant stimulus produced by the tree. Moeck (1981) conducted rigorous field experiments to determine whether the bark beetles colonizing penderosa pine in California were using primary attraction. He cut logs from trees, treated them anaerobically to induce production of chemicals associated with this type of stress. He also treated some living trees with the herbicide, cacodylic acid, to ensure rapid death, and froze the bases of other trees with solid CO_2 to interrupt the upward flow of water and so simulate drought stress. Trees were covered from the ground into the crown with fine mesh metal screen to prevent boring by landing beetles which would have produced secondary attractants and nullified the experiment.

In none of his experiments could Moeck demonstrate that primary attraction is used in host selection by the three principal species in the area: *Dendroctonus brevicomis* (the western pine beetle), *D. ponderosae* (the mountain pine beetle) and *Ips paraconfusus* (the California five-spined engraver beetle). What Moeck's results did show, however, was that beetles landed indiscriminantly on healthy and stressed trees at about one beetle a day on each tree. Theoretically only one beetle is needed to initiate mass attack, since as soon as it releases pheromone, landing rates would increase markedly on that tree. In Moeck's study, landing rates on trees that became attacked by *D. brevicomis* increased to up to 800 beetles a day.

The evidence thus seems to indicate that beetles of these three species land on host trees and non-host trees at random. Only after landing and close contact with the host does the beetle select or reject the tree. This is corroborated by Elkinton and Wood (1980) who found that *I. paraconfusus* selected their host, ponderosa pine, in preference to a non-host, white fir, only after boring through the outer bark. Male beetles bored in white fir only until they reached the phloem, whereas they bored extensively in ponderosa pine. However, if confined in artificial entrance tunnels made in white fir, male *I. paraconfusus* produced pheromone capable of attracting conspecifics, though in much lower quantities than normal (Elkinton *et al.*, 1980). The apparent waste of energy in boring through the bark of an unsuitable tree before rejecting it may be necessary since the outer bark consists of dead cells and may not contain enough stimuli to permit distinction. Also, beetles on the surface of trees are subject to attack by predators so that there may be a distinct selective advantage in rapid initiation of boring. Beetles invariably seek out bark fissures in which to bore, so the distance and energy required to reach the phloem may be minimal.

The decision to colonize or to leave their host tree, *Pinus radiata*, is also made by *Ips grandicollis*, in Australia, only after the beetles have landed and bored into the bark (Witanachchi and Morgan, 1981). In resistant trees, the exudation of resin leads to retreat of the beetles without pheromone production.

In many species, rapid resin flow, which covers the surfaces exposed by boring and hardens round the wound, as well as flushing beetles from their boring holes, is a very obvious resistance mechanism of the tree. Equally important, however, may be the hypersensitive reaction induced in the tree by beetle attack. In resistant trees, the infection of pathogenic fungi introduced along with beetle attacks is localized by a rapidly developing necrosis and wound reaction which isolates the pathogen within the localized attack area (Berryman, 1972). If the pathogen overcomes this reaction, the tree may be sufficiently weakened that further beetle attacks will be more successful. Alternatively, very large numbers of attacks on a resistant tree may cause enough damage to overcome it whatever the resistance level.

Besides pathogenic fungi which they may transport on the body surface, beetles often carry one or more species of fungi in specialized chambers or mycangia. In *D. brevicomis*, the mycangial fungi are *Ceratocystis nigracarpa* and an unidentified basidiomycete, and the pathogenic fungus is *Ceratocystis minor*. The mycangial fungi clearly gain an advantage from the beetle in being transported between hosts, but the reciprocal advantage is less clear. Paine (1981) has now shown that inoculation of mycangial fungi into ponderosa pine can increase water stress, so that these fungi may aid in killing trees and enhance beetle success.

12.3 PHEROMONE PRODUCTION

Once the first, or first few, beetles have bored into a suitable host tree they begin to release pheromone which attracts both sexes of the same species. In polygamous species, such as *Ips*, males attack first and both sexes respond. Males initiate new attacks, females locate the entrance holes of established males, enter, mate and then construct galleries along which the eggs are laid. In monogamous species such as *Dendroctonus*, the females attack first. In both cases, as more beetles arrive at the tree and start producing pheromone, the attraction of the host increases until the whole tree or log is colonized. The focus for attraction may spread to neighboring trees as beetles land on those and so create a group of killed trees. Some species, especially those that attack living trees, seem to be able to release pheromone as soon as they have selected a suitable host tree, which may stimulate more rapid colonization and help to overcome the tree's resistance. Those species that colonize dead timber may need to feed before pheromone is released.

The first identification of a bark beetle aggregation pheromone was from *I. paraconfusus* (then *I. confusus*). The pheromone was isolated by collecting several kilograms of the frass (boring dust and faeces) produced by males boring in ponderosa pine. The pheromone is released from the hind guts of the males along with the faeces. Repeated extraction of the frass in benzene, followed by fractionation, resulted in the identification of a three-component pheromone; ipsenol, ipsdienol and *cis*-verbenol (Silverstein *et al.*, 1966) (Fig. 12.1). The combination of the three compounds is synergistic, little or no attraction being elicited by any of them alone. The isolation procedure was monitored by a bioassay based on the upwind walking response of beetles to sources of odour in a multiple-choice olfactometer (Wood *et al.*, 1966). Although the beetles fly to attacked trees in the field, controlled flight responses in the laboratory are exceedingly difficult to obtain. However, the walking response also monitors an upwind orientation behavior, and compounds isolated using it have later proved attractive in the field (Wood *et al.*, 1968).

Isolation and identification techniques have since become more sophisticated. Extraction of frass did lead to identification of the pheromone of *D. brevicomis*, but other pheromones have been isolated by condensation of the air from around logs containing boring male beetles (e.g., *I. pini*, Browne *et al.*, 1974; Birch *et al.*, 1980a) (Fig. 12.2), by extraction from the hind-guts of male beetles (e.g., *I. grandicollis* Vité and Renwick, 1971), or by the absorption of pheromone-laden air on a substrate (such as Porapak®) and its later extraction by solvent (e.g., *Scolytus multistriatus*, Pearce *et al.*, 1975).

The choice of method for bioassaying for activity during pheromone isolation has long been a contentious issue. Field bioassays are agreed to be the most reliable method but are very dependent on the availability of field sites, sufficient beetles and weather conditions. Rigorous tests are also exceedingly expensive to run, so that some sort of laboratory assay is usually essential to monitor chemical procedures until pure compounds are isolated. For the identification of the pheromone of *I. pini* in California as one enantiomer of ipsdienol (Birch *et al.*, 1980a) (Fig. 12.2) a laboratory bioassay was essential because the enantiomers were very difficult to prepare and available only in μg quantities. Only after a strong positive response in the laboratory was it worth testing the pure material in the field.

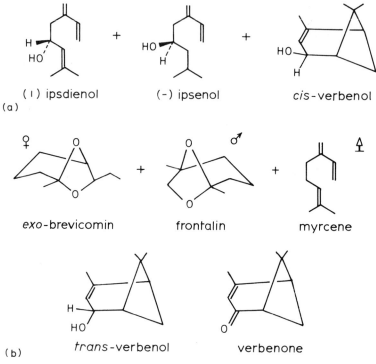

Fig. 12.1 Aggregation pheromones of: (a) *Ips paraconfusus*, three male-produced synergistic components — (+)2-methyl-6-methylene-2,7-octadien-4-ol (*ipsdienol*); (−)2-methyl-6-methylene-7-octen-4-ol (*ipsenol*); and *cis*-verbenol. (b) *Dendroctonus brevicomis*, three attractive components, *exo*-7-ethyl-5-methyl-6,8-dioxabicyclo-[3.2.1] octane (*exo*-brevicomin) from the female; 1,5-dimethyl-6,8-dioxabicyclo-[3.2.1] octane (frontalin) from the male; and myrcene released from the host tree. Verbenone and *trans*-verbenol, released after a male locates the female, interrupts response to the attractant blend.

Fig. 12.2 Schematic outline of procedures leading to the identification of (−)ipsdienol as the aggregation pheromone of *Ips pini* in California (Browne *et al.*, 1974; Birch *et al.*, 1980a). All the volatile compounds produced by male beetles boring in ponderosa pine were liquefied in a cold trap. After distilling off the air at low temperature and removing the water, the resulting concentrate was fractionated by gas chromatography (GC). The activity of the condensate and all GC fractions at each step were monitored using a bioassay based on the upwind walking response of beetles to an attractive odor source. Repeated GC and assays showed that only one isolated compound, ipsdienol, had consistent activity, but synthetic (racemic) ipsdienol failed to elicit any positive responses. When the enantiomers of ipsdienol were resolved to 95 + % enantiomeric purity, the beetles responded positively to (−)ipsdienol in the laboratory, whereas (+)ipsdienol interrupted the response to (−), explaining the early lack of activity from synthetic racemic ipsdienol. In the field, (−)ipsdienol was competitive with boring male *I. pini* in attractiveness to flying beetles.

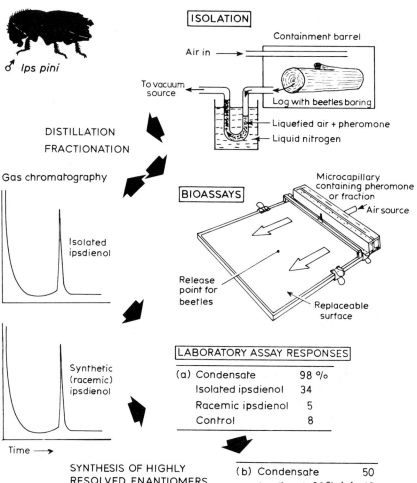

♂ *Ips pini*

ISOLATION

Air in →

Containment barrel

To vacuum source

Log with beetles boring

Liquefied air + pheromone

Liquid nitrogen

DISTILLATION
FRACTIONATION

Gas chromatography

Isolated ipsdienol

BIOASSAYS

Microcapillary containing pheromone or fraction

Air source

Release point for beetles

Replaceable surface

Synthetic (racemic) ipsdienol

Time →

LABORATORY ASSAY RESPONSES

(a) Condensate	98 %
Isolated ipsdienol	34
Racemic ipsdienol	5
Control	8

(b) Condensate	50
Ipsdienol 96% (−)	48
Ipsdienol 88% (−)	10

SYNTHESIS OF HIGHLY
RESOLVED ENANTIOMERS
OF IPSDIENOL

TRAP CATCHES IN THE FOREST

Trap bait	Beetles trapped (mean)
25 ♂♂ boring *I. pini*	60
Racemic ipsdienol	1
(+) ipsdienol	0
(−) ipsdienol	83
Control	0

PHEROMONE

HO

H

(R) (−) IPSDIENOL

The identification of the aggregation pheromone of *D. brevicomis* as the synergistic mixture of two bicyclic ketals, *exo*-brevicomin and frontalin, and the terpene alcohol myrcene (Bedard *et al.*, 1969; Bedard *et al.*, 1980b) (Fig. 12.1) raised the issue of definition. *Exo*-brevicomin produced by females and frontalin by male beetles satisfy the conspecific criterion for a pheromone. Myrcene, however, is released by the host tree, ponderosa pine, when it is attacked, and the combination forms a highly attractive blend. Like most definitions, that of pheromone breaks down if it is applied too inflexibly. It is most straightforward to call the three-component attractive blend the pheromone of *D. brevicomis*. In view of the many similar terpenes and derivatives being released by trees and beetles in the forest environment, the use of blends has the effect of increasing specificity. Only one combination of compounds functions as the aggregation pheromone. Others may convey other information or be part of the background chemical 'noise' which is ignored.

Many of the pheromone components of bark beetles are chiral compounds, and the different enantiomers have very different biological activities (Silverstein, 1979). Thus, it is $(1R,5S,7R)$-$(+)exo$-brevicomin and $(1S,5R)(-)$frontalin that synergise with myrcene in the pheromone of *D. brevicomis* (Fig. 12.1) (Wood *et al.*, 1976); the opposite enantiomers are inactive. Similarly, *I. grandicollis* responds to $(S)(-)$ipsenol and not to $(R)(+)$ipsenol (Vité *et al.*, 1976). The pheromone of *I. paraconfusus* was identified as $(S)(-)$ipsenol, predominantly the $(S)(+)$enantiomer of ipsdienol, and $(S)(+)cis$-verbenol (Silverstein *et al.*, 1966), and that of *I. pini* to be $(R)(-)$ipsdienol (Birch *et al.*, 1980a) (Fig. 12.2). However, in *I. pini* the activity of $(R)(-)$ipsdienol is completely nullified or interrupted by as little as 3% of $(S)(+)$ipsdienol (Birch *et al.*, 1980a) (Fig. 12.2; Table 12.2). A similar interruption of response by *Gnathotrichus retusus* to its pheromone $(S)(+)$sulcatol (6-methyl-5-hepten-2-ol) is achieved by small percentages of $(R)(-)$sulcatol (Borden *et al.*, 1980). The high behavioral resolution in response to enantiomeric pheromones implies that mechanisms of synthesis and perception are also chiral in nature.

Pheromone production and release are controlled by physiological and environmental variables, such as age, hormonal levels and circadian rhythmicity (Borden, 1974). In most species, pheromone production occurs in the hindgut region although the precise site of biosynthesis is still unclear. There has been much speculation as to whether beetles accumulate and release compounds from their host tree, whether they convert host compounds in their tissues or extracellularly in the hindgut to pheromones, or indeed whether the pheromone components are actually generated by micro-organisms in the beetles' gut rather than by the beetles. All of these suggestions have some basis.

Some species incorporate host compounds into the pheromone blend, such as myrcene with *D. brevicomis*. Vité *et al.* (1972) distinguished between species like *D. brevicomis* and *D. frontalis* (southern pine beetle) which apparently contain pheromone in their hindguts when they emerge and release it on contact with a suitable host tree, and others, such as most *Ips* species, which only produce pheromone after entering the new host and release it with their frass.

Knowledge of the biosynthesis of bark beetle pheromones has been largely conjectural, based on simple metabolic pathways from host terpenes. However, the biosynthetic routes of the pheromone components of *I. paraconfusus* are now known (Fig. 12.3). Both sexes produced *cis*-verbenol when exposed to (−)α-pinene (Renwick *et al.*, 1976), and *trans*-verbenol from (+)α-pinene. When exposed to the vapour of myrcene, ipsdienol and ipsenol appeared in the hindguts of male beetles, but not in females (Hughes, 1974; Hughes and Renwick, 1977; Byers *et al.*, 1979), whereas no pheromone was detected in the guts of beetles not exposed to myrcene. Deuterium-labelling techniques have now confirmed that myrcene is converted in male *I. paraconfusus* to ipsdienol and ipsenol (Hendry *et al.*, 1980). Thus, all three known pheromone components are obtained by simple oxidation of host plant chemicals.

This picture may be more complex than it appears. Brand *et al.* (1975) isolated a bacterium, *Bacillus cereus* from the hindgut of *I. paraconfusus* which could convert α-pinene to *cis*- and *trans*-verbenol *in vitro*. Male *I. paraconfusus* fed the antibiotic streptomycin, no longer synthesized ipsdienol and ipsenol on exposure to myrcene, although ingestion and absorption of myrcene still occurred (Byers and Wood, 1981b). This implies that bacteria are involved in the sex-specific conversion of myrcene to the pheromone components, although the lack of effect of these antibiotics on *cis*-verbenol synthesis is unexplained. In *D. frontalis*, the mycangial fungus carried by the female is capable of oxidising *trans*-verbenol to verbenone (Brand *et al.*, 1976).

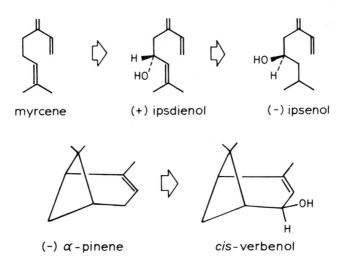

myrcene   (+) ipsdienol   (−) ipsenol

(−) α-pinene   *cis*-verbenol

Fig. 12.3 Biosynthesis of the components of the aggregation pheromone of *Ips paraconfusus* by simple conversion from the host tree terpenes myrcene and α-pinene.

Thus, symbiotic fungi, bacteria, and possibly other micro-organisms, may be far more involved in pheromone production than previously thought.

The production of ipsdienol and ipsenol in *I. paraconfusus* appears to be under hormonal control (Hughes and Renwick, 1977; Borden *et al.*, 1969). Pheromone production, normally stimulated by feeding, could also be induced by treatment with juvenile hormone. In *I. cembrae*, which attacks living larch trees, ipsdienol and ipsenol are produced on contact with myrcene vapour, without feeding, but production of a third component (3-methyl-3-buten-1-ol) appears to be controlled hormonally (Renwick and Dickens, 1979). The extent of hormonal stimulation of pheromone release is, however, still equivocal.

The sensitivity of pheromone perception in bark beetles appears to be generally typical of the very high sensitivity shown by many insects to their pheromones (Seabrook, 1978; Payne, 1979; see also Mustaparta, Chapter 2). In general, beetles have a lower threshold for their own pheromonal components than to other terpenes and related compounds (Mustaparta *et al.*, 1979; Dickens, 1981). In *I. pini*, receptor cells specialized for pheromone perception respond minimally, if at all, to host compounds, and vice versa (Mustaparta *et al.*, 1979, 1980). However, the behavioral interruption of the response of *I. pini* to (−)ipsdienol by a few percent of (+)ipsdienol or by ipsenol must be due to processing in the central nervous system of the insect and not to any interaction of compounds at the antennal receptor sites. Similarly, the synergistic effect of a pheromone blend is due to central integration.

The low thresholds of perception for some compounds indicate the ability of the beetle to perceive the compounds at a long distance from their source. A wide concentration range from theshold to 'saturation' indicates that the particular compound might be used in distance orientation, e.g., ipsdienol in *I. pini* (Mustaparta *et al.*, 1979, 1980) and (−)*cis*-verbenol in *I. typographus* (Dickens, 1981). Conversely higher thresholds and narrower response ranges might indicate compounds used for close-range orientation.

Male *I. paraconfusus* have a greater overall antennal sensitivity to (−)ipsdienol produced by *I. pini* than to their pheromonal component (+)ipsdienol (Light and Birch, 1982). Females, however, show a greater and more typical response to their pheromonal enantiomer.

Thus, the chirality of the receptor as well as production mechanisms has also been established, and the sensitivities of the antennal receptors to pheromonal and host compounds parallels that of the behavior. Apparent anomalies, such as the antennal response of male *I. paraconfusus* to (−)ipsdienol, also have a behavioral rationale (see Section 12.6).

Although it is often tacitly assumed that the pheromone of a species is a constant, this is not necessarily true. There is little evidence that the pheromone of polyphagous species varies from one host to another, but the pheromone of *I. pini*, which attacks pines and spruces throughout N. America, does vary geographically. Eastern and western populations of *I. pini* respond preferentially to aggregation pheromone produced by males from their own populations, but very little to the pheromone from the other population (Table 12.1) (Lanier *et al.*, 1972). The chemical basis for this difference is that *I. pini* from the west produces and responds to (−)ipsdienol as their pheromone, whereas *I. pini* from the east produces a 65 : 35 ratio of (+):(−)ipsdienol and responds more strongly to racemic ipsdienol in the field (Lanier *et al.*, 1980). The response of western *I. pini* is completely interrupted by a small proportion of (+)ipsdienol (Birch *et al.*, 1980a), whereas eastern *I. pini* are not attracted to western *I. pini* males because no (+)ipsdienol is released by them.

Table 12.1 Geographic variation in production and response to aggregation phero- mone by *Ips pini* in California, Idaho and New York, and response of the clerid predator, *Enoclerus lecontei*, to *I. pini* aggregation pheromone in California and Idaho. Responses are given as trap catches in the field and are shown as a percentage of the response of local beetles to males from that locality. (Data from Lanier *et al.*, 1972)

I. pini treatment	% Response of *I. pini* (and *E. lecontei*) in:		
	California	Idaho	New York
California males	100 (100)	77 (161)	1
Idaho males	115 (77)	100 (100)	3
New York males	9 (416)	61 (461)	100

This short summary of pheromone production has been largely limited to pheromones of three species in N. California, and even then has not considered factors such as release rates of pheromones, and variation in these, or inter- action of components at antennal receptor sites. There are many similar ques- tions and only a few answers from a relatively few species.

12.4 AGGREGATION AND ITS TERMINATION

Production and release of pheromone attracts other conspecific beetles of both sexes and aggregation or mass attack is underway. In *Ips*, males attack first and additional males as well as females arrive. Males initiate new boring holes. Females locate established males, enter their boring holes and mate. In *Den- droctonus*, pheromone release is modified as a tree is colonized. The female arrives first and releases *exo*-brevicomin and myrene, which preferentially attract males. Male arrivals release frontalin near female entrance holes and the blend attracts both males and females in equal numbers. The release of ver- benone and *trans*-verbenol then deters further arrivals (Wood, 1982).

Two questions about aggregation behavior appear to be paramount. One is what stimuli beetles use to orient to trees under attack, then to locate the entrance hole to gain entry and to mate. The second question concerns the mechanism of terminating or shifting attacks to different parts of the tree, or to different trees, since there is clearly an optimal density above which the space available for larvae to feed and mature is sharply reduced.

It seems likely that bark beetles orient anemotactically in flight to attractive trees (see Cardé, Chapter 5), though there is no experimental evidence to support this. The upwind walking response of beetles to a source of pheromone in the laboratory indicates that they can follow an odor plume to its source, at least while walking (Wood *et al.*, 1966; see also Bell, Chapter 4). Increasing concentration of pheromone near a source may be used for final orientation. There appears to be a maximum pheromone concentration above which beetles do not continue orientation, but rather land and initiate new attacks. This explains why trees are often killed in groups around one susceptible tree that is attacked first.

Visual and chemical stimuli may be integrated. Thus, *D. brevicomis* responded in much higher numbers to pheromone traps associated with a vertical silhouette similar to a standing tree trunk, than to traps without a silhouette (Tilden, 1976). Conversely, *D. ponderosae*, which frequently attacks fallen timber, could be caught in greater numbers on horizontal traps.

Auditory stimuli are also important. In *Ips*, females landing on an attractive log search along the bark until they find attractive frass and then locate the male entrance hole. Males normally block the hole and females gain entrance by pushing against them while stridulating. Stridulation probably carries species-specific information (Barr, 1969). In *I. pini*, the mean number of females accepted is three, and for females to enter a hole containing a male and three females they must stridulate longer than those outside holes with fewer females (Swaby and Rudinsky, 1976). Attraction in this species declines as more females enter a gallery, but there is no evidence whether this is simply due to curtailment of pheromone production (probable) or to release of an inhibitory compound (unlikely because of the slow decrease in attraction).

The spacing of attacks over a log surface is fairly even, but how the males determine where to bore when they arrive at a tree is not known. The decision may be a function of pheromone concentration or simply visual distance from neighbouring frass piles. In any case, the final result is a log with attacks all over its surface which has ceased to be attractive to other beetles.

In *Dendroctonus* species, stridulation may trigger the release of pheromones that deter further arrivals. Thus arriving male *D. pseudotsugae* (Douglas fir beetle) stridulate outside the females' entrance hole, stimulating her to release 3,2-methyl-cyclohexenone (3,2-MCH) (Rudinsky *et al.*, 1973). Females also stridulate, and this appears to stimulate males also to release 3,2-MCH and 3,3-MCH, both of which inhibit further beetle responses (Rudinsky *et al.*, 1976; Libbey *et al.*, 1976). If the effect of these compounds is local they could provide the spacing mechanism which prevents fresh attacks next to active entrance holes.

In *D. brevicomis*, verbenone and *trans*-verbenol are produced by both sexes and inhibit response to the aggregation pheromone (Wood, 1972; Bedard *et al.*, 1980a). The precise nature and timing of the stimuli triggering release of each behavior are still unclear. There is some evidence that verbenone is released simultaneously with frontalin by males (Browne *et al.*, 1979), that release of verbenone inhibits the response of both sexes and is released by males after locating the female (Renwick and Vité, 1970), and also that a sonic stimulus from females triggers the release of verbenone by males (Rudinsky *et al.*, 1976). The gradual decrease in attraction could be explained as well by a reduction in release of frontalin and *exo*-brevicomin as by increased release of verbenone and *trans*-verbenol (Byers and Wood, 1980). Bedard *et al.* (1980a) suggest that *trans*-verbenol may increase response to the pheromone at a distance, but as the concentration of it and verbenone increase, response is increasingly interrupted so that incoming beetles colonize different areas of the tree or adjacent trees.

12.5 INTERSPECIFIC INTERACTIONS

Pheromones are not 'reserved' for conspecifics. Like any other chemicals released into the environment, they contain information that can be interpreted by any organism that detects them. A pine tree under attack by bark beetles soon becomes a community of species, some competing for the same primary resource, and others parasitic or predaceous on the primary consumers.

The responses of parasites and predators of bark beetles to the pheromones of their hosts are well known, but only in a few instances have they been defined through experimentation. Pheromones interpreted and exploited by parasites and predators are also acting as kairomones, because the adaptive benefit accrues to the receiver rather than the emitter (Borden, 1977; Brown *et al.*, 1970), i.e., the kairomonal response results in predators or parasites arriving at trees simultaneously with large numbers of their prey.

Trapping on the surface of pines infested with *D. brevicomis* in California, Stephen and Dahlsten (1976) collected over 100 species of insects. Predators of larval and adult *D. brevicomis*, such as the beetles *Enoclerus lecontei* and *Temnochila chlorodia* and the predaceous fly *Medetera aldrichii*, arrive during and shortly after the mass arrival of *D. brevicomis*. *T. chlorodia* responds very specifically to *exo*-brevicomin from female *D. brevicomis*, although its response decreases when *trans*-verbenol and verbenone are also present (Bedard *et al.*, 1969; Bedard *et al.*, 1980b). *E. lecontei*, which is a signfiicant predator of *D. brevicomis* (Berryman, 1970), responds to host tree volatiles, but not significantly to the pheromone of *D. brevicomis*, although it does respond to the pheromone of *I. paraconfusus*, another of its prey species (Lanier *et al.*, 1972; Wood, 1972). The pteromalid parasite, *Tomicobia tibialis*, also responds to pheromones or other odors associated with boring male *I. paraconfusus* (Rice, 1969).

As with *D. brevicomis*, many predatory and parasitic insects arrive at trees synchronously with *D. frontalis* (Camors and Payne, 1973), several of these species also being attracted to chemicals either produced by *D. frontalis* (e.g. the clerid *Thanasimus dubius* to frontalin) (Vité and Williamson, 1970; Dixon and Payne, 1979), or by the tree (Dixon and Payne, 1980).

Other parasites attack late-instar scolytid larvae. How these species locate their prey from a distance is unknown, but one species, the braconid wasp *Coeloides brunneri*, detects larvae through the bark by their slightly elevated temperature (Richerson and Borden, 1972).

It is clearly advantageous for parasites and predators to respond to scolytid pheromones and to synchronize their arrival at a tree with large numbers of their host species. No parasites or predators appear to have exploited the sex pheromones of Lepidoptera by responding to them*, and it seems that only where the responding insect is likely to encounter aggregations of prey individuals in the vicinity of a pheromone source, has this behavior been a selective advantage.

Where several species of bark beetle co-exist and also use the same host tree species for breeding, one species may exploit the pheromone of another and so colonize a host more efficiently, or conversely, it may avoid the pheromone of another species and so avoid potential hybridization or detrimental competition for a resource. Wood (1970) suggested that the specificity of pheromones in the Scolytidae might be one mechanism maintaining breeding isolation between sympatric species. This has been substantiated for sympatric species of *Ips*, and besides breeding isolation, competition for host resources is also mediated *via* chemical communication. Two mechanisms have been discovered: (i) non-attraction to the pheromone of another species, and (ii) interruption of a potential response to host material occupied by another species.

The sibling species *I. paraconfusus*, *I. montanus* and *I. confusus* are broadly cross-attractive (Lanier and Wood, 1975), and their ranges are contiguous (parapatric) in California: *I. paraconfusus* breeding principally in ponderosa pine west of the Sierra Nevada crest and up to 5000 ft elevation, *I. montanus* in western white pine at higher elevations, and *I. confusus* in the semi-desert piñon pine areas of the state. In general, closely related species such as these, which are cross-attractive, are not sympatric, whereas species from different groups, such as *I. paraconfusus* and *I. pini* in N. California, are not at all cross-attractive and co-exist in the same area and often the same host pine, in this case ponderosa pine.

Specificity of communication between *I. paraconfusus* and *I. pini* is based not only on the non-attraction that would be expected from species of different groups, but also on mutual interruption of the other's response (Birch and Wood, 1975). Two components of the male pheromone of *I. paraconfusus*, (−)ipsenol and (+)ipsdienol, completely interrupt the response of *I. pini* to its own pheromone (Birch *et al.*, 1977; Birch *et al.*, 1980a). Similarly, (−)ipsdienol, the pheromone of *I. pini*, interrupts the response of *I. paraconfusus* to its

---

* However, mature female bolas spiders, *Mastophora* sp., attract males of the fall armyworm, *Spodoptera frugiperda*, with a volatile substance that apparently mimics the female sex attractant of this species (Eberhard, 1977).

pheromone (Light and Birch, 1979) (Table 12.2). Thus, these components have at least two effects; as components of pheromones, at the same time as interrupting the responses of a second species. Since the interruption is at least immediately advantageous to the sender, this action is allomonal. The advantages to the emitter of these allomones are: (i) that they reserve a particular piece of host material for the first colonising species, and (ii) that they minimize the possible reduction in reproductive capacity if the two species colonize a log together. There is some indication that the normal even distribution of galleries in a log fully colonized by either *I. paraconfusus* or *I. pini* alone, is disrupted when both species are present, resulting in less than optimal use of the available phloem and consequent reduction in brood (Birch, 1978).

Interspecific interactions are not restricted to these two species. In the same environment, *D. brevicomis* frequently attacks ponderosa pine, infesting the lower and mid-trunk, while *I. paraconfusus* and *I. pini* attack the top and larger limbs. The gallery systems of *Ips* rarely overlap with those of *D. brevicomis*, even though later-instar larvae of *D. brevicomis* feed in the bark, and *I. paraconfusus* and *I. pini* larvae develop entirely in the phloem. *D. brevicomis* is attracted to low concentrations of the pheromone blend of *I. paraconfusus*, but not to higher (or lower) concentrations. Higher levels interrupt the response of *D. brevicomis* to its own pheromone. Conversely, *I. paraconfusus* is neither attracted nor interrupted by the aggregation pheromone of *D. brevicomis*, but verbenone, released primarily after male beetles arrive and mate, effectively interrupts the response of *I. paraconfusus* (Byers and Wood, 1980, 1981a).

I. latidens also infests ponderosa pine in N. California and responds to two components of the *I. paraconfusus* pheromone: ipsenol and *cis*-verbenol. Addition of the third component, ipsdienol (racemic), eliminates the response

Table 12.2 Mutual interruption of response to their aggregation pheromones by *Ips paraconfusus* and *I. pini*. Treatments are male beetles of one or both species boring in ponderosa pine, with, in some treatments, one or more enantiomers or racemic ipsenol and ipsdienol. Responses are given as a percentage of the response to conspecifics. (Data from: Birch and Wood, 1975; Light and Birch, 1979; Birch *et al.*, 1980a)

Treatment	% response	
	I. paraconfusus	*I. pini*
50 ♂♂ *I. paraconfusus*	100	0
50 ♂♂ *I. pini*	0	100
50 ♂♂ *I. paraconfusus* + 50 ♂♂ *I. pini*	1	15
50 ♂♂ *I. paraconfuses* + (−)ipsdienol	7	*
50 ♂♂ *I. pini* + racemic ipsdienol	*	8
50 ♂♂ *I. pini* + (+)ipsdienol	*	0
(−)ipsdienol (95.6% E.E.)	*	138
(+)ipsdienol (88.2% E.E.)	*	2

* Not tested

of *I. latidens* (Wood *et al.*, 1967). However, *I. latidens* appears to be attracted by (+)ipsdienol (Birch *et al.*, 1980a). Although both sets of field data are tenuous, it is possible to speculate that *I. latidens* is attracted by ipsenol, *cis*-verbenol and (+)ipsdienol, but is interrupted by (−)ipsdienol. It would thus be deterred from attacking trees simultaneously with *I. pini* (whose pheromone is (−)ipsdienol) or *I. paraconfusus* (which maintains a percentage of (−)ipsdienol in its pheromone blend).

There are similar complex interspecific relationships among other species groups. In E. Texas, the four species, *D. frontalis*, *I. avulsus*, *I. calligraphus* and *I. grandicollis*, often attack the same loblolly pines (*P. taeda*) and partition them such that *D. frontalis* occupies primarily the lower trunk, *I. avulsus*, the upper trunk, both overlapping with *I. calligraphus* and *I. grandicollis* infests primarily the larger branches in the crown (Birch *et al.*, 1980b; Paine *et al.*, 1981; Švihra *et al.*, 1980). There is some mutual interruption of responses among these species, but also some enhancement. For example, *D. frontalis* and *I. grandicollis* mutually interrupt each other's response, but *I. grandicollis* increases the response of *I. avulsus* to conspecific males. Depending on the effect, the response evoked by the chemical components in these interactions can be labelled pheromonal, allomonal or kairomonal (Birch *et al.*, 1980b; Hedden *et al.*, 1976; Vité *et al.*, 1978). The net effects are that interspecific facilitation during attack favors the survival of all species in rapidly over-whelming the resistance mechanisms of a tree, and that specificity or competitive exclusion (which may operate close to or on the surface of a tree) results in the delineation of separate breeding areas. Similar highly specific behavioral interactions are shown by ambrosia beetles colonizing the xylem of dead trees or lumber. The pheromone of *Gnathotrichus sulcatus* is a 65:35 blend of (+):(−)sulcatol. *G. sulcatus* is not attracted to *G. retusus*, which produces only (+)sulcatol, and the (−)sulcatol in the pheromone of *G. sulcatus* interrupts the response of *G. retusus* to its pheromone (Borden *et al.*, 1980).

Studies such as these indicate that host colonization behavior and the partitioning of available resources for colonization are the outcome of complex interactions, many of which are chemically mediated. If the evolution of the system is to be understood or if chemical communication is to be effectively exploited in forest management, the aggregation behavior of individual species must be investigated in the natural system rather than in isolation.

12.6 HOST SELECTION PARADIGM

It should now be clear that the initial stimuli involved in host selection may be initially directional (primary attractants) or non-directional (random landing) depending on the species, but that highly directional stimuli (pheromones) take over once a tree is attacked. Using the example of *I. paraconfusus*, initial positive host selection seems to occur only after landing and feeding, but negative selection, i.e., that a host is already colonized and unsuitable, may act over a distance (Fig. 12.4).

This distinction clarifies the surprising differential sensitivity of male and female *I. paraconfusus* to components of their pheromone and to that of *I. pini*. Males and females are equally sensitive to their natural pheromone and to its component, (+)ipsdienol, over a wide range of concentrations. However, females are significantly more sensitive to pheromonal (+)ipsdienol than to allomonal (−)ipsdienol (the pheromone of *I. pini*), whereas males are more sensitive to allomonal (−)ipsdienol than to their own pheromonal (+)ipsdienol (Light and Birch, 1982). There is a clear adaptive advantage for males to locate new host material quickly and to avoid resources occupied by *I. pini*, unless nothing else is available, since if the two species do co-colonize a resource, the reproductive potentials of both are reduced. This premium on selection of unoccupied resources underlies the high sensitivity of males to allomonal ipsdienol. The interruption of response in *I. paraconfusus* by verbenone from *D. brevicomis* may have a similar basis.

Other species are deterred from material infested by *I. paraconfusus* in the same way that *I. paraconfusus* is deterred from attacking occupied resources. Throughout colonization, until the progeny leave the host, there is constant interaction with other species, particularly parasites and predators of parent adults, larvae and emerging adult beetles.

During the evolution of bark beetles, their hosts and associated fungi, parasites, predators and organisms, the composition of bark beetle pheromones will probably have been modified as other species evolved responses to components of the pheromone and exerted selection pressures on the beetle population. There is only circumstantial evidence that this might have happened. For example, in the experiments on widely separated populations of *I. pini* (Lanier *et al.*, 1972), California beetles attracted more local *I. pini* than did New York beetles, and vice versa in New York. However, New York beetles in California attracted far more of the local predator *E. lecontei* than did the local *I. pini*. California and New York *I. pini* use different ratios of the enantiomers of ipsdienol as their pheromones. The high resolution of (−)ipsdienol in California populations versus the blend in New York supports the idea of co-evolution of chemical systems of predator and prey: production of and response to (+)ipsdienol being eliminated in California by a predator which had evolved specific responses to it. Perhaps in the absence of *E. lecontei* in New York there was no pressure to resolve the blend.

The small proportion of (−)ipsdienol in the pheromone blend of *I. paraconfusus* might similarly be advantageous in spite of it being the pheromone of its competitor, *I. pini*, because (−)ipsdienol also appears to deter *I. latidens*. The exploitation of pheromone components by predators need not necessarily select for different attractive blends but could result in selection for components that interrupt the predator response. For example, the highly specific response of *T. chlorodia* to *exo*-brevicomin from female *D. brevicomis* is interrupted by

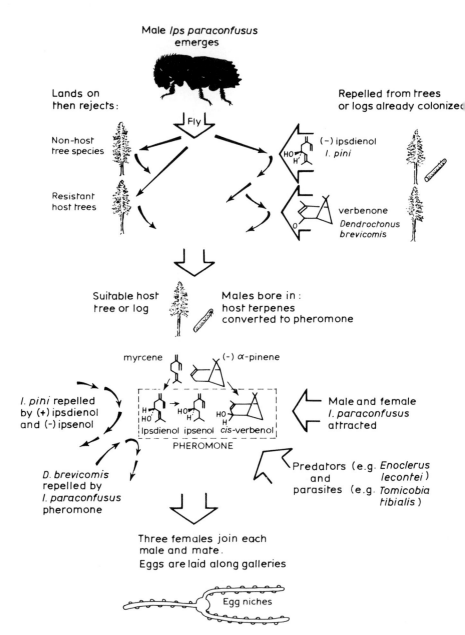

Male *Ips paraconfusus* emerges

Lands on then rejects:

Non-host tree species

Resistant host trees

Fly

Repelled from trees or logs already colonized

(−) ipsdienol
I. pini

verbenone
Dendroctonus brevicomis

Suitable host tree or log

Males bore in: host terpenes converted to pheromone

myrcene
(−) α-pinene

I. pini repelled by (+) ipsdienol and (−) ipsenol

Ipsdienol   ipsenol   *cis*-verbenol
PHEROMONE

Male and female *I. paraconfusus* attracted

D. brevicomis repelled by *I. paraconfusus* pheromone

Predators (e.g. *Enoclerus lecontei*) and parasites (e.g. *Tomicobia tibialis*)

Three females join each male and mate.
Eggs are laid along galleries

Egg niches

Fig. 12.4 Behavioral paradigm for host selection in *Ips paraconfusus* from adult emergence to construction of galleries and egg laying.

the release of verbenone and *trans*-verbenol on mating (Bedard *et al.*, 1980b). Once mating has occurred it is clearly advantageous to deter predators as quickly as possible, when no longer necessary to attract further males.

This chapter has described what has been learned about bark beetle aggregation behavior in the 15 years since the first pheromone was identified from *I. paraconfusus*. It has also indicated how little we yet know about many of the behavioral aspects of aggregation. Our ability to make highly resolved enantiomers of pheromones opens the way to analyse these behaviors more precisely. Similarly, the biosynthesis of pheromones from host compounds, the orientation mechanisms used by beetles in flight, the pheromone production of individual beetles, and their interaction with sonic and visual stimuli can now all be used to understand the many behavioral steps taken by individual beetles from eclosion to mating and egg laying, which will result in a composite picture of tree colonization built up from information rather than intuitive assumptions.

Finally, the impetus for much of this work has come from the potential use of pheromones in forest pest management. Aggregation pheromones have been used in attempting to mass-trap some species (e.g. *D. brevicomis*, *S. multistriatus*, *I. typographus* and ambrosia beetles) and also, in conjunction with trap trees, to reduce populations of other species, (e.g., *D. pseudotsugae* and *S. multistriatus*). Chemicals which interrupt intra- or inter-specific responses have been used to deter attacks (e.g., in *D. frontalis*, *D. brevicomis*, *D. pseudotsugae* and *I. pini*). Such studies are covered by Borden (1977), Roelofs (1979) and in Mitchell (1981). There is still a long way to go before pheromones, allomones and kairomones are used routinely in forest pest management and other directions could still be explored. The response of predators such as *T. chlorodia* to pheromones could be exploited to increase bark beetle mortality rather than being considered as an undesirable effect of the use of pheromones. Not only should the chemical communication system within a species be intimately known in order to manipulate behavior effectively, but the effect of manipulating one species on the complex of other species in that environment should also be understood. The more that is known of the chemicals and behaviors they evoke prior to large-scale use, the more likely it is that control measures will succeed. There is tremendous scope for original work at all levels.

ACKNOWLEDGMENT

I would like to thank M. L. Birch, K. F. Haynes, D. M. Light, and T. D. Paine for their stimulating comments on early versions of this chapter.

REFERENCES

Barr, B. A. (1969) Sound production in Scolytidae (Coleoptera) with emphasis on the genus *Ips. Can.Ent.*, **101**, 636–72.

Bedard, W. D., Tilden, P. E., Wood, D. L., Silverstein, R. M., Brownlee, R. G. and Rodin, J. O. (1969) Western pine beetle: field response to its sex pheromone and a synergistic host terpene myrcene. *Science*, **164**, 1284–5.

Bedard, W. D., Tilden, P. E., Lindahl, Jr., K. Q., Wood, D. L. and Rauch, P. A. (1980a) Effects of verbenone and *trans*-verbenol on the response of *Dendroctonus brevicomis* to natural and synthetic attractant in the field. *J. Chem. Ecol.*, **6**, 997–1013.

Bedard, W. D., Wood, D. L., Tilden, P. E., Lindahl, Jr., K. Q., Silverstein, R. M. and Rodin, J. O. (1980b) Field responses of the western pine beetle and one of its predators to host- and beetle- produced compounds. *J. Chem. Ecol.*, **6**, 625–41.

Berryman, A. A. (1970) Evaluation of insect predators of the western pine beetle. In: *Studies on the Population Dynamics of the Western Pine Beetle*, Dendroctonus brevicomis *LeConte* (*Coleoptera: Scolytidae*) (Stark, R. W. and Dahlsten, D. L., eds). University of California Press, Berkeley.

Berryman, A. A. (1972) Resistance of conifers to invasion by bark beetle-fungus associations. *Bioscience*, **22**, 598–602.

Birch, M. C. (1978) Chemical communication in pine bark beetles. *Am. Sci.*, **66**, 409–19.

Birch, M. C. and Wood, D. L. (1975) Mutual inhibition of the attractant pheromone response by two species of *Ips* (Coleoptera: Scolytidae). *J. Chem. Ecol.*, **1**, 101–13.

Birch, M. C., Light, D. M. and Mori, K. (1977) Selective inhibition of response of *Ips pini* to its pheromone by the (S)-$(-)$-enantiomer of ipsenol. *Nature*, **270**, 738–9.

Birch, M. C., Light, D. M., Wood, D. L., Browne, L. E., Silverstein, R. M., Bergot, B. J., Ohloff, G., West, J. R. and Young, J. C. (1980a) Pheromonal attraction and allomonal interruption of *Ips pini* in California by the two enantiomers of ipsdienol. *J. Chem. Ecol.*, **6**, 703–17.

Birch, M. C., Švihra, P., Paine, T. D. and Miller, J. C. (1980b) Influence of chemically mediated behavior on host tree colonization by four cohabiting species of bark beetles. *J. Chem. Ecol.*, **6**, 395–414.

Borden, J. H. (1974) Aggregation pheromones in the Scolytidae. In: *Pheromones* (Birch, M. C., ed.) North-Holland, Amsterdam.

Borden, J. H. (1977) Behavioral responses of Coleoptera to pheromones, allomones and kairomones. In: *Chemical Control of Insect Behavior: Theory and Application* (Shorey, H. H. and McKelvey, Jr., J. J., eds). Wiley, New York.

Borden, J. H., Nair, K. K. and Slater, C. E. (1969) Synthetic juvenile hormone: induction of sex pheromone production in *Ips confusus*. *Science*, **166**, 1626–7.

Borden, J. H., Handley, J. R., McLean, J. A., Silverstein, R. M., Chong, L., Slessor, K. N., Johnston, B. D. and Schuler, H. R. (1980) Enantiomer-based specificity in pheromone communication by two sympatric *Gnathotrichus* species (Coleoptera: Scolytidae). *J. Chem. Ecol.*, **6**, 445–56.

Brand, J. M., Bracke, J. W., Markovetz, A. J., Wood, D. L. and Browne, L. E. (1975) Production of verbenol pheromone by a bacterium isolated from bark beetles. *Nature*, **254**, 136–7.

Brand, J. M., Bracke, J. W., Britton, L. N., Markovetz, A. J. and Barras, S. J. (1976). Bark beetle pheromones: production of verbenone by a mycangial fungus of *Dendroctonus frontalis*. *J. Chem. Ecol.*, **2**, 195–9.

Brown, Jr., W. L., Eisner, T. E. and Whittaker, R. H. (1970) Allomones and kairomones: transpecific chemical messengers. *Bioscience*, **20**, 21–2.

Browne, L. E., Birch, M. C. and Wood, D. L. (1974) Novel trapping and elution systems for airborne insect pheromones. *J. Insect Physiol.*, **20**, 183–93.

Browne, L. E., Wood, D. L., Bedard, W. D., Silverstein, R. M. and West, J. R. (1979) Quantitative estimates of the attractive pheromone components, exo-brevicomin, frontalin and myrcene, of the western pine beetle in nature. *J. Chem. Ecol.*, **5**, 397–414.

Byers, J. A., Wood, D. L., Browne, L. E., Fish, R. H., Piatek, B. and Hendry, L. B. (1979) Relationship between a plant compound, myrcene, and pheromone production in the bark beetle *Ips paraconfusus. J. Insect Physiol.*, **25**, 477–82.

Byers, J. A. and Wood, D. L. (1980) Interspecific inhibition of the response of the bark beetles, *Dendroctonus brevicomis* and *Ips paraconfusus* to their pheromones in the field. *J. Chem. Ecol.*, **6**, 149–64.

Byers, J. A. and Wood, D. L. (1981a) Interspecific effects of pheromones on the attraction of the bark beetles, *Dendroctonus brevicomis* and *Ips paraconfusus* in the laboratory. *J. Chem. Ecol.*, **7**, 9–18.

Byers, J. A. and Wood, D. L. (1981b) Antibiotic-induced inhibition of pheromone synthesis in a bark beetle. *Science*, **213**, 763–4.

Camors, Jr., F. B. and Payne, T. L. (1973) Sequence of arrival of entomophagous insects to trees infested with the southern pine beetle. *Env. Ent.*, **2**, 267–70.

Dickens, J. C. (1981) Behavioral and electrophysiological responses of *Ips typographus* L. to potential pheromone components. *Physiol. Ent.*, **6**, 251–61.

Dixon, W. N. and Payne, T. L. (1979) Aggregation of *Thanasimus dubius* on trees under mass-attack by the southern pine beetle. *Env. Ent.*, **8**, 178–81.

Dixon, W. N. and Payne, T. L. (1980) Attraction of entomophagous and associate insects of the southern pine beetle to beetle- and host tree-produced volatiles. *J. Georgia Ent. Soc.*, **15**, 378–89.

Eberhard, W. G. (1977) Aggressive chemical mimicry by a bolas spider. *Science*, **198**, 1173–5.

Elkinton, J. S. and Wood, D. L. (1980) Feeding and boring behavior of the bark beetle *Ips paraconfusus* (Coleoptera: Scolytidae) on the bark of host and non-host tree species. *Can. Ent.*, **112**, 589–601.

Elkinton, J. S., Wood, D. L. and Hendry, L. B. (1980) Pheromone production by the bark beetle, *Ips paraconfusus*, in the nonhost, white fir. *J. Chem. Ecol.*, **6**, 979–87.

Hedden, R., Vité, J. P. and Mori, K. (1976). Synergistic effect of a pheromone and a kairomone on host selection and colonization by *Ips avulsus. Nature*, **261**, 696–7.

Hendry, L. B., Piatek, B., Browne, L. E., Wood, D. L., Byers, J. A., Fish, R. H. and Hicks, R. A. (1980) *In vivo* conversion of a labelled host plant chemical to pheromones of the bark beetle *Ips paraconfusus. Nature*, **284**, 485.

Hughes, P. R. (1974) Myrcene: a precursor of pheromones in *Ips* beetles. *J. Insect Physiol.*, **20**, 1271–5.

Hughes, P. R. and Renwick, J. A. A. (1977) Neural and hormonal control of pheromone biosynthesis in the bark beetle, *Ips paraconfusus. Physiol. Ent.*, **2**, 117–23.

Lanier, G. N., Birch, M. C., Schmitz, R. F. and Furniss, M. M. (1972) Pheromones of *Ips pini* (Coleoptera: Scolytidae): variation in response among three populations. *Can. Ent.*, **104**, 1917–23.

Lanier, G. N. and Wood, D. L. (1975) Specificity of response to pheromones in the genus *Ips* (Coleoptera: Scolytidae). *J. Chem. Ecol.*, **1**, 9–23.

Lanier, G. N., Classon, A., Stewart, T., Piston, J. J. and Silverstein, R. M. (1980) *Ips pini*: the basis for interpopulational differences in pheromone biology. *J. Chem. Ecol.*, **6**, 677–87.

Libbey, L. M., Morgan, M. E., Putnam, T. B. and Rudinsky, J. A. (1976) Isomer of antiaggregative pheromone identified from male Douglas fir beetle: 3-methylcyclohex-3-en-1-one. *J. Insect Physiol.*, **22**, 871–3.

Light, D. M. and Birch, M. C. (1979) Inhibition of the attractive pheromone response in *Ips paraconfusus* by (*R*)-(−)-ipsdienol. *Naturwissenchaften*, **66**, 159–60.

Light, D. M. and Birch, M. C. (1982) Bark beetle enantiomeric chemoreception: greater sensivity to allomone than pheromone. Naturwissenchaften, **69**, 243–5.

Mitchell, E. R. (ed.) (1981) *Management of Insect Pests with Semiochemicals*. Plenum Press, New York.

Moeck, H. A. (1970) Ethanol as the primary attractant for the ambrosia beetle *Trypodendron lineatum* (Coleoptera: Scolytidae). *Can. Ent.*, **102**, 985–95.

Moeck, H. A. (1981) Host selection behavior of bark beetles (Coleoptera: Scolytidae) attacking *Pinus ponderosa*, with special emphasis on the western pine beetle, *Dendroctonus brevicomis*. *J. Chem. Ecol.*, **7**, 49–83.

Mustaparta, H., Angst, M. E. and Lanier, G. N. (1979) Specialization of olfactory cells to insect- and host-produced volatiles in the bark beetle *Ips pini* (Say). *J. Chem. Ecol.*, **5**, 109–23.

Mustaparta, H., Angst, M. E. and Lanier, G. N. (1980) Receptor discrimination of enantiomers of the aggregation pheromone ipsdienol in two species of *Ips*. *J. Chem. Ecol.*, **6**, 689–701.

Paine, T. D. (1981) *Aspects of the Physiological and Ecological Relationships between the Western Pine Beetle*, Dendroctonus brevicomis, *and associated Fungi*. PhD Thesis University of California, Davis.

Paine, T. D., Birch, M. C. and Švihra, P. (1981) Niche breadth and resource partitioning by four sympatric species of bark beetles (Coleoptera: Scolytidae) *Oecologia*, **48**, 1–6.

Payne, T. L. (1979) Pheromone and host odor perception in bark beetles. In: *Neurotoxicology of Insecticides and Pheromones* (Narahashi, T., ed.) Plenum Press, New York.

Pearce, G. T., Gore, W. E., Silverstein, R. M., Peacock, J. W., Cuthbert, R. A., Lanier, G. N. and Simeone, J. B. (1975) Chemical attractants for the smaller European elm bark beetle *Scolytus multistriatus* (Coleoptera: Scolytidae). *J. Chem. Ecol.*, **1**, 115–24.

Renwick, J. A. A. and Vité, J. P. (1970) Systems of chemical communication in *Dendroctonus*. *Cont. Boyce Thompson Inst. Plant Res.*, **24**, 283–92.

Renwick, J. A. A., Hughes, P. R. and Krull, I. S. (1976) Selective production of *cis*- and *trans*-verbenol from (–)- and (+)-α-pinene by a bark beetle. *Science*, **191**, 199–201.

Renwick, J. A. A. and Dickens, J. C. (1979) Control of pheromone production in the bark beetle, *Ips cembrae*. *Physiol. Ent.*, **4**, 377–81.

Rice, R. E. (1969) Response of some predators and parasites of *Ips confusus* (Le-C.) (Coleoptera: Scolytidae) to olfactory attractants. *Cont. Boyce Thompson Inst. Plant Res.*, **24**, 189–94.

Richerson, J. V. and Borden, J. H. (1972) Host finding behavior in *Coeloides brunneri* (Hymenoptera: Braconidae). *Can. Ent.*, **104**, 1877–81.

Roelofs, W. L. (ed.) (1979) *Establishing Efficacy of Sex Attractants and Disruptants for Insect Control*. Entomological Society of America, College Park, Md.

Rudinsky, J. A., Morgan, M., Libbey, L. M. and Michael, R. R. (1973) Sound production in Scolytidae: 3-methyl-2-cyclohexen-1-one released by the female Douglas fir beetle in response to male sonic signal. *Env. Ent.*, **2**, 505–9.

Rudinsky, J. A., Ryker, L. C., Michael, R. R., Libbey, L. M. and Morgan, M. E. (1976) Sound production in Scolytidae: female sonic stimulus of male pheromone release in two *Dendroctonus* beetles. *J. Insect Physiol.*, **22**, 1675–81.

Seabrook, W. D. (1978) Neurobiological contributions to understanding insect pheromone systems. *A. Rev. Ent.*, **23**, 471–85.

Silverstein, R. M. (1979) Enantiomeric composition and bioactivity of chiral semiochemicals in insects. In: *Chemical Ecology: Odour Communication in Animals* (Ritter, F. J., ed.). Elsevier/North-Holland, Amsterdam.

Silverstein, R. M., Rodin, J. O. and Wood, D. L. (1966) Sex attractants in frass produced by male *Ips confusus* in ponderosa pine. *Science*, **154**, 509–10.

Stephen, F. M. and Dahlsten, D. L. (1976) The arrival sequence of the arthropod complex following attack of *Dendroctonus brevicomis* in ponderosa pine. *Can. Ent.*, **108**, 282–304.

Švihra, P., Paine, T. D. and Birch, M. C. (1980) Interspecific olfactory communications in southern pine beetles. *Naturwissenschaften.*, **67**, 518–19.

Swaby, J. A. and Rudinsky, J. A. (1976) Acoustic and olfactory behaviour of *Ips pini* (Say) (Coleoptera: Scolytidae) during host invasion and colonization. *Z. Angew. Ent.*, **81**, 421–32.

Tilden, P. E. (1976) *Behavior of* Dendroctonus brevicomis *near Source of Synthetic Pheromones in the field*. MS thesis, University of California, Berkeley.

Vité, J. P. and Williamson, D. L. (1970) *Thanasimus dubius*: prey perception. *J. Insect Physiol.*, **16**, 233–9.

Vité, J. P. and Renwick, J. A. A. (1971) Population aggregating pheromone in the bark beetle, *Ips grandicollis. J. Insect Physiol.*, **17**, 1699–704.

Vité, J. P., Bakke, A. and Renwick, J. A. A. (1972) Pheromones in *Ips* (Coleoptera: Scolytidae): occurrence and production. *Can. Ent.*, **104**, 1967–75.

Vité, J. P. and Francke, W. (1976) The aggregation pheromones of bark beetles: progress and problems. *Naturwissenschaften.*, **63**, 550–5.

Vité, J. P., Hedden, R. and Mori, K. (1976) *Ips grandicollis*: field response to the optically pure pheromone. *Naturwissenshaften.*, **63**, 43–4.

Vité, J. P., Ohloff, G. and Billings, R. F. (1978) Pheromone chirality and integrity of aggregation response in southern species of the bark beetle *Ips* sp. *Nature*, **272**, 817–18.

Witanachchi, J. P. and Morgan, F. D. (1981) Behaviour of the bark beetle, *Ips grandicollis*, during host selection. *Physiol. Ent.*, **6**, 219–23.

Wood, D. L. (1970) Pheromones of bark beetles. In: *Control of Insect Behavior by Natural Products* (Wood, D. L., Silverstein, R. M. and Nakajima, M., eds) Academic Press, New York.

Wood, D. L. (1972) Selection and colonization of ponderosa pine by bark beetles. In: *Insect/Plant Relationships* (van Emden, H. F., ed.) *Symp. R. Ent. Soc. Lond.*, **6**. Blackwell, Oxford.

Wood, D. L. (1982) The role of pheromones, kairomones, and allomones in the host selection and colonization behavior of bark beetles. *A. Rev. Ent.*, **27**, 411–46.

Wood, D. L., Browne, L. E., Silverstein, R. M. and Rodin, J. O. (1966) Sex pheromones of bark beetles. I. Mass production, bio-assay, source, and isolation of the sex pheromones of *Ips confusus* (Le C.). *J. Insect Physiol.*, **12**, 523–36.

Wood, D. L., Stark, R. W., Silverstein, R. M. and Rodin, J. O. (1967) Unique synergistic effects produced by the principal sex attractant compounds of *Ips confusus* (Le Conte). *Nature*, **215**, 206.

Wood, D. L. Browne, L. E., Bedard, W. D., Tilden, P. E., Silverstein, R. M. and Rodin, J. O. (1968) Response of *Ips confusus* to synthetic sex pheromones in nature. *Science*, **159**, 1373–4.

Wood, D. L., Browne, L. E., Ewing, B., Lindahl, K., Bedard, W. D., Tilden, P. E., Mori, K., Pitman, G. B. and Hughes, P. R. (1976) Western pine beetle: specificity among enantiomers of male and female components of an attractive pheromone. *Science*, **192**, 896–8.

13

Sexual Communication with Pheromones

Ring T. Cardé and Thomas C. Baker

13.1 INTRODUCTION

A critical event in sexual reproduction is location or recruitment of a mate. In a number of insect groups, the necessary movements in time and space are often mediated by pheromones. One sex may recruit the other, or both sexes may be attracted to the chemical emitters. Aggregation may be viewed as the end result of movement reactions that reduce the distance between individuals in their environment. Such clustering may be brought about by a combination of attraction and arrestment, which are themselves not orientation mechanisms but rather end results, i.e., displacements, caused by movement reactions (Kennedy, 1978). For sex- and aggregation-pheromone communication, we define attraction as the net displacement of one individual toward the chemical source. Conversely, arrestment is the lack of net displacement toward or away from the chemical source. Both displacement phenomena may be viewed as part of a continuum caused by pheromone mediation of quite disparate movement reactions, such as orthokinesis, klinotaxis and anemotaxis (see Bell, Chapter 4 and Cardé, Chapter 5). That attraction and arrestment are only outcomes, not mechanisms, does not diminish the heuristic value of these terms; they are a capsule summary of the change in spacing between an individual and the chemical source. To an organism responding to sex pheromone, proximate cues and orientation mechanisms notwithstanding, such outcomes are the result of evolutionary selection.

13.2 DIVERSITY OF COMMUNICATION SYSTEMS

Among the insect groups in which chemical attraction is a major means of

Chemical Ecology of Insects. Edited by William J. Bell and Ring T. Cardé
© 1984 Chapman and Hall Ltd.

sexual recruitment are the moths (Lepidoptera), many Coleoptera, Hymenoptera, Orthoptera, Diptera and Homoptera. In these groups, females are predominantly the emitters and males the receivers. Hence, the reduction in spacing is due primarily to chemically mediated movements of males. Where attraction of females by males does occur, interesting ecological and evolutionary situations are being discovered that involve sexual selection and parental investment in gametes and offspring. The following examples of chemical communication in a number of insect groups illustrates the variety of these chemical communication systems.

Within the Lepidoptera, only in a few cases discovered so far do male-emitted chemicals attract females. The vast majority of aggregation is initiated by females who release volatiles from a gland typically located at the end of their abdomen. In some species, a sort of dual system is used, in which both female and male-emitted chemicals cause aggregation. As discussed in more detail later in this chapter, male oriental fruit moths, *Grapholitha molesta*, chemically attract females from a few centimeters away after they have been attracted from long distances by the females' pheromone (Baker and Cardé, 1979b). In the salt marsh caterpillar, *Estigmene acraea*, males hang from plants just after sunset and inflate huge coremata at the end of their abdomens (Willis and Birch, 1982). A yet-to-be-identified pheromone released from these organs apparently attracts other males which land and evert their coremata, resulting in the formation of an aggregation of sexually displaying males, or lek. Females are attracted to a male or males in this group, one of which then mates with her. Later in the night, females attract males with a blend of (Z,Z)-9,12-octadecadienal,(Z,Z,Z)-9,12,15-octadecatrienal, and (Z,Z)-3,6-*cis*-9,10-epoxyheneicosadiene (Hill and Roelofs, 1981). In the pyralid wax moths, *Galleria melonella* and *Achroia grisella*, males emit pheromone from costal wing glands to lure flying females, although in the latter species the chemicals, *n*-undecanal and (Z)-11-octadecenal, apparently only 'prime' females to respond to auditory cues generated by males' vibrating wings (Dahm *et al.*, 1971). In *G. melonella* it is not clear whether the chemicals, *n*-nonanal and *n*-undecanal (Roller *et al.*, 1968; Leyrer and Monroe 1973), attract females from a distance or merely cause arrestment of randomly flying individuals (Finn and Payne, 1977). However, in another pyralid species, *Eldana saccharina*, the African sugarcane borer, an unidentified pheromone emitted from male wing glands causes females to walk rapidly and search the grassy vegetation until they locate the male and touch his abdominal hairpencils (Zagatti *et al.*, 1981). Vanillin and *p*-hydroxybenzaldehyde have been identified from these hairpencils, which function during courtship to aid in females' acceptance of males during copulatory attempts (Zagatti, 1981).

In the Coleoptera, most knowledge of sexual communication pertains to bark beetles, family Scolytidae (discussed in detail by Birch, Chapter 12), where one of the sexes, depending on the species, initiates aggregation by boring into host trees. The blend of pheromone plus host-tree volatiles attracts predominantly members of the opposite sex. In the curculionid species *Anthonomus grandis*, the boll weevil, a male-emitted blend of two terpene alcohols plus two terpene aldehydes (Tumlinson *et al.*, 1969) results primarily in the attraction of females, although a close-range, female-emitted pheromone plus cotton plant volatiles attracts principally males (Hedin *et al.*, 1979). Much work has also been done in the dermestid genus *Trogoderma*, in which solitary females release either (*E*) or (*Z*), 14-methyl, 8-hexadecenal or a blend of both from an intersegmental abdominal gland (Cross *et al.*, 1976). Males are not only attracted preferentially to a precise geometrical isomeric blend, but also respond specifically to the correct optical enantiomers (Silverstein *et al.*, 1980). In the Scarabaeidae, female Japanese beetles, *Popilla japonica* release the sex pheromone (*Z*)-5-(1-decenyl)dihydro-2(3H)-furanone, attracting males from many tens of meters away (Tumlinson *et al.*, 1977). A small amount of the (*S*) form added to the pure (*R*) form of this compound reduces the attraction of males, and therefore optical configuration also plays a role in pheromone channel partitioning in this family.

For hymenopterans, much is known about chemically mediated aggregation in the sawflies, subfamily Symphyta. Female-emitted pheromone attracts males from some distance. In the species *Neodiprion lecontei*, *N. sertifer* and *N. pinetum*, females emit, and males respond most strongly to, the acetate ester of 3,7-dimethyl pentadecan-2-ol (Jewett *et al.*, 1976), whereas in *Diprion similis* the propionate ester is used. Communication channel segregation is thus achieved between the latter species and the others by functional moiety; however, *N. lecontei* and *N. pinetum* males both preferentially respond to the same (*S*,*S*,*S*) optical isomer of the acetate compound (Kraemer *et al.*, 1979, 1981). It is unclear how partitioning is accomplished, except that there are some host-tree preference differences between these species. Presumably, optical differences in pheromone structure could play a role in isolating some of the other members of this genus that apparently use this acetate in sexual communication. In the subfamily Apocrita, a variety of sexual communication systems is employed. Queens of some primitive ant species in the subfamilies Myrmicinae, Myrmeciinae, and Ponerinae (such as *Rhytidoponera metallica*) raise their abdomens in a typical calling posture and release an unidentified sex pheromone from a tergal gland to attract males (Hölldobler and Haskins, 1977). More advanced species such as the socially parasitic *Harpagoxenus sublaevis*, *Doronomyrmex pacis* and *Leptothorax kutteri* also employ female-emitted pheromone for sexual communication (Buschinger and Alloway, 1979). Queen honey bees emit sex pheromone from their mandibular glands, which initiates mate-finding by drones (males), evokes upwind flight and heightens visual orientation to the female flying several meters above the ground. A mandibular gland chemical, (*E*)-9-oxo-2-decenoic acid has been implicated as a sex pheromone component (Boch *et al.*, 1975). Bumblebee queens also produce from their mandibular glands a sex pheromone that evokes copulatory attempts from males (Van Honk *et al.*, 1978). Females are arrested on vegetation at the sites of mandibular gland deposits made by males (Svensson and Bergstrom, 1979), who patrol particular routes along their chemically marked territory.

Most work on sexual communication in the Orthoptera has centered on the cockroaches, including many elegant studies on orientation covered in Chapter 4 by Bell. Female American cockroaches, *Periplaneta americana* emit sex pheromone, (1*Z*,5*E*)-1,10(14)-diepoxy-4(15),5-germacradien-9-one from their body surfaces (Persoons *et al.*, 1979; Adams *et al.*, 1979); this chemical elicits increased rates of locomotion, upwind movement, and wing-raising by males (Tobin *et al.*, 1981). Wing-raising attracts females from a few centimeters away, and in the German cockroach, *Blattella germanica* is elicited by the female's cuticular constituent 3*S*,11*S*-dimethyl-2-nonacosanone (Nishida *et al.*, 1979). In both species, the male accomplishes copulation while the female is arrested and feeding on the male's dorsal abdominal cuticule. Hence, a 'dual' system of sexual aggregation is found also in the order Orthoptera. Both males and females take part in a chemical dialog that reduces spacing and results in mating.

In the Diptera, a wide variety of sexual aggregating systems is known. A female sciarid, *Bradysia impatiens*, attracts males from at least 1 m downwind using a pheromone evidently released from the thorax and legs (Alberts *et al.*, 1981). Cuticular hydrocarbons such as the blend of (*Z*)-9-tricosene and branched alkanes of 28–30 carbons from female houseflies (*Musca domestica*) cause males to fly and to land on fly-like models (Uebel *et al.*, 1976). Such cuticular pheromones are also known in the stable fly (*Stomoxys calcitrans*) (Sonnet *et al.*, 1979) and face fly (*Musca autumnalis*) (Uebel *et al.*, 1975). Among tephritid fruit flies, a variety of systems is used, some in which females attract males with pheromone, and others involving males attracting females, either by isolated or group emission (leks), depending on the species (Prokopy, 1980). Lekking males are also found in some tropical *Drosophila* species, in which pheromone apparently attracts females into the group whereupon one male may mate with her (Spieth, 1968).

Investigations of homopteran sexual communciation have focused on the economically important scale insects and mealybugs. Female diaspidid scales are sedentary and release blends of terpenoid-type compounds such as (3*S*,6*R*)-3-methyl-6-isopropenyl-9-decen-1-yl acetate plus (3*S*,6*R*)-3-methyl-6-isopropenyl-3,9-decadien-1-yl acetate by the California red scale, *Aonidiella aurantii* (Roelofs *et al.*, 1978; Gieselmann *et al.*, 1980). Such blends attract the vagile, flying males from several meters away. Mealybug females of the family Coccidae also produce terpenoid compounds that attract males, such as 2,6-dimethyl-1,5-heptadien-3-ol acetate by the Comstock mealybug, *Pseudococcus comstocki* (Bierl-Loenhardt *et al.*, 1980). The preferential attraction of male scale insects and mealybugs to only the 'correct' geometrical and optical isomers (Gieselmann *et al.*, 1979; Bierl-Leonhardt *et al.*, 1981) indicates myriad possibilities for maintaining separate communication channels by sympatric species in this group.

These have been brief descriptions of sexual communication in a variety of groups. In the remainder of our contribution we will define the major selective forces that shape chemical sexual communication systems: environmental conditions, competition for an exclusive communication channel (a function of the signal-to-noise ratio), reproductive isolation, stabilizing selection, and sexual selection. These factors could be explored with any of the aforementioned insect groups, but we will use moth species as examples because of the comparative wealth in our current understanding of their chemical communication systems. A related discussion for the bark beetles is found in Chapter 12.

13.3 SELECTIVE FORCES

13.3.1 Competition and reproductive isolation

A major ecological paradigm contends that interspecific competition plays a major role in the structuring of communities; this belief has been applied to the analyses of the partitioning of pheromone communication channels in moths (e.g., Comeau, 1971; Roelofs and Cardé, 1974; Cardé *et al.*, 1977; Greenfield and Karandinos, 1980). If the competing species are closely related, distinct pheromone communication channels also provide premating isolation mechanisms that prevent hybridization (Roelofs and Cardé, 1974; Cardé *et al.*, 1977). Although these two, related selective agents provide ample hypotheses to explain distinct communication channels, a rigorous demonstration of the importance of either competition or reproductive isolation in the initial evolution and continued maintenance of these distinct channels is not so readily forthcoming. Autecological constraints can be expected to influence these communication systems, particularly in their daily temporal patterning. The avoidance of avian and chiropteran predation, flight energetics in various air temperatures and insolations and dispersal of pheromone in different wind fields (see Elkinton and Cardé, Chapter 3) can all be expected to offer substantial effects.

In the following examples, description of the communication channels and their degree of overlap between sibling and congeneric species, as well as species in different families, will be straightforward, although the elucidation of these phenomena is by no means complete even in these cases. The selective forces apparently shaping these patterns will be presented as hypotheses. There may be several plausible explanations for differences among the communication channels of the species, and it may be difficult to test experimentally any hypothesis. It could be argued that we rarely observe direct competition between species for a communication channel in nature because the selective effects of competition are so powerful. Character displacement, in the form of a narrower partitioning of the pheromone communication channel where competing species overlap, would provide direct support of the competition paradigm. Insight into the importance of these factors can be gained by experimental manipulation of the natural blend of synthetic pheromone components or by emitting the pheromone at uncharacteristic times or emission rates and documenting the effects upon behavioral response. Similarly, observing the effects of natural fluctuations in environmental conditions upon the animal's communication success may reveal the influence of autecological factors in shaping the communciation channel.

The preponderance of attractant pheromones identified to date for various moth species are even-numbered 10–18 carbon-chain acetates, alcohols and aldehydes, typically with one or two double bonds along the chain. Of course, some moth families employ quite different pheromones, as in the long-chain epoxides and ketones in the Lymantriidae. Yet other major families (e.g., Sphingidae and Geometridae) currently remain with few identified attractants; these groups probably possess unique structures.

Now although some species evidently use only a single chemical for attraction, most species use a blend of two to four related chemicals to create a distinct communication channel. Moths in phylogenetically distant groups may 'share' the same attractant components. (Z)-11- and (E)-11-tetradecenyl acetates, for example, are used by two co-existing moths in North America, the tortricid, *Argyrotaenia velutinana* and the pyralid, *Ostrinia nubilalis* (Klun *et al.*, 1973; Kochansky *et al.*, 1975). These species nonetheless maintain non-overlapping chemical channels by employing slightly different ratios of these chemicals (93:7 and 97:3, respectively) and through the use of an additional component, dodecyl acetate, emitted by *A. velutinana* (Roelofs *et al.*, 1975).

One of the most thoroughly investigated attractant pheromone systems in any group of moths are the attractants in the Tortricidae. The Tortricidae is generally divided into two subfamilies, the Oleuthreutineae, which are characterized by somewhat food plant-specialized, internally feeding larvae and the Torticineae, which are primarily folivorous. In species occurring worldwide (especially North America, Europe and Japan), much is known of the chemical structures of their pheromones. But our knowledge of the tortricids of northeastern North America is particularly detailed, principally due to the efforts of Wendell Roelofs and his colleagues. A typical attractant pheromone in the Tortricidae is comprised of one to four compounds of either 12 or 14 carbon-chain length and possesses either an acetate, alcohol or aldehyde moiety. The compounds eliciting upwind flight typically contain one or two double bonds between the C_7 to the terminal position. A major difference among species in the two subfamilies is the nearly exclusive use of 12-carbon-chain attractants by oleuthreutines and 14-carbon-chain attractants by tortricines (Roelofs and Comeau, 1971; see the discussion by Roelofs and Brown, 1982).

Within the tortricines of northeastern North America many species overlap broadly in their spatial and temporal distributions, and thus we may surmise that the chemical communication channel is partitioned in some fashion. Indeed cross-communication, in which a female lures a non-conspecific mate, appears negligible (Comeau, 1971), despite the relatively few structures employed by the species examined to date.

The partitioning of the chemical channel in tortricines feeding on apple (*Malus*) in this region is especially instructive because in a given locality most of the species emerge some three to five weeks after the late spring optimal feeding period for the larvae. Because of the initial larval dispersal from the egg mass and the solitary leaf-rolling habits of the larvae, adults of many species emerge over the entire tree at the same time (Chapman and Lienk, 1971). One might expect that differing daily times of sexual recruitment would offer an effective partitioning mechanism.

The importance of mating periodicity in creation of an exclusive communication channel between some moth species has long been recognized (Rau and Rau, 1929; see discussion in Roelofs and Cardé, 1974). For example, discrete, non-overlapping daily mating rhythms appear to be primary barriers to cross-attraction in three saturniid moths in South Carolina. *Callosamia promethea* is active from about 10.00 to 15.00–16.00 h; *C. securifera* from the latter time until dusk and *C. angulifera* from dusk to about midnight (Collins and Weast, 1961; Ferguson, 1971–2). Among the *Hemileuca* species of the California Sierra Nevada mountains, specificity also seems to be based in part upon different diel rhythms of attraction (Collins and Tuskas, 1979). But among the apple-feeding tortricines in eastern North America exclusive rhythms of mating appear relatively unimportant. First, if the average times of attraction of males to synthetic pheromone (and where data are available to females) are calculated for a typical June evening (Table 13.1), then it is apparent that the daily rhythms of attraction in these species are so broadly overlapping as to be ineffective partitioning mechanisms. Second, the rhythms shift according to the daily temperature conditions: female calling and male response are altered by the current temperature conditions. This plasticity evidently allows optimization of flight times so that for these tortricines mating occurs late at night when temperatures are warm and early in the evening (or even before dusk) when temperatures are cool (Comeau, 1971; Cardé *et al.*, 1975a; Comeau *et al.*, 1976). An example of such daily alterations in time of attraction for *Archips argyrospilus* in New York is given in Fig. 13.1.

The effect of energetics upon the rhythms of sexual activity would be expected to be most pronounced in small moths, i.e., those possessing a high surface area to volume ratio (Comeau, 1971, Cardé *et al.*, 1975a). Among the nocturnal species, we might expect (at least in temperate regions) that large species would fly later in the night than small ones, as seems to be the case in central New York (Comeau, 1971).

Among moths a large proportion of the day-flying species are either mimics (e.g., male *C. promethea* mimics the butterfly *Battus phelinor*, and sesiids are wasp mimics), or distasteful and aposomatically colored as in many arctiids. Thus, daytime flight among temperate and tropical moths is not typical, unless the species is comparatively well protected against predation.

Table 13.1 Attractant blend ratio and inhibitors*

Species	Z11–14:Ac	E11–14:Ac	Z9–14:Ac	12:Ac	Z11–14:OH	E11–14:OH	Adult seasonal distribution†	Regression for attraction in June–July‡	Typical activity time in June–July§	Structure references¶
Archips argyrospilus	60* ~	40* ~	4* ~	200*	—	—	June to early July	$10.46 + 0.63T_{2050}$	22.5	Cardé *et al.*, 1977
Archips mortuanus	90 ~	10 ~	1 ~	200			June to early July	$14.98 + 0.43T_{2220}$	22.3	Cardé *et al.*, 1977
Archips cervasivoranus	20* ~	80* ~	~	—			July	$20.00 + 0.27T_{2415}$	24.5	Roelofs *et al.*, 1980
Archips semiferanus	30* ~	70* ~					June to early July	$18.18 + 0.25T_{1215}$	22.4	Miller *et al.*, 1976
Argyrotaenia velutinana	90* ~	10* ~		150*	—		April; late June to mid July; August to early September	$17.45 + 0.22T_{2300}$	21.1	Roelofs *et al.*, 1975
Christoneura rosaceana	95* ~	5* ~			5*		June to early July; August	$16.74 + 0.40T_{2300}$	23.4	Hill and Roelofs, 1979
Christoneura fractivittana	—				100		June	$17.65 + 0.25T_{2200}$	21.9	Roelofs and Comeau, 1970
Pandemis limitata	90* ~	—	10* ~	—			June			Roelofs *et al.*, 1976
Platynota idaeusalis	—	50*			—	50*	June; August			Hill *et al.*, 1974
Platynota flavedana	—			15*	~2	85*	June; August			Hill *et al.*, 1977

* Compounds asterisked have been identified as present in either the female's abdominal tip or her effluvium. The remaining compounds have been determined as attractants or antagonists of attraction (−) by empirical screening in the field

† Flight periods for the lower Hudson Valley of New York according to Chapman and Lienk (1971)

‡ Regression equations for the hour of attraction as modified by the temperature at the mean time of attraction for this species. See Comeau (1971) and Comeau *et al.* (1976). All times are Eastern Daylight Time

§ Peak time of attraction for a typical June evening where the temperature is 23°C at 1800, 19°C at 2000, 17°C at 2200, 16.5°C at 2400, and 15.5°C at 0200

¶ Additional references to compound identifications and biological activities are given in these references

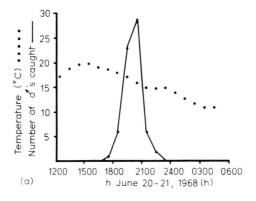

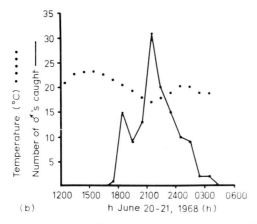

Fig. 13.1 Diel rhythms of male *Archips argyrospilus* attraction to female pheromone extract on (a) a typical June night and on (b) an unusually warm evening (after Comeau, 1971).

The sibling *Archips* species *argyrospilus* and *mortuanus* exhibit nearly identical times of attraction to caged virgin females (Fig. 13.2); specificity in attraction is mediated largely by the ratio of the (Z) and (E) isomers of 11-tetradecenyl acetate. The same pattern of chemical blend specificity holds for the other species (Table 13.1) and evidently it is enhanced by the antagonistic effects upon attraction by the pheromone components emitted by the *other* species. For example, *Pandemis limitata* utilizes the same proportion of (Z)-11- and (Z)-9-tetradecenyl acetates as *A. mortuanus*, but males of *P. limitata* are not lured to this blend when the two additional blend components for *A. mortuanus*, (E)-11-tetradecenyl and dodecyl acetates, are added.

Apart from energetic and predation considerations, the pattern of the

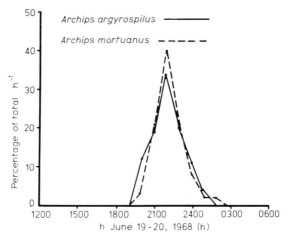

Fig. 13.2   Periodicities of attraction of the sibling species male *Archips argyrospilus* and
A. mortuanus to female pheromone extracts of these species (after Comeau, 1971).
Specificity in attraction is achieved mainly by differences in the ratio of (*Z*) to (*E*) isomers
of 11-tetradecenyl acetate (Table 13.1).

pheromone dispersal can be modified greatly by atmospheric conditions. First,
wind speed and dispersion coefficients, for example, on average vary with time
of day (Fig. 13.3; see Chapter 3) and such daily patterns modify the dimensions
of the active space. Second, sending and receiving individuals may alter their
patterns of emission and response in accord with the current conditions.
Trichoplusia ni females call in bouts the duration of which are adjusted from
about 20 min at wind velocities of <0.1 m sc^{-1} to 5 min at velocities of 3 m sc^{-1}
(Kaae and Shorey, 1972). But the overall daily timing of sexual activity undoubtedly is influenced by daily cycles of fluctuation in atmospheric conditions
and their modification of the patterns of pheromone dispersion.

A quite different partitioning of the sex communication channel evidently
occurs in the day-active Sesiidae (Greenfield and Karandinos, 1979). Most of
their data on attraction specificity is based on male attraction to synthetic lures;
for, in nearly all of the species studied, the identities of the natural pheromone
system remains unknown. Presumably, however, they are identical or very
similar to the empirically determined attractants. Greenfield and Karandinos
found that year of emergence (some sesiids have a 2-year life cycle) or habitat
preference are not involved in isolation; instead the species are partitioned by a
combination of diel and seasonal differences and the ratio of the (*Z,Z*) and
(*E,Z*)-3,13-octadecadienyl acetate attractants. In 93% of the species pairs
examined, a communicational channel (niche) overlap of less than 5% was
achieved by partitioning along a *single* channel dimension (chemical, seasonal
or diel), with the remaining 7% of species pairs being isolated by a combination
of factors (Greenfield and Karandinos, 1979).

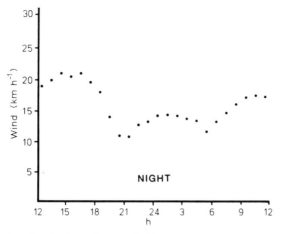

Fig. 13.3   Mean hourly wind speeds recorded in Geneva, New York during the last half of July, 1968 (after Comeau, 1971).

13.3.2 Variation in communication systems

As suggested earlier, the view that either reproductive isolation or interspecific competition dictate the variance of the communication channel may be illuminated by examining the variance of features of the channel in localities where the number of competitive species (and hence the intensity of competition) differ. Where there are few species we might expect the channel to be comparatively broad; conversely, in places where there were many species potentially employing overlapping channels, we might expect relatively narrow tuning. For example, *A. argyrospilus* in New York seems to be more acutely tuned to the proportion of the Z9-14:Ac, in its blend than populations in British Columbia (Cardé *et al.*, 1977). Although the number of tortricines on apple is similar in the two localities (Chapman and Lienk, 1971; Mayer and Beirne, 1974a, b), in New York a sibling species, *A. mortuanus*, competes for the same channel (Table 13.1). Most of the specificity of attraction comes from using different proportions of the 11-14:Ac's. In British Columbia, male *A. argyrospilus* were attracted to mixtures lower in E11-14:Ac or 12:Ac or higher in Z9-14:Ac than in New York (Roelofs *et al.*, 1974; Cardé *et al.*, 1977), suggesting character displacement (a narrowing of the communication channel) in the New York population due to competition with *A. mortuanus*.

Geographical variation in male response also appears to be documented in the noctuid *Agrotis segetum*. The attractant components Z5-10:Ac, Z7-12:Ac and Z9-14:Ac and their blends that evoke male catch appear to differ in Denmark, Switzerland, France and Hungary (Arn *et al.*, 1982). Geographical differences in pheromone production and response occur in *Ostrinia nubilalis*, the European corn borer, and will be considered in Section 13.3.6.

Interpretation of the published attraction spectra of different populations of a given species is obscured by several factors. First, most traps employed in these studies ensnare responding animals on a sticky surface; as a trap's capacity is approached, it retains proportionally fewer responders than traps baited with less active treatments (see Cardé, 1979). Second, trap (treatment) interaction is dependent upon the intertrap distance and the position of treatments (Wall and Perry, 1978). The same experimental design (such as a Latin Square with a constant intertrap distance) should be employed in comparative field trials. Third, general population activity levels seem to influence the apparent variance of response spectra (e.g., Baker and Cardé, 1979a), possibly by altering the number of trap visitations per animal. These factors mean that variances of response spectra among different populations, particularly when test conditions vary and non-saturating traps are employed, must be compared cautiously.

We know even less about the ratio of *emitted* pheromone components. The variance in ratio of the Z11 and E11-14:Ac components extracted from the pheromone gland has been determined for *A. velutinana* (Miller and Roelofs, 1980), and for various strains of *O. nubilalis* (see Section 13.3.6), but as yet we do not know if these ratios remain constant over time in individuals or if the ratios in the pheromone gland match the ratios actually released.

13.3.3 Physiochemical constraints upon blends

One interesting feature of the multichemical attractant systems is the dissimilarity among species in the effect of modifications of the optimal blend ratio. Relatively subtle alterations of the optimal blend for some species greatly decreases the field trap catches. For example, a 5% alteration in the optimum ratio of Z11-14:Ac to E11-14:Ac drastically lowers the male catch in *A. velutinana* (Klun *et al.*, 1973; Roelofs *et al.*, 1975). However, the addition of 12:Ac in widely varying proportions (1:5 to 2:1) to the 9:1 Z11-14:Ac, E11-14:Ac attractant combination increases the trap catch ten-fold over the attractant alone (Roelofs *et al.*, 1975). Similar cases of noncritical ratios are given in Table 13.1.

The *raison d'être* of these divergent blends may be explained by the relative volatility of the components. Those compounds differing in the position or geometrical configuration of the double bond but which have identical carbon-chain lengths and functionalities possess essentially identical vapor pressures at different environmental temperatures. Such components would emanate from a pheromone gland and diffuse within the active space at nearly the same rates. (Most models of pheromone dispersion in wind do not consider molecular diffusion as a major factor in dispersion (see Elkinton and Cardé, Chapter 3); but in still air, over relatively short distances differences in molecular diffusivity may be important (see Bradshaw and Howse, Chapter 15)). Ratios would remain constant throughout the active space, provided that the components are transported at equal rates to the surface of the pheromone gland.

Blend components with either different functional moieties or chain lengths would evaporate at slightly different rates at typical environmental temperatures (e.g., 16–30°C). Hence, the ratio of components emitted from the gland would vary with temperature and thus preclude the use of such compounds in very precise ratios.

13.3.4 Parental investment and stabilizing selection

Features of the chemical communication system are also likely to be molded by the probabilities of mate finding under different population densities. In the absence of any other pressures, stabilizing selection (selection for the population norm) should prevail, favoring (in species with a multichemical pheromone) those males most sensitive to the blend emitted by the majority of females, and those females emitting the blend to which males are most sensitive. Individuals varying from the norm should have a reduced probability of finding a mate, especially at moderate to low population densities, even in the absence of competition with other species. At high densities, however, discrimination by females for 'better' males may also mold the system.

It is possible that lepidopterous sex pheromone systems, in which, as a rule, females are the emitters, arose originally due to the disparity in parental investment between the two sexes (Trivers, 1972; Thornhill, 1979). Female moths produce fewer gametes than males, but furnish each egg with a large supply of nutrients, representing a larger 'investment' in each potential offspring than males give each sperm cell. Females' production of offspring is limited by the number of eggs they can produce, whereas male offspring production depends upon how many matings males can procure. Females therefore become a limiting resource competed for among males. Males able to detect and locate females more rapidly and from greater distances than other males would be at a reproductive advantage, and volatile chemicals emanating from the females' body surfaces might be likely cues originally utilized for detection by males. Thornhill (1979) has proposed just such a scenario for the origin of sex pheromones in *Bittacus* and *Panorpa* scorpionflies (Mecoptera). In these groups, however, males are a limiting resource for reproductive success since they provide nutrients needed for females' egg production, in the form of proteinaceous saliva balls or prey items. Accordingly, these mecopteran males' chemical emanations are detected from a distance by females who fly upwind and locate males and their gifts.

In the Lepidoptera where females are limiting, why, among all the potential volatiles arising from females' bodies would such a narrow range of chemicals be selected to enable early detection by males of each species? Stabilizing selection must be part of the explanation, as males that can detect what the majority of females are producing should be able to locate more females and presumably gain more matings than other males, assuming, of course, an equal probability of courtship success among those males locating females. In addition, it would seem that males should be under pressure to produce not only more receptors and a decoding mechanism tuned to the norm of female emissions, but also to detecting the emissions of 'abnormal' females on the extremes of the distribution. One of the costs of such broad tuning would be an increase in chemical background noise from the environment, thus effectively reducing detection sensitivity through a decrease in the signal:noise ratio. Time and energy lost to responding to airborne volatiles from non-conspecifics could be another cost of the broader tuning, bringing about a decline in male reproductive success.

Once males began tuning their receptors to emissions common to most conspecific females, presumably selection should have favored those females producing greater quantities of these volatiles, as such females would be most likely to attract males and succeed in mating, especially at low population densities. However, the amounts of lepidopterous sex pheromone emitted by females, usually of the order of 1–100 billionths of a gram per hour, are miniscule compared to emissions of courtship compounds by males (Birch, 1974) or of defensive compounds by insects of both sexes.

For a given species there is a characteristic rate of pheromone release as well as a range (minimum to maximum) of pheromone concentrations that elicit normal behavioral responses. Indisputably, of course, there will be variation in the Q (emission rate) and K (threshold of response) features of individuals, although at present we have only fragmentary information on how much variation exists or its genetic basis. Leaving aside the important question of biosynthetic capability, it can be suggested that an increase in Q (and therefore the generated active space) would seem to confer an advantage to an emitter in locating a mate. In sparse populations, particularly, a female emitting a higher than average rate of pheromone should have an increased probability of luring a mate over typical females. Why, then, does there appear to be such a range of release rates among the various moth species and why are many of these rates so low? The difference between *G. molesta* which releases at about 2 ng h^{-1} (Baker *et al.*, 1980) and *T. ni* which emits at up to 1 μg h^{-1} (Bjostad *et al.*, 1980) cannot be attributed to the differences in gland surface area alone.

Part of this disparity could relate to the *demographics* of calling females and responsive males. In populations in which males are sparse, selection should favor high Q values. However, females releasing pheromone with a higher than average Q may lure mates possessing on average an above-normal K. These responders could enter the active space relatively close to the female and assuming that the higher K is heritable, her sons could be less sensitive to pheromone. Also, responders possessing a normal K might respond to pheromone from such high Q females at optimal concentrations occurring far from the source and then exhibit less than optimal performance of orientation and mating behaviors in the abnormally high concentration of pheromone near the female. This would be most likely in species possessing an upper threshold limit such as *G. molesta* (Baker and Roelofs, 1981).

A corollary to the hypothesis that the limiting sex is competed for by the non-limiting sex is that the limiting sex can afford to be 'choosy' or discriminating (Trivers, 1972; Thornhill, 1979). Possibly, one of the reasons for the small quantities of pheromone emitted by female moths of some species is that under average-to-high population densities the parsimonious emission of pheromone by (discriminating) females selects for males having lowest K's and most acute mate-finding abilities (Greenfield, 1981). At the same time this would put further pressure on the males' sensory systems to optimize their tuning by narrowing the bandwidth (variance in the receptor's response spectrum) and amplifying the signal (number of receptors), thereby improving the signal:noise ratio. Hence, the narrow range of pheromone blend ratios and low emission rates often utilized by Lepidoptera could be the result of selective pressures relating to reproductive (communication) success and parental invest-ment, and not reproductive isolation *per se*, resource partitioning of the communication channel, or other pressures such as avoidance of detection by predators, as in some beetles (see Birch, Chapter 12).

These examples show that, in addition to the posited interspecific effects of competition for an exclusive communication channel, other factors including energetics, predation, pheromone dispersion, physiochemical constraints, and stabilizing selection, may dictate the design of these signals. We will now consider in detail the sexual communication system of the oriental fruit moth, *G. molesta*, as an example of the interplay of the chemical and non-chemical signals in sexual communication and the potential role of sexual selection.

13.3.5 Integration of chemical and other cues in the sexual behavior of the oriental fruit moth

Sexual communication in the oriental fruit moth is initiated by the female, who, a few hours before dusk, raises her wings and abdomen and releases four compounds from a gland near her extended ovipositor. Three of these com-pounds elicit behavioral changes in males: (Z)-8-dodecenyl acetate (Z8-12:Ac) (Roelofs *et al.*, 1969), (E)-8-dodecenyl actate (E8-12:Ac), and (Z)-8-dodecenyl alcohol (Z8-12:OH) (Cardé *et al.*, 1979) (Fig. 13.4). The fourth compound,

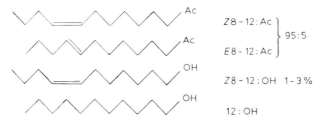

Fig. 13.4   Pheromone components emitted by female *Grapholitha molesta*, the Oriental fruit moth.

dodecyl alcohol (12:OH) seems to increase males' reactions only if Z8-12-OH is at or lower than normal levels. The first three compounds appear to act in concert to elicit movement reactions resulting in attraction to the female (Baker and Cardé, 1979b). It is clear that the ratio of these components is crucial to optimal attraction, as demonstrated in field experiments with synthetic sources. When more or less than 6% E8-12:Ac relative to Z8-12:Ac is present, the number of approaches of males toward the source diminishes significantly (Baker and Cardé, 1979b), as reflected in reduced trap captures of males to these off-ratios (Baker *et al.*, 1981a) (Fig. 13.5). Wind tunnel observations support the field results, and indicate that reduced upwind flights to the source in response to too little E8-12:Ac is due to the tendency of males to lose contact with the plume during erratic, rapid velocity flights. The reduction in source location by males flying upwind to an excess of E8-12:Ac is characterized by arrestment within the plume at some distance downwind of the source, followed by upward flight out of the plume. Too little or too much Z8-12:OH added to the optimal Z8:12:Ac mixture also reduces attraction of males, the optimal percentage being ca. 3% relative to Z8-12:Ac. Not surprisingly, those ratios that elicit optimal attraction of males to synthetic sources are nearly identical to those emitted by females. Interestingly, Z8-12:OH appears important for the reproductive isolation of the oriental fruit moth from a congener, *G. prunivora*, the lesser appleworm. This component reduces captures of *G. prunivora* males when added to the two acetates while it simultaneously increases attraction of *G. molesta*.

Female *G. molesta* release this blend of components at a rate of ca. 2 ng h^{-1} (Baker *et al.*, 1980). Trapping experiments indicate that rubber septum dispensers releasing the synthetic blend at between 1 and 12 ng h^{-1} capture the maximum number of males, demonstrating that emission rates close to the females' are optimal. Emission rates higher or lower than these reduced captures (Fig. 13.5), but for different reasons. The lower concentrations are not sufficient to elicit upwind flight from males more than a few meters on average from the source. Conversely, super-normal rates evoke upwind flight from more than 80 m away but result in within-plume arrestment more than a meter away from the source (Baker and Roelofs, 1981). More detailed observations in a wind tunnel confirmed that only an intermediate range of dosages and (*E*)-(*Z*) ratios resulted in attraction to the source without prematurely causing arrestment.

Once the male has landed near a female, these same pheromone components elicit close approach to the female by walking while wing-fanning. Of all the incomplete combinations of components, only the two acetates are sufficient to evoke some close approaches, but inclusion of Z8-12:OH in the blend increases close approaches and courtship behaviors dramatically.

The most striking behavior seen when the complete blend is present is the hairpencil display of courtship. The natural blend of chemicals alone, however, is not all that is needed before males will display at a female. The visual cues

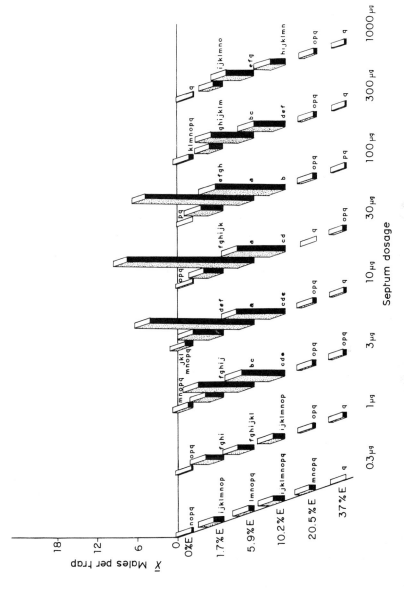

Fig. 13.5 Captures of male *Grapholitha modesta* in traps baited with different combinations of (Z)-8-dodecenyl and (E)-8-dodecenyl acetates in different dosages (Baker *et al.*, 1981a).

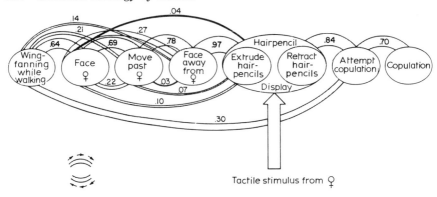

Fig. 13.6   The conditional probabilities of male and female *Grapholitha molesta* court-ship behaviors occurring in cases where the female delivers a tactile stimulus to the male by hitting her head into his abdominal tip (Baker and Cardé, 1979a).

from a sitting female (or artificial model) need to be coupled with the presence of pheromone (Baker and Cardé, 1979a). However, apparently males are not able to discriminate between potential mates on the basis of visual cues alone, for a large red rubber septum is very effective at evoking hairpencil displays. If the model is not within, or at the source of, the pheromone plume, displays will not be evoked as frequently.

When the visual and chemical stimuli are correct, a fairly rigidly 'fixed action pattern' of male behaviors is evoked, one of which is the hairpencil display (Fig. 13.6) (Baker and Cardé, 1979a). Chemical, anemotactic (wind), and possibly visual cues from the hairpencil display now elicit movement by the female, over a few cm to the male's abdominal tip. The predominant cues effecting female attraction are chemicals released by the rhythmically extruded and retracted hairpencils that give them a pleasant herb-like odor to the human nose. Four hairpencil compounds (Fig. 13.7) have been identified: ethyl-*trans*-cinnamate (I), (−)-mellein (II), methyl-jasmonate (III), and methyl 2-epijasmonate (IV). The combination of I and IV is behaviorally active in attracting females, unlike many other combinations of the four compounds (Baker *et al.*, 1981b). Wind of ca. 90 cm sc^{-1} generated by the male's vibrating wings propels these compounds to the female, possibly imparting directionality to the signal. Although typically the male walks upwind of the female before displaying, he can display from any position because his wind-producing ability would confer indepen-dence from the ambient wind. His wind also might allow the female to use anemotaxis to orient more accurately to him. Visual cues from the male's body, including the rhythmic extension of the light-colored hairpencils, do not seem to affect the accuracy of the female's orientation, although the contribution of the hairpencils' visual cues to attraction cannot yet be dismissed entirely.

The final stimulus in the aggregation process, as one might expect, is tactile; the attracted female touches the tip of the male's abdomen, and experimental

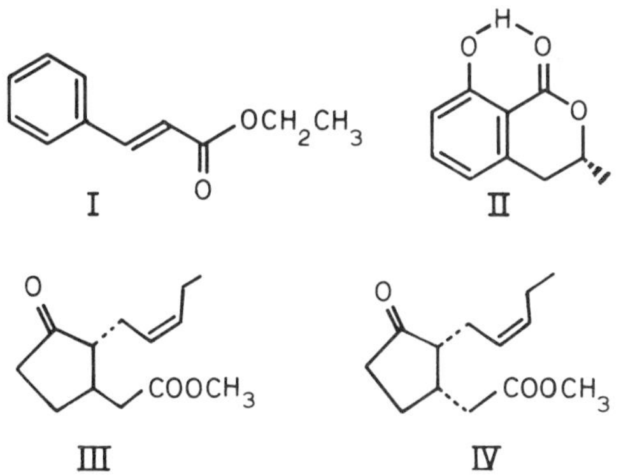

Fig. 13.7 Compounds present in male *Grapholitha molesta* hairpencils: ethyl-*trans*-cinnamate (I), (−)-mellein (II), methyl-jasmonate (III), and methyl 2-epijasmonate (IV) (Baker *et al.*, 1981b).

manipulations confirm that it is this touch that evokes the male's copulatory attempt (Baker and Cardé, 1979a). Only those males that display and induce females to touch their abdomens copulate. However, the situation is quite different if two or more males arrive near a female at about the same time (Baker, 1983). If a male is displaying when a second male approaches, the latter male may attempt to copulate (often successfully) with the female without first displaying. Such late-arriving males appear to 'sneak' in for a copulatory attempt under the cover of the first male's chemical barrage. Other alterations in the normal male courtship sequence also occur, including late-arriving males being attracted to first-arrivals' displays, and the touches they deliver causing misdirected copulatory attempts by displaying males and reduced mating success. Two may display simultaneously toward one female, whereupon she will 'choose' one male over the other by touching his hairpencils. Simultaneous displaying occurs infrequently, but even for sequentially arriving males, females sometimes choose late arrivals after not responding to earlier males' displays. Females fail to respond to males that are prevented experimentally from everting their hairpencils, even though the rest of their courtship sequence is otherwise unaltered.

Because females control whether copulation will occur, it is possible that female-choice sexual selection could have been responsible for the evolution of the hairpencil organs, the courtship pheromone blend, and the rigidly fixed sequence of movements that comprise the display (Baker and Cardé, 1979a). As discussed above, as the limiting sex, females can exercise discrimination in selecting mates, considering their larger parental investment in potential offspring. However, here the benefit received by females choosing males with 'better' displays would be the mating advantage conferred upon their sons in the presence of discriminating females in the next generation. Protandry (early seasonal emergence of male adults), or a skewed sex ratio with a predominance of males, are two conditions that could result in a proportion of the male population being excluded from mating, but in the oriental fruit moth a third situation effectively creates a shortage of females during the mating period. Most females mate only once whereas males can mate an average of more than once per evening (Dustan, 1964).

Thus, some males could acquire a disproportionate share of matings at the expense of other males. For the highly rapid, directional selection (selection for an extreme) of male courtship traits to proceed, first a small percentage of discriminating females and males with an appropriate trait must be present in the population. Initially, such characters may confer an advantage in *reproductive* rather than courtship success (Thornhill, 1979). After that, sexual selection could proceed on its own to produce more extreme male scent dissemination structures and increasingly selective females.

13.3.6 Speciation

The most commonly accepted view of speciation in sexually reproducing animals assumes that the diverging populations are geographically isolated from one another for many generations, during which time differing selective pressures acting upon the two populations cause them to acquire substantial genetic differences. If they have not thus acquired effective premating reproductive isolating mechanisms and hybrids are less fit, then it is posited that barriers to hybridization could develop during periods of secondary contact between the populations when traits, such as differences in the sex pheromone communication channel, would be accentuated. Such an allopatric model is most compatible with multicomponent pheromone communication systems (cf. Shorey, 1970) that have the ready possibility of modification of the communication system. This traditional allopatric view of speciation is currently being challenged in two ways. First, many suggest that speciation may occur in 'quantum' or rapid steps. Rapid speciation is viewed as most likely to occur in isolated or peripheral segments of a species' distribution. A second model holds that new species can arise, presumably also quite rapidly, within the general distribution of the species. The latter process is termed sympatric speciation and has been most cogently advocated by Bush (1975).

As with the previous discussions of the selective forces that mold the communication channel, it is difficult to arrive at either definitive proofs or absolute refutations of these models. Notwithstanding such difficulties, several cases provide illuminating evidence of the potential contribution of sex pheromones to the speciation process. In many of the bark beetles it is known that the aggregation pheromone is biosynthesized from precursors obtained by the feeding adult beetles from the phloem of the host trees. In *Ips paraconfusus* the pheromone component *cis*-verbenol is biosynthesized from the host terpene $(-)$-pinene. Ingestion of the opposite isomer produces the opposite optical form of the pheromone (Renwick *et al.*, 1976). Clearly this suggests that beetles colonizing a tree species that contains a 'wrong' precursor either alone or in combination with the correct precursor would possess a new communication system. If this occurred in individual animals colonizing inappropriate hosts, however, we cannot readily assume that this process will lead to a rapid evolution of a new pheromone, because there would have to be numerous males and females *responsive* to this new communication system if the colonizing beetles are to attack a tree successfully (see Chapter 12). Thus, the rapid alteration of pheromone communication channel of bark beetles through colonization of new hosts containing novel precursor materials may be limited.

Ips pini from eastern and western North America produce and respond preferentially to pheromone produced by their own aggregating populations (Lanier *et al.*, 1980; see Section 11.3). These differences presumably arose in isolation, character displacement, or possibly clinal divergence.

Roelofs and Comeau (1969) hypothesized that rapid evolution of new pheromone communication channels could occur within relatively few generations. The rare female homozygous (recessive trait) for a novel pheromone component or blend would lure the rare male homozygous (recessive trait) for the ability to perceive the same pheromone. The resulting progeny, of course, would have a normal phenotype. If, however, some of these siblings mated, as might be feasible in very small populations, then recombination of traits in males responsive to the novel system and females releasing it could carry this new communication system to fixation within a relatively few generations. If either the production or response traits were not completely recessive, or if the traits were carried on the X-chromosome, then the model would be more appealing.

The occurrence of an insect's pheromone trait on the X-chromosome (in the Lepidoptera the female is the heterogametic sex) has been demonstrated for the major 'species-recognition' pheromone in *Colias eurytheme*, sulfer butterfly, by Grula and Taylor (1979). Another signal used by males in courtship is the ultraviolet reflection pattern of the wing, which is also transmitted on the X-chromosome. In contrast to the evidently close linkage of courtship signals in *C. eurytheme*, it was found that genes controlling a multicomponent courtship pheromone system in *Drosophila melanogaster* were dispersed throughout the genome. Such an independent assortment of loci would enhance the diversity of genotypes and promote the outbreeding found with the negative assortative mating system of this species (Averhoff and Richardson, 1976). That genes controlling a species-specific communication system should be inherited as a co-adapted complex in *C. eurytheme* (Grula and Taylor, 1979) substantiates the theoretical analyses of Alexander (1962) and O'Donald (1962). The documented systems in *C. eurytheme* and *D. melanogaster* raise 'the interesting possibility that insect communication systems have both variable and invariable components and that different sets of genes with different modes of inheritance give rise to this dichotomy' (Grula and Taylor, 1979).

Populations of *O. nubilalis* in North America (which were introduced from unknown European localities several times in the early 1900s) vary in their production and response to the ratio of the two pheromone components, Z11- and E11-14: Ac. Populations from Iowa and New York, for example, produce and are most attracted to 97:3 and 4:96 ratios, respectively, of the two acetates (Klun *et al.*, 1973; Kochansky *et al.*, 1976). Isozyme studies of males attracted to these two blends in central Pennsylvania suggest that these strains are not panmictic (Cardé *et al.*, 1975b, 1978; Harrison and Vawter, 1977). In some areas such as Maryland and North Carolina captures of males in traps occur across the spectrum of attractant blends from the predominantly (*Z*) to (*E*) isomer mixes (Klun, 1975; Kennedy and Anderson, 1980). The so-called Z and E strains both occur in Europe, but evidently not in the same localities, so that an allopatric origin of these strains is the most plausible explanation. The Klun and Maini (1979) study of the genetics of pheromone production and laboratory response in hybrid crosses of these strains indicates that the major features of these traits are controlled by simple Mendelian inheritance. As White (1978 p. 332–3) notes,

'An important unresolved question in speciation theory is whether the primary role in initiating speciation is usually a premating isolating mechanism, such as a pheromonal or bio-acoustic difference between the two diverging populations, or a postmating one, whether due to chromosomal rearrangements or gene mutation. Only a series of investigations on suitable cases in various stages of speciation, in which both the ethological and the cytogenetic factors are intensively studied, will resolve this basic question. The answer will not necessarily be the same for all cases or for all groups of organisms. In all but a minute number of instances closely related species differ in karyotype. But, at the same time, the great majority of closely related animal species differ in courtship behavior. Thus, most studies of sibling species are likely to be uninformative on this point, since both kinds of differences will exist. The cases that may help to resolve this problem will be those few instances of semispecies or sibling species that differ in a single respect only.'

13.4 CONCLUSION

During the past 20 years the burgeoning study of pheromone communication has resulted in the characterization of the chemistry and behavioral responses in numerous insect species. Despite the rapid accumulation of identified pheromones and a description of the behaviors elicited, for a large number of the pheromone systems, we do not as yet know all of the chemical components involved, nor do we have a thorough understanding of the orientation mechanisms that are involved in many pheromone responses. The selective forces that shape these communication systems largely remain uncharted.

The rhythms of daily communication cycles and the proximate environmental cues regulating their expression have been described for a large number of temperate insects, and these general patterns can be expected to be similar for other insect species. But how the timing of these events relates to ultimate selection factors, such as flight energetics of the responder, predation, optimal conditions for atmospheric transmission of the pheromone message, light and wind conditions favorable for successful orientation, and numerous other factors, has scarcely been documented. The paradigm of interspecific partitioning of the communication channel appears to be a likely explanation for differences among closely allied species in chemical blends, the daily timing of reproductive activity, and other differences in the communication channel. However, rigorous proof of the importance of these selective forces in the creation and maintenance of such partitioning is lacking. Future studies, if they are to resolve these questions, will need to separate the effect of environmental factors that favor particular daily patterns of communication from divergences resulting from interspecific competition for discrete communication channels. In part, distinctions between the effects of environmental and species partitioning forces may become evident by careful description and comparison of the communication channels of given species in different localities where the number of species sharing pheromone components varies and therefore the degree of competition differs.

Stabilizing or normalizing selection would act to keep characteristics of the communication channel at some intermediate level, as in the ratio of two pheromone components. Unfortunately, we have little information on individual variation in production of and response to ratios of components. Sexual selection and choice of a partner by the non-resource limited sex may be powerful agents, promoting new and elaborate courtship behaviors and pheromones. The seemingly advantageous strategy of increasing the rate of pheromone emission in the calling sex, thereby increasing the active space and the opportunity for luring a mate, may have disadvantageous consequences. This practice may tend to result in females attracting males that have a higher threshold of response and thus their male offspring would be less successful in mate location. To test these hypotheses it will be necessary to have descriptions of natural variation in the communication channel and to determine its genetic basis.

The degree to which pheromones are involved in speciation will probably remain speculative, inasmuch as allopatric speciation generally is not amenable to experimentation. It is clear that a bark beetle could alter its aggregation pheromone simply by selecting a host tree species that provides a precursor for a novel pheromone. If speciation occurs primarily in an allopatric mode, then changes in the pheromone communication channel that effect reproductive isolation may come about or at least be reinforced after initial changes in the communication system that occurred in isolation.

The present attempt to summarize the selective forces molding the pheromone systems in insects has emphasized moth examples. This restriction has limited our exploration of phenomena such as aggregation, lekking, male-produced attractants, and the effects of plant hosts upon sexual behavior, all of which appear to occur infrequently in this group. McNeil and Turgeon (1982) have suggested that both temperature and larval food source modify the rapidity of the onset of sexual behavior (including pheromone emission) in the cutworm moth *Pseudaletia unipuncta*. This in turn determines the reproductive success of the fall generation and the likelihood of wide-scale epidemics occurring in the ensuing year. Except for this example (and of course the phenomenon of mass host attack in bark beetles), interactions between population dynamics and pheromone-mediated sexual behavior have been little explored in the non-social insects.

Much remains to be unravelled if we are to understand how environmental factors, the community, and intraspecific mating success shapes the pheromone communication channel. The exciting prospect of defining and testing these hypotheses is one of the future challenges in elucidating the chemical ecology of insects.

ACKNOWLEDGMENT

We thank R. Charlton for a valuable critique of this chapter.

REFERENCES

Adams, M. A., Nakanishi, K., Still, W. C., Arnold, E. V., Clardy, J. and Persoons, C. J. (1979) Sex pheromone of the American cockroach: absolute configuration of periplanone-B. *J. Am. Chem. Soc.*, **101**, 2495−8.

Alberts, S. A., Kennedy, M. K. and Cardé, R. T. (1981) Pheromone-mediated anemotactic flight and mating behavior of the sciarid fly *Bradysis impatiens*. *Env. Ent.*, **10**, 10−15.

Alexander, R. D. (1962) Evolutionary change in cricket accoustical communication. *Evolution*, **16**, 443−67.

Arn, H., Baltensweiler, W., Bues, R., Buser, H. R., Esbjerg, P., Guerin, P., Mani, E., Rausher, S., Szocs, G. and Toth, M. (1982) Refining lepidopteran sex attractants. Les médiateurs chimiques agissant sur le comportement des insects. *INRA Coll.*, **7**, 261−5.

Baker, T. C. (1982) Variations in male oriental fruit moth courtship patterns due to male competition. *Experientia*, **39**, 112−4.

Baker, T. C. and Cardé, R. T. (1979a) Courtship behavior of the oriental fruit moth (*Grapholitha molesta*): experimental analysis and consideration of the role of sexual selection in the evolution of courtship pheromones in the Lepidoptera. *Ann. Ent. Soc. Am.*, **72**, 173−88.

Baker, T. C. and Cardé, R. T. (1979b) Analysis of pheromone-mediated behavior in male *Grapholitha molesta*, the oriental fruit moth (Lepidoptera: Tortricidae). *Env. Ent.*, **8**, 956−68.

Baker, T. C., Cardé, R. T. and Miller, J. R. (1980) Oriental fruit moth pheromone component release rates measured after collection by glass surface adsorption. *J. Chem. Ecol.*, **6**, 749−58.

Baker, T. C., Mayer, W. and Roelofs, W. L. (1981a) Sex pheromone dosage and blend specificity of response by oriental fruit moth males. *Ent. exp. appl.*, **30**, 269–79.

Baker, T. C., Nishida, R., and Roelofs, W. L. (1981b) Close-range attraction of female oriental fruit moths to herbal scent of male hairpencils. *Science*, **214**, 1359–61.

Baker, T. C. and Roelofs, W. L. (1981) Initiation and termination of oriental fruit moth male response to pheromone concentrations in the field. *Env. Ent.*, **10**, 211–8.

Bierl-Leonhardt, B. A., Moreno, D. S., Schwarz, M., Gorster, H. S., Plimmer, J. R. and DeVilbiss, E. D. (1980) Identification of the pheromone of the Comstock mealybug. *Life Sci.*, **27**, 399–402.

Bierl-Leonhardt, B. A., Moreno, D. S., Schwarz, M., Fargerlund, J. and Plimmer, J. R. (1981) Isolation, identification, and synthesis of the sex pheromone of the citrus mealybug, *Planococcus citri* (Risso). *Tetrahedron Leters*, **22**, 389–92.

Birch, M. (1974) Aphrodisiac pheromones in insects. In: *Pheromones* (Birch, M., ed.) pp. 115–34. North-Holland Publications, Amsterdam.

Bjostad, L. B., Gaston, L. K. and Shorey, H. H. (1980) Temporal pattern of sex pheromone release by female *Trichoplusia ni*. *J. Insect Physiol.*, **26**, 493–8.

Boch, R., Shearer, D. A. and Young, J. C. (1975) Honey bee pheromones: field tests of natural and artificial queen substance. *J. Chem. Ecol.*, **1**, 133–48.

Buschinger, A. and Alloway, T. M. (1979) Sexual behaviour in the slave-making ant, *Harpagoxenus canadensis* (M. R. Smith) and sexual pheromone experiments with *H. canadensis*, *H. americanus* (Emery), and *H. sublaevis* (Nylander) (Hymenoptera: Formicidae). *Z. Tierpsychol.*, **49**, 113–19.

Bush, G. L. (1975) Modes of animal speciation. *A. Rev. Ecol. Syst.*, **6**, 339–64.

Cardé, A. M., Baker, T. C. and Cardé, R. T. (1979) Identification of a four-component sex pheromone of the female oriental fruit moth. *J. Chem. Ecol.*, **5**, 423–7.

Cardé, R. T. (1979) Behavioral responses of moths to female-produced pheromones and the utilization of attractant-baited traps for population monitoring. In: *Movement of highly mobile insects: concepts and methodology in research* (Rabb, R. L. and Kennedy, G. G., eds) pp. 286–315. North Carolina State University Press.

Cardé, R. T., Baker, T. C. and Roelofs, W. L. (1975a) Moth mating periodicity: temperature regulates the circadian gate. *Experientia*, **31**, 46–8.

Cardé, R. T., Kochansky, J., Stimmel, J. F., Wheeler, Jr., A. G. and Roelofs, W. L. (1975b) Sex pheromones of the European corn borer *Ostrinia nubilalis: cis-* and *trans-* responding males in Pennsylvania. *Env. Ent.*, **4**, 413–4.

Cardé, R. T., Cardé, A. M., Hill, A. S. and Roelofs, W. L. (1977) Sex pheromone specificity as a reproductive isolating mechanism among the sibling species *Archips argyrospilus* and *A. mortuanus* and other sympatric tortricine moths (Lepidoptera: Tortricidae). *J. Chem. Ecol.*, **3**, 71–84.

Cardé, R. T., Roelofs, W. L., Harrison, R. G., Vawter, A. T., Brussard, P. F., Mutuura, A. and Monroe, E. (1978) European corn borer: pheromone polymorphism or sibling species. *Science*, **199**, 555–6.

Chapman, P. J. and Lienk, S. E. (1971) Tortricid fauna of apple in New York (Lepidoptera: Tortricidae): including an account of apples' occurrence in the state, especially as a naturalized plant. *Spec. Publ. N.Y.S. Agr. Exp. Sta.*, Geneva, New York.

Collins, M. M. and Weast, R. D. (1961) *Wild Silk Moths of the United States*. Saturniinae. Collins Radio Company, Cedar Rapids, Iowa.

Collins, M. M. and Tuskes, P. M. (1979) Reproductive isolation in sympatric species of dayflying moths (*Hemileuca*: Saturniidae). *Evolution*, **33**, 728–33.

Comeau, A. (1971) *Physiology of Sex Pheromone Attraction in Tortricidae and other Lepidoptera (Heterocera)*. PhD Thesis. Cornell University, Ithaca, New York.

Comeau, A., Cardé, R. T. and Roelofs, W. L. (1976) Relationship of ambient temperatures to diel periodicities of sex attraction in six species of Lepidoptera. *Can. Ent.*, **108**, 415–18.

Cross, J. H., Byler, R. C., Cassidy, Jr., R. F., Silverstein, R. E., Greenblatt, R. E., Burkholder, W. E., Levinson, A. R. and Levinson, H. Z. (1976) Porapak-Q collection of pheromone components and isolation of (Z)- and (E)-14-methyl-8-hexadecenal, sex pheromone components from the female of four species of *Trogoderma* (Coleoptera: Dermestidae). *J. Chem. Ecol.*, **2**, 457–68.

Dahm, K. H., Meyer, D., Finn, W. E., Reinhold, V. and Roller, H. (1971) The olfactory and auditory mediated sex attraction in *Achroia grisella* (Fabr.). *Naturwissenschaften*, **58**, 265–66.

Dustan, G. G. (1964) Mating behaviour of the oriental fruit moth, *Grapholitha molesta* (Busck) (Lepidoptera: Olethreutidae). *Can. Ent.*, **96**, 1087–93.

Ferguson, D. C. (1971–72) Bombycoidae (Saturniidae) In: *The Moths of America North of Mexico* (Dominick *et al.*, eds). Fasc. 20.2.

Finn, W. E. and Payne, T. L. (1977) Attraction of greater wax moth females to male-produced pheromones. *Southw. Ent.*, **2**, 62–5.

Gieselmann, M. J., Rice, R. E., Jones, R. A. and Roelofs, W. L. (1979) Sex pheromone of the San Jose scale. *J. Chem. Ecol.*, **5**, 891–900.

Gieselmann, M. J., Henrick, C. A., Anderson, R. J., Moreno, D. S. and Roelofs, W. L. (1980) Responses of male California red scale to sex pheromone isomers. *J. Insect Physiol.*, **26**, 179–82.

Greenfield, M. D. (1981) Moth sex pheromones: an evolutionary perspective. *Fla. Ent.*, **64**, 4–17.

Greenfield, M. D. and Karandinos, M. G. (1979) Resource partitioning of the sex communication channel in clearwing moths (Lepidoptera: Sesiidae) of Wisconsin. *Ecol. Mon.*, **49**, 403–26.

Grula, J. W. and Taylor, O. R. (1979) The inheritance of pheromone production in the sulfur butterflies *Colias eurytheme* and *C. philodice*. *Heredity*, **42**, 359–71.

Harrison, R. G. and Vawter, A. T. (1977) Allozyme differences between pheromone strains of the European corn borer, *Ostrinia nubilalis*. *Ann. Ent. Soc. Am.*, **70**, 717–20.

Hedin, P. A., McKibben, G. H., Mitchell, E. B. and Johnson, W. L. (1979) Identification and field evaluation of the compounds comprising the sex pheromone of the female boll weevil. *J. Chem. Ecol.*, **5**, 617–27.

Hill, A., Cardé, R., Comeau, A., Bode, W. and Roelofs, W. (1974) Sex pheromones of the tufted apple bud moth (*Platynota ideausalis*). *Env. Ent.*, **3**, 249–52.

Hill, A., Cardé, R., Bode, W. and Roelofs, W. (1977) Sex pheromone components of the varigated leafroller moth, *Platynota flavedana*. *J. Chem. Ecol.*, **3**, 369–76.

Hill, A. S. and Roelofs, W. L. (1979) Sex pheromone components of the oblique-banded leafroller moth *Choristoneura rosaceana*. *J. Chem. Ecol.*, **5**, 3–11.

Hill, A. S. and Roelofs, W. L. (1981) Sex pheromone of the saltmarsh caterpillar moth, *Estigmene acrea*. *J. Chem. Ecol.*, **7**, 655–68.

Hölldobler, B. and Haskins, C. P. (1977) Sexual calling behavior in primitive ants. *Science*, **195**, 793–4.

Jewett, D. M., Matsumura, F. and Coppel, H. C. (1976) Sex pheromone specificity in the pine sawflies: interchange of acid moieties in an ester. *Science*, **192**, 51–3.

Kaae, R. S. and Shorey, H. H. (1972) Sex pheromones of noctuid moths. XXVII. Influence of wind velocity on sex pheromone releasing behavior of *Trichoplusia ni* females. *Ann. ent. Soc. Am.*, **65**, 436–40.

Kennedy, J. S. (1978) The concepts of olfactory 'arrestment' and 'attraction'. *Physiol. Ent.*, **3**, 91–8.

Kennedy, G. G. and Anderson, T. E. (1980) European corn borer trapping in North Carolina with various sex pheromone component blends. *J. Econ. Ent.*, **73**, 642–6.

Klun, J. A. (1975) Insect sex pheromones: intraspecific pheromonal variability of *Ostrinia nubilalis* in North America and Europe. *Env. Ent.*, **4**, 891–4.

Klun, J. A., Chapman, D. L., Mattes, K. C., Wojtkowski, P. W., Beroza, M. and Sonnet, P. E. (1973) Insect sex pheromones: minor amount of opposite geometrical isomer critical to attraction. *Science*, **181**, 661–3.

Klun, J. A. and Maini, S. (1979) Genetic basis of an insect chemical communication system: the European corn borer. *Env. Ent.*, **8**, 423–6.

Kochansky, J., Cardé, R. T., Liebherr, J. and Roelofs, W. L. (1975) Sex pheromone of the European corn borer, *Ostrinia nubilalis* (Lepidoptera: Pyralidae) in New York. *J. Chem. Ecol.*, **1**, 225–31.

Kraemer, M., Coppel, H. C., Matsumura, F., Kikukawa, T. and Mori, K. (1979) Field responses of the white pine sawfly, *Neodiprion pinetum*, to optical isomers of sawfly sex pheromones. *Env. Ent.*, **8**, 519–20.

Kraemer, M. E., Coppel, H. C., Matsumura, F., Wilkinson, R. C. and Kikukawa, T. (1981) Field and electro-antennogram responses of the red-headed pine sawfly, *Neodiprion lecontei* (Fitch), to optical isomers of sawfly sex pheromones. *J. Chem. Ecol.*, **7**, 1063–72.

Lanier, G. N., Classon, A., Stewart, T., Piston, J. J. and Silverstein, R. M. (1980) *Ips pini*: the basis for interpopulational differences in pheromone biology. *J. Chem. Ecol.*, **6**, 677–87.

Leyrer, R. L. and Monroe, R. E. (1973) Isolation and identification of the scent of the moth, *Galleria mellonella*, and a re-evaluation of its sex pheromone. *J. Insect Physiol.*, **19**, 2267–71.

McNeil, J. N. and Turgeon, J. J. (1982) Pheromone biology in the population dynamics of *Pseudaletia unipuncta* (Haw) (Lepidoptera; Noctuidae), a sporadic pest. Les médiateurs chimiques agissant sur le comportement des insects. *INRA Coll.*, **7**, 215–24.

Mayer, D. F. and Beirne, B. P. (1974a) Aspects of the ecology of apple leaf rollers (Lepidoptera: Tortricidae) in the Okanagan Valley, British Columbia. *Can. Ent.*, **106**, 349–52.

Mayer, D. F. and Beirne, B. P. (1974b) Occurrence of apple leaf rollers (Lepidoptera: Tortricidae) and their parasites in the Okanagan Valley, British Columbia. *J. Ent. Soc. Br. Col.*, **71**, 22–5.

Miller, J. R., Baker, T. C., Cardé, R. T. and Roelofs, W. L. (1976) Re-investigation of oak leaf roller sex pheromone components and the hypothesis that they vary with diet. *Science*, **192**, 140–3.

Miller, L. J. and Roelofs, W. L. (1980) Individual variation in sex pheromone component ratios in two populations of the redbanded leafroller moth, *Argyrotaenia velutinana*. *Env. Ent.*, **9**, 359–63.

Nishida, R., Kuwahara, Y., Fukami, H. and Ishii, S. (1979) Female sex pheromone of the German cockroach, *Blattella germanica* (L.) (Orthoptera: Blatellidae), responsible for male wing-raising: IV. The absolute configuration of the pheromone, 3,11-dimethyl-2-nonacosanone. *J. Chem. Ecol.*, **5**, 289–97.

O'Donald, P. (1962) The theory of sexual selection. *Heredity*, **17**, 541–52.

Persoons, C. J., Verwiel, P. E. J., Talman, E. and Ritter, F. J. (1979) Sex pheromone of the American cockroach, *Periplaneta americana*: isolation and structure elucidation of periplanone-B. *J. Chem. Ecol.*, **5**, 221–36.

Prokopy, R. J. (1980) Mating behavior of frugivorous Tephritidae in nature. *Proc. Symp. Fruit Fly Probl. Nat. Inst. Agric. Sci. Japan*, pp. 37–46.

Rau, P. and Rau, N. L. (1929) The sex attraction and rhythmic periodicity in the giant saturniid moths. *Trans. Acad. Sci. St. Louis*, **26**, 83–221.

Renwick, J. A. A., Hughes, P. R. and Krull, I. S. (1976) Selective production of *cis*- and *trans*-verbenol from (−)- and (+)-pinene by a bark beetle. *Science*, **191**, 199–201.

Roelofs, W. L. and Comeau, A. (1969) Sex pheromone specificity: Taxanomic and evolutionary aspects in Lepidoptera. *Science*, **165**, 398–400.

Roelofs, W. L., Comeau, A. and Selle, R. (1969) Sex pheromone of the oriental fruit moth. *Nature*, **224**, 723.

Roelofs, W. L. and Comeau, A. (1970) Lepidopterous sex attractants discovered by field screening tests. *J. Econ. Ent.*, **63**, 969–74.

Roelofs, W. L. and Comeau, A. (1971) Sex attractants in Lepidoptera. *Proc. 2nd Int. Cong. Pest. Chem.*, pp. 91–114.

Roelofs, W. L. and Cardé, R. T. (1974) Sex pheromones in the reproductive isolation of lepidopterous species. In: *Pheromones* (Birch, M. C., ed.) pp. 96–114. North Holland, Amsterdam.

Roelofs, W., Hill, A., Cardé, R., Tette, J., Madsen, H. and Vakenti, J. (1974) Sex pheromone of the fruit tree leafroller moth, *Archips argyrospilus*. *Env. Ent.*, **3**, 747–51.

Roelofs, W., Hill, A. and Cardé, R. (1975) Sex pheromone components of the redbanded leafroller, *Argyrotaenia velutinana* (Lepidoptera: Tortricidae). *J. Chem. Ecol.*, **1**, 83–9.

Roelofs, W., Cardé, A., Hill, A. and Cardé, R. (1976) Sex pheromones of the threelined leafroller, *Pandemis limitata*. *Env. Ent.*, **5**, 649–52.

Roelofs, W. L., Gieselmann, M., Cardé, A., Tashiro, H., Monero, D. S., Henrick, C. A. and Anderson, R. J. (1978) Identification of the California red scale sex pheromone. *J. Chem. Ecol.*, **4**, 211–24.

Roelofs, W. L., Tamhankar, A. J., Comeau, A., Hill, A. S. and Taschenberg, E. F. (1980) Moth activity periods and identification of the sex pheromone of the uglynest caterpillar, *Archips cerasivoranus*. *Ann. ent. Soc. Am.*, **73**, 631–4.

Roelofs, W. L. and Brown, R. L. (1982) Pheromones and the evolutionary relationships of the Tortricidae. *A. Rev. Ecol. Syst.*, **13**, 395–42.

Roller, H., Biemann, K., Bjerke, J., Norgard, D. and McShan. W. (1968) Sex pheromones of the pyralid moths. I. Isolation and identification of the sex attractant of *Galleria mellonella* L. (greater wax moth). *Acta Ent. Bohemoslov.*, **65**, 209–11.

Shorey, H. H. (1970) Sex pheromones of Lepidoptera. In: *Control of insect behavior by natural products* (Wood, D. L., Silverstein, R. M. and Nakajima, M., eds) pp. 249–84. Academic Press, New York.

Silverstein, R. M. Cassidy, R. F., Burkholder, W. E., Shapas, T. J., Levinson, H. Z., Levinson, A. R. and Mori, K. (1980) Perception by *Trogoderma* species of chirality and methyl branching at a site far removed from a functional group in a pheromone component. *J. Chem. Ecol.*, **6**, 911–17.

Sonnet, P. E., Uebel, E. C., Lusby, W. R., Schwarz, M. and Miller, R. W. (1979) Sex pheromone of the stable fly: identification, synthesis, and evaluation of alkenes from female stable flies. *J. Chem. Ecol.*, **5**, 353–61.

Spieth, H. T. (1968) Evolutionary implication of sexual behavior in *Drosophila*. *Evol. Biol.*, **2**, 157–93.

Svensson, B. G. and Bergstrom, G. (1979) Marking pheromones of *Alpinobombus* males. *J. Chem. Ecol.*, **5**, 603–15.

Thornhill, R. (1979) Male and female sexual selection and the evolution of mating strategies in insects. In: *Sexual Selection and Reproductive Competition in Insects* (Blum, M. S. and Blum, N.A., eds) pp. 81–121. Academic Press, New York.

Trivers, R. L. (1972) Parental investment and sexual selection. In: *Sexual Selection and the Descent of Man, 1871–1971*, pp. 136–79. Aldine, Chicago.

Tobin, R. T., Seelinger, G. and Bell, W. J. (1981) Behavioral responses of male *Periplaneta americana* to periplanone B, a synthetic component of the female sex pheromone. *J. Chem. Ecol.*, **7**, 969–79.

Tumlinson, J. H., Hardee, D. D., Gueldner, R. C., Thompson, A. C., Hedin, P. A. and Minyard, J. P. (1969) Sex pheromones produced by male weevils: isolation, identification, and synthesis. *Science*, **166**, 1010–12.

Tumlinson, J. H., Klein, M. G., Doolittle, R. E., Ladd, T. L. and Proveaux, A. T. (1977) Identification of the female Japanese beetle sex pheromone: inhibition of male response by an enantiomer. *Science*, 197, 789–92.

Uebel, E. C., Sonnet, P. E., Miller, R. W. and Beroza, M. (1975) Sex pheromone of the face fly, *Musca autumnalis* De Geer (Diptera: Muscidae). *J. Chem. Ecol.*, 1, 195–202.

Uebel, E. C., Sonnet, P. E. and Miller, R. W. (1976) House fly sex pheromone: enhancement of mating strike activity by combination of (*Z*)-9-tricosene with branched saturated hydrocarbons. *Env. Ent.*, 5, 905–8.

Van Honk, C. G. J., Velthuis, H. H. W. and Roseler, P.-F. (1978) A sex pheromone from the mandibular glands in bumblebee queens. *Experientia*, 34, 838–9.

Wall, C. and Perry, N. J. (1978) Interactions between pheromone traps for the pea moth, *Cydia nigricana* (F.) *Ent. Exp. Appl.*, 24, 155–62.

White, M. D. J. (1978) *Modes of Speciation*. W. H. Freeman, San Francisco.

Willis, M. A. and Birch, M. C. (1982) Male lek formation and female calling in a population of the arctiid moth *Estigmene acraea*. *Science*, 218, 168–70.

Zagatti, P. (1981) Micro-comportements Induits par les Pheromones Sexuelles chez quelques Lepidopteres revageurs des cultures en milieu sahelien. PhD Thesis, University of Pierre and Marie Curie, Paris.

Zagatti, P., Kunesch, G. and Morin, N. (1981) La vanilline, constituant majoritaire de la secretion aphrodisiaque emise par les androconis du male de la pyral de la canne a sucre: *Eldana saccharina* (Wlk.) (Lepidoptere, Pyralidae, Galleriinae). *CR Acad. Sci. Paris*, 292, 633–5.

Sociochemicals

14

Sociochemicals of Bees

Richard M. Duffield, James W. Wheeler
and George C. Eickwort

14.1 INTRODUCTION

The aculeate Hymenoptera include approximately 14 000 species of ants, 16 000 species of wasps and 20 000 species of bees world-wide (Wilson, 1971; Michener, 1974). Such insects exhibit diversity both in the chemistry of their exocrine secretions and their behaviors. These compounds range from simple alkanes to terpenoids and triglycerides; social behaviors range from strictly solitary to the complex caste systems of honey bees and stingless bees.

This chapter focuses on the chemical ecology of bees. Collaboration among behaviorists, field biologists, systematists, and chemists has led to a surge of publications on the natural product chemistry of these insect exocrine secretions. This has been stimulated by those with practical interests in apiculture and medicine (bee and wasp stings kill more people annually in the USA than do snake bites).

The importance of bees in pollination is well known. It may be possible to manipulate wild bees by their sociochemicals to increase pollination. For example, exocrine products from the alfalfa-pollinating alkali bee, *Nomia melanderi*, may be used to induce increased nesting at artificial sites near alfalfa fields.

Determining the chemistry of different exudates over a wide range of species, genera, families, and orders provides additional data for the systematist. Mandibular and Dufour's secretions of *Andrena* are species-specific. Colletid, halictine, and nomiine bees all produce lactones in their Dufour's glands, indicating a common ancestry.

The recent surge in publications has been facilitated by the development of

Chemical Ecology of Insects. Edited by William J. Bell and Ring T. Cardé
© 1984 Chapman and Hall Ltd.

new analytical techniques and advances in instrument sensitivity (gas chromatographic-mass spectrometric analysis, GC–MS). It is possible to analyze a single Dufour's gland. Mandibular gland analyses require more specimens both because the gland is generally smaller, and also because loss prior to chemical analysis can be greater. Mandibular compounds are generally of lower molecular weight and thus more volatile.

Chemical investigations have surpassed studies which use chemicals or glandular extracts in behavioral studies. Field testing of compounds is slow and difficult. Although it is relatively easy to trap Southern pine beetles with pheromone, similar procedures employing standard compounds found in the exocrine extracts of bees have not been successful. The identification of compounds has little value unless their significance to the biology of the bees can be determined with quantitative behavioral assays.

At present there is no review of the sociochemicals of all bees. This chapter's value is providing a concise, organized data base for the behaviorist, field biologist, and chemist alike. We have also tried to identify areas where our knowledge is especially weak and where future research might be most fruitful.

14.2 CLASSIFICATION OF BEES

Bees comprise the superfamily Apoidea, a highly successful offshoot from the sphecoid wasps that provision nests with plant products instead of arthropod prey. Most bees, like their sphecoid relatives, dig nests in the soil where the larva develops in its own cell on a mixture of pollen and concentrated nectar. Unlike wasps, most bees line their cells with glandular secretions. Like wasps, some bees excavate rotten wood or pithy stems or use natural cavities for nests. Some species construct free-standing nests of mud, mortar, resin, or wax. The typical bee is non-social, making and provisioning the nest alone and not interacting with her developing offspring. However, aggregations of nests are common and some species in all major families form communal nests in which several females share a burrow but all lay eggs. True social behavior, in which some forego reproduction and function solely as workers, has evolved independently in the Apidae, Anthophoridae, and Halictidae. The evolution of cleptoparasitic or 'cuckoo behavior', in which a species lays its eggs in the cells of another species and the larva consumes the provisions intended for the host, has occurred many times in most families.

The 20 000 species of bees are currently placed in 11 families (Fig. 14.1). The apoid families are traditionally grouped into 'short-tongued' and 'long-tongued' bees (Fig. 14.1), with the latter considered to have evolved from the former. This taxonomic summary is based largely on Michener (1974, 1979), updated as noted by more recent studies. Table 14.1 summarizes the classification of genera of bees whose glandular products have been chemically analyzed.

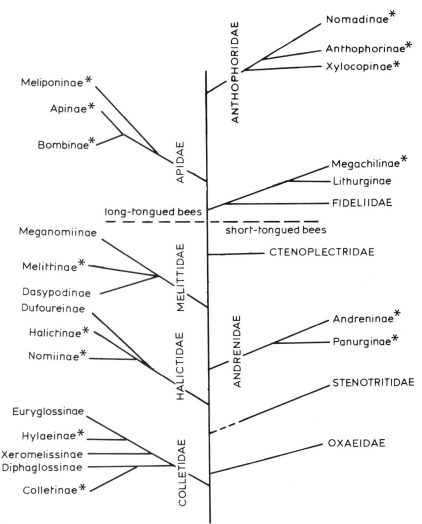

* represents taxa for which chemical data are available

Fig. 14.1 Dendrogram showing relationships among the major groups of bees (modified from Michener, 1974).

Phylogenetic relationships among the families of short-tongued bees are poorly understood. The large, cosmopolitan family Colletidae has traditionally been considered the most primitive family of bees (Michener, 1974), mostly because of the bilobed glossa which superficially resembles that of sphecid wasps. However, McGinley (1980) has shown that the colletid glossa is a specialized structure, used for applying the nest cell lining. All colletid bees are non-social, lining their cells with a cellophane-like secretion and filling them

Table 14.1   Genera of bees with identified exocrine secretions

COLLETIDAE	*Andrena*	Nomadinae
Colletinae	Panurginae	Nomadini
Colletes	*Panurginus*	*Nomada*
Hylaeinae	MELITTIDAE	Xylocopinae
Hylaeus	Melittinae	Xylocopini
HALICTIDAE	*Melitta*	*Xylocopa*
Nomiinae	MEGACHILIDAE	Ceratinini
Nomia	Megachilinae	*Ceratina*
Halictinae	*Megachile*	*Pithitis*
Augochlora	*Osmia*	APIDAE
Augochlorella	ANTHOPHORIDAE	Meliponinae
Augochloropsis	Anthophorinae	*Trigona*
Agapostemon	Eucerini	*Lestrimelitta*
Dialictus	*Melissodes*	Bombinae
Evylaeus	*Svastra*	Bombini
Lasioglossum	Anthophorini	*Bombus*
Halictus	*Anthophora*	*Psithyrus*
ANDRENIDAE	Centridini	Apinae
Andreninae	*Centris*	*Apis*

with semiliquid provisions. The Euryglossinae and Hylaeinae are unique among non-parasitic bees in lacking pollen-collecting hairs, carrying pollen instead in the crop. The Hylaeinae and Xeromelissinae nest in hollow stems and other cavities, while members of the other subfamilies are typical soil-nesters. The family is best represented in the southern hemisphere, especially Australia.

The Halictidae is a large, cosmopolitan family, that includes cleptoparasitic, wood-dwelling, and typical soil-nesting species. The Halictinae or 'sweat bees' are among the most common bees throughout the world. This subfamily is of particular interest because primitive social behavior has repeatedly evolved within it and all gradients of sociality occur among its species. The Nomiinae is well represented in the Old World tropics and south temperate regions, with some holarctic species. They can be solitary, communal, or perhaps weakly social. The Dufoureinae is a relatively small group of non-social bees, typically specialists for pollen on particular plants (oligolectic), and are best represented in the holarctic region.

The Andrenidae is a large family of short-tongued bees, common in most parts of the world except Australia, Indonesia, and most of southeast Asia. All are soil-nesting, non-parasitic, and non-social, although communal nesting is common. The Panurginae is most diverse in the western hemisphere, especially South America, and is most abundant in xeric habitats. Most species are oligolectic and are thought to coat their provision masses with glandular secretions. The Andreninae is most diverse in the holarctic and its species are among the most common visitors to spring flowers.

There are also four small families among the short-tongued bees. The Stenotritidae (once a subfamily of Colletidae) is limited to Australia. The Oxaeidae (once a subfamily of Andrenidae) is principally neotropical. Both have soil-nesting, non-social species. The Ctenoplectridae, until recently in the Melittidae (Michener and Greenberg, 1980), is now considered to share a common ancestor with the long-tongued bees. It is paleotropical; its members nest in wood and, like some Melittinae (*Macropis*), provision nests with plant oils. Some species are cleptoparasitic. The glands have been described morphologically only in Oxaeidae, and nothing is published concerning the chemistry of any glandular secretions.

The Melittidae, another small family of short-tongued bees, may be the taxon from which the Ctenoplectridae and long-tongued bees evolved (Michener and Greenberg, 1980). All species are soil-nesting, non-social, and non-parasitic. The Melittinae and Dasypodinae are most diverse in Africa, with representatives in the holarctic. Meganomiinae are African (Michener, 1981).

The long-tongued bee families share a set of specialized morphological features that attest to their common evolutionary origin. The Fideliidae is a small group of soil-dwelling desert bees in southern Africa and Chile, sometimes considered as a subfamily of Megachilidae. Nothing is known of the morphology of the glands or the chemistry of their secretions. The Megachilidae is a large, cosmopolitan family of distinctive non-social bees, in which pollen is carried on the abdominal venter instead of the hind legs. They line their cells with foreign materials such as leaves, plant down, or resin, apparently without using glandular secretions. Many species nest in wood or stems, use preformed cavities, or construct free-standing cells of soil and saliva or resin. The Lithurginae is a small but cosmopolitan group, best represented in temperate South America, while the Megachilinae is well represented throughout the world.

The Anthophoridae is another large cosmopolitan family. The Xylocopinae, large and small carpenter bees, nest in wood or stems, except for the palearctic *Proxylocopa* which nests in soil. Extended parental care is common among carpenter bees and Old World Ceratinini have evolved primitive social behavior. The Anthophorinae, a diverse group of generally large, fuzzy, fast-flying bees, usually nest in soil, but some in wood. The subfamily has been split into 13 tribes, of which three, the Eucerini, Anthophorini, and Centridini, contain species with analyzed glandular secretions. The Nomadinae are all clepto-parasites. The subfamily contains about 12 tribes, of which the Nomadini has been analyzed chemically. All subfamilies are cosmopolitan, although the Xylocopinae is best represented in the tropics and the other two subfamilies are poorly represented in Australia.

The Apidae, which includes all of the highly social bees, carry pollen–nectar mixtures in specialized corbiculae on the hind tibiae and construct free-standing nest cells of secreted wax, resin, or other foreign materials. The Meliponinae, until recently considered a tribe of the Apinae, are the stingless, highly social bees in the Old and New World tropics. The Bombinae includes two tribes: the Euglossini, the brilliant metallic 'orchid bees' of tropical America, whose males pollinate orchids and whose females are solitary or communal; and the Bombini, the social bumble bees, principally holarctic with neotropical and oriental representatives. The Apinae includes honey bees (*Apis*), originally oriental, palearctic, and African, but spread by humans throughout the world.

14.3 GLANDULAR SOURCES OF EXOCRINE SECRETIONS

This chapter considers only those exocrine glands whose secretions influence either the behavior or survival of other individuals of the same or different species. Glands that secrete nest materials and larval food are also included.

The head has glands whose contents are secreted via the mouthparts. Labial glands open on the proboscis near the base of the glossa, but the secretory cells are mostly or entirely in the thorax. Mandibular glands open at the bases of the mandibles. Hypopharyngeal glands open at the base of the proboscis. Many short-tongued bees possess facial foveae between the eyes and antennae that appear to be glandular.

The thorax seems an unlikely part of the body to contain glands, but there are several associated with the legs. Some male *Centris* have glands in their enlarged hind femora and tibiae, and similar glands should be sought in enlarged legs of other male bees. Male euglossine bees have glands in their enlarged hind tibiae. Tarsal glands occur in the last tarsomere of legs of Apidae and may exist in other bees.

The abdomen contains large glands associated with the female reproductive system. The venom gland empties its secretions into the base of the sting, as does the Dufour's gland. Koshevnikov glands occur on the basal apparatus of the sting, and other secretory cells on the sting also produce pheromones, at least in *Apis*. The sterna and terga and intersegmental membranes of the abdomen are richly endowed with glands. Some of these are hypertrophied to form Nasanov glands and wax glands in Apidae.

A rich literature exists on the morphology and histology of apoid exocrine glands. Since the honey bee, *Apis mellifera*, is the best studied of all insects, it is not surprising that more is known about its glands than those of any other bee. Still, Pain (1973) estimates that there may be 31 pheromones in honey bees, of which only 13 had been identified by 1973. Older studies of glandular structure in *A. mellifera* are summarized in Snodgrass (1956); more recent studies, especially of fine structure, will be cited in this chapter.

Comparative studies of bee glands began with the classic papers of Bordas (1895) and Heselhaus (1922). More recently, Cruz Landim (1967) conducted a landmark study on the mandibular, labial, and hypopharyngeal glands of 78 species, representing all major families except Melittidae. Lello (1971a, b, c, d, 1976; Kerr and Lello, 1962) and Pessotzkaja (1929) examined the venom and Dufour's glands in all major families. Following early comparative studies by Heselhaus (1922) and Jacobs (1924), Altenkirch (1962) conducted a detailed study of the abdominal, tergal, sternal, intersegmental, and sting glands of 66 species of European bees, representing all major families. Michener (1974, p. 12–14) has conveniently summarized the best studied glands in bees. Earlier studies are cited in the above comprehensive papers; more recent studies are cited below.

14.4 CEPHALIC GLANDS

14.4.1 Labial glands

(a) Morphology

Besides those species studied by Cruz Landim (1967), labial glands have been described in *Megachile* (Graf, 1967) and *Peponapis pruinosa* (Mathewson, 1965) and other Eucerini (Graf, 1970). The ultrastructure of these glands has been described for *Melipona, Xylocopa, Megachile, Colletes* (Cruz Landim, 1968), and male *Bombus* (Ågren, 1975, 1979; Ågren *et al.*, 1979).

The paired thoracic labial glands have ducts which unite to form a median salivary duct to carry the secretion through the head to the opening at the salivary syringe, a chamber on the anterior surface of the proboscis at the bases of the glossa and paraglossae. The secretion is presumably conveyed by the paraglossae to the back of the glossa, by which it can be mixed with the food or licked onto a substrate.

Species of Apidae have an additional pair of labial glands situated in the posterior region of the head, whose ducts join the median salivary duct. These cephalic glands are clusters of globular acini, very different in shape and histology from the thoracic glands. They are absent in most other bees, although the median salivary duct itself is often lined by cells that appear to be secretory. In *Nomia*, an expansion of the salivary duct in the head appears to be glandular. This expansion is long and branched in *Megachile*, and long and ribbon-like in *Melissoptila* and *Melissodes* but not in other Eucerini. These cephalic portions are absent in drone honey bees, but are better developed in males than in females of Eucerini, *Megachile*, and some *Bombus*. The columnar or palisade secretory cells that comprise the thoracic and cephalic glands lack tubules and so the secretion apparently passes directly across the cuticle lining the glands.

In honey bees, the thoracic labial glands produce a watery saliva that dissolves dry sugar, but the major source of invertase is the hypopharyngeal glands. In *Melipona quadrifasciata* the thoracic gland secretion is proteinaceous, whereas the cephalic labial gland secretion is a lipid, as it is in *Trigona* and *Apis* (Cruz Landim, 1968, and references therein). We will not consider the chemistry of the non-pheromone-producing labial glands.

(b) Chemistry

The major work on the labial gland pheromones has been done on bumble bees, namely 25 species in *Bombus* and *Psithyrus*. From the cephalic extracts, these secretions appear to be species-specific and found in the males only. Since Stein (1963a, b) identified farnesol in cephalic extracts of *B. terrestris*, aliphatic straight-chain hydrocarbons, alcohols, aldehydes, acetates, and ethyl esters have been demonstrated (Table 14.2). Both saturated and unsaturated, odd-numbered hydrocarbons ranging from 21–27 are present in all extracts. Acyclic mono-, sesqui-, and diterpenoid alcohols and acetates are also detected. Geranylgeranyl acetate and ethyl dodecanoate are found in six species, whereas geranylgeraniol and hexadecenol are found in five species. Some contain only one compound in addition to the hydrocarbons, but several species contain six to eight compounds and one contains fourteen.

It should be noted that the compounds listed in Table 14.2 were assumed to come from labial glands, but this has been verified only in several species. Some of the compounds, identified in head extracts, could be from mandibular glands. (See discussion of *Bombus* mandibular glands.)

(c) Function

Males of most bumble bee species patrol flight paths up to several hundred meters in a habitat that differs from species to species. In the morning the males land on objects and lick them, marking them with cephalic pheromones. Other males are attracted to these chemical markers, and so a flight path may be patrolled by several males. Later in the day, males cease marking, but still approach the marks. Such flight paths in temperate *Bombus* contain fewer than 30 males, but a flight path of tropical *B. pullatus* contained an estimated 460 to 720 males (Stiles, 1976).

Analysis of labial gland extracts of several species has demonstrated that the labial glands are the source of these chemical markers (Kullenberg *et al.*, 1973; Ågren *et al.*, 1979). These secretions are spread over the body during grooming behaviors: thus after a male has rested on an object, it smells of the secretion. Presumably, receptive females are attracted to these chemical markers where mating can occur. This has rarely been observed, although this is not surprising if females mate only immediately after emergence.

Labial gland secretions may be mixed with those of the Dufour's gland by *Colletes* females, the enzymes in the saliva hastening polymerization of the lactones to form the cellophane-like cell linings. A bee first licks its sting area, imbibing the Dufour's gland secretion containing the lactones, then spreads the mixture with its glossa over the cell wall. The hypertrophied labial glands of female *Hylaeus*, in contrast to the smaller Dufour's gland, suggests that the labial gland is the principal source of the silk-like cell lining in that colletid (Batra, 1980). Mason bees (Megachilinae: *Hoplitis, Chalicodoma*) produce copious saliva from large thoracic labial glands. This saliva is mixed with dry soil to produce a mortar for constructing their free-standing nests. These nests are remarkably strong and weatherproof (Eickwort, 1975). Chemical analysis of the labial gland secretions used in cell construction has never been reported for any bee but would be worthwhile.

Table 14.2 Volatile components identified in the cephalic (labial) gland extracts of bumble bees

Compound (mol. wt.)

Species	Geraniol (154)	Citronellol (156)	Geranyl acetate (196)	Citronellyl acetate (198)	Farnesenes (204)	Farnesol (222)	2,3-Dihydrofarnesol (224)	2,3-Dihydrofarnesal (222)	Farnesyl acetate (264)	2,3-Dihydrofarnesyl acetate (266)	Geranylgeraniol (290)	Geranylcitronellol (292)	Geranylgeranyl acetate (332)	1-Dodecanol (186)	1-Tetradecanol (214)	Tetradecanal (212)	1-Hexadecanol (242)	1-Hexadec-?-enol (240)	Hexadecanal (240)	1-Octadecanol (268)	1-Eicosenol (296)	Tetradecyl acetate (256)	Hexadecyl acetate (284)	Hexadecenyl acetate (282)	Octadecyl acetate (312)	Octadecenyl acetate (310)	Eicosyl acetate (340)	Eicosenyl acetate (338)	Docosyl acetate (368)	Ethyl decanoate (200)	Ethyl dodecanoate (228)	Ethyl tetradecanoate (256)	Ethyl tetradecenoate (254)	Ethyl hexadecenoate (282)	Ethyl hexadienoate (280)	Ethyl hexatrienoate (278)	Ethyl octadecanoate (312)	Ethyl octadecenoate (310)	Ethyl octadecadienoate (308)	Ethyl octadecatrienoate (306)	Hydrocarbons sat. and unsat.	References
Bombus																																										
B. agrorum	X	—	X	—	—	—	—	—	—	—	—	—	—	—	X	—	X	7M	—	—	—	—	—	—	—	—	—	—	—	—	—	—	—	—	—	—	—	—	—	—	—	5,7
B. callumanus	—	—	—	—	X	M	M	X	X	—	X	X	M	—	—	—	X	—	—	—	—	—	—	—	—	—	—	X	—	—	—	—	—	—	—	—	—	—	—	—	—	7
B. cingulatus	—	—	M	M	—	—	M	—	—	—	—	—	—	X	—	—	—	—	—	—	—	—	—	—	—	—	—	—	—	—	—	—	—	—	—	—	—	—	—	—	X	8
B. derhamellus	—	—	—	—	—	—	—	—	—	—	—	—	—	—	—	—	—	—	—	—	—	—	—	—	—	—	—	—	—	—	—	—	—	—	—	—	—	—	—	—	M	5
B. hortorum	—	—	—	—	—	—	M	M	—	X	—	X	M	—	—	—	X	7X	—	X	—	—	—	—	—	—	X	—	—	—	—	X	—	—	—	—	—	—	—	—	X	6,7
B. hypnorum	—	—	—	—	—	—	—	—	—	—	—	—	—	—	—	—	X	—	—	—	—	—	—	—	—	—	—	—	—	—	—	—	—	—	—	—	—	—	—	—	—	8
B. jonellus	—	—	—	—	—	—	—	—	—	—	X	—	—	—	X	—	X	9M	—	—	—	—	—	—	—	—	—	—	—	—	—	—	—	—	X	—	—	—	—	—	X	2,8
B. lapidarius	—	—	X	—	X	—	—	—	—	—	X	—	—	—	—	—	X	—	—	—	—	—	X	X	—	—	X	—	—	—	—	—	—	—	—	—	—	—	—	—	X	5,7
B. lapponicus (blond)	—	—	X	—	X	—	—	—	—	—	—	—	—	—	—	—	X	—	—	—	—	—	—	—	—	—	—	—	—	—	—	—	—	—	—	—	—	—	—	—	X	4
B. lapponicus (dark)	—	—	X	—	—	—	—	—	—	—	—	—	—	—	—	—	X	—	—	—	—	—	—	—	—	—	—	—	—	—	—	—	—	—	—	—	—	—	—	—	X	8
B. lapponicus (unspec.)	—	—	X	—	X	—	—	—	—	X	—	X	—	—	—	—	X	—	—	—	—	—	X	—	X	—	X	X	—	X	M	X	—	X	X	X	X	X	X	X	X	3,7
B. lucorum (dark)	—	—	—	—	X	—	—	—	—	—	X	X	X	—	—	—	X	—	—	—	—	—	—	—	—	—	—	—	—	X	X	X	X	X	X	X	X	X	X	X	X	3,7
B. lucorum (light)	—	—	—	—	—	—	—	—	—	—	—	—	X	—	—	—	—	—	—	—	—	M	—	X	—	—	—	—	—	X	M	X	—	X	—	—	—	—	—	—	X	5
B. lucorum (unspec.)	—	—	—	—	—	—	—	—	—	—	—	—	—	—	—	—	—	—	—	—	—	—	—	X	—	M	—	—	—	—	—	—	—	—	—	—	—	—	—	—	—	7
B. muscorum	—	—	—	M	—	M	M	—	—	M	—	—	M	—	—	—	X	M	—	M	—	—	—	—	—	—	—	—	—	X	M	—	—	—	—	—	—	—	—	—	X	7
B. patagiatus	—	—	X	—	X	—	—	—	M	—	X	X	M	—	—	—	X	5M	—	X	—	—	X	X	—	—	—	X	—	X	M	—	—	X	—	—	—	—	—	—	M	5,7,8
B. pratorum	X	—	X	—	X	M	—	—	M	—	—	—	—	—	—	—	X	—	—	—	—	—	X	X	—	—	—	—	—	X	—	—	—	X	—	—	—	—	—	—	X	8
B. scandinavicus	—	—	X	—	X	—	—	—	—	—	—	—	—	—	—	—	X	—	—	—	—	—	—	—	—	—	—	—	—	—	—	—	—	—	—	—	—	—	—	—	X	7
B. sorocensis	—	—	X	—	—	—	—	—	—	—	—	—	—	—	—	X	—	—	—	—	—	M	—	—	—	—	X	X	—	—	X	M	X	—	—	—	—	—	—	—	X	7
B. sporadicus	—	—	—	—	—	—	—	—	—	—	—	—	—	—	—	—	—	—	—	—	—	—	—	—	—	—	—	—	—	—	—	—	—	—	—	—	—	—	—	—	X	7
B. subterraneus	—	—	—	—	—	M	—	M	—	X	—	X	—	—	X	X	X	—	—	—	—	—	—	—	X	—	—	—	X	X	X	X	X	X	—	—	—	—	—	—	—	7
B. terrestris	—	—	—	—	—	—	M	—	—	—	—	—	M	—	X	—	X	—	—	M	—	—	—	—	—	—	—	—	—	—	X	X	X	—	—	—	—	—	—	—	X	1,3,7
Psithyrus																																										
P. barbatellus	—	—	—	—	—	—	—	—	—	—	—	—	—	—	X	X	—	—	X	—	—	—	—	X	X	—	—	—	—	—	—	—	—	—	—	—	—	—	—	—	—	7
P. bohemicus	—	X	—	—	—	—	—	—	—	—	—	—	—	—	X	X	X	5M	X	X	X M	—	X	—	X	—	—	—	—	—	—	—	—	—	—	—	—	—	—	—	—	7
P. campestris	—	—	—	—	—	—	—	—	—	—	—	—	—	—	X	X	X	—	—	X	M	—	—	—	—	—	—	—	—	—	—	—	—	—	—	—	—	—	—	—	—	7
P. globosus	—	—	—	—	—	—	—	—	—	—	—	—	M X	—	—	—	X	—	—	—	—	—	—	—	—	—	—	—	—	—	—	9X	—	—	—	—	—	—	—	—	—	7
P. rapestris	—	—	—	—	—	—	M	—	—	—	—	X	—	—	X	—	X	—	—	—	—	M	—	X	—	—	—	—	—	X	X	—	X	—	—	—	—	—	—	—	X	7
P. silvestris	—	—	—	—	—	—	—	—	—	—	—	—	—	—	X	X	X	M	—	—	—	—	—	—	—	—	—	—	—	—	—	X	X	—	—	—	—	—	—	—	X	7

M = major, X = present, — = absent
References: (1) Bergström *et al.*, 1968; Ståhlberg-Stenhagen, 1970; (2) Bergström and Svensson, 1973b; (3) Bergström *et al.*, 1973a; (4) Bergström and Svensson, 1973b; (5) Calam, 1969; (6) Kullenberg *et al.*, 1973; (7) Kullenberg *et al.*, 1970; (8) Svensson and Bergström, 1977
5, 7, 9 indicates position of double bond

14.4.2 Mandibular glands

(a) Morphology

Nedel (1960) described in detail the mandibular gland of *Apis mellifera* and compared it to those of Meliponinae, *Bombus*, *Hylaeus*, *Andrena* and *Osmia*. The secretory cycle of the gland in worker honey bees was described by Costa Leonardo (1980). In addition to those species studied by Cruz Landim (1967), glands have been described for *Peponapis pruinosa* (Mathewson, 1965) and other Eucerini (Graf, 1970). Fine structural analyses have been conducted on the mandibular glands of male *Bombus terrestris* (Stein, 1962) and workers of the cleptoparasitic stingless bee *Lestrimelitta limao* (Cruz Landim and Camargo, 1970).

The mandibular gland consists of a sac-like reservoir that is partially or completely lined by secretory cells. The reservoir opening in the mesal junction of the mandible with the head is controlled by a valve that permits the bee to regulate the discharge of the secretion when the mandible is opened.

The secretory cells are 'Leydig' or 'ampulla' glands, common throughout the Hymenoptera. Each resembles a balloon on a string, having an 'intracellular' ductule entering a cavity lined by microvilli. The ductule continues out of the secretory cell to form the 'string' that connects each secretory cell independently to the reservoir. Electron micrographs show these 'unicellular' glands are actually multicellular, as the ductule is a product of one or more cells separate from the secretory cell (Ågren, 1979).

Most of the short-tongued bees and some long-tongued bees have simple mandibular glands, with the secretory cells covering the entire reservoir. The glands are small in Colletidae, Panurginae, and Halictinae, and moderate to large in Nomiinae, Dufoureinae, and Oxaeidae. In most of the long-tongued bees the secretory cells are restricted to part of the reservoir. The gland varies in size from a very small lobe situated on top of the mandible to a huge structure extending to the top of the eyes. In Anthophorinae, Xylocopinae, Megachilidae, and many Meliponinae, the gland is bilobed, with secretory cells restricted to one lobe or at the ends of both lobes. In Eucerini, the secretory cells form a separate gland, connected to one lobe of the reservoir by a long duct. The Meliponinae have varied glands, which are especially large in *Trigona tataira*, *postica* and *xanthotricha*.

Size may differ among the sexes or castes of the same species. Glands of male *Centris decolorata* and *dirrhoda*, but not *C. fasciata*, are much larger than those of their respective females (Raw, 1975). In *Apis mellifera*, the gland extends to the base of the antenna in the worker, to the top of the eye in the queen, but is a tiny structure in the drone.

(b) Chemistry

Colletidae

Mandibular gland secretions of the Colletidae contain blends of monoterpenes. Linalool dominates the mandibular gland secretions of *Colletes* (Bergström and

Tengö, 1978; Hefetz *et al.*, 1979a; Cane and Tengö, 1981). Bergström and Tengö estimate that the mandibular gland of *C. cunicularius* contains approximately 10 µg in the female and less in the male. Both isomers of citral (geranial and neral) have been found in several (but not all) species of *Colletes* (Bergström and Tengö, 1978; Hefetz *et al.*, 1979a), and in *Hylaeus* (Blum and Bohart, 1972; Bergström and Tengö, 1973; Duffield *et al.*, 1980). Both of these compounds are common natural products, particularly citral, and therefore not ideal as specific volatile signals.

Halictidae
No compounds have been identified in the mandibular gland secretions of bees belonging to the Halictinae and Nomiinae. In the Dufoureinae, *Dufourea novaeangliae* has well-developed mandibular glands with a citral-like odor.

Andrenidae
The first report of mandibular gland secretions of *Andrena* was that of Tengö and Bergström (1976a). Extracts of eight species contained citral as well as a homologous series of ketones. Later, Tengö and Bergström (1977) found a series of saturated and unsaturated primary and secondary alcohols as well as acetates and butanoates. The secretions are species-specific and these differences can be discerned by the human nose (Tengö and Bergström, 1977). In almost all of the species investigated, both male and female head extracts were identical. Individual differences have yet to be determined. Unusual among the compounds reported in the mandibular gland extracts of *Andrena* is the recent identification of eight spiroketals (7-M.W.-184; 1-M.W.-198) from *A. wilkella* and *A. haemorrhoa* (Francke *et al.*, 1980). Several of these multioxygenated compounds are new natural products for insects.

The mandibular glands of both male and female *Panurginus potentillae* (Panurginae) contain citral (Duffield *et al.*, 1983) while *Panurginus atramontensis* contains 8-acetoxy-2,6-dimethyloct-2-en-1-al (Wheeler, Avery, Birmingham and Duffield, unpublished results).

Megachilidae
No components have been identified in the mandibular gland secretions of bees in this family.

Anthophoridae
The mandibular gland chemistry and function of *Nomada* (Nomadinae) will be discussed in Section 14.7 on parasitic bees.

The chemistry of mandibular gland secretions has been studied in two genera of Anthophorinae. Batra and Hefetz (1979) showed that the secretions of male *Melissodes denticulata* are dominated by a homologous series of saturated and unsaturated butanoates ranging from C_{12} to C_{18} (even carbon compounds only). However, none of these compounds has been detected in female head extracts, so that it is possible that the chemicals actually came from the cephalic labial glands, which are better developed in males. In contrast, the secretions of

Centris adani are dominated by a variety of monoterpenoid alcohols, acids, and acetates (Vinson *et al.*, 1982). Other species of *Centris* show no detectable mandibular gland secretions.

Wheeler *et al.* (1976) showed that the secretions of *Xylocopa hirsutissima* (Xylocopinae) are dominated by *cis*-2-methyl-5-hydroxyhexanoic acid lactone. In addition, traces of benzaldehyde, *p*-cresol, benzoic acid, and vanillin were detected.

Mandibular gland exudates of small carpenter bees, *Ceratina* (Ceratinini), vary considerably among species. Secretions of *C. strenua* contain traces of geranyl, farnesyl, and neryl acetates while those of *C. calcarata* contain traces of nerolic and geranic acids. *Ceratina cucurbitina* males exhibit a trace of 1-hexadecyl acetate. Pentadecane is a major component of all three species. They also show a series of saturated and unsaturated hydrocarbons (Wheeler *et al.*, 1977). *Pithitis smaragdula*, another ceratinine species native to southern Asia, has secretions containing salicyaldehyde, several terpenoid acetates, ethyl hexadecanoate and hydrocarbons (Hefetz *et al.*, 1979b).

Apidae

Geranial and neral have been reported as major volatile constituents in two studies of stingless bees (Blum *et al.*, 1970; Crewe and Fletcher, 1976) while a series of 2-alkanols and 2-alkanones have been found in other species (Blum, 1971; Luby *et al.*, 1973). Several of the alkanols and alkanones suggested by Luby *et al.* were reported as tentative identifications and the experimental details of the species studied by Blum (1971) have never been published. Eleven new world species of *Trigona* studied by us (Wheeler, unpublished) show these alkanols and alkanones, other alcohols, two spiroketals, and a series of aliphatic esters. Benzaldehyde and 2-heptanone, common constituents of previously investigated species, were absent in these 11 species.

Kullenberg (1956) reported the identification of butanoic acid from the heads of bumble bees. Mandibular gland extracts of queens, males, and workers of *Bombus lapidarius* were later shown to contain butanoic acid, a homologous series of odd numbered 2-alkanones ranging from C_5 to C_{13}, and a number of minor components including 2-nonanol, 2-undecanol, geraniol, and citronellol. Butanoic acid was found in eight species of *Bombus* and two of *Psithyrus* (Cederberg, 1977).

The chemistry of the cephalic secretions of bumble bees needs clarification. Cederberg (1977) reports geraniol from the mandibular glands while Kullenberg *et al.* (1970), and Svensson and Bergström (1977) report geraniol from male labial gland extracts. Similarly, citronellol has been reported from both sources.

The chemistry of queen and worker mandibular gland secretions of the honey bee, *Apis mellifera*, has been reviewed by numerous authors (Wilson, 1971; Blum and Brand, 1972; Gary, 1974; Blum, 1976; Boch *et al.*, 1979; Costa Leonardo, 1980), so further review here is unnecessary. The distinction between worker bees producing (*E*)-10-hydroxy-2-decenoic acid, a component of 'bee milk', and queen bees producing (*E*)-9-oxo-2-decenoic acid, the 'queen phero-mone', has been blurred by the discovery that laying workers of the Cape race (*A. m. capensis*) produce queen pheromone (Crewe and Velthuis, 1980). The biosynthetic capabilities of the two castes are apparently similar.

(c) Function

A widespread and perhaps primitive function of the mandibular gland secretions is defense. Bees typically release these chemicals when they are roughly handled. The secretions include chemicals that are common arthropod defens-ive compounds. When crickets are treated with the mandibular gland com-pounds of *Ceratina*, attack by ants is deterred (Wheeler *et al.*, 1977). 2-Heptanone in *Apis* worker mandibular glands acts as an alarm pheromone in conjunction with sting gland secretions.

Terpenes and other compounds present in mandibular gland secretions have been established as fungistatic agents (Cole and Blum, 1975) and may function in this manner in the nest cells.

Many ground-nesting bees nest in dense aggregations, and chemical cues may be responsible for attracting more bees to an existent aggregation. Mandibular gland secretions are often suggested as a source of these cues, but there has been no experimental confirmation. Likewise, many male bees form 'sleeping' aggregations on plant stems and the like. Pheromonal cues seem obvious, with the mandibular glands most frequently suggested as the source.

The attractant role of mandibular gland secretions has been best demon-strated in *Colletes*, which nest in dense aggregations. Males emerge before females and patrol the nest site, aggregating above pre-emergent bees (both males and females) and digging them out (Cane and Tengö, 1981). The mandibular gland secretion of *C. cunicularius* contains only linalool. Cane and Tengö, using velvet rectangles with synthetic linalool, demonstrated significant male attraction and, by analyzing odors collected from digging bees, demon-strated that the emerging bees secrete linalool. Female *Colletes* are also attracted to netted females swung about a nest site (Hefetz *et al.*, 1979a; Cane and Tengö, 1981). Hefetz *et al.* found that *C. thoracicus* is not attracted to citral or linalool alone. When presented in the ratio of the natural secretion (3:1:1; linalool, neral, geranial), both sexes were attracted. It is difficult to repeat these biological assays because the experimental protocols were not adequately described. Duffield (unpublished) placed filter paper discs treated with 50 μl of Hefetz *et al.*'s attractant solution 1 m upwind of a dense nesting site of *C. thoracicus* above which numerous males were flying and did not attract bees, although males appeared to increase activity. Analysis of male mandibular glands indicated only linalool with no detectable citral (Wheeler, unpublished).

Mandibular gland chemicals are released by males of diverse groups of bees while they are patrolling for females. The secretions are believed to attract females, although this function as a sex pheromone has never been conclusively demonstrated (summarized in Eickwort and Ginsberg, 1980). Males of many species of *Andrena* patrol areas similar to, but smaller than, the flight paths of bumble bees, also marking vegetation with cephalic secretions that attract other males and presumably females (Tengö, 1979). As with bumble bees, the areas differ among species, but unlike bumble bees, the *Andrena* males do not follow particular flight paths within the patrolled areas. Velvet rectangles soaked with synthetic mandibular gland compounds were used by Tengö to demonstrate that males are attracted to the scent marks in the patrolled areas. The straight-chain alkanols are potent behavior-releasing chemicals, but the blend of components in the mandibular gland secretions has a significantly higher activity than the individual components.

Some territory-defending males also release mandibular gland secretions. In *Xylocopa hirsutissima*, the male secretion (which is apparently spread over the venter of his abdomen), when applied to a dummy, causes a territory owner to respond aggressively to the dummy (Velthuis and Camargo, 1975). Males of some species of *Centris* mark vegetation within their territories with mandibular gland chemicals. Others employ chemicals produced by leg glands (see following section) (Frankie *et al.*, 1980; Vinson *et al.*, 1982). Mandibular gland secretions are deposited on grass stems by *C. adani* in a rough parabola upwind of a male's perch. If the scent-marked grass stems are experimentally removed, the territorial male leaves the site. Frankie *et al.* (1980) observed mating in these territories and hypothesized that the scent markers attract females. During mating, males mark females with mandibular gland secretion, which may reduce the attractiveness of the female to other males.

The use of mandibular gland secretions in female bees as sex pheromones is frequently hypothesized, but rarely demonstrated. In *Xylocopa sulcatipes*, territorial males recognize females by their mandibular gland pheromones (Velthuis and Gerling, 1980). Honk *et al.* (1978) showed that mandibular gland secretions of virgin bumble bee queens attract males. The use of 'queen pheromone', the mandibular gland pheromone, as a long-range sex attractant by virgin honey bee queens is well known and need not be reviewed here. Queen pheromone also indirectly regulates the production of new queens by influencing workers, which also occurs in *Bombus terrestris* (Honk *et al.*, 1980).

Many stingless bees (Meliponinae) use mandibular gland secretions to indicate the location of food. When encountering a rich food source, *Meliponula bocandei* deposits secretions nearby that attract other foragers. *Trigona cupira* workers follow a scout bee returning to a rich food source; the scout seems to be enveloped in a strong odor that may create an aerial odor trail. In many other species of *Trigona*, the scout bees mark objects along the route from the food source to the nest. These scouts follow the odor marks as they lead new foragers to the food (summarized in Michener, 1974, pp. 154–5, from the studies of Kerr and Lindauer). At high concentrations the mandibular secretions act as alarm pheromones, inducing attack by workers (Blum *et al.*, 1970; Luby *et al.*, 1973). Trail pheromones deposited on flowers by aggressive *Trigona* species cause alarm behavior, including flight and defensive postures, in non-aggressive *T. fulviventris* foragers (Johnson, 1980).

Lestrimelitta limao, a 'robber bee', pilfers the nests of various *Trigona*. Scouts of *L. limao* mark a trail to the intended target *Trigona* nest with their mandibular secretions. The main components were identified by Blum (1966) as the citral isomers. During the raid, the *L. limao* evacuate their mandibular glands in or at the raided nest. This attracts more *L. limao* and also disturbs workers of the raided species, breaking down a cohesive defense (Blum *et al.*, 1970). The alarm pheromones of *L. limao* and *T. pectoralis* also act as allomones for other species of *Trigona*, inducing defensive behavior that might be effective against these actual and potential nest robbers (Weaver *et al.*, 1975).

While stingless bees would seem to be defenseless, such is not the case. Many species are nasty biters and efficiently chase humans from their nests. *Trigona tataira* is notoriously effective because the secretion from its hypertrophied mandibular glands produces a terrible 'burning', earning it the name of 'fire bee' (Michener, 1974, pp. 212–17).

14.4.3 Other cephalic glands

(a) Hypopharyngeal glands

The hypopharyngeal glands have been surveyed by Cruz Landim (1967). Their fine structure in *A. mellifera* has been documented by numerous authors (Ortiz-Picon and Diaz-Flores, 1972; Cruz Landim and Hadek, 1969; and references therein). They consist of a pair of glands that open on the hypopharyngeal plate. In most bees these glands consist of numerous Leydig cells individually connected by their ductules to the hypopharyngeal plate. In some Halictinae and in *Megachile*, *Centris*, *Xylocopa* and Apidae, ductules from the secretory cells on each side join a common duct that in turn opens laterally on the hypopharyngeal plate. The glands are long in workers of Bombini, Apinae and Meliponinae. In the latter two taxa, the extended glands are much longer than the body of the bee. In *A. mellifera* 550 Leydig glands may join each duct. Hypopharyngeal glands are absent in queen and drone *Apis*.

The function of the hypopharyngeal glands is known only in the honey bee. In 'nurse' worker bees, they secrete a portion of the 'bee milk', the food given to larvae which is a complex mixture of lipids, vitamins and proteins. They also secrete invertase and an enzyme that oxidizes glucose to an acid (Cruz Landim and Hadek, 1969).

(b) Foveal glands

Female Colletidae and Andrenidae possess a pair of depressed areas, densely covered by short setae, on the face between the eyes and antennae. These areas, the facial foveae, are lined by enlarged epidermal cells that appear to be glandular (Heselhaus, 1922; Nedel, 1960). The foveae are small or absent in males. Nothing is known of the chemistry or function of their secretions.

14.5 THORACIC GLANDS

The principal glands of the thorax are actually the thoracic portions of the labial glands, described in Section 14.4.1. Other glands occur in the legs. In many male *Centris*, the hind femora and tibiae are greatly enlarged. A large gland with two lobes, one tibial and one femoral, fills most of the space in those podites not occupied by muscles (Stort and Cruz Landim, 1965; Williams *et al.*, 1982). Both lobes join a common duct, that (according to Stort and Cruz Landim) runs through the tarsus to exit on the arolium of the pretarsus or (according to Williams *et al.*) exits apically on the tibia. The gland in the femur is an elongate sac lined by secretory cells, apparently without ductules.

Male 'orchid bees' (Euglossini) have greatly enlarged hind tibiae, each with a groove into which they rub exudates from orchids and other sources. The morphology and fine structure of these tibiae have been described for *Euglossa cordata*, *Euplusia violacens* and *Eulaema nigrita* (Cruz Landim *et al.*, 1965; Sakagami, 1965). The seta-lined groove opens into a cuticular cavity that fills most of the tibia, so that in transverse section the tibia roughly resembles two concentric circles joined by the groove. Cuticular projections fill most of the inner cavity, giving it a 'spongy' texture that appears to adsorb the materials entering through the groove. Cruz Landim *et al.* suggest that the cells lining the inner cavity may draw ingredients from the adsorbed liquid across the cavity walls into the space between the cavity and the tibial integument. There the fluid is exposed to large, glandular cells near the groove that may resorb it and pass it across the cuticle to the outside of the tibia. Presumably one or more of these cellular layers chemically modifies the substance, as the floral components alone do not attract female bees.

Male *Euglossa imperialis* and *Eulaema meriana* perch on tree trunks in their territories and display in such a manner that release of a sex attractant pheromone from the tibiae is likely (Kimsey, 1980). Although the chemicals from orchids that attract male euglossine bees have been characterized, the chemical structure of the pheromone released by the males has not been determined, nor, indeed, has its presence been proven. This is a crucial missing link in the 'orchid bee' story.

The last tarsomere in all legs of both sexes and all castes of *Apis*, *Bombus*, *Trigona* and *Melipona* contains a large gland, called the tarsal or Arnhart's gland (Cruz Landim and Staurengo, 1965). The tarsal gland is a secretory cell-lined sac that fills most of the tarsomere and empties via the arolium of the pretarsus.

In honey bees the tarsal gland produces an oily exudate, the 'footprint pheromone', that is deposited wherever a bee walks. Queens produce 13 times more pheromone than do workers, and with mandibular gland secretion it inhibits queen cell construction by workers (Lensky and Slabezki, 1981). Worker honey bees may recognize a hive entrance by the footprint pheromone deposited on it by other workers (Butler *et al.*, 1969).

Although males of some species of *Centris* mark territories with chemicals released by the mandibular glands (Vinson *et al.*, 1982), males of those species with enlarged hind legs use secretions applied to substrates with the hind legs. These femoral/tibial secretions contain a homologous series of ketones with pentadecanone dominating, acetates and ethyl myristate (Williams *et al.*, 1982).

14.6 ABDOMINAL GLANDS

14.6.1 Dufour's gland

(a) Morphology

In addition to the species considered by Pessotzkaja (1929) and Lello (1971a, b, c, d, 1976; Kerr and Lello, 1962), the Dufour's gland has been described in *Colletes* and *Hylaeus* (Batra, 1980), *Peponapis pruinosa* (Mathewson, 1965), and *Anthophora abrupta* (Norden *et al.*, 1980). In general appearance the sac-like gland is flattened, milky white, with a smooth or convoluted surface. Lining the reservoir are secretory cells which lack ductules and empty directly across the cuticular intima into the reservoir. This reservoir, which has thin spine-like cuticular projections lining the lumen, narrows to a duct which opens via a valve in the sting chamber below the sting bulb so the secretion does not enter the sting shaft.

The Dufour's gland is large, filling most of the dorsal abdomen, and elongate to somewhat horseshoe-shaped in Colletinae, Dufoureinae and Andrenidae. In *Colletes* the gland represents 10% of the live weight of the bee and may contain 3–4 µl of secretion (Duffield, unpublished). It is similar in shape but smaller in *Hylaeus*. The sac is wider and more crescent-shaped in *Nomia*, *Megalopta* (Halictinae) and *Hesperapis* (Melittidae), whereas in *Melitta* it is an immense bilobed structure with numerous fingerlike projections. In other Halictinae the gland is more similar in shape to that of *Colletes* but is smaller and bilobed. The Dufour's gland of *Oxaea* is small and egg-shaped. It is narrow and ribbon-shaped in most Megachilinae, Anthophorinae and Xylocopinae. In Megachilinae it varies in size from virtually vestigial in some *Osmia* to longer than the abdomen in *Megachile terrestris*. Length also varies among the Anthophoridae, with the longest glands in Xylocopini and some *Exomalopsis* (Anthophorinae). In *Anthophora* the gland is immense and wide, with many fingerlike projections. The gland in Bombinae is small in the worker, and disproportionately larger in queens. It is narrow in *Apis* and, in *A. mellifera*, is shorter than the venom gland.

The Meliponinae lack a functional sting. Lello (1976) considers the single gland opening at the end of the abdomen to be homologous with the Dufour's gland. The gland varies considerably in size and shape. In *Meliponula bocandei*, *Trigona staudingeri* and *T. varia* it is a large, wide, U-shaped sac, occupying about half the abdomen, but it is more pear-shaped in other species with large glands. Many species of *Trigona* and *Melipona* have a short to vestigial ribbon-shaped gland in the workers, but well-developed ribbon or pear-shaped glands in the queens. Two species of *Trigona* have anomalous Dufour's glands: two equal-sized elliptic lobes united by a common duct in *T. muelleri*; and a chain of transparent spherules united by a thin duct in *T. ferruginea*.

(b) Chemistry

Colletidae
Macrocyclic lactones were first identified in the Halictidae by Andersson *et al.* (1966) (see Table 14.4). Bergström (1974) showed that the secretions of

Table 14.3   Chemical components found in the Dufour's glands of the Colletidae

Subfamily, genus and species	16-Hexadecanolide (254)	18-Octadecanolide (282)	18-Octadecenolide (280)	20-Eicosanolide (310)	20-Eicosenolide (308)	21-Henicosanolide (324)	22-Docosanolide (338)	24-Tetracosanolide (366)	Hydrocarbons	Methyl esters	Ethyl esters	n-Octadecanal (268)	n-Eicosanal (296)	Biogeographical realm	Reference
Colletinae															
Colletes															
C. cunicularius	X	X	X	X	—	—	—	—	X	—	—	—	—	P	1
	—	X	X	X	—	—	—	—	—	—	—	—	—	P	2
C. cunicularius celticus	—	X	X	X	X	X	X	—	X	—	—	—	—	P	3
C. inaequalis	—	X	—	X	—	—	—	—	—	—	—	—	—	H	4
C. succinctus	—	X	X	X	—	—	—	—	X	—	—	—	—	P	3
C. thoracicus	X	X	X	X	—	X[a]	X	—	X	—	—	X	X	H	4
C. validus	—	X	—	X	—	—	—	—	X	—	—	—	—	H	4
Hylaeinae															
Hylaeus															
H. modestus	—	X	—	X	—	—	X	X	X	X	X	—	—	H	5

X = present, — = absent
References: (1) Bergström, 1974; (2) Cane, 1981; (3) Albans *et al.*, 1980; (4) Hefetz *et al.*, 1979c; (5) Duffield *et al.*, 1980
H = holarctic, P = palearctic
a = methyl-branched 20-eicosanolide

Colletes cunicularis are dominated by saturated and unsaturated macrocyclic lactones as well as hydrocarbons (Table 14.3). Subsequent analyses of five additional species revealed an additional lactone (Hefetz *et al.*, 1979c; Albans *et al.*, 1980; Cane, 1981). Secretions of both palearctic and holarctic species have macrocyclic lactones and hydrocarbons as major components. In addition to these, *Colletes thoracicus* has *n*-octadecanal and *n*-eicosanal (Hefetz *et al.*, 1979c). Extensive analyses of the Dufour's secretions of *C. cunicularius celticus* also showed di- and trihydroxymonocarboxylic acids, dicarboxylic acids, hydrocarbons and a C_{21} lactone in addition to the even carbon-containing lactones (Albans *et al.*, 1980).

The Dufour's gland secretions of *Hylaeus modestus* are similar to those of *Colletes* and are dominated by macrocyclic lactones. The secretions of *H. modestus* exhibit only saturated macrocyclic lactones as well as methyl and ethyl esters (Duffield *et al.*, 1980).

Halictidae

The Dufour's gland secretions of species representing nine genera of the Halictinae and Nomiinae are similar to those of the Colletidae; all are dominated at least in part by macrocyclic lactones (Table 14.4). The chemistry of the Dufour's gland secretions of North American and European species of the Halictinae have been investigated by Bergström (1974), Hefetz *et al.* (1978), Bergström and Tengö (1979), and Duffield *et al.* (1981). They have identified a series of even numbered ($C_{16}-C_{26}$), saturated and unsaturated macrocyclic lactones. Duffield *et al.* (1981) also reported the presence of trace amounts of a homologous series of eight esters containing branched C_5-alkenols and fatty acids in several species.

The Dufour's gland secretions of four species of *Nomia* (Table 14.4) exhibit even numbered, saturated macrocyclic lactones and isopentenyl esters similar to those found in the Halictinae. The major difference between secretions of *Nomia* and Halictinae is the presence of the isopentenyl esters as major components of the blend in *Nomia* (Duffield *et al.*, 1982).

Andrenidae

The chemistry of the Dufour's gland secretions of the Andrenidae was first studied in *Andrena* by Bergström and Tengö (1974); the exudates contain a variety of terpenes including farnesene, farnesol, geraniol and a series of farnesyl and geranyl esters with either geranyl octanoate or farnesyl hexanoate dominant (Table 14.5). Later studies (Tengö and Bergström, 1975; Tengö and Bergström, 1978; Fernandes *et al.*, 1981; Duffield *et al.*, unpublished) of holarctic and palearctic species have added many additional compounds. The work of Fernandes *et al.* (1981) suggested three different classes of *Andrena* secretions. Some species produce terpenoid esters as the dominant compounds, either farnesyl hexanoate or geranyl octanoate. The second type produces a series of straight-chained esters, dominated by either hexanoates or octanoates. The third type secretes a mixture of two types of components: farnesyl and straight-chain esters.

Table 14.4 Compounds identified in the Dufour's gland extracts of the Halictidae

Subfamily, genus and species	Compound (mol. wt.)															Biogeographical realm	Reference
	18-Octadecanolide (282)	18-Octadecenolide (280)	20-Eicosanolide (310)	20-Eicosenolide (308)	22-Docosanolide (338)	22-Docosenolide (336)	24-Tetracosanolide (366)	24-Tetracosenolide (364)	26-Hexacosanolide (394)	3-Methyl-2(3)-buten-1-yl tetradecanoate (296)	3-Methyl-2(3)-buten-1-yl hexadecanoate (324)	3-Methyl-2(3)-buten-1-yl octadecanoate (352)	3-Methyl-2(3)-buten-1-yl eicosanoate (380)	3-Methyl-2(3)-buten-1-yl docosanoate (408)	3-Methyl-2(3)-buten-1-yl tetracosanoate (436)		
Halictinae																	
Augochlora																	
A. pura pura	—	—	X	X	M	—	—	—	—	—	—	—	tr	tr	—	H	1
A. pura mosieri	—	—	X	—	M	—	X	—	—	—	—	—	tr	tr	—	H	1
Augochlorella																	
A. striata	—	—	X	X	M	—	X	—	—	—	—	tr	tr	tr	—	H	1
Augochloropsis																	
A. metallica	M	X	X	X	X	X	X	X	X	—	—	—	—	—	—	H	1
Agapostemon																	
A. sericeus	M	—	X	X	X	—	X	—	—	—	—	tr	—	—	—	H	1
A. splendens	M	—	M	X	X	—	—	—	—	—	—	—	—	—	—	H	2
A. texanus	M	—	X	—	X	—	—	—	—	—	—	tr	tr	tr	—	H	1
Dialictus																	
D. coeruleus	X	—	X	X	X	—	—	—	—	—	tr	tr	tr	tr	—	H	1
D. cressonii	X	X	M	X	X	X	X	—	—	—	—	tr	—	—	—	H	1
D. laevissimus	X	—	X	—	X	—	—	—	—	—	—	—	—	—	—	H	2

Compound occurrence in Dufour's gland secretions

Species	1	2	3	4	5	6	7	8	9	10	11	12	13	Dist.	Ref.
D. lineatulus	X	—	X	X	X	—	—	—	—	—	—	—	—	H	2
D. nymphalis	X	—	X	X	X	—	—	—	—	—	—	—	—	H	2
D. pilosus	M	—	X	X	X	—	—	—	—	—	—	tr	—	H	1
D. rohweri	X	X	M	X	X	—	—	—	X	—	—	—	—	H	1
D. tamiamensis	X	—	X	X	X	—	—	—	X	—	—	—	—	H	2
D. versatus	X	X	M	X	—	—	—	—	—	tr	tr	tr	—	H	2
Evylaeus															
E. albipes[a]	X	X	X	X	—	—	—	—	—	—	—	—	—	P	3
E. calceatus[a]	X	X	X	X	—	—	—	X	—	—	—	—	—	P	4
E. quebecensis	X	—	X	X	X	—	—	—	—	—	—	—	—	H	2
E. truncatus	X	—	X	?	X	X	—	—	—	—	—	—	—	H	1
Lasioglossum															
L. coriaceum	M	X	X	X	X	X	X	?	—	tr	tr	tr	—	H	1
L. fuscipenne	M	—	X	X	X	X	X	—	—	tr	tr	tr	—	H	1
L. leucozonium	X	—	X	—	—	—	—	—	—	—	—	—	—	H	2
Halictus															
H. confusus	—	—	M	X	X	—	X	—	—	tr	tr	tr	—	H	1
H. ligatus	X	—	X	M	M	X	X	X	—	—	—	—	—	H	1
H. parallelus	—	—	M	X	X	X	X	—	—	—	—	—	—	H	1
H. rubicundus	X	X	X	M	X	X	X	—	—	—	—	—	—	H	5
Nomiinae															
Nomia															
N. heteropoda	tr	—	X	tr	—	—	—	X	—	—	M	X	X	H	6
N. nevadensis	—	tr	tr	M	M	tr	—	M	tr	—	M	M	tr	H	6
N. triangulifera	—	tr	tr	M	M	—	—	M	—	—	M	M	X	H	6
N. t. tetrazonata	tr	—	—	—	—	—	—	—	—	—	M	M	—	H	6

M = major component, X = present, tr = trace, — = absent
References: (1) Duffield *et al.*, 1981; (2) Hefetz *et al.*, 1978; (3) Cane, 1981; (4) Bergström, 1974; (5) Bergström and Tengö, 1979; (6) Duffield *et al.*, 1982
H = holarctic, P = palearctic
a = 16-hexadecanolide present

407

Table 14.5 shows that with few exceptions each species appears to contain a unique blend of terpenoid esters as defined by its molecular composition and/ or the relative concentration of each component. Such a tendency to species-specificity is not suggested in the Colletidae or Halictidae, the only other families for which there are enough examples to draw any conclusions.

Citronellyl citronellate and another minor component, citronellyl geranate, have been identified in the Dufour's gland secretion of *Panurginus potentillae* but *P. atramontensis* has only the latter compound (Duffield *et al.*, 1983). These two compounds have been found previously in the seventh sternal glands of the European hornet, *Vespa crabro* (Wheeler *et al.*, 1982).

Melittidae

The chemistry of the Dufour's gland secretions have been studied in two palearctic species of *Melitta* by Tengö and Bergström (1976b). Both species exhibit a homologous series of even numbered alkyl butanoates ranging from C_{12} to C_{20}. The major component in each is hexadecyl butanoate. Several acetates and unsaturated alcohols were also detected.

Megachilidae

Although there are no published reports on the chemistry of the Dufour's gland secretions of members of the Megachilidae, several research groups are currently investigating the Dufour's gland secretions of them. *Megachile* Dufour's extracts contain complex mixtures of triglycerides; those of some *Osmia* have a homologous series of hydrocarbons (Wheeler *et al.*, unpublished results).

Anthophoridae

The chemistry of the Dufour's gland of the few species of Anthophoridae investigated is very heterogeneous. In the Anthophorinae, terpenoid and straight-chain acetates have been found in one species of *Melissodes* (Eucerini) (Batra and Hefetz, 1979). The secretions of *Svastra*, another Eucerini, are the most complex of all the bee extracts examined in total number of compounds present (Duffield *et al.*, unpublished results). The esters are like those found in some species of *Andrena* (Fernandes *et al.*, 1981), although each mixture of the same molecular weight is more complex and more molecular weights are represented. A series of triglycerides have been identified in the Dufour's gland extracts of *Anthophora abrupta* by Norden *et al.* (1980). They also reported triglycerides in the subgenus *Clisodon* but the specific compounds were not identified.

The chemistry of Dufour's glands in *Nomada*, the only nomadine genus investigated, is discussed in Section 14.7. In the Xylocopinae, saturated and unsaturated hydrocarbons dominate the Dufour's gland extracts of the large carpenter bee, *Xylocopa virginica* (Vinson *et al.*, 1978), which also contains methyl esters of fatty acids.

Table 14.5 Chemical components found in the Dufour's gland extracts of *Andrena* (Andrenidae)

Compound (mol. wt.)

Subgenus and species	Geraniol (154)	Farnesol (222)	Geranyl hexanoate (252)	Geranyl octanoate (280)	Geranyl decanoate (308)	Geranyl dodecanoate (336)	trans-α-Farnesene (204)	trans, trans-β-Farnesene (204)	Farnesyl acetate (264)	Farnesyl butanoate (292)	Farnesyl hexanoate (320)	Farnesyl octanoate (348)	Farnesyl decanoate (376)	Geranylgeranyl octanoate (416)	Octyl hexanoate (228)	Decyl hexanoate (256)	Octyl octanoate (256)	Dodecyl hexanoate (284)	Decyl octanoate (284)	Tetradecyl hexanoate (312)	Dodecyl octanoate (312)	Hexadecyl hexanoate (340)	Tetradecyl octanoate (340)	Octadecyl hexanoate (368)	Hexadecyl octanoate (368)	Eicosyl hexanoate (396)	Octadecyl octanoate (396)	Dodecyl tetradecanoate (396)	Docosyl hexanoate (424)	Eicosyl octanoate (424)	Tetradecyl tetradecanoate (424)	Hexanoic acid (116)	Biogeographical realm	Reference
Andrena																																		
A. clarkella	M	—	—	—	—	—	—	—	—	—	—	—	—	—	—	—	—	—	—	—	—	—	—	—	—	—	—	—	—	—	—	—	P	4
A. fucata	—	—	—	M	X	—	X	X	—	—	M	X	—	—	—	—	—	—	—	—	—	—	—	—	—	—	—	—	—	—	—	—	P	4
A. helvola	X	—	tr	M	X	—	X	tr	—	—	—	tr	—	—	—	—	—	—	—	—	—	—	—	—	—	—	—	—	—	—	—	—	P	1
A. longifacies	—	—	tr	M	tr	—	—	—	tr	—	—	tr	—	—	—	—	—	—	—	—	—	—	—	—	—	—	—	—	—	—	—	—	H	2
A. milkwaukeensis	—	—	—	M	tr	—	—	—	tr	—	—	X	—	—	—	—	—	—	—	—	—	—	—	—	—	—	—	—	—	—	—	—	H	3
A. mandibularis	—	—	—	M	—	—	—	—	—	—	—	X	—	—	—	—	—	—	—	—	—	—	—	—	—	—	—	—	—	—	—	—	H	3
A. praecox	—	—	—	M	—	—	—	—	—	—	—	tr	—	—	—	—	—	—	—	—	—	—	—	—	—	—	—	—	—	—	—	—	P	4
A. tridens	—	—	X	M	X	X	—	tr	—	—	tr	tr	tr	—	—	—	—	—	—	—	—	—	—	—	—	—	—	—	—	—	—	—	H	3
Biareolina																																		
A. haemorrhoa	—	X	—	—	—	—	X	X	X	X	tr	X	—	—	—	—	—	—	—	—	—	—	—	—	—	—	—	—	—	—	—	—	P	1
Callandrena																																		
A. gardineri	—	—	—	—	—	—	—	X	tr	tr	M	—	tr	—	—	—	—	—	—	—	—	—	—	—	—	—	—	—	—	—	—	—	H	3
Charitandrena																																		
A. hattorfiana	—	—	—	—	—	—	—	tr	tr	tr	M	X	tr	—	—	—	—	—	—	—	—	—	—	—	—	—	—	—	—	—	—	—	P	4
Chrysandrena																																		
A. fulvago	—	—	—	—	—	—	—	—	—	—	M	—	—	—	—	—	—	—	—	—	—	—	—	—	—	—	—	—	—	—	—	—	P	4
Cnemidandrena																																		
A. denticulata	X	X	X	X	—	—	X	X	X	X	M	—	—	—	—	—	—	—	—	—	—	—	—	—	—	—	—	—	—	—	—	—	P	1
Conandrena																																		
A. bradleyi	—	—	—	M	—	—	—	—	—	tr	—	X	—	—	—	—	—	—	—	—	—	—	—	—	—	—	—	—	—	—	—	—	H	2
Euandrena																																		
A. bicolor	X	X	X	—	—	—	X	X	X	X	M	—	—	—	—	—	—	—	—	—	—	—	—	—	—	—	—	—	—	—	—	—	P	1
Hoplandrena																																		
A. carantonica	—	—	—	—	—	—	—	—	—	—	M	—	—	—	—	—	—	—	—	—	—	—	—	—	—	—	—	—	—	—	—	—	P	4

(continued)

Table 14.5—continued

Subgenus and species	Geraniol (154)	Farnesol (222)	Geranyl hexanoate (252)	Geranyl octanoate (280)	Geranyl decanoate (308)	Geranyl dodecanoate (336)	trans-α-Farnesene (204)	trans, trans-β-Farnesene (204)	Farnesyl acetate (264)	Farnesyl butanoate (292)	Farnesyl hexanoate (320)	Farnesyl octanoate (348)	Farnesyl decanoate (376)	Geranylgeranyl octanoate (416)	Octyl hexanoate (228)	Decyl hexanoate (256)	Octyl octanoate (256)	Dodecyl hexanoate (284)	Decyl octanoate (284)	Tetradecyl hexanoate (312)	Dodecyl octanoate (312)	Hexadecyl hexanoate (340)	Tetradecyl octanoate (340)	Octadecyl hexanoate (368)	Hexadecyl octanoate (368)	Eicosyl hexanoate (396)	Octadecyl octanoate (396)	Dodecyl tetradecanoate (396)	Docosyl hexanoate (424)	Eicosyl octanoate (424)	Tetradecyl tetradecanoate (424)	Hexanoic acid (116)	Biogeographical realm	Reference
Iomelissa																																		
A. violae	—	—	—	—	—	—	tr	tr	—	X	M	X	—	—	—	—	—	—	—	—	—	—	—	—	—	—	—	—	—	—	—	—	H	2
Larandrena																																		
A. miserabilis	—	—	—	X	—	—	—	—	—	—	—	M	—	—	—	—	—	—	—	—	—	—	—	—	—	—	—	—	—	—	—	—	H	3
Leucandrena																																		
A. erythronii	—	—	—	tr	—	—	tr	tr	tr	tr	M	X	tr	—	tr	tr	tr	tr	tr	tr	tr	tr	tr	—	tr	—	—	—	—	—	—	—	H	3
A. placida	—	—	—	—	—	—	—	—	—	—	—	tr	—	—	—	—	—	—	—	—	—	—	—	—	—	—	—	—	—	—	—	—	H	3
Melandrena																																		
A. carlini	—	—	—	—	—	—	tr	tr	tr	tr	M	tr	tr	—	—	—	—	—	—	—	—	—	—	—	—	—	—	—	—	—	—	—	H	3
A. confederata	—	—	—	—	X	—	tr	—	tr	tr	M	X	X	—	—	—	—	—	—	—	—	—	—	—	—	—	—	—	—	—	—	—	H	3
A. hilaris	X	—	X	X	X	—	X	X	—	—	M	tr	—	—	—	—	—	—	—	—	—	—	—	—	—	—	—	—	—	—	—	—	H	2
A. nigroaenea	—	X	X	X	—	—	tr	tr	X	—	M	X	—	—	—	—	—	—	—	—	—	—	—	—	—	—	—	—	—	—	—	—	P	1
A. nivalis	—	—	—	—	—	—	—	tr	—	tr	M	X	tr	—	—	—	—	—	—	—	—	—	—	—	—	—	—	—	—	—	—	—	H	3
A. pruni	—	—	—	—	—	—	tr	tr	tr	tr	M	tr	tr	—	—	—	—	—	—	—	—	—	—	—	—	—	—	—	—	—	—	—	H	3
A. vaga	—	—	—	—	—	—	—	—	—	—	M	—	—	—	—	—	—	—	—	—	—	—	—	—	—	—	—	—	—	—	—	—	P	4
A. vicina	—	—	—	—	—	—	tr	tr	tr	tr	M	—	—	—	—	X	tr	X	tr	X	tr	—	—	—	—	—	—	—	—	—	—	—	H	3
Opandrena																																		
A. cressoni	—	—	—	—	—	—	—	—	—	—	M	tr	tr	tr	—	—	—	—	—	—	—	—	—	—	—	—	—	—	—	—	—	—	H	3
Parandrena																																		
A. andrenoides	—	—	—	—	—	—	—	—	—	—	M	—	—	—	—	—	—	—	—	—	—	—	—	—	—	—	—	—	—	—	—	—	H	3
Plastandrena																																		
A. bimaculata	—	—	—	X	—	—	—	—	—	—	M	X	—	—	—	—	—	—	—	—	—	—	—	—	—	—	—	—	—	—	—	—	P	4
A. carbonaria	X	X	X	X	X	—	—	—	—	—	M	—	—	—	—	—	—	—	—	—	—	—	—	—	—	—	—	—	—	—	—	—	P	1
A. crataegi	—	—	—	—	—	—	tr	tr	—	—	M	X	—	—	—	—	—	—	—	—	—	—	—	—	—	—	—	—	—	—	—	—	H	2
A. tibialis	—	—	—	—	—	—	—	—	—	—	M	—	—	—	—	—	—	—	—	—	—	—	—	—	—	—	—	—	—	—	—	—	P	4

Compound (mol. wt.)

Taxon		Type	Ref.
Ptilandrena			
	A. erigeniae	H	2
Rhacandrena			
	A. robertsonii	H	2
Scaphandrena			
	A. arabis	H	2
Scrapteropsis			
	A. fenningeri	H	3
	A. ilicis	H	3
	A. imitatrix	H	3
	A. w-scripta	H	2
Simandrena			
	A. nasonii	H	3
Taeniandrena			
	A. gelri	P	4
	A. ovatula	P	4
	A. russula	P	5
	A. wilkella	H	2
Thysandrena			
	A. bisalicis	H	3
Trachandrena			
	A. ceanothi	H	2
	A. forbesii	H	2
	A. hippotes	H	3
	A. mariae	H	2
	A. nuda	H	3
	A. rugosa	H	3
	A. spiraena	H	3
Tylandrena			
	A. perplexa	H	3
Unassigned to subgenus			
	A. flexa	H	3

M = major compound, X = present, tr = trace, — = absent

References: (1) Bergström and Tengö, 1974; (2) Duffield, LaBerge and Wheeler, unpublished data; (3) Fernandes *et al.*, 1981; (4) Tengö and Bergström, 1975; (5) Tengö and Bergström, 1978

H = holarctic, P = palearctic

411

(c) Chemical notes

Although it is known that unsaturated lactones accompany their saturated counterparts in halictid and colletid Dufour's glands, the position of the double bond in these large rings is unknown in all but one case (Albans *et al.*, 1980). Whether there are single unsaturated lactones or mixtures of positional and/or geometrical isomers is also unknown. Similarly, the position of the double bond in the hexadecenol and octadecenol found in melittid bees has not been established. In many cases these unsaturated compounds represent a small percentage of the saturated and/or total volatiles. Until a specific function is found for the Dufour's gland secretion, the investment of time necessary to specify the position of the double bond or to synthesize these isomers does not appear warranted.

A similar problem exists with the triglycerides found in *Megachile* and *Anthophora*. While it is apparent from their mass spectra (which show appropriate acylium ions RCO^+) that these compounds are mixtures of triglycerides, it is an entirely different matter to show which mixtures of specific triglycerides are present. Work in our laboratory suggests that these Dufour's gland products in megachilid bees are complex mixtures of isomeric triglycerides.

While the esters (and lactones) mentioned above may (or may not) be mixtures, the farnesyl and geranyl esters found in *Andrena* are not. They appear to all be of all-*trans*-farnesol and devoid of esters of the other farnesol isomers. Similarly the geranyl esters are free of their neryl counterparts. The specific isomer of dihydrofarnesyl acetate from *Melissodes* is unknown although 2,3-dihydro-6-*trans*-farnesol has been found in a *Bombus jonellus* cephalic gland (Bergström and Svensson, 1973a).

In many cases the esters and lactones listed above are accompanied by variable amounts of hydrocarbons (C_{19} and higher) which may or may not be saturated. It is unknown whether these hydrocarbons are specific glandular products or artifacts of the extraction process as the higher fatty acids are thought to be. In most cases mentioned above the Dufour's gland products were obtained by extraction of the dissected gland with an organic solvent. Thus some trace chemicals listed as representative of the secretion may be indicative only of the total extraction procedure.

While hydrocarbons containing odd numbers of carbon are commonplace in both mandibular and Dufour's gland secretions of these bees, these same compounds are common ant natural products and do not appear to have a function other than as a carrier for the low-molecular-weight materials. Although they have been reported by many workers and have been listed by us in various tables, any function has yet to be documented and we believe their importance is minimal.

Similarly, the variation in relative concentrations of compounds between species may or may not be significant. Only now is work being done on individual glands of a single species to determine the variation within one species.

The blends in *Andrena* appear to be species-specific but it remains to be determined how much these blends vary within each species.

(d) Function

The large Dufour's gland of many species has long been suggested as the source of the liquid which, when applied to cell walls, hardens to form the linings observed in the Colletidae, Halictidae, Andrenidae, and Anthophoridae (*Anthophora*) (Semichon, 1906). These linings, which may either thin, cellophane-like and transparent (Colletidae) or waxy (other families), probably function (i) to provide chemical cues for nesting, (ii) to maintain humidity control, (iii) to defend against microbial infection, and/or (iv) as a food source.

Recently, the chemical composition of the cell wall linings of several bees has been shown to be the same as some components found in the Dufour's gland: Colletidae – *Colletes* (Hefetz *et al.*, 1979c; Albans *et al.*, 1980; Cane, 1981); Halictidae – *Augochlora* (Duffield *et al.*, 1981), *Evylaeus* (Cane, 1981); Andrenidae – *Andrena* (Cane, 1981); and Anthophoridae – *Anthophora* (Norden *et al.*, 1980). The cellophane-like linings of *Colletes* are particularly interesting because the material is a polymer of the Dufour's secretion. Although the polymerization process is not understood, Albans *et al.* (1980) suspect that the thoracic labial glands release an enzyme which is applied as the female spreads the Dufour's secretion with her mouth parts. This enzyme polymerizes the Dufour's secretion. Similar involvement of labial gland secretions is probable for halictids (Batra, 1968) and *Anthophora abrupta* (Norden *et al.*, 1980).

The maternal Dufour's gland secretions/cell wall linings may also provide a food source for developing larvae. Norden *et al.* (1980) contend that *Anthophora abrupta* larvae have an unusual behavior; they eat the provisions which contain diglycerides provided by the Dufour's gland secretion and then eat the white cell lining which also contains the diglycerides. Duffield *et al.* (1981) have also shown that the major lactone components found in the Dufour's gland of *Augochlora pura* (Halictinae) are in the pollen/nectar provisions. These observations lead to an interesting speculation. The farnesyl esters of *Andrena* could be hydrolyzed and oxidized when consumed by the larvae to yield farnesenic acids, known precursors of juvenile hormone (JH) (Reibstein *et al.*, 1979). Both the farnesyl esters and JH do have a *trans, trans* configuration. Thus the mother bee, by providing terpenoid esters used in the synthesis of JH, may have an influence on her offspring's development.

The secreted cell linings may have pheromonal functions. Halictine foragers returning to a nest are attracted to freshly lined cells, and sweat bees avoid breaking into completed cells when excavating new cells, even under crowded conditions (Batra, 1968; May, 1973). However, May was not able to demonstrate that cell linings alone deter bees from accidentally digging into cells. The Dufour's gland secretions are also hypothesized to act as aggregation pheromones, inducing other females to nest nearby. They may also attract males to the nest site and enable females to recognize their nests (Bergström and Tengö, 1974; Tengö and Bergström, 1975; Hefetz *et al.*, 1979c). None of these pheromonal functions have been demonstrated experimentally with synthetic chemicals.

Other functions of the Dufour's gland have been little investigated, despite the fact that the glands are well developed in many Megachilidae, Xylocopinae, and Apidae, where their production of cell lining secretions is not likely. *Xylocopa virginica texana* females scent-mark flowers that they visit for nectar, and females are repulsed from such flowers for up to 10 minutes. Extracts of the Dufour's gland applied experimentally to flowers duplicated the repulsion (Frankie and Vinson, 1977; Vinson *et al.*, 1978). There is a remarkable absence of data and even speculation concerning other Dufour's gland functions in these long-tongued bees, even for the remarkable glands of stingless bees.

14.6.2 Venom glands

(a) Morphology

The venom gland has been described along with the Dufour's gland by Lello (1971a, b, c, d, 1976; Kerr and Lello, 1962), Pessotzkaja (1929), and Mathewson (1965). Its fine structure in *Apis mellifera* was described by Cruz Landim and Kitajima (1966) and Bridges (1977). The basic structure is similar in all bees and resembles that in other aculeate Hymenoptera (Blum and Hermann, 1969). The venom sac is a reservoir whose duct enters the sting capsule and deposits the secretion in the basal bulb of the sting shaft. The movement of the basal valves of the sting shaft drives the venom to the tip of the sting (Snodgrass, 1956; Maschwitz and Kloft, 1971). The secretory cells surround a long, narrow duct that enters the distal end of the venom sac. This tubular secretory portion, called the free filaments by Blum and Hermann, bifurcates in most bees. The venom sac is lined by muscles that presumably force the venom from the reservoir into the sting. The venom sac itself is largely free of secretory cells except for a cluster near the entrance of the free filaments. These may represent the convoluted gland described by Blum and Hermann. The secretory cells of the free filaments and venom sac have ductules.

The structure of the venom gland varies only slightly. The venom sac can be ovoid to cylindrical. The gland is especially well developed in Bombini, in which the two branches of the free filaments branch multiply. In many Megachilinae and in *Nomada*, the free filament is unbranched. Both the sting and the venom gland are small in *Hesperapis carinata* (Melittidae), and the free filaments are unbranched. Lello (1971a) stated that *Andrena andrenoides* and *A. erythrogaster* have atrophied stings and no venom glands, but Radović and Hurd (1980) found well-developed stings in these species. Our dissections indicate that some species of *Andrena* have unusually small stings, although all parts are present. The stingless bees (Meliponinae) have reduced, non-functional stings

(Radović, 1981) and lack any trace of a venom gland. The median gland, interpreted as a venom sac by Kerr and Lello (1962), is apparently a modified Dufour's gland (Lello, 1976).

(b) Function

The stings of bees are used exclusively in defense and the venoms are of considerable human importance. In some other aculeate Hymenoptera, venom gland secretions also function as pheromones (Maschwitz and Kloft, 1971), but this has not been demonstrated in any bee. Bee venoms are complex mixtures of proteins, histamine, serotonin, and acetylcholine (Maschwitz and Kloft, 1971) and will not be reviewed here.

14.6.3 Sting glands

(a) Morphology

The Koshevnikov glands open laterally on the sting. They have been surveyed by Altenkirch (1962) and their fine structure in *A. mellifera* was described by Hemstedt (1969). They consist of a set of Leydig cells connected by their ductules to the membrane between the eight abdominal hemitergite (stigmal plate) and the first valvifer (quadrate plate). The gland is present in all bees examined by Altenkirch except *Hylaeus*, *Nomada* and *Heriades* (Megachilinae). The Koshevnikov gland is especially well developed in *Apis*, where the Leydig cells lie close together and the membrane into which they empty forms a reservoir. These glands produce an attractant pheromone in honey bee queens that has not been chemically analyzed.

The epidermis forms a glandular palisade layer on parts of the sting apparatus. Such a layer is especially well developed on the membrane that connects the dorsal and ventral corners of the sting sheath in *Andrena*, *Colletes*, *Panurginus* (Panurginae), *Anthophora* and *Crocisa* (Anthophorinae). The sting sheath itself bears a well-developed gland of this type in honey bee workers.

Sting glands, whose exact location in the basal apparatus is unknown, produce an alarm pheromone that is stored in a setose membrane connecting the second valvifers (oblong plates) in workers of all *Apis* species (summarized in Maschwitz and Kloft, 1971; Gary, 1974; Michener, 1974).

(b) Chemistry

The first volatile associated with alarm, isoamyl acetate (Boch *et al.*, 1962), was later found in varying amounts in all four species of *Apis* (Morse *et al.*, 1967). Sting apparatus extracts of *A. mellifera* (Blum *et al.*, 1978; Pickett *et al.*, 1982; Collins and Blum, 1982) contain additional compounds including a series of acetates and several alcohols.

The major volatile in sting extracts of *A. dorsata* and *A. florea* is 2-decen-1-yl acetate (Veith *et al.*, 1978) with, tentatively, 1-octyl acetate. *A. cerana*, *A. dorsata*, and *A. florea* also have at least six volatile components which have not been identified. Isoamyl acetate alone will not initiate stinging behavior in *A. mellifera* (Free and Simpson, 1968), but it does recruit workers. None of the species reacts to octyl acetate alone although it is present in all four. Pickett *et al.* (1982) showed that (Z)-ll-eicosen-1-ol plus isopentyl acetate elicits the full alarm response of *A. mellifera*. Collins and Blum (1982) also demonstrated alarm responses to 2-nonanol, *n*-butyl acetate, benzyl acetate and isopentyl acetate. 2-Decen-1-yl acetate can attract *A. dorsata* and *A. florea* by itself, but it is more attractive when mixed with isoamyl acetate (Koeniger *et al.*, 1979). The most long-lasting apid defensive behavior is exhibited by *A. dorsata*, which actually follows a sting victim (Lindauer, 1956), perhaps in part because its 2-decen-1-yl acetate is less volatile and more persistent than isoamyl acetate.

14.6.4 Abdominal integumental glands

(a) Morphology

The abdominal terga and sterna and the intersegmental membranes of all bees are richly endowed with glands, most of which are small (Heselhaus, 1922; Jacobs, 1924; and Altenkirch, 1962). Their morphology and fine structure have been described by the following: intersegmental glands of *Nomia* by Youssef (1968, 1969, 1975); *Xylocopa pubescens* by Gerling *et al.* (1979); tergal Nasanov's gland in *A. mellifera* by Renner (1960) and Belik (1979); wax glands of the honey bee summarized in Snodgrass (1956), and Sanford and Dietz (1976); and the Nasanov's and wax glands surveyed in the Apidae by Cruz Landim (1963, 1967).

Intersegmental glands occur in the membrane connecting abdominal sterna III–VII in female Andrenidae, Colletidae, Halictinae, and Anthophoridae. The glands are located just over the anterior-lateral corners of the sterna, where they are overlapped by the preceding terga. Where well formed, the glands consist of numerous clusters of Leydig cells connected by their ductules to peripheral pouches of a distinct reservoir, which narrows to form a duct joining the intersegmental membrane. The glands do not occur in all intersegmental membranes in all species. A distinct reservoir was found only in *Andrena, Panurginus, Nomia* and *Xylocopa*.

The surfaces of most terga and sterna of all bees may also bear Leydig cells and limited areas of secretory epithelium (palisade epithelium of Altenkirch). In most female bees, the palisade layers are restricted to the parts of the terga surrounding the spiracles, but all Megachilinae are unique in having a well-developed secretory epidermis on the anterior portion of sternum VII. This gland is greatly enlarged to form a pair of huge sacs, extending anteriorly to reach the center of sternum V, in *Anthidium manicatum*. The anterior portions of some or all of sterna III–VI contain a well developed glandular lining in *Xylocopa* and *Ceratina* (Jacobs, 1924; Daly, 1966), which in *Ceratina* is a palisade layer that secretes wax-like scales.

Table 14.6 Volatile compounds identified in extracts of the sting apparatus of *Apis* species

Species	n-Butyl acetate (116)	Isoamyl acetate (130)	Isoamyl alcohol (88)	n-Hexyl acetate (144)	n-Octyl acetate (172)	2-Nonanol (144)	n-Decyl acetate (200)	2-Decenyl acetate (198)	Benzyl acetate (150)	Benzyl alcohol (108)	(Z)-11-Eicosen-1-ol (296)	Hydrocarbons	Unidentified compounds	Reference
Apis														
A. mellifera	—	X	—	—	—	—	—	—	—	—	—	—	—	1
	—	X	—	—	—	—	—	—	—	—	—	—	—	2
	X	X	X	X	X	X	X	—	X	X	—	X	—	3
	—	X	—	—	—	—	—	—	—	—	X	—	—	6
	—	X	X	X	X	X	—	X	X	X	—	—	—	7
A. cerana	—	X	—	—	—	—	—	—	—	—	—	—	—	2
	—	X	—	—	X	—	—	—	—	—	—	—	X	4
A. dorsata	—	X	—	—	—	—	—	—	—	—	—	—	—	2
	—	X	—	—	—	—	—	X	—	—	—	—	X	4
	—	X	—	—	X	—	—	X	—	—	—	—	—	5
A. florea	—	X	—	—	—	—	—	—	—	—	—	—	—	2
	—	X	—	—	—	—	—	X	—	—	—	—	X	4
	—	—	—	—	X	—	—	X	—	—	—	—	—	5

X = present, — = absent
References: (1) Boch *et al.*, 1962; (2) Morse *et al.*, 1967; (3) Blum *et al.*, 1978; (4) Koeniger *et al.*, 1979; (5) Veith *et al.*, 1978; (6) Pickett *et al.*, 1982; (7) Collins and Blum, 1982

In female *Bombus*, worker *Apis*, and queen (but not worker) Meliponinae, the thin anterior margin of tergum VII is the site of the Nasanov's or 'scent' gland. It is composed of 500–600 closely packed Leydig cells. Abdominal terga III–V of queen *A. mellifera* also possess complexes of Leydig cells (Renner and Baumann, 1964).

(b) Chemistry

Nasanov's gland produces citral, geraniol, nerol, (*E,E*)-farnesol, geranic acid and nerolic acid in *A. mellifera* (Shearer and Boch, 1966; Butler and Calam, 1969; Boch and Shearer, 1963; Pickett *et al.*, 1980). A worker honey bee releases the secretion by stretching her abdomen, exposing the gland which is normally hidden by tergum VI, and dispersing the chemicals by wing-fanning. The secretion is an attractant pheromone in species of *Apis*, and is especially used to attract workers to unscented food and to water, as well as in swarming.

The function and chemistry of the secretions of this gland in queen stingless bees and in bumble bees is unknown.

Except for the Nasanov's gland, the chemistry of the secretions of abdominal integumental glands is uninvestigated. The tergal glands of queen honey bees produce a pheromone that supplements the mandibular gland secretion as a close-range attractant. It attracts young workers and maintains the 'court' and, in the mating flight, attracts drones at close range and stimulates copulation (Vierling and Renner, 1977; Renner and Vierling, 1977).

(c) Function

The functions of integumental glands in solitary bees are unproven; these glands are potentially exciting subjects for chemical–behavioral research. Some glands may provide some components of cell linings. Altenkirch (1962) hypothesized that the secretions of large intersegmental glands of *Anthidium* help form chambers in the plant pubescence used to line nest cavities. Gerling *et al.* (1979) hypothesized that the proteins and mucopolysaccharides secreted by the intersegmental glands in *Xylocopa pubescens* are spread on the burrow walls to make a waterproof lining. The wax-like scales secreted by the sternal glands of *Ceratina* may also be used to line burrows (Daly, 1966), an analog or possible homolog of the use of wax by Apidae.

14.6.5 Wax glands

(a) Morphology

The wax glands of Apidae are highly developed integumental glands of the palisade type. They are located on the anterior portions of the abdominal sterna and/or terga, where the cuticle forms thin, shining 'wax mirrors'. Wax glands occur on tergum VI of female *Euglossa cordata*, terga and sterna III–VI of *Bombus* females (and apparently males), on terga III–VI of meliponine queens and workers, and on sterna III–VI of *Apis* workers, but not queens. The glandular layer is lined by fat cells (trophocytes) and oenocytes that apparently synthesize the wax or its precursors, the epithelial glandular cells themselves transporting these chemicals through the cuticle. The columnar glandular cells lack ductules, and transport through the cuticle may be via pore and wax canals. Wax accumulates as scales on the surfaces of the mirrors; these scales are groomed from the abdomen with the hind legs.

(b) Function

The wax is a complex mixture of esters, fatty acids, and hydrocarbons. We will not review its structure. Its function in producing nest cells is well known, and beeswax is used by humans. Apidae thus use a different set of glands to create homeostatic conditions for their developing young than do those bees that line cells with Dufour's gland secretions. Wax is well-adapted for the construction of free-standing cells, which in turn enable the social bees to feed efficiently and otherwise care for large numbers of young, in ways not possible for a bee that individually excavates cells in the soil. Thus the evolution of wax glands may have been a necessary precursor to highly eusocial behavior in bees.

14.7 HOST–PARASITE RELATIONSHIPS

The relationship between cleptoparasitic bees and their hosts is poorly understood. There are approximately 5000 species of cleptoparasitic bees, yet host bees are known for less than 5% of them. In most cases it is not known whether a species utilizes one host or several.

The chemical basis for several cleptoparasitic relationships has been described (Tengö and Bergström, 1975, 1976b, 1977). The host (*Andrena* or *Melitta*) Dufour's secretions are similar to the cephalic secretions of the male (but not the female) cleptoparasite (*Nomada* sp.). For example, in two host–parasite pairs, *Andrena haemorrhoa–Nomada bifida* and *A. carantonica–N. marshamella*, the secretions are dominated by all-*trans*-farnesyl hexanoate. In two other pairs, *A. helvola–N. panzeri* and *A. clarkella–N. leucophthalma*, the Dufour's and cephalic secretions are dominated by geranyl octanoate. Finally, with bees of the genus *Melitta* and their *Nomada* cleptoparasites, the exocrine products are alkyl butanoates. Tengö and Bergström (1977) note that this odor correspondence does not occur in all *Andrena–Nomada* pairs.

Tengö and Bergström (1977) state that *Nomada* males and females produce relatively large amounts of cephalic secretions (1 mg per individual); yet chemically are quite different. We (RMD and JWW), however, find that for several *Nomada* species the male and female mandibular gland secretions are both dominated by a compound with a molecular weight of 170 which is not a farnesyl or geranyl ester. This compound is not present in the Dufour's gland of the *Andrena* hosts.

Alcock (1978) observed *Nomada* males scent-marking grass stems, presumably with a mandibular gland secretion, and hypothesized that this secretion attracts females. He suggested that the females are responding to odor cues of the host nest site that emanate from the host's Dufour's gland secretion, which are mimicked by the *Nomada* male's pheromone. In copulation, male *Nomada* spray the mandibular gland secretion over the female, and Tengö and Bergström (1977) hypothesized that the female is then able to enter a host nest and avoid aggression from the host female. They provide no behavioral data to substantiate this hypothesis.

The orientation mechanism of cleptoparasitic bees to the host nests also remains enigmatic. Cleptoparasitic bees typically 'investigate' host burrows with their antennae before entering, suggesting that chemical cues may be important. This topic remains a prime area for research. A major stumbling block is not obtaining chemicals but rather the design of a quantitative behavioral assay that can be repeated easily in the field without disrupting or destroying the nesting site.

As mentioned in the section on the Andrenidae, it is interesting to note that the Dufour's gland blends of *Andrena* are virtually species-specific. One might argue that this specificity enables an *Andrena* to recognize its own species when it forms nesting aggregations. However, one aspect that may have been selected for chemical specificity is the extent to which *Andrena* are parasitized by *Nomada*. If the Dufour's gland secretions of *Andrena* also act as chemical signals for *Nomada* searching for host nests, this parasitic pressure could generate the chemical diversity seen in *Andrena*, because it would be advantageous for the host to alter its 'chemical profile'.

14.8 CONCLUSIONS

In the field of chemical ecology, the bees provide potential research material virtually unrivaled in other insect groups. This potential results from (i) their diverse exocrine glands, primarily in the head and abdomen; (ii) the wide range in their social, nesting, and mating behaviors; (iii) the large number of species, many of which are relatively common; (iv) the variety of cleptoparasitic species; and (v) the diversity of their foraging strategies.

This chapter may give the impression that the exocrine chemistry of bees is well established. The contrary is closer to the truth. There are numerous subfamilies of bees and numerous glandular systems (e.g., the facial foveal glands, the tarsal glands, the abdominal integumental glands, even the Dufour's glands of most long-tongued bees, especially Meliponinae) for which we have no chemo-behavioral data (Fig. 14.1). Although the chemistries of the exocrine products and the morphology of the glands are much better established than are the behavioral significance of the secretions, there are still significant gaps in the chemical data. These include the following: (i) possible sexual differences in exocrine products; (ii) quantification of exocrine components, which is particularly important to insure that bioassays are conducted with components of the correct relative concentrations; (iii) biosynthesis of these components; (iv) possible age differences; and (v) possible seasonal differences.

It also must be noted that chemical analyses should use rigorous criteria for identifying unknown natural products. It is not sufficient to identify a component on the basis of its mass spectrum alone, even with the extensive mass spectra files now available; gas-chromatographic retention times and comparisons of mass spectra with standards are also necessary. When optical isomers are possible (e.g., linalool in *Colletes*), no attempt has been made to see which enantiomer is active.

Another 'chemical' gap in this research area is that of the supply of compounds for behavioral testing. Not all compounds are commercially available, so chemists usually synthesize milligram quantities for their identifications, often leaving insufficient quantities available for behavioral bioassays. Thus larger scale syntheses of many exocrine compounds are necessary to establish the behavioral correlates of these chemistries.

A greater obstacle to establishing the behavioral and ecological significance of these sociochemicals is the lack of reproduceable, quantitative behavioral bioassays. With the possible exception of the bumble bee work in Sweden (Kullenberg and colleagues), the design of such bioassays has not been successful. Behavioral research deserves a larger percentage of the overall effort if the increasing catalog of exocrine chemistries is to have any significance in our understanding of the chemical ecology of bees.

REFERENCES

Ågren, L. (1975) Fine structure of the thoracic salivary gland of male *Bombus lapidarius* L. (Hymenoptera, Apidae). *Zoon*, **3**, 19–31.

Ågren, L. (1979) A morphological approach to the chemical communication system in Hymenoptera, Aculeata. *Acta Univ. Upsal.*, **501**.

Ågren, L., Cederberg, B. and Svensson, B. G. (1979) Changes with age in ultrastructure and pheromone content of male labial glands in some bumble bee species (Hymenoptera, Apidae). *Zoon*, **7**, 1–14.

Albans, K. R., Aplin, R. T., Brehcist, J., Moore, J. F. and O'Toole, C. (1980) Dufour's gland and its role in secretion of nest cell lining in bees of the genus *Colletes* (Hymenoptera: Colletidae). *J. Chem. Ecol.*, **6**, 549–64.

Alcock, J. (1978) Notes on male mate-locating behavior in some bees and wasps of Arizona (Hymenoptera: Anthophoridae, Pompilidae, Sphecidae, Vespidae). *Pan-Pac. Ent.*, **54**, 215–25.

Altenkirch, G. (1962) Untersuchungen über die Morphologie der abdominalen Hautdrüsen einheimischer Apiden (Insecta, Hymenoptera). *Zool. Beitr.*, **7**, 161–238.

Andersson, C. O., Bergström, G., Kullenberg, B. and Ställberg-Stenhagen, S. (1966) Studies on natural odouriferous compounds I. Identification of macrocyclic lactones as odouriferous components of the scent of the solitary bees *Halictus calceatus* Scop. and *Halictus albipes* F. *Ark. Kemi.*, **26**, 191–8.

Batra, S. W. T. (1968) Behavior of some social and solitary halictine bees within their nests: a comparative study (Hymenoptera: Halictidae). *J. Kansas ent. Soc.*, **41**, 120–33.

Batra, S. W. T. (1980) Ecology, behavior, pheromones, parasites and management of the sympatric vernal bees *Colletes inaequalis, C. thoracicus* and *C. validus. J. Kansas ent. Soc.*, **53**, 509–38.

Batra, S. W. T. and Hefetz, A. (1979) Chemistry of the cephalic and Dufour's gland secretions of *Melissodes* bees. *Ann. ent. Soc. Am.*, **72**, 514–15.

Belik, M. Y. (1979) On the sub-microscopic organization of the cells of the Nasanov gland. *Pchelovodstvo no.*, **1**, 16–8 (in Russian).

Bergström, G. (1974) Studies on natural odouriferous compounds X. Macrocyclic lactones in the Dufour gland secretion of the solitary bees *Colletes cunicularius* L. and *Halictus calceatus* Scop. (Hymenoptera, Apidae). *Chem. Scr.*, **5**, 39–46.

Bergström, G., Kullenberg, B., Ställberg-Stenhagen, S. and Stenhagen, E. (1968) Studies on natural odoriferous compounds. II. Identification of 2,3-dihydrofarnesol as the main component of the marking perfume of male bumble-bees of the species *Bombus terrestris* L. *Arkiv. Kemi*, **28**, 453–69.

Bergström, G., Kullenberg, B. and Ställberg-Stenhagen, S. (1973) Studies on natural odoriferous compounds. VII. Recognition of two forms of *Bombus lucorum* L. (Hymenoptera, Apidae) by analysis of the volatile marking secretions from individual males. *Chem. Scr.*, **3**, 3–9.

Bergström, G. and Svensson, B. G. (1973a) 2,3-Dihydro-6,*trans*-farnesol: main component in the cephalic marker secretion of *Bombus jonellus* K. (Hym., Apidae) males. *Zoon, Suppl.*, **1**, 61–5.

Bergström, G. and Svensson, B. G. (1973b) Studies on natural odoriferous compounds. VIII. Characteristic marking secretions of the forms *lapponicus* and *scandinavicus* of *Bombus lapponicus* Fabr. (Hymenoptera, Apidae). *Chem. Scr.*, **4**, 231–8.

Bergström, G. and Tengö, J. (1973) Geranial and neral as main components in cephalic secretions of four species of *Prosopis* (Hym., Apidae). *Zoon, Suppl.*, **1**, 55–9.

Bergström, G. and Tengö, J. (1974) Studies on natural odoriferous compounds IX. Farnesyl- and geranyl esters as main volatile constituents of the secretion from Dufour's gland in 6 species of *Andrena* (Hymenoptera, Apidae). *Chem. Scr.*, **5**, 28–38.

Bergström, G. and Tengö, J. (1978) Linalool in the mandibular gland secretion of *Colletes* bees (Hymenoptera, Apoidea). *J. Chem. Ecol.*, **4**, 437–49.

Bergström, G. and Tengö, J. (1979) C_{24}-, C_{22}-, C_{20}-, and C_{18}- macrocyclic lactones in Halictidae bees. *Acta Chem. Scand. Ser. B*, **29**, 390.

Blum, M. S. (1966) Chemical releasers of social behavior-VIII. Citral in the mandibular gland secretion of *Lestrimelitta limao* (Hymenoptera: Apoidea: Melittidae). *Ann. ent. Soc. Am.*, **59**, 962–4.

Blum, M. S. (1971) Dimensions of Chemical Sociality. *Chemical Releasers in Insects*, **3**, 147–62.

Blum, M. S. (1976) Pheromonal communication in social and semisocial insects. In: *Proceedings Symposium on Insect Pheromones and Their Application*. pp. 49–60. Nat. Inst. Agric. Sci., Tokyo, Japan.

Blum, M. S. and Hermann, H. R. (1969) The hymenopterous poison gland: probable functions of the main glandular elements. *J. Ga. ent. Soc.*, **4**, 23–8.

Blum, M. S., Crewe, R. M., Kerr, W. E., Keith, L. H., Garrison, A. W. and Walker, M. M. (1970) Citral in stingless bees: isolation and functions in trail-laying and robbing. *J. Insect Physiol.*, **16**, 1637–48.

Blum, M. S. and Bohart, G. E. (1972) Neral and geranial: identification in a colletid bee. *Ann. ent. Soc. Am.*, **65**, 274–5.

Blum, M. S. and Brand, J. M. (1972) Social insect pheromones: their chemistry and function. *Am. Zool.*, **12**, 553–76.

Blum, M. S., Fales, H. M., Tucker, K. W. and Collins, A. M. (1978) Chemistry of the sting apparatus of the worker honeybee. *J. Apic. Res.*, **17**, 218–21.

Boch, R., Shearer, D. A. and Stone, B. C. (1962) Identification of iso-amyl acetate as an active component in the sting pheromone of the honey bee. *Nature*, **195**, 1018–20.

Boch, R. and Shearer, D. A. (1963) Production of geraniol by honey bees of various ages. *J. Insect Physiol.*, **9**, 431–4.

Boch, R., Shearer, D. A. and Shuel, R. W. (1979) Octanoic acid and other volatile acids in the mandibular glands of the honeybee and in royal jelly. *J. Apic. Res.*, **18**, 250–3.

Bordas, M. L. (1895) Appareil glandulaire des Hyménoptères. *Ann. Sci. Nat. (Zool.) Ser. 7*, **19**, 1–362.

Bridges, A. R. (1977) Fine structure of the honey bee (*Apis mellifera* L.) venom gland and reservoir. A system for the secretion and storage of a naturally produced toxin. *Micros. Soc. Can.*, **4**, 50–1.

Butler, C. G. and Calam, D. H. (1969) Pheromones of the honeybee – the secretion of the Nassanoff pheromones of the worker. *J. Insect Physiol.*, **15**, 237–44.

Butler, C. G., Fletcher, D. J. C. and Watler, D. (1969) Nest-entrance marking with pheromones by the honeybee *Apis mellifera* L., and by a wasp, *Vespula vulgaris* L. *Anim. Behav.*, **17**, 142–7.

Calam, D. H. (1969) Species and sex-specific compounds from the heads of male bumblebees (*Bombus* spp). *Nature*, **221**, 856–7.

Cane, J. H. (1981) Dufour's gland secretion in the cell linings of bees (Hymenoptera: Apoidea). *J. Chem. Ecol.*, **7**, 403–10.

Cane, J. H. and Tengö, J. O. (1981) Pheromonal cues direct mate-seeking behavior of male *Colletes cunicularius* (Hymenoptera: Colletidae). *J. Chem. Ecol.*, **7**, 427–36.

Cederberg, B. (1977) Chemical basis for defense in bumble bees. *Proc. VIII Int. Cong. I.U.S.S.I.*, Wageningen, Netherlands, p. 77.

Cole, L. K. and Blum, M. S. (1975) Antifungal properties of the insect alarm pheromones, citral, 2-heptanone and 4-methyl-3-heptanone. *Mycologia*, **67**, 701–8.

Collins, A. M. and Blum, M. S. (1982) Bioassay of compounds derived from the honeybee sting. *J. Chem. Ecol.*, **8**, 463–70.

Costa Leonardo, A. M. (1980) Estudos morfológicos do ciclo secretor das glândulas mandibulares de *Apis mellifera* L. (Hymenoptera, Apidae). *Rev. Bras. Ent.*, **24**, 143–51.

Crewe, R. M. and Fletcher, D. J. C. (1976) Volatile secretions of two Old World stingless bees. *S. Af. J. Sci.*, **72**, 119–20.

Crewe, R. M. and Velthuis, H. H. W. (1980) False queens: A consequence of mandibular gland signals in worker honeybees. *Naturwissenschaften*, **67**, 467–9.

Cruz Landim, C. da (1963) Evolution of the wax and scent glands in the Apinae (Hymenoptera: Apidae). *J. N.Y. ent. Soc.*, **71**, 2–13.

Cruz Landim, C. da (1967) Estudo comparativo de algumas glândulas das abelhas (Hymenoptera, Apoidea) e respectivas implicações evolutivas. *Arq. Zool. S. Paulo*, **15**, 177–290.

Cruz Landim, C. da (1968) Histoquímica e ultraestrutura das glândulas salivares das abelhas (Hymenoptera, Apoidea). *Arq. Zool. S. Paulo*, **17**, 113–66.

Cruz Landim, C. da and Staurengo, M. A. (1965) Glande tarsale des abeilles sans aiguillon. *Proc. V Int. Cong. I.U.S.S.I.*, pp. 219–25.

Cruz Landim, C. da, Stort, A. C., Costa Cruz, M. A. de and Kitajima, E. W. (1965) Órgão tibial dos machos de Euglossini. Estudo ao microscópio óptico e electrônico. *Rev. Brasil. Biol.*, **25**, 323–41.

Cruz Landim, C. da and Kitajima, E. W. (1966) Ultraestrutura do aparelho venenífero de *Apis* (Hymenoptera, Apidae). *Men. Inst. Butantan Simp. Int.*, **33**, 701–10.

Cruz Landim, C. da and Hadek, R. (1969) Ultrastructure of *Apis mellifera* hypopharyngeal gland. *Proc. VI Int. Cong. I.U.S.S.I.*, pp. 121–30.

Cruz Landim, C. da and Camargo, I. J. B de (1970) Light and electron microscope studies of the mandibular gland of *Lestrimelitta limao* (Hym., Meliponidae). *Rev. Brasil. Biol.*, **30**, 5–12.

Daly, H. V. (1966) Biological studies on *Ceratina dallatorreana*, an alien bee in California which reproduces by parthenogenesis (Hymenoptera: Apoidea). *Ann. ent. Soc. Am.*, **59**, 1138–54.

Duffield, R. M., Fernandes, A., McKay, S., Wheeler, J. W. and Snelling, R. R. (1980) Chemistry of the exocrine secretions of *Hylaeus modestus* (Hymenoptera: Colletidae) *Comp. Biochem. Physiol.*, **67B**, 159–62.

Duffield, R. M., Fernandes, A., Lamb, C., Wheeler, J. W. and Eickwort, G. C. (1981)

Macrocyclic lactones and isopentenyl esters in the Dufour's gland secretion of halictine bees (Hymenoptera: Halictidae). *J. Chem. Ecol.*, **7**, 319–31.

Duffield, R. M., LaBerge, W. E., Cane, J. H. and Wheeler, J. W. (1982) Exocrine secretions of bees. IV. Macrocyclic lactones and isopentenyl esters in the Dufour's gland secretions of *Nomia* bees (Hymenoptera: Halictidae). *J. Chem. Ecol.*, **8**, 535–43.

Duffield, R. M., Harrison, S. E., Maglott, D., Ayorinde, F. O. and Wheeler, J. W. (1983) Exocrine secretions of bees V. Terpenoid esters in the Dufour's secretions of *Panurginus* bees (Hymenoptera: Andrenidae). *J. Chem. Ecol.*, **9**, 277–83.

Eickwort, G. C. (1975) Nest-building behavior of the mason bee *Hoplitis anthocopoides* (Hymenoptera: Megachilidae). *Z. Tierpsychol.*, **37**, 237–54.

Eickwort, G. C. and Ginsberg, H. S. (1980) Foraging and mating behavior in Apoidea. *A. Rev. Ent.*, **25**, 421–46.

Fernandes, A., Duffield, R. M., Wheeler, J. W. and LaBerge, W. E. (1981) Chemistry of the Dufour's gland secretions of North American andrenid bees (Hymenoptera: Andrenidae). *J. Chem. Ecol.*, **7**, 453–63.

Francke, W., Reith, W., Bergström, G. and Tengö, J. (1980) Spiroketals in the mandibular glands of *Andrena* bees. *Naturwissenschaften*, **67**, 149–50.

Frankie, G. W. and Vinson, S. B. (1977) Scent marking of passion flowers in Texas by females of *Xylocopa virginica texana* (Hymenoptera: Anthophoridae). *J. Kansas ent. Soc.*, **50**, 613–25.

Frankie, G. W., Vinson, S. B. and Coville, R. E. (1980) Territorial behavior of *Centris adani* and its reproductive function in the Costa Rican dry forest (Hymenoptera: Anthophoridae). *J. Kansas ent. Soc.*, **53**, 837–57.

Gary, N. E. (1974) Pheromones that affect the behavior and physiology of honey bees. In: *Pheromones* (Birch, M. C., ed.) pp. 200–21. North-Holland Publishing Co., Amsterdam.

Gerling, D., Orion, T. and Ovadia, M. (1979) Morphology, histochemistry and ultrastructure of the yellow glands of *Xylocopa pubescens* Spinola (Hymenoptera: Anthophoridae). *Int. J. Insect Morph. Embryol.*, **8**, 123–34.

Graf, V. (1967) Nota sôbre a glândula salivar da cabeça em *Megachile* (Megachilidae – Apoidea). *Dusenia*, **8**, 131–3.

Graf, V. (1970) Nota sôbre a ocorrência das glândulas salivares da cabeça em *Melissoptila* (Anthophoridae, Apoidea). *Bol. Univ. Federal Paraná, Zool.*, **3**, 281–8.

Hefetz, A., Blum, M. S., Eickwort, G. C. and Wheeler, J. W. (1978) Chemistry of the Dufour's gland secretion of halictine bees. *Comp. Biochem. Physiol.*, **61B**, 129–32.

Hefetz, A., Batra, S. W. T. and Blum, M. S. (1979a) Linalool, neral and geranial in the mandibular glands of *Colletes* bees – an aggregation pheromone. *Experientia*, **35**, 319–20.

Hefetz, A., Batra, S. W. T. and Blum, M. S. (1979b) Chemistry of the mandibular gland secretion of the Indian bee *Pithitis smaragdula*. *J. Chem. Ecol.*, **5**, 753–8.

Hefetz, A., Fales, H. M. and Batra, S. W. T. (1979c) Natural polyesters: Dufour's gland macrocyclic lactones form brood cell laminesters in *Colletes* bees. *Science*, **204**, 415–7.

Hemstedt, H. (1969) Zum Feinbau der Koshewnikowschen Drüse bei der Honigbiene *Apis mellifica* (Insecta, Hymenoptera). *Z. Morphol. Ökol. Tiere*, **66**, 51–72.

Heselhaus, F. (1922) Die Hautdrüsen der Apiden und verwandter Formen. *Zool. Jahrb. Anat.*, **43**, 369–464.

Honk, C. G. J. van, Velthuis, H. H. W. and Röseler, P. F. (1978) A sex pheromone from the mandibular glands in bumble-bee queens. *Experientia*, **34**, 838–9.

Honk, C. G. J. van, Velthuis, H. H. W., Röseler, P. F. and Malotaux, M. E. (1980) The mandibular glands of *Bombus terrestris* queens as a source of queen pheromones. *Ent. exp. appl.*, **28**, 191–8.

Jacobs, W. (1924) Das Duftorgan von *Apis mellifica* und ähnliche Hautdrüsenorgane sozialer und solitärer Apiden. *Z. Morph. Ökol. Tiere*, **3**, 1–80.

Johnson, L. K. (1980) Alarm response of foraging *Trigona fulviventris* (Hymenoptera: Apidae) to mandibular gland components of competing bee species. *J. Kansas ent. Soc.*, **53**, 357–62.

Kerr, W. E. and Lello, E. de (1962) Sting glands in stingless bees – a vestigial character (Hymenoptera: Apidae). *J. N. Y. ent. Soc.*, **70**, 190–214.

Kimsey, L. S. (1980) The behaviour of male orchid bees (Apidae, Hymenoptera, Insecta) and the question of leks. *Anim. Behav.*, **28**, 996–1004.

Koeniger, N., Weiss, J. and Maschwitz, U. (1979) Alarm pheromones of the sting in the genus *Apis. J. Insect Physiol.*, **25**, 467–76.

Kullenberg, B. (1956) Field experiments with chemical sexual attractants on aculeate Hymenoptera males. I. *Zool. Bidrag Uppsala*, **31**, 253–354.

Kullenberg, B., Bergström, G. and Ställberg-Stenhagen, S. (1970) Volatile components of the marking secretion of male bumblebees. *Acta Chem. Scand.*, **24**, 1481–3.

Kullenberg, B., Bergström, G., Bringer, B., Carlberg, B. and Cederberg, B. (1973) Observations on scent marking by *Bombus* latr. and *Psithyrus* lep. males (Hym., Apidae) and localization of site of production of the secretion. *Zoon, Suppl.*, **1**, 23–30.

Lello, E. de (1971a) Adnexal glands of the sting apparatus of bees: anatomy and histology, I. (Hymenoptera: Colletidae and Andrenidae). *J. Kansas ent. Soc.*, **44**, 5–13.

Lello, E. de (1971b) Adnexal glands of the sting apparatus of bees: anatomy and histology, II. (Hymenoptera: Halictidae). *J. Kansas ent. Soc.*, **44**, 14–20.

Lello, E. de (1971c) Anatomia e histologia das glândulas do ferrão das abelhas. III. Hymenoptera: Megachilidae e Melittidae. *Ciência e cultura*, **23**, 253–8.

Lello, E. de (1971d) Glândulas anexas ao aparelho de ferrão das abelhas: anatomia e histologia. IV. Hymenoptera, Anthophoridae. *Ciência e cultura*, **23**, 765–72.

Lello, E. de (1976) Adnexal glands of the sting apparatus in bees: anatomy and histology, V. (Hymenoptera: Apidae). *J. Kansas ent. Soc.*, **49**, 85–99.

Lensky, Y. and Slabezki, Y. (1981) The inhibiting effect of the queen bee (*Apis mellifera* L.) foot-print pheromone on the construction of swarming queen cups. *J. Insect Physiol.*, **27**, 313–23.

Lindauer, M. (1956) Über die Verstandigung bei indischen Bienen. *Z. vergl. Physiol.*, **38**, 521–57.

Luby, J. M., Regnier, F. E., Clarke, E. T., Weaver, E. C. and Weaver, N. (1973) Volatile cephalic substances of the stingless bees, *Trigona mexicana* and *Trigona pectoralis. J. Insect Physiol.*, **19**, 1111–27.

McGinley, R. J. (1980) Glossal morphology of the Colletidae and recognition of the Stenotritidae at the family level (Hymenoptera: Apoidea). *J. Kansas ent. Soc.*, **53**, 539–52.

Maschwitz, U. W. J. and Kloft, W. (1971) Morphology and function of the venom apparatus of insects – bees, wasps, ants and caterpillars. In: *Venomous Animals and their Venoms*, Vol. 3 (Bücherl, W. and Buckley, E. E., ed.) pp. 1–60. Academic Press, New York.

Mathewson, J. A. (1965) The internal morphology of the eastern cucurbit bee, *Peponapis pruinosa* (Hymenoptera: Apoidea). *J. Kansas ent. Soc.*, **38**, 209–33.

May, D. G. K. (1973) Factors contributing to recognition of the brood cells of a solitary sweat bee, *Augochlora pura* (Hymenoptera, Halictidae). *J. Kansas ent. Soc.*, **46**, 301–10.

Michener, C. D. (1974) *The Social Behavior of the Bees*. Harvard University Press, Cambridge.

Michener, C. D. (1979) Biogeography of the bees. *Ann. Missouri Bot. Gard.*, **66**, 277–347.

Michener, C. D. (1981) Classification of the bee family Melittidae with a review of species of Meganomiinae. *Contr. Am. Ent. Inst.*, **18**, 1–135.

Michener, C. D. and Greenberg, L. (1980) Ctenoplectridae and the origin of long-tongued bees. *Zool. J. Linn. Soc.*, **69**, 183–203.

Morse, R. A., Shearer, D. A., Boch, R. and Benton, A. W. (1967) Observations on alarm substances in the genus *Apis*. *J. Apic. Res.*, **6**, 113–8.

Nedel, J. O. (1960) Morphologie und Physiologie der Mandibeldrüse einiger Bienen-Arten (Apidae). *Z. Morph. Ökol. Tiere*, **49**, 139–83.

Norden, B., Batra, S. W. T., Fales, H. M., Hefetz, A. and Shaw, G. J. (1980) *Anthophora* bees: unusual glycerides from maternal Dufour's glands serve as larval food and cell lining. *Science*, **207**, 1095–7.

Ortiz-Picón, J. M. and Díaz-Flores, L. (1972) Étude structurale, optique et electronique, des glandes hypophyaryngiennes de *Apis mellifera*. *Trabajos Inst. Cajal Investigaciones Biol.*, **64**, 225–40.

Pain, J. (1973) Pheromones and Hymenoptera. *Bee World*, **54**, 1–14.

Pessotzkaja, K. (1929) Die Rolle des Drüsenapparates in der instinktiven Tätigkeit der Apiden. I. Die Funktion der alkalischen Drüse. (in Russian) *Leningrad Obshch. Estest. Trudy*, **59**, 21–46.

Pickett, J. A., Williams, I. H., Martin, A. P. and Smith, M. C. (1980) Nasanov pheromone of the honey bee, *Apis mellifera* L. (Hymenoptera: Apidae) Part I. Chemical characterization. *J. Chem. Ecol.*, **6**, 425–34.

Pickett, J. A., Williams, I. H. and Martin, A. P. (1982) (Z)-11-eicosen-1-ol, an important new pheromonal component from the sting of the honey bee, *Apis mellifera* L. (Hymenoptera, Apidae). *J. Chem. Ecol.*, **8**, 163–75.

Radović, I. T. (1981) Anatomy and function of the sting apparatus of stingless bees (Hymenoptera: Apidae: Apinae). *Proc. Ent. Soc. Wash.*, **83**, 269–73.

Radović, I. T. and Hurd, Jr., P. D. (1980) Skeletal parts of the sting apparatus of selected species in the family Andrenidae (Apidae: Hymenoptera). *Proc. Ent. Soc. Wash.*, **82**, 562–7.

Raw, A. (1975) Territoriality and scent marking by *Centris* males (Hymenoptera, Anthophoridae) in Jamaica. *Behaviour*, **54**, 311–21.

Reibstein, D., Law, J. H., Bowlus, S. B. and Katzenellenbogen, J. A. (1979) Enzymatic synthesis of juvenile hormone in *Manduca sexta*. In: *The Juvenile Hormones* (Gilbert, L. I., ed.) pp. 131–46. Plenum Press, New York.

Renner, M. (1960) Das Duftorgan der Honigbiene und die physiologische Bedeutung ihres Lockstoffes. *Z. vergl. Physiol.*, **43**, 411–68.

Renner, M. and Baumann, M. (1964) Über Komplexe von subepidermalen Drüsenzellen (Duftdrüsen?) der Bienenkönigin. *Naturwissenschaften*, **51**, 68–9.

Renner, M. and Vierling, G. (1977) Die Rolle des Taschendrüsenpheromons beim Hochzeitsflug der Bienenkönigin. *Behav. Ecol. Sociobiol.*, **2**, 329–38.

Sakagami, S. F. (1965) Über den Bau der männlichen Hinterschiene von *Eulaema nigrita* Lepeletier (Hymenoptera, Apidae). *Zool. Anz.*, **175**, 347–54.

Sanford, M. T. and Dietz, A. (1976) The fine structure of the wax gland of the honey bee (*Apis mellifera* L.). *Apidologie*, **7**, 197–207.

Semichon, M. L. (1906) Recherches morphologiques et biologiques sur quelques melliferes solitaires. *Bull. Sci. Fr. Belg. (Paris)*, **40**, 281–442.

Shearer, D. A. and Boch, R. (1966) Citral in the Nassanoff pheromone of the honeybee. *J. Insect Physiol.*, **12**, 1513–21.

Snodgrass, R. E. (1956) *Anatomy of the Honey Bee*. Comstock Publ. Ass., Ithaca, New York.

Ställberg-Stenhagen, S. (1970) The absolute configuration of terrestrol. *Acta Chem. Scand.*, **24**, 348–60.

Stein, G. (1962) Über den Feinbau der Mandibeldrüse von Hummelmännchen. *Z. Zellforsch. mikrosk. Anat.*, **57**, 719–36.

Stein, G. (1963a) Über den Sexuallockstoff von Hummelmännchen. *Naturwissenschaften*, **50**, 305.

Stein, G. (1963b) Untersuchungen über den Sexuallockstoff der Hummelmännchen. *Biol. Zentralblatt*, **82**, 343–9.

Stiles, E. W. (1976) Comparison of male bumblebee flight paths: temperate and tropical (Hymenoptera: Apoidea). *J. Kansas ent. Soc.*, **49**, 266–74.

Stort, A. C. and Cruz Landim, C. da (1965) Glândulas dos apêndices locomotores do género *Centris* (Hymenoptera, Anthophoridae). *Bol. Inst. Angola* nos. 21/23, 5–14.

Svensson, B. G. and Bergström, G. (1977) Volatile marking secretions from the labial gland of north European *Pyrobombus* D. T. males (Hymenoptera, Apidae). *Insectes Sociaux*, **24**, 213–24.

Tengö, J. (1979) Odour-released behaviour in *Andrena* male bees (Apoidea, Hymenoptera). *Zoon*, **7**, 15–48.

Tengö, J. and Bergström, G. (1975) All-*trans* farnesyl hexanoate and geranyl octanoate in the Dufour gland secretion of *Andrena* (Hymenoptera: Apidae). *J. Chem. Ecol.*, **1**, 253–68.

Tengö, J. and Bergström, G. (1976a) Comparative analyses of lemon-smelling secretions from heads of *Andrena* F. (Hymenoptera, Apoidea) bees. *Comp. Biochem. Physiol.*, **55B**, 179–88.

Tengö, J. and Bergström, G. (1976b) Odor correspondence between *Melitta* females and males of their nest parasite *Nomada flavopicta* K. (Hymenoptera: Apoidea). *J. Chem. Ecol.*, **2**, 57–65.

Tengö, J. and Bergström, G. (1977) Cleptoparasitism and odor mimetism in bees: Do *Nomada* males imitate the odor of *Andrena* females? *Science*, **196**, 1117–9.

Tengö, J. and Bergström, G. (1978) Identical isoprenoid esters in the Dufour's gland secretions of North American and European *Andrena* bees (Hymenoptera: Andrenidae). *J. Kansas ent. Soc.*, **51**, 521–6.

Veith, H. J., Weiss, J. and Koeniger, N. (1978) A new alarm pheromone (2-decen-1-yl acetate) isolated from the stings of *Apis dorsata* and *Apis florea* (Hymenoptera: Apidae). *Experientia*, **34**, 423.

Velthuis, H. H. W. and Camargo, J. M. F. de (1975) Further observations on the function of male territories in the carpenter bee *Xylocopa* (*Neoxylocopa*) *hirsutissima* Maidl (Anthophoridae, Hymenoptera). *Neth. J. Zool.*, **25**, 516–28.

Velthuis, H. H. W. and Gerling, D. (1980) Observations on territoriality and mating behaviour of the carpenter bee *Xylocopa sulcatipes*. *Ent. exp. appl.*, **28**, 82–91.

Vierling, G. and Renner, M. (1977) Die Bedeutung des Sekretes der Tergittaschendrüsen für die Attraktivität der Bienenkönigin gegenüber jungen Arbeiterinnen. *Behav. Ecol. Sociobiol.*, **2**, 185–200.

Vinson, S. B., Frankie, G. W., Blum, M. S. and Wheeler, J. W. (1978) Isolation, identification and function of the Dufour's gland secretion of *Xylocopa virginica texana* (Hymenoptera: Anthophoridae). *J. Chem. Ecol.*, **4**, 315–23.

Vinson, S. B., Williams, H. J., Frankie, G. W., Wheeler, J. W., Blum, M. S. and Coville, R. F. (1982) Mandibular glands of male *Centris adani*, (Hymenoptera: Anthophoridae): Their morphology, chemical constituents, and function in scent marking and territorial behavior. *J. Chem. Ecol.*, **8**, 319–27.

Weaver, N., Weaver, E. C. and Clarke, E. T. (1975) Reactions of five species of stingless bees to some volatile chemicals and to other species of bees. *J. Insect Physiol.*, **21**, 479–94.

Wheeler, J. W., Evans, S. L., Blum, M. S., Velthuis, H. H. W. and Camargo, J. M. F. de (1976) *cis*-2-Methyl-5-hydroxyhexanoic acid lactone in the mandibular gland secretion of a carpenter bee. *Tetrahedron Letters*, **45**, 4029–32.

Wheeler, J. W., Blum, M. S., Daly, H. V., Kislow, C. J. and Brand, J. M. (1977)

Chemistry of mandibular gland secretions of small carpenter bees (*Ceratina* spp.). *Ann. ent. Soc. Am.*, **70**, 635–6.

Wheeler, J. W., Ayorinde, F. O., Greene, A. and Duffield, R. M. (1982) Citronellyl citronellate and citronellyl geranate in the European hornet (*Vespa crabro*) (Hymenoptera: Vespidae). *Tetrahedron Letters*, **23**, 2071–2.

Wilson, E. O. (1971) *The Insect Societies*. Harvard University Press, Cambridge.

Williams, H. J., Vinson, S. B., Frankie, G. W., Coville, R. E. and Ivie, G. W. (1982) Description, chemical contents and function of the tibial gland in *Centris nitida* and *Centris trigonoides subtarsata* (Hymenoptera: Anthophoridae) in the Costa Rica dry forest. *J. Kansas ent. Soc.* (in press).

Youssef, N. N. (1968) Musculature, nervous system and glands of pregenital abdominal segments of the female of *Nomia melanderi Ckll. (Hymenoptera, Apoidea). J. Morph.*, **125**, 205–18.

Youssef, N. N. (1969) Musculature, nervous system and glands of metasomal abdominal segments of the male of *Nomia melanderi* Ckll. (Hymenoptera, Apoidea). *J. Morph.*, **129**, 59–80.

Youssef, N. N. (1975) Fine structure of the intersegmental membrane glands of the sixth abdominal sternum of female *Nomia melanderi* (Hymenoptera, Apoidea). *J. Morph.*, **146**, 307–24.

15

Sociochemicals of Ants

J. W. S. Bradshaw and P. E. Howse

15.1 INTRODUCTION

Our knowledge of the chemical ecology of ants has changed considerably in the last 10 years. Whereas, previously, much emphasis was placed upon the chemical characterization of glandular secretions, recently many studies have concentrated on the adaptive significance of chemically mediated behavior. In this chapter, we attempt to show how chemical signals are used by ants in relationships with their nest-mates, members of other colonies and species of ants, and other organisms. Since chemical signals have characteristics very different from other modes of communication, we have included some discussion of mechanism where relevant.

The ants are classified as a single family, Formicidae, and the most useful divisions for comparative purposes are subfamilies (Fig. 15.1). Members of widely separated subfamilies frequently occupy similar niches in different parts of the world, and we have therefore tended to emphasize functional aspects rather than taxonomic divisions. The use of chemical data in the systematics of ants (and termites) is fraught with difficulties (Howse and Bradshaw, 1980), not least because of the complexity of glandular secretions of ants, and an exploration of the role of these secretions in behavior and ecology must now be a primary aim.

15.2 EXOCRINE GLANDS

Many of the chemicals and secretions we will discuss originate in exocrine glands, that is, glands which discharge outside the body of the ant. A full

Chemical Ecology of Insects. Edited by William J. Bell and Ring T. Cardé
© 1984 Chapman and Hall Ltd.

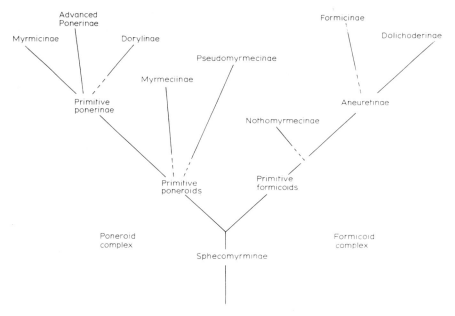

Fig. 15.1   A hypothetical phylogenetic diagram of the Formicidae, including the major subfamilies (after Taylor, 1978).

discussion of the anatomy of these glands is outside the scope of this chapter; recent reviews of gland structure include those of Blum and Hermann (1978a, b), Hölldobler and Engel (1978) Kugler (1978) and Jessen *et al.* (1979). In a great many species, the functions of these glands are not known. The positions of a number of glands are shown in Figs 15.2 and 15.3, which are intended to illustrate the similarities and differences between the species chosen.

The most important exocrine structures in the head are the *mandibular glands*, which are paired, and discharge their secretions on to the mesal side of the mandibles. These secretions are normally associated with alarm and defence (Sections 15.3 and 15.4). Other cephalic glands, such as the maxillary, salivary and post-pharyngeal (Vinson *et al.*, 1980) have digestive functions. In the thorax, the main glands are the *metapleural* or *metathoracic*. These are evidently multi-functional (Maschwitz, 1974; Brown, 1968; Schildknecht and Koob, 1971), but their exact role has only been established for a very few species. The majority of the ecologically important exocrine glands discharge near to the tip of the abdomen, dispensing venoms, territorial markers, trail pheromones, recruitment pheromones and probably many other chemical messages as yet uncategorized. The venom apparatus itself consists of the *poison gland* and *Dufour's gland*, which discharge through the sting in all subfamilies except the Formicinae, where the sting is lost and is replaced by a short duct, the acidopore, leading to a poison funnel ringed with bristles. A third structure, associated with the venom apparatus, is the *sting-sheath gland* (Bazire-Bénazet and Zylberberg, 1979).

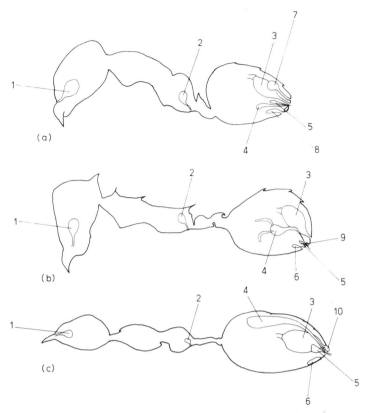

Fig. 15.2   Diagrams of some exocrine glands of worker ants in three subfamilies. (a) Dolichoderinae (*Iridomyrmex*). (b) Myrmicinae (*Atta*). (c) Formicinae (*Oecophylla*). (1) Mandibular gland, (2) metapleural or metathoracic gland, (3) rectal sac, (4) poison gland, (5) Dufour's gland, (6) sternal glands, (7) pygidial or anal gland, (8) Pavan's (sternal) gland, (9) sting sheath gland, (10) rectal gland.

A number of glands are derived from the intersegmental membranes of the gastral tergites. These have recently been classified by Hölldobler and Engel (1978) and Kugler (1978). The glands associated with the membrane between abdominal terga 6 and 7 are now known as *pygidial glands*. Large pygidial glands are found in most Dolichoderinae, where they are also known as anal glands, and in a number of species scattered throughout the other subfamilies. When enlarged, their function is generally to produce a defensive secretion. The *post-pygidial gland*, associated with the membrane between abdominal terga 7 and 8, is occasionally large, but of unknown function.

On the ventral surface of the abdomen, several different types of *sternal gland* have been discovered. Some of these are intersegmental, such as the two in *Leptogenys* (Fig. 15.3); others, including Pavan's gland, the source of the trail pheromone in many Dolichoderine species, are associated with a conspicuous

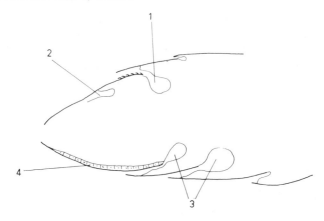

Fig. 15.3   Diagram showing some intersegmental glands in the abdomen of *Leptogenys*. (1) Pygidial gland, (2) postpygidial gland, (3) sternal glands, (4) glandular epithelium (after Hölldobler and Engel, 1978).

palisade epithelium on the 7th sternum. A further type of sternal gland, found in *Oecophylla longinoda* (Hölldobler and Wilson, 1978), consists of single gland cells discharging into cuticular cups on the outer surface of the 7th sternite.

Many of these intersegmental glands are associated with complex cuticular structures, which are presumably used in disseminating their secretion. Some of these structures are vividly illustrated in Hölldobler *et al.* (1976), Hölldobler and Wilson (1978), Hölldobler and Engel (1978) and Jessen *et al.* (1979).

Finally, the *rectum* may also be a source of chemical signals, with the faecal material acting as a 'carrier' (Hölldobler and Wilson, 1977a). In *Oecophylla longinoda*, a separate *rectal gland* is formed from an infolding of the rectal wall. This gland is the source of the trail pheromone, and when in use rests on the bristles of the acidopore.

In this brief survey we have tried to provide a basic description of the positions of the most important glands found in the bodies of ants. The functions of these glands in behaviour and ecology will form the subject of much of the rest of this chapter.

15.3 ALARM AND DEFENSE

The vast majority of formicid species use chemical secretions for defensive purposes. However, within many groups of ants, the proteinaceous venom typical of other Hymenoptera has been superceded by mixtures of structurally simpler organic compounds, and in many cases a communicative function has been added to the defensive or offensive properties. In this section we will concentrate upon the role of alarm communication within colonies, and leave intercolony and interspecies effects aside. However, it must be pointed out that these secretions have presumably become adapted for a number of roles, including the integration of colony defence, aggressive interactions with other colonies and species of ants, both by toxins and more subtle communicative channels, defence against invertebrates and vertebrate predators, and in some cases the capture of living prey. Some features of the secretion may relate to only one of these areas, others may be multifunctional, and very detailed examination is needed to elucidate fully the roles of every compound found in even a single secretion.

15.3.1 Venoms

Ant venoms have been reviewed comprehensively by Blum and Hermann (1978a, b), and will not be discussed in detail here, apart from the most recent work. Venoms from the subfamilies Ponerinae, Myrmicinae, Dorylinae and Myrmeciinae are produced in the poison gland, and disseminated via the sting. They are normally proteinaceous, although several myrmicines produce alkaloids. Several specialized venoms have been found in ponerine species. Both *Harpegnathus saltator* and *Leptogenys chinensis* are able to paralyse their prey, and store it alive in the nest as a food reserve (Maschwitz *et al.*, 1979). Two species of *Pachycondyla* can spray their venom to form long white threads of foam which impede, but do not kill, other ants (Maschwitz *et al.*, 1981).

In species from the subfamily Formicinae, the sting is lost and the poison gland contains formic acid. This, together with the contents of Dufour's gland, is either sprayed at an attacker, or introduced into a wound made by the mandibles (Blum and Hermann, 1978a). The risk of autotoxicity is reduced by minimal accumulation of free acid in the poison gland cells, and a cuticular intima which lines the gland reservoir (Hefetz and Blum, 1978). The hydrocarbons and acetates commonly found in Dufour's gland greatly enhance the penetration of formic acid into the bodies of insects, probably *via* the tracheal system (Lofqvist, 1977). Ketones, such as 2-tridecanone, are also found in Dufour's gland of many formicine ants (Blum and Hermann, 1978a), and are themselves toxic to a range of insect species (Williams *et al.*, 1980).

The function of the poison gland in species of the subfamilies Dolichoderinae and Aneuretinae is not known, nor have any chemicals been characterized from this source. The main defense secretion is produced by hypertrophied pygidial (anal) glands containing cyclopentanoid monoterpenes and other low-molecular-weight organic compounds (Blum and Hermann, 1978b). Many of these compounds are toxic or repellent; the cyclopentyl ketones in the anal glands of *Azteca* ants are repellent to *Solenopsis* species (Wheeler *et al.*, 1975), and interactions of this kind may be partly responsible for the great success of dolichoderine ants in the American tropics.

Enlarged pygidial glands are also found in a few species outside the Dolichoderinae and Aneuretinae (Kugler, 1978). This gland in *Pheidole biconstricta* produces a two-phase secretion; one comprises volatile material and is repellent to other ants, while the other is yellowish and viscous and forms a sticky defense secretion (Kugler, 1979). Release of the whole secretion is apparently rare, but the volatile fraction can be emitted alone by partial opening of the pygidium. In addition to being repellent, the volatile part also acts as an alarm pheromone, and other components of aggressive behavior are released by the poison gland secretion.

15.3.2 Alarm signals

So-called 'alarm pheromones' have been investigated extensively in all the major subfamilies of ants, and it is now certain that the possession of volatile chemical signals for communication in aggressive or defensive situations is an almost universal trait among ants. However, the composition of the secretions, their exocrine sources, and the behavior exhibited by responding nest-mates, vary considerably from one species to another. The latter should be interpreted in terms of the overall ecology of the species. For example, a typical component of the reaction in many species is frenzied behavior, involving an increased rate of locomotion. However, the myrmicine *Zacryptocerus varians*, whose small colonies live in mangrove trees, reacts to disturbances by standing motionless and flattening its body to the substrate (Olubajo *et al.*, 1980). Frenzied alarm would probably result in a proportion of ants accidentally dropping from the trees and drowning in the water below. The relationship between colony size and the form of the alarm behavior has been discussed for species in the subfamily Formicinae (Wilson and Regnier, 1971).

One aspect of alarm communication which is difficult to study is the amount and timing of release of secretions, unless they are readily visible, as is the bright orange mandibular gland secretion of *Calomyrmex spendidus* (Brough, 1978), or readily detected by the human nose, as are the alkyl sulphides produced by *Paltothyreus* species (Casnati *et al.*, 1967). Lack of certainty surrounding the generation of the alarm signal is a problem when the total analysis of alarm communication is attempted (Pasteels, 1975). A further problem facing the investigator is the sheer number of compounds found in many secretions, particularly from Dufour's glands (Blum and Hermann, 1978a). Presumably this diversity relates to the multiple functions of each secretion, listed above, together with biosynthetic considerations; it has even been suggested, without much evidence, that some of the compounds are biosynthetic intermediates. However, at this time there has been no complete ethological, ecological and biochemical analysis of any of the more complex secretions.

(a) Simple alarm pheromones

In cases where the composition of the secretion is comparatively simple, a reasonably complete picture can be obtained. For example, several species of ponerine ants produce dimethyl alkyl pyrazines in their mandibular glands, and utilize these for alarm communication. Detailed study of the behavior of

Odontomachus troglodytes demonstrates that of the four compounds produced by workers, three, 2,6-dimethyl-3-butyl pyrazine, and its 3-pentyl and 3-hexyl homologues all release the same behavior pattern in worker ants. These ants are alerted, approach the source with mandibles opened, where some of them attack. The fourth compound, 2,6-dimethyl-3-ethyl pyrazine, has no detectable communicative role between workers, even though it is not the least abundant of the four compounds in the whole secretion. There is no qualitative difference between the behavior patterns released by the three other compounds in the secretion, and the workers' response to the whole secretion can be explained in terms of any one of these single components (Longhurst *et al.*, 1978).

Similar systems, where one, or a small number of compounds release identical behavior patterns, are apparently common among ants, even when the total composition of the secretion is more complex. In leaf-cutting ants of the genus *Atta* (Myrmicinae) there are frequently large numbers of compounds in the mandibular glands, which are sources of alarm/defense secretions (Blum *et al.*, 1968; Schildknecht, 1976). Oxygenated monoterpenes, such as geranial, neral, geraniol and citronellol, present in large quantities, are primarily defensive and do not affect behavior. The alarm pheromone in every species investigated is 4-methyl-3-heptanone (the (+)-isomer in at least *Atta texana* and *Atta cephalotes*; Riley *et al.*, 1974a). In *Atta texana*, this releases attraction at a concentration of 3×10^9 molecules cm^{-3}, and 'alarm,' fast running with open mandibules, at approximately 10 times this concentration (Moser *et al.*, 1968). This gives the ant which releases the secretion the possibility of modulating the response. A minor disturbance could be signalled by the release of a small amount of the secretion, and nearby workers would be attracted, but not alarmed. As the stimulus escalates, so the release of more secretion could result in the attraction of more nest-mates, and the addition of the alarm response as the vapour concentration of pheromone builds up. This could escalate still further if several workers discharged their mandibular glands in the same area at the same time. Such a system thus contains a degree of flexibility, while still relying on a single chemical releaser.

Further in-built sophistication results directly from the two different behavioral thresholds for attraction and alarm. To describe this it is necessary to know something of the way in which chemical signals are transmitted through the air. At close range, and in comparatively still air, the predominant mode of transmission is by molecular diffusion (Bossert and Wilson, 1963; see Cardé and Elkinton, Chapter 3). The 'active spaces' for attraction and alarm, the volumes within which the concentration of chemical is at or above the behavioral threshold, will both be hemispheres if the chemical is released instantaneously on a plane surface, and their radii R at any one time t will be given by

$$R_{(t)} = \sqrt{\{4Dt \log [2Q/K(4\pi Dt)^{3/2}]\}},$$

where D is the diffusion coefficient of the compound, Q is the number of molecules of the pheromone emitted and K is the threshold concentration.

Several conclusions can be drawn from this equation, as applied to the alarm pheromone of *Atta texana*. The first is that as the amount of chemical emitted increases (i.e., Q increases), so at any one time the radii of the active spaces increase, and so does the total *duration* of the signal, given by the equation, derived from that above

$$t_{\text{fadeout}} = \frac{0.126}{D} \, (Q/K)^{2/3}.$$

Furthermore, the values of K for alarm and attraction differ by a factor of 10. Thus for any value of t and Q, the active space for attraction will be greater than that for alarm (Fig. 15.4), and this will result in the responding ants receiving a sequential message. They are first stimulated by low concentrations of $(+)$-4-methyl-3-heptanone, and are attracted toward the source; as they move in this direction they will encounter higher concentrations of the same compound, and when the ants cross the second threshold, they will be alarmed.

The predictions of this model have to be considered in terms of the assumptions made. In the context of ants fighting off an intruder, it is likely that the chemical signal will be released over a period of time, or several times, or both, whereas the model assumes a single, instantaneous emission. This will result in distorted, overlapping signals, but the principle of different active spaces for the two behavioral reactions will still hold. A more serious problem is presented by local air movements, which will tend to move 'pockets' of air containing high concentrations of the pheromone, away from the source, much faster than if they had been transmitted by molecular diffusion. Under these conditions

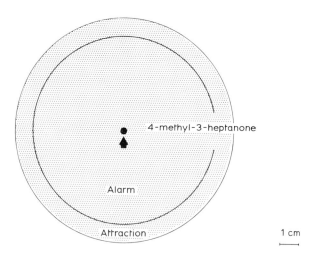

Fig. 15.4   Predicted active spaces of 4-methyl-3-heptanone for workers of *Atta texana*, 20 s after deposition of 10% of mandibular gland contents, at the point arrowed, on a flat surface in still air. The area of attraction is shaded (compare Figs 15.5 and 15.6) (calculated from data in Moser *et al.*, 1968).

attraction to the source does not occur (the gradient of concentration would be far too shallow to perceive), and only the alarm reaction will be observed (Bradshaw, 1981).

(b) Multicomponent alarm pheromones

There are several cases known where, within a complex secretion, two or more compounds occur, which each release different elements of behavior. Given that each of these compounds may have different release rates, behavioral threshold concentrations and diffusion coefficients, it is inevitable that the active spaces of each compound will spread at different rates. In such cases it is essential to analyse the whole system in as much detail as possible, in order to elucidate the complete chemical message and its ecological significance.

The poison glands of *Myrmicaria* species contain not the typical proteinaceous venom of most other myrmicine ants, but large quantities of monoterpene hydrocarbons, such as β-myrcene, β-pinene and limonene (Brand *et al.*, 1974; Longhurst *et al.*, 1983). The secretion is emitted from the tip of the sting in discrete droplets, and in addition to their defensive role, several of the compounds act as releasers of alarm behavior in workers. Since in this case single droplets of secretion can be collected and analysed, the exact amounts of each component can be measured, as can the behavioral threshold concentrations.

In the quantities present in a single droplet from *M. eumenoides*, all three of the monoterpene hydrocarbons mentioned will alert and attract workers, while limonene alone also causes the ants to circle 1–2 cm around the source. The behavior observed in response to a single droplet of secretion is essentially the sum of these responses, a sequence of alerting, attraction and circling. When the sizes of the active spaces are calculated, it is β-pinene which spreads furthest and fastest, and the alerting and attraction properties of the other two compounds are apparently redundant; the only observable function of limonene in the whole secretion is the circling behavior (Fig. 15.5). However, the proportions of these and other monoterpenes in the whole secretion vary between different species and colonies of *Myrmicaria* (Longhurst, unpublished), and may provide species- and colony-specific odours.

In the field, workers of *M. eumenoides* will place a single droplet of secretion on potential prey items. Following this marking, sister workers are attracted toward the prey and help with its immobilization. The circling behavior is observed briefly, but the ants quickly grasp the prey in their mandibles; presumably the prey provides visual cues which override the response to limonene. The net result is that each worker becomes more or less displaced from the original site of attack, apparently giving a more effective deployment of attackers around the prey item.

Larger quantitites of secretion are used in more aggressive interactions, for example with termite soldiers or small lizards. The poison gland, which contains the equivalent of about five droplets, may be almost completely voided in these circumstances, and then responding ants are frequently further alarmed

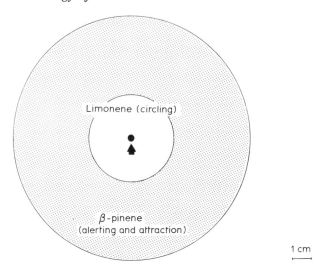

Fig. 15.5   Predicted active spaces of the two behaviorally important components in the poison gland secretion of *Myrmicaria eumenoides*, 40 s after deposition, at the point arrowed, on a flat surface.

and repelled close to the source, by high concentrations of all three of the chemicals. This results in the ants running some distance from the source, and there recruiting any workers they encounter. Thus there is direct evidence of a two-step system available to worker ants, which can be used as the situation dictates.

Examples of such complex communication systems are not confined to the more advanced subfamilies. The mandibular gland secretion of workers of the ponerine *Bothroponera soror*, which is used to attract other workers to prey items, produces a similarly complex message (Longhurst *et al.*, 1980).

(c) Multisource alarm pheromones

In the previous example, communication is provided by the mandibular glands and venom by the poison gland. In some aggressive species, several glands may be used for alarm communication. This apparent redundancy is probably quite common among ants, with, for example, both the mandibular glands and Dufour's gland contributing to 'alarm' communication. Rarely is there any overlap between the compounds found in the two glands, and it seems unlikely that two distinct signalling systems would persist in so many species unless they were used for different purposes, even though the reactions of responding workers may be similar to both.

The most complex alarm communication system to be investigated to date is that of the African weaver ant *Oecophylla longinoda* (Formicinae), where no less than four glands are employed. One, the sternal gland, is used for short-range recruitment of nestmates to disturbances or prey items, and will be discussed in Section 15.5.6 together with the other types of recruitment in this species. The remaining sources are the mandibular glands, and the poison gland complex, comprising the poison and Dufour's glands. The mandibular glands contain over 30 compounds, of which only four have a direct role in alarm communication. The secretion elicits a complex pattern of behavior in major workers. All ants within a range of 5–10 cm are alerted, increasing their rate of locomotion, making short fast runs with frequent changes of direction, with the mandibles held open and the antennae raised. Ants up to 5 cm away from the source are attracted directly toward it where they may circle with mandibles spread, before biting at or near the source. Although this sequence of behavior is similar to that described for *Myrmicaria eumenoides*, each stage is under the control of a different compound in the secretion (Bradshaw *et al.*, 1975, 1979a).

Alerting is released by hexanal, attraction and probably the circling by 1-hexanol, and biting by the less volatile 3-undecanone and 2-butyl-2-octenal. The way in which the active spaces of these compounds interact is shown in Fig. 15.6; since that for hexanal is the largest, some alerted ants will run more

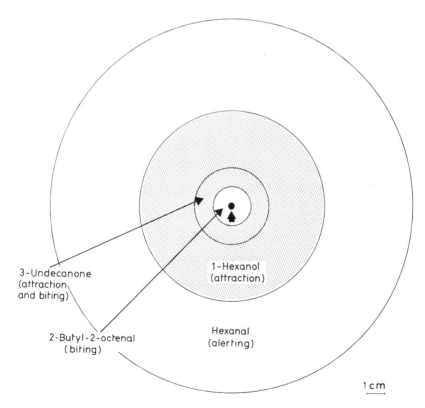

Fig. 15.6 Active spaces of four components of the mandibular gland secretion of major workers of *Oecophylla longinoda*, 20 s after deposition, at the point arrowed, on a flat surface.

or less away from the source of the chemicals, and not approach the source immediately. Workers within about 5 cm are attracted to the source, where they bite, presumably using a mixture of chemical, visual and tactile cues (Fig. 15.7). This is the opposite order of events to that in *Atta texana*, where workers are first attracted and then alarmed (Fig. 15.4). This may reflect a difference in strategies between the phytophageous *Atta* and the highly territorial, predatory *Oecophylla*. The advantage of the arrangement in *Oecophylla* becomes apparent when the enemy, marked with mandibular gland secretion, manages to move a few cm, where it is likely to encounter ants alerted by hexanal, which are stimulated to attack by the less volatile biting markers contaminating its cuticle.

The venom, produced by the poison gland and Dufour's gland discharging simultaneously through the poison funnel, releases similar, but not identical, aggressive behavior in nearby major workers. The main differences lie in the intensity of the reaction, which is higher, and in the type of biting, which is rapidly repeated in response to the venom but prolonged in response to the mandibular gland secretion. The two major components of each gland, formic acid from the poison gland, and *n*-undecane from Dufour's gland, act synergistically to elicit attraction followed by attack near the source. There is no evidence in this case for less volatile compounds playing a crucial role in short-range behavior.

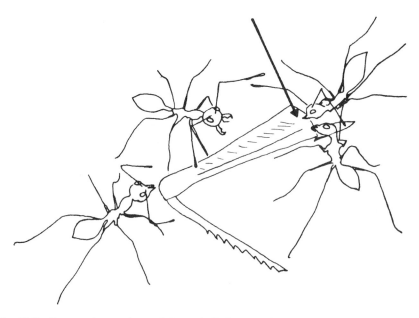

Fig. 15.7   Four major workers of *Oecophylla longinoda*, responding to the excised leg of a locust which has been contaminated with mandibular gland secretion at the cut end (arrowed). The worker on the left is being attracted by 1-hexanol (compare with Fig. 15.6), the one in the centre is about to attack near to the source, where the two workers on the right are already biting (drawn from a photograph).

Release of either or both secretions is usually preceded or accompanied by a characteristic threat display, which presumably adds visual cues to the chemicals. The ant raises its gaster to the vertical, opens its mandibles, and jerks back and forth in an attempt to grasp the enemy with its mandibles (Höll-dobler, 1979a; Bradshaw *et al.*, 1979c). If an ant succeeds in grasping its adversary, it may spread its legs wide apart and pull backwards, or continue to rotate the gaster forward until the poison funnel is above the back of the head, when the venom may be sprayed. Prey is normally immobilized by a number of workers pulling in opposite directions, and when biting, fluid from the mandibular glands may be seen on the upper surfaces of the mandibles. The precise contexts in which two 'alarm' communication systems are used are not readily distinguished, at least in the laboratory, but the use of venom may be restricted by two considerations. First, worker ants contaminated by their own venom are rapidly paralysed, and are therefore at risk when spraying, and second, prey items may be less palatable if contaminated with formic acid. It may therefore be less costly for the workers to use the mandibular gland secretion, plus the sternal gland for recruitment, as described by Hölldobler (1979a), whenever possible, and reserve the use of venom for highly aggressive interactions.

(d) Caste, age and individual variations in chemistry and behavior

Most studies of alarm communication have concentrated on the workers, but, in most nests, males, females and several types of worker are present. Frequently, each of these castes responds differently to alarm pheromones, and may also produce its own characteristic secretion.

Males

Male ants are often more frail than the workers, and hence ready perception of alarm pheromones should be of advantage to them. The alkyl pyrazines which communicate alarm between workers of *Odontomachus troglodytes* (see above) release a quite different behavior in males, which retreat and hide under any suitable object nearby (Longhurst *et al.*, 1978). Alarm pheromones identical to those found in workers are found in males of some species, not in others. In cases where the mandibular glands are used by workers for this purpose, the same or a similar secretion often occurs in males (Crewe, 1973; Brand *et al.*, 1973; Bradshaw *et al.*, 1979b).

Variations between and within the female castes

In the female castes, there are several ways in which production and perception of alarm substances can vary. There may be quantitative and qualitative differences between morphologically distinct castes, such as queens, virgin female reproductives, major workers, minor workers, soldiers, etc., and there may also be differences between age-classes, associated with age-dependent polyethism (division of labor), or even between individuals of the same age and size class. Differences in responsiveness of individual workers are frequently evident when testing alarm substances (Blum, 1977). Some of these differences may be accounted for by the immediate experience and environment of individuals. For example, leaf-carrying workers of *Atta texana* are rather unresponsive to their alarm pheromone (Moser *et al.*, 1968). Workers of *Crematogaster scutellaris* are similarly unresponsive to their pheromone if they are carrying food, but their threshold decreases with their degree of starvation (Leuthold and Schlunegger, 1973). There is also evidence for variations with age. In *Myrmica rubra*, the age classes of workers can be assessed by their degree of pigmentation, and older and younger foragers differ both in their production of mandibular gland alarm pheromones and in their response (Cammaerts-Tricot, 1974a). Social effects on the aggressiveness of individual workers have also been observed (Cole, 1981; Le Roux and Le Roux, 1979).

The various worker castes frequently carry out different roles in defence. In many species there is a specialized soldier caste, whose behavior and exocrine chemistry may differ markedly from those of other workers (e.g., Law *et al.*, 1965). Workers of the African weaver ant *Oecophylla longinoda* are dimorphic; the major workers forage and defend the nests and territory, while the minor workers serve as nurses to the brood, and are rarely seen outside the nests (Weber, 1949). The blend of compounds in the mandibular glands differs markedly between the two castes (Bradshaw *et al.*, 1979b); all the compounds present in minor workers are alcohols, and three of the four compounds most important in alarm communication between majors are absent, namely hexanal, 3-undecanone and 2-butyl-2-octenal. The only abundant compound held in common is 1-hexanol, and the secretion in minor workers is largely made up of the monoterpene nerol. Both nerol and 1-hexanol are highly repellent to minor workers, and they are thus displaced from sources of both minor and major worker secretions. Nerol is a powerful attractant and arrestant for major workers.

Thus with two chemically distinct castes, there are four possible combinations of secretion and response (although only three are evident here). As the number of castes increases, so the situation becomes more complex; there are apparently three chemically distinct castes of the Asian species *O. smaragdina*, leading to nine possible combinations (Howse and Bradshaw, 1980). If these are combined with the developmental and social factors described above, one achieves an extremely elaborate system of both signals and behavior within a colony. The concept of a single alarm pheromone eliciting a stereotyped response in all colony members within range is therefore a gross oversimplification.

15.4 AGGRESSION AND COMPETITION BETWEEN COLONIES AND SPECIES

Encounters between ants of different colonies or species are frequently characterized by aggression, although in the field such encounters are minimized by various types of territorial and avoidance behavior. Interactions between ant colonies can be mediated at a number of levels, including recognition of individual workers, specific interactions involving chemical secretions and behavior patterns, resource partitioning by means of trails, and territorial markers. In areas where ants are dominant such competition has greatly influenced the whole invertebrate fauna.

15.4.1 Colony and species recognition

Worker ants from different colonies of the same species, or different species, are often able to recognize one another rapidly and may attack one another, or one or both may attempt to escape. The basis of this recognition is not fully understood, but is probably largely based upon 'surface pheromones', compounds with very shallow active spaces which surround each ant, or may be perceived on contact. These odours are probably complex, and are therefore difficult to study after extraction. Most studies have therefore focused on the reactions of ants one to another. For example, the responses of workers of *M. rubra* to intruders have been characterized (de Vroey and Pasteels, 1978) and then used to show that workers of this species can distinguish between members of other colonies of their own species, and between these and members of some other sympatric species (de Vroey, 1979). Foreign *Myrmica rubra* are seized for prolonged periods, but rather infrequently, and are very rarely damaged; a group of other species, *Myrmica sabuleti, Tetramorium caespitum, Lasius flavus* and *L. niger*, all induce different patterns of response. For example, only *Lasius flavus* induce recruitment behavior in the *M. rubra*, while *L. niger* elicit no aggressive responses at all. In some species with small colonies, recognition at the individual level may be possible. The workers within a nest of *Leptothorax allardycei* form a dominance hierarchy, and prior to displays of dominance, touch one another with their antennae (Cole, 1981). This suggests that these ants can recognize one another by means of surface pheromones.

Surface pheromones may be transferred from one individual to another by mutual grooming, and thereby become a 'colony odour', common to all members of the colony. The components may either be innate, or acquired from the environment. The relative importance of these sources has been investigated in the leaf-cutting ant *Acromyrmex octospinosus* (Jutsum *et al.*, 1979). Colonies given the same forage material in the laboratory show less aggression than those put on different materials, although there is a latency of about seven days before fighting disappears in the first case, or appears in the second. If colonies are split into two equal halves, and put on to different diets, the degree of investigation of workers from the other half increases, but they are never attacked. These experiments suggest that although odours acquired from the diet are important, they cannot entirely over-ride the intrinsic odours. Three species of leaf-cutting ants, *Acromyrmex octospinosus*, *Atta cephalotes* and

Atta sexdens, all attack intruders from alien colonies or species, but recognition at the species level takes place when the ants are separated by 0.5–1.5 cm, whereas the corresponding range for intraspecies recognition is 0–0.8 cm (Jutsum, 1979). This implies a difference in the chemical signals involved.

Interspecies recognition between *Acromyrmex octospinosus* and *Atta cephalotes* probably depends on compounds originating in the head (Bradshaw, Baker and Howse, unpublished). Excised gasters, thoraces and legs of *A. cephalotes* are treated as rubbish by workers of *A. octospinosus*, but heads are immediately and repeatedly attacked. Workers of *A. cephalotes* will attack all parts of the body of *A. octospinosus*, but the head is attacked most frequently and persistently. The main source of volatile compounds in these heads is the mandibular glands, the contents of which vary from one species to another within the Attini (Blum *et al.*, 1968; Crewe and Blum, 1972). It is possible that, on alien territory, workers release alarm substances from the mandibular glands which alert nearby nest mates, and that these chemicals induce recognition in the other species. The compounds on the cuticle of *A. octospinosus* which elicit attack may be traces of mandibular gland material absorbed into the cuticular wax, or they may originate in a separate exocrine structure.

One such possibility is the metapleural gland, which Brown (1968) has proposed as the source of colony and species odours. This gland is lost or considerably reduced in species which need to enter the nests of other colonies or species, such as male army ants, workers of *Polyergus* slave-makers and queens of social parasites.

Workers of *Pheidole dentata* can evidently recognize one of their major enemies, *Solenopsis invicta*, since a single *Solenopsis* worker will elicit a specific alarm-recruitment behavior. In this case, compounds in both the venom and on the body surface of the *Solenopsis* provide cues for the recognition (Wilson, 1975).

The available evidence suggests that surface pheromones mediating recognition can come from a variety of sources, and it may not be possible to make generalizations about what is a very heterogeneous phenomenon.

15.4.2 Competitive interactions

Apart from the use of venoms, which are normally general-purpose defense secretions, several species of ants have been shown to use specific secretions to compete with particular species. Examples are only likely to arise from field studies, or laboratory studies conducted in semi-natural conditions, and this phenomenon is, therefore, not particularly well documented.

(a) Competition at food sources

A number of small, mass-recruiting species of ants are able to repel larger species by means of glandular secretions. *Monomorium pharaonis*, *Solenopsis fugax* (Hölldobler, 1973) and *M. minimum* (Adams and Traniello, 1981) use their poison glands, and *Iridomyrmex pruinosus* and *Forelius foetidus* their anal glands (Hölldobler *et al.*, 1978). For example, workers of *M. minimum*, if they encounter workers of other species while they are feeding, raise their gasters to the vertical and exude a droplet of fluid from the tip of the sting. Intruders which become contaminated with this secretion withdraw, and groom vigorously (Adams and Traniello, 1981). The European thief ant, *Solenopsis fugax*, preys on the brood of other ants. The workers dig tunnels into the brood chambers and then recruit a large number of nest mates using a Dufour's gland-trail (Hölldobler, 1973). The poison gland contains the alkaloid (*E*)-2-butyl-5-heptylpyrrolidine, which is both highly repellent to workers of other species (e.g., *Lasius flavus*) and also inhibits their pickup of contaminated brood (Blum *et al.*, 1980). By the time the repellent has evaporated, the raid is complete and the *Solenopsis* have retreated with their booty.

(b) Chemical 'propaganda' in slave-making ants

True slave-making, or *dulosis*, occurs in only three ant tribes, the Tetramoriini and Leptothoracini from Myrmicinae, and the Formicini from Formicinae. All prey on closely related species or genera, and the trait is likely to have evolved several times. However, many of the features of slave-raids are held in common between more than one tribe of slave-makers (Buschinger *et al.*, 1980). The basic pattern is as follows: scout ants discover a suitable colony of the slave species, nest-mates are recruited and the slave species is attacked. Once the workers from the target colony have been killed or expelled, the brood, with preference given to the pupae and large larvae, is transported back to the slave-makers' nest, where it is reared, and the resulting workers become part of the slave-maker colony.

In several species, the use of brute force has been substituted by chemical signals. So-called 'propaganda substances' are known from *Formica subintegra* and *F. pergandei*, from *Harpagoxenus* spp. and probably from *Polyergus* spp. (Buschinger *et al.*, 1980). Workers of *Harpagoxenus americanus* only rarely attack the workers of the raided colony. Instead, they are able to induce panic in the target workers, and any of the defenders which do attack are subsequently attacked by their own nest-mates (Alloway, 1979). The propaganda substance is probably disseminated *via* the sting. Additionally, in the brood transport stage, workers of *H. sublaevis* smear the captured brood with their Dufour's gland secretion. This makes the brood unattractive to its own workers (*Leptothorax* spp.), and probably reduces the risk of successful rescue attempts (Buschinger *et al.*, 1980); the propaganda substance and brood repellent may be identical. What is most remarkable is that slave *Leptothorax* workers accompanying *H. americanus* raiders are apparently unaffected by the propaganda substances, unlike their conspecifics in the raided colony (Alloway, 1979).

The chemical nature of the *Harpagoxenus* substances is unknown, but the main components from *Formica pergandei* and *F. subintegra* are decyl, dodecyl and tetradecyl acetates. These substances originate in the hypertrophied

Dufour's gland, and are discharged in such quantities that the workers of the slave species are disoriented and unable to organize a defense of their brood (Regnier and Wilson, 1971).

The metabolic cost to the raiders of releasing such large amounts of chemicals must be high, and may balance the advantage of reduced casualties. The main advantage of chemical warfare may be conservation of resources in the proximity of the slave-raiders' nest: workers from the raided colony are not killed and can return to their nest to rear the younger brood left behind by the slave-makers.

15.4.3 Imprinting

The success of dulotic ants depends upon the ability of workers of the host species to accept the brood of the slave-maker. Jaisson (1975) investigated this phenomenon in the European species, *Formica polyctena*, and found that there was a critical two-week period during which newly emerged workers learn the features of cocoons to which they are exposed. *F. polyctena* workers that had experienced only the cocoons of other species accepted them as their own and treated those of their own species as they would prey. Recognition depends upon a combination of visual and chemical stimuli. It persists after a six-month period, but as in vertebrate imprinting, to which it is clearly analogous, learning cannot take place after the first two weeks.

Jaisson (1980) went on to show that early experience could also control environmental preferences in ants. Young *Camponotus vagus* and *Formica polyctena* workers were kept for three to five weeks in artificial nests containing thyme. They are normally repelled by the odour of thyme and there is some indication that it is toxic to the ants over a long period. In a choice chamber, the ants exposed to thyme then selected a part of the chamber containing thyme. The possibility remains that response to environmental odours and pheromones may depend in some measure on such learning processes.

The reactions of ants to pupae has been developed as a taxonomic tool for ants of the *Formica rufa* group (Rosengren and Cherix, 1981). Presented with pupae from different species, or from geographically separated populations, workers showed varying 'preferences' for their own pupae alongside those of other populations. These preferences were correlated with morphological characters.

15.4.4 Territories

Many ant colonies are associated with more or less well-defined territories, areas which are defended against other colonies of the same species, and frequently against competing species as well. The degree of territoriality is correlated with the number of queens found in a colony. In most polygynous species (those with two or more egg-laying queens), there is little aggression between workers of adjacent colonies; indeed in some cases all the nests in a particular area can be thought of as one giant colony, with free interchange of brood, workers and even queens between different nests. In these cases intra-specific territories will be ill-defined or absent, although interspecific territories are still possible. Vigorous, well-defined territoriality is largely restricted to monogynous colonies (those with only one egg-laying queen) (Hölldobler and Wilson, 1977b; Hölldobler, 1979b). The territorial strategies of various species have been reviewed comprehensively (Wilson, 1971; Hölldobler, 1979b; Höll-dobler and Lumsden, 1980).

Memorized visual cues probably form a considerable part of the basis for territorial behavior, but recently, for several species, chemical cues have been demonstrated which can be termed territorial markers because they are both colony-specific, and are deposited on the substratum by the ants. They may not induce any overt response, but merely modify existing behavior patterns, and hence may be more widespread than the few documented examples suggest.

One recent study has indicated that trails can be colony-specific, and may thereby form part of territorial marking. Recruitment trails of *Lasius neoniger*, which originate in the hindgut, consist of an ephemeral stimulatory component and a more durable orientation cue. It is the latter which is colony-specific, but apart from workers showing a preference for their nestmates' trails, no overt behavior is observed when workers encounter foreign trails; no aggressive or aversive reaction is evident (Traniello, 1980).

Aggressive behavior is released by the territorial pheromone of *Oecophylla longinoda*, however. This pheromone is located in the drops of brown rectal sac fluid which the workers deposit over the territorial surface. They are particu-larly active in marking any fresh areas which they may encounter, but estab-lished territory is also marked constantly, albeit at a greatly reduced rate. Workers which encounter spots of rectal sac fluid produced by members of another colony inspect them frequently with their antennae, and often adopt threat postures in the complete absence of any foreign ants. Moreover, the presence of their own territorial marks enables workers to recruit faster than workers on alien territory (Hölldobler and Wilson, 1977a).

The use of rectal sac material in this way cannot be widespread among ants, as most ants defecate in confined areas away from the nest. Workers of *Myrmica rubra* exploring new territory mark the ground with the tip of the abdomen, but in this case the marker, which is only partly species-specific within the genus *Myrmica* (Cammaerts *et al.*, 1981), originates in Dufour's gland (Cammaerts *et al.*, 1977). Spots of rectal sac material have no effect on territorial behavior of *Atta cephalotes*, and in this species the territorial marker apparently originates in the valves gland which contains both species-specific and colony-specific components. Their effects probably last for only about one hour after deposition (Jaffe *et al.*, 1979). Leaf-cutting ants do not defend permanent territories in the field, but change their foraging sites quite frequently. A relatively emphemeral territorial marker would suit this strategy better than a long-lasting signal. The latter would give inaccurate information much of the time and could lead to increased levels of aggression between adjacent colonies.

15.4.5 The ant mosaic

In many habitats, particularly in the tropics, the most abundant ants are dominants, that is, they exclude other dominants, and normally have semi-permanent territories. It is not yet known whether the interspecific territory boundaries are marked by exocrine chemicals, but this seems entirely possible. Since interspecies fighting is probably no more common than fighting between colonies of the same species, it is likely that mechanisms exist to reduce direct aggression. The importance of the dominant ants in determining the insect fauna within their territories has been stressed (e.g., Leston, 1978) particularly with reference to tropical tree crops, where the presence or absence of various pests is often strongly associated with particular dominants.

15.5 RECRUITMENT BEHAVIOR AND FOOD RETRIEVAL

The term 'recruitment', like the term 'alarm' has become diffuse as the diversity and complexity of social insect behavior has emerged. In this review we will take recruitment to mean the inducement of workers to move from one area to another, either to retrieve food or to move to new territory or a new nest site. The chemical signals involved are usually, but not always, deposited on the sub-stratum, and may therefore also constitute an orientation signal. The two possible functions of 'trail pheromones', recruitment of nest mates and orien-tation to food or nest, need to be carefully separated to estimate the importance of these chemical signals in the ecology of each species.

Recruitment communication has been comprehensively reviewed by Höll-dobler (1978). Oster and Wilson (1978) have classified foraging behavior into five broad types, depending on the degree of co-operation and communication between workers.

15.5.1 Chemical signals – contact and airborne

Some ants use specialized chemical signals (as distinct from alarm pheromones) which are not deposited on the ground as trails. The most primitive form of recruitment is thought to be tandem-running (Hölldobler, 1978); each scout can only recruit one nest-mate, which has to keep in close antennal contact with the scout in order to be guided to the food-source. In most cases that have been investigated, tactile signals from the follower keep the leader motivated, but the follower is attached by both mechanical stimuli, and also a surface pheromone (Maschwitz *et al.*, 1974; Hölldobler *et al.*, 1974). In the ponerine *Pachycondyla obscuricornis*, the tandem-running pheromone originates in the pygidial glands, the contents of which contaminate the whole body surface, including the head (Hölldobler and Traniello, 1980b). Scouts of several species of *Leptothorax*, in a 'tandem-calling' position, release their poison gland secretion from the tip of the sting as an airborne signal which attracts nest-mates prior to tandem-running (Fig. 15.8a) (Möglich *et al.*, 1974; Buschinger and Winter, 1977; Möglich, 1979). The same secretion helps to bind the follower to the leader, and if the pair is broken the leader halts and re-assumes the calling posture until it is re-established (Möglich *et al.*, 1974).

In these cases, orientation of scouts is probably achieved without the use of pheromones. However, recruitment by tandem-running in *Camponotus sericeus* is guided by an orientation trail previously laid by scouts from their hind-guts (Hölldobler *et al.*, 1974).

Windborne chemical signals for recruitment, which are transmitted in a manner comparable with that of sex pheromones (see Section 15.6, and Cardé, Chapter 5), are known in *Novomessor* species. When scouts discover large food sources, they release their poison gland secretions into the air, and other

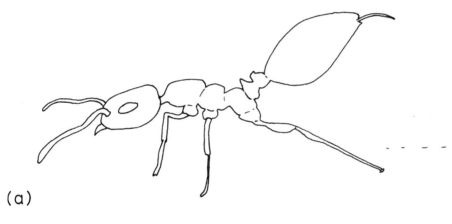

(a)

(b)

Fig. 15.8 Contrasting postures adopted by two myrmicine ants when releasing their recruitment pheromones *via* the sting. (a) *Leptothorax acervorum* in 'calling' position (after Möglich *et al.*, 1974); (b) *Solenopsis saevissima* laying an odour trail (after Wilson, 1971).

foragers are attracted from as far as 100 cm downwind (Hölldobler *et al.*, 1978). The same secretion can also be used to produce a short-lived chemical trail back to the nest, and the whole response can be modulated by the use of vibration signals produced by stridulation (Markl and Hölldobler, 1978).

15.5.2 Chemical trails

Many species use chemical trails for orientation to food sources and new nest sites; some, but not all, of these species also use components of the trail to stimulate recruitment.

(a) Group recruitment

Camponotus socius, like the congeneric *C. sericeus*, utilizes a hind-gut trail for orientation, but instead of recruiting one worker at a time by tandem-running, scouts perform a 'waggle' display inside the nest. This alerts nest-mates, and a group of them will leave the nest, following the scout (Hölldobler, 1971a). The presence of the leader is essential for prolonged trail-following, although all the recruited ants use the trail for orientation. A modification of this method is seen in many raiding species (see Section 15.5.4).

In several species, the presence of a leader is not essential. Workers of *C. pennsylvanicus* lay hind-gut orientation trails, and recruit nest-mates with a waggle display, but the recruited ants follow the trail independently of the scouts (Traniello, 1977). Some workers will even follow the trail without witnessing a waggle display, indicating an increased reliance on chemical signals over the previous example. Other intermediates between group recruitment and mass recruitment have been demonstrated in *Formica fusca* (Moglich and Hölldobler, 1975), *Polyergus lucidus* (Talbot, 1967), and others (see Hölldobler, 1978).

(b) Mass recruitment

In what may be the most advanced form of recruitment in social insects (Wilson, 1971; Hölldobler, 1978), chemical signals control both recruitment and orientation. Little is yet known about the way in which the two messages are separated, if at all. In *Myrmica rubra*, the poison gland trail is orientating and the Dufour's gland trail, laid on top of the poison gland trail by a scout returning to the prey, is activating (Cammaerts-Tricot, 1974b); such a system evidently retains some element of group recruitment (Fig. 15.9). In other cases, the same secretion has long-lasting and short-lived components, of which only the short-lived stimulates recruitment; this is evidently the case in *Atta texana*, where the volatile component, methyl 4-methylpyrrole-2-carboxylate, both recruits and orientates ants, but there are also one or more less volatile components (Tumlinson *et al.*, 1972).

Many species utilizing trunk-trails employ mass recruitment, including species of *Pogonomyrmex* (Hölldobler, 1976), *Pheidole* (Hölldobler and

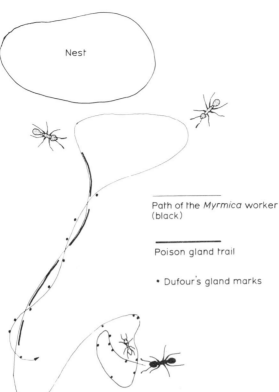

Fig. 15.9 Aggressive recruitment in *Myrmica rubra*. A worker (the black ant) runs around an intruder (*Lasius flavus*, the white ant), depositing the Dufour's gland secretion, then stings, and returns towards the nest laying a poison gland trail. In the nest area she alerts nearby workers (shaded), and returns to the intruder, depositing streaks of attractive Dufour's gland material on top of the poison gland trail (after Cammaerts-Tricot, 1974b).

Möglich, 1980) and *Atta* (Jaffé and Howse, 1979). By contrast, *Solenopsis saevissima* utilizes the same system for its comparatively ephemeral trails (Wilson, 1962). On finding large sources of food, ants return to the nest laying a trail from Dufour's gland (Fig. 15.8b). This is activating as well as orientating. The first and subsequent ants to follow the trail to the bait also lay their own trails back to the nest. Since the number of workers recruited is directly proportional to the concentration of the trail pheromone, the initial build-up at the food is exponential. Thereafter the number reaches a limit, partly because of crowding; only those which have contacted the food will lay trails. The quality and quantity of food is communicated largely by the proportion of workers which lay trails (Wilson, 1962).

A more sophisticated system is available to the leaf-cutting ant *Atta cephalotes*

(Jaffé and Howse, 1979). The first scouts to find leaf material recruit other workers exclusively, and do not join in leaf-cutting for up to an hour. Scouts recruit nest-mates both by the trail pheromone and by tactile displays. As in *Solenopsis*, the degree of recruitment is modulated by the strength of the trail, but in the case of *Atta* this is not dependent on the number of recruiting ants, so that individuals apparently regulate the amount of trail pheromone they deposit, based upon the information they have about the quality and quantity of food available.

15.5.3 Chemistry and lifetime of trail pheromones

Compared to alarm pheromones, very few trail pheromones have been characterized chemically; this reflects the extremely low levels of compound present in the trail secretions of most species investigated. The confirmed structures are shown in Fig. 15.10. In many cases only a single compound has been identified, usually one which stimulates both recruitment and orientation. These compounds are normally stored and applied in highly dilute aqueous solution, and

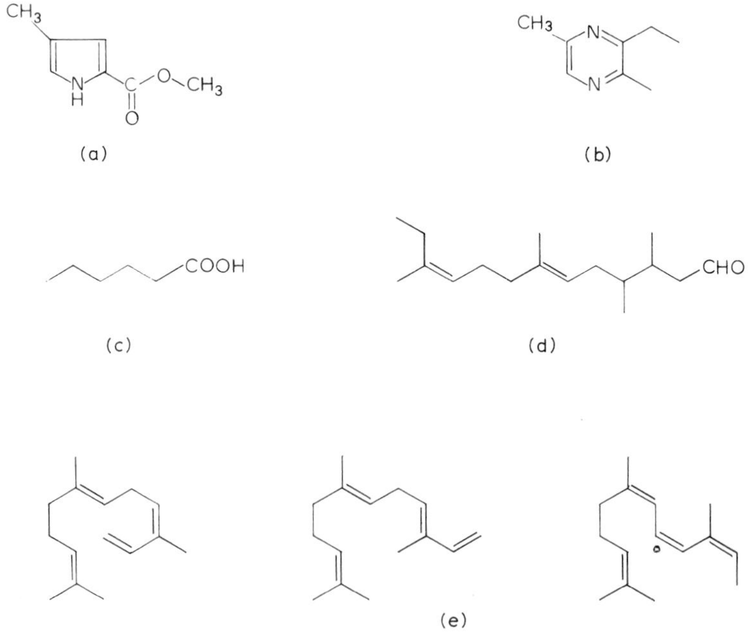

Fig. 15.10   Trail pheromones identified from myrmicine ants, indicating the diversity of structures encountered. (a) *Atta texana* (Tumlinson *et al.*, 1972) and *A. cephalotes* (Riley *et al.*, 1974b); (b) *A. sexdens rubropilosa* (Cross *et al.*, 1979) and *Myrmica* spp. (Evershed *et al.*, 1981); (c) *Lasius fuliginosus* (Huwyler *et al.*, 1975); (d) *Monomorium pharaonis* (Ritter *et al.*, 1977); (e) *Solenopsis invicta* (Williams *et al.*, 1981; Vandermeer *et al.*, 1981).

are usually only part of a complex mixture of organic compounds. The evaporation of trail pheromones is therefore a complex process, depending on humidity, temperature, the blend of compounds in the secretion, and, probably most important, the nature of the substratum on which they are deposited. For example, the persistence of trails of *Lasius fuliginosus* depends upon the volume of secretion discharged, and the porosity of the surface. Old, inactive trails can be re-activated after days of disuse simply by moistening the substratum with water (Hangartner, 1967). Measurements of the active spaces of trail pheromones have therefore been largely empirical, and based upon the whole secretion rather than the individual components.

The durability of the trail pheromone is probably related to the foraging requirements of each species, although, environmental factors are important in determining the volatilization of a deposited signal. The trail laid by a single worker of *Solenopsis saevissima* remains active for about 100 s on a glass surface (Fig. 15.11), and if the food is much more than 50 cm from the nest, a trail-laying worker is unable to return to the food with the nest-mates it has recruited, before the first-laid, outermost part of the trail has evaporated below the threshold (Wilson, 1962). Glass is, of course, an artificial substrate from which the trail pheromone evaporates and dissipates much more quickly than from a natural substrate, but these experiments demonstrate the ephemeral nature of the trail pheromone of this species, a feature necessary to minimize the possibility of inaccurate information from the trail. The lifetime of the trail in terms of a mass response was judged by removing the food after foraging had become established. The time of maximum overshoot, when the greatest number of workers were present in the food area, ranged from 90 to 420 seconds, and increased with the length of the trail (Wilson, 1962). Artificial trails of methyl 4-methylpyrrole-2-carboxylate have a lifetime of about 300 seconds on a wooden surface for *Atta cephalotes*, as do those drawn with poison sac extract, although the latter also contains a much less volatile signal which releases orientation but not recruitment (Jaffe and Howse, 1979).

Orientation trails of other species last considerably longer than minutes, although the recruitment component may decline quite rapidly. Trails of the slave-raider *Polyergus lucidus*, which practices group recruitment, last for approximately two hours (Talbot, 1967); nests which have been raided are no longer available as sources of slaves, and a longer-lived orientation signal would be counter-productive. *Oecophylla smaragdina* which maintains permanent territories, has trails which last for about three days under humid tropical conditions (Jander and Jander, 1979), and species which maintain permanent trunk-trails may have still more durable chemical cues, although memory is undoubtedly important.

15.5.4 Raiding behavior of *Megaponera*, *Pachycondyla* and *Leptogenys*

Megaponera foetens is a large African ponerine ant which feeds almost

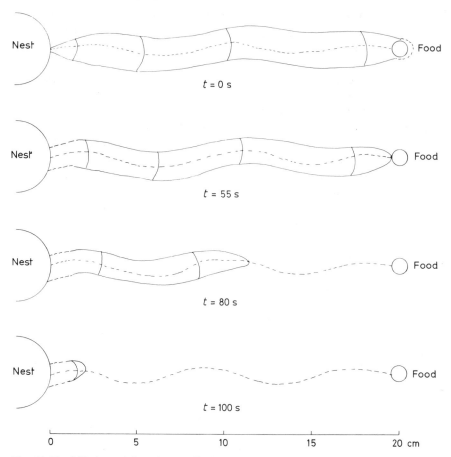

Fig. 15.11   Lifetime of the odour trail of *Solenopsis saevissima*, laid on glass. From the time the worker returns to the nest ($t = 0$), the trail contracts as the pheromone evaporates (from Wilson, 1971).

exclusively on termites of the sub-family Macrotermitinae (fungus-growers). Individual ants, usually major workers forage as 'scouts' (Fletcher, 1973) searching an area up to 95 m from the nest. Their prey in the area around Mokwa, Nigeria, consists mainly of *Macrotermes bellicosus* and *Odontotermes* spp.

These species forage under the cover of soil 'sheeting' constructed from regurgitated soil particles or faeces cemented together with a salivary secretion. A scout ant that contacts fresh moist sheeting searches no longer, but returns to the nest and recruits a trail of sister workers which may consist of 20 to several 100 individuals. Longhurst and Howse (1978) showed that the ants responded to sheeting of *Macrotermes* and *Odontotermes*, less than four hours old. The presence or absence of termites beneath sheeting had no effect on the ants. Ants did respond, however, to extracts of the head or thorax of termites and to extracts of fresh soil sheeting, suggesting that they are using a kairomone of salivary origin, possibly the 'cement pheromone' described by Bruinsma (1979) (see Howse, Chapter 16).

The scout ant returns to the nest and re-emerges after a short interval at the head of a trail of workers. The ants follow the same path as the returning scout and appear to re-inforce the trail with pheromone from their own extruded stings (Longhurst and Howse, 1979b). When they reach the prey site, the workers spread out around the end of the trail. Major workers break open the soil sheeting, but only the minor workers enter the termite galleries and capture termites, which they assemble in piles around the entrances to the galleries. After about 10 minutes, the attack ceases and the major workers gather up the piles of termites in their jaws, return to the end of the recruitment trail, and march back along the trail to the nest.

The trail pheromone is produced by the poison gland (Longhurst *et al.*, 1979a), and in the field lasted for a maximum of three hours. There is also evidence that chemical stimuli attract the ants from the end of the trail to the sites where major soldiers have broken through the termite sheeting. Workers respond immediately to freshly crushed heads or excised mandibular glands of sister workers, biting into the substrate around them and digging with their forelegs. Approach is controlled by dimethyl disulphide and dimethyl trisulphide present in the mandibular glands, and digging by a further unidentified component. Dimethyl disulphide will also attract ants away from foraging trails. Crushed gasters provoke high intensity alarm; followed by approach, and mandibular grasping. Possibly, some components of the abdominal glands influence prey capture and transport. The ants are able to detect termite soldiers at a distance by odour cues (discussed further in Howse, Chapter 16) and they then approach, bite, and sting soldiers of certain species, although they may be repelled by defensive secretions of others.

The trail pheromone is also used by the winged males, which, after a short dispersal flight, search on the ground and follow recruitment trails back to the nest (Longhurst and Howse, 1979a) where they presumably encounter the wingless females of the alien colony. The trail pheromone is also probably used in emigration to new nests, and in marking the entrance to a new nest (Longhurst and Howse, 1979b). The various factors involved in the chemical ecology of *M. foetens* are summarized in Fig. 15.12.

Regional variations apparently exist in the behavior of *Megaponera*. Studies in a different biotope in the Ivory Coast (Levieux, 1966) indicated that there was no fixed scout ant, and that major workers at the head of the column could be removed without disrupting the raid. Ants attacked the first termites they encountered. In Kinshasa (Collart, 1927) South Africa (Fletcher, 1973) and in Mokwa (*loc. cit.*) removal of the scout ant resulted in the workers milling around and then returning to the nest. In South African colonies, the scout ant is the only major worker in the front part of the party, but in Mokwa colonies it

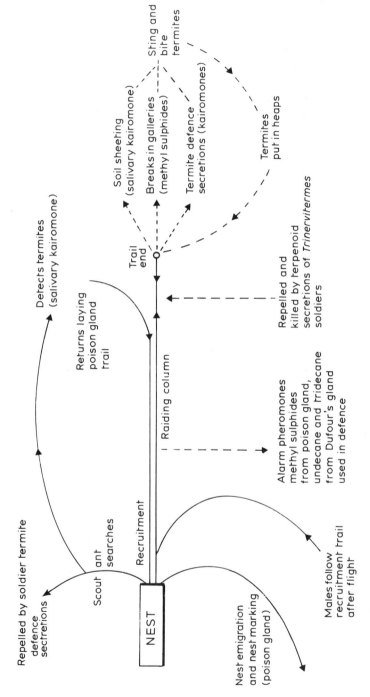

Fig. 15.12 Diagram to show the involvement of chemical stimuli in the raiding behavior of *Megaponera foetens* (see text for references).

is accompanied by 5–12 other major workers. Differences in foraging behavior may be related to prey'density and differences in prey species in different biotopes.

The neotropical species *Pachycondyla* (= *Termitopone*) *laevigata* shows raiding behavior analogous to that of *Megaponera* (Wheeler, 1936; Schneirla, 1971; Hölldobler and Traniello, 1980a). A scout ant or group of scouts stings termites it encounters and returns to the nest laying a trail from the pygidial gland (not the poison gland). The trail elicits rapid recruitment of workers which follow the one or more scouts in a single line. Schneirla (1971) counted a trail of 576 individuals which, after attacking termites in a stump, piled them in caches in three or four locations before transporting them in a single column to the nest. In both *Megaponera* and *Pachycondyla* there is apparently no information transferred on the richness of the food source; in *P. laevigata* recruitment can be triggered by a single termite (Hölldobler and Traniello, 1980a), which usually will indicate the presence of a foraging group or a nest that has been opened.

The genus *Leptogenys* is of great interest for the diversity of hunting and raiding techniques that the ants employ. *Leptogenys attenuata* and *L. nitida* in S. Africa are predators of terrestrial isopods, which, like termites, exist in numbers in particular sites. Both solitary foraging and group raiding with a scout ant occurs (Fletcher, 1971, 1973), but *L. nitida* workers co-operate in transport of the relatively large woodlice along the trail to the nest. Similar behavior occurs in the oriental species, *L. diminuta* (Maschwitz and Mühlenberg, 1975) in which a scout that has found prey returns to the nest laying a trail with her sting. Large items are transported co-operatively to the nest by the raiding column that results. *L. binghami*, in contrast, usually hunts alone for termites and woodlice.

The most complex retrieval behavior is seen in *Leptogenys ocallifera* (Maschwitz and Mühlenberg, 1975). This species sets up long-lasting trunk trails which ants occupy continuously, some workers acting as guards alongside the trail. The preferred prey are earthworms and termites, which are also the most abundant protein sources. If a worker finds a worm, she runs to the trunk trail and recruits perhaps hundreds of workers which sting the worm and dig, helping to free it from the hole and bring it to the surface. It is then cut into pieces and transported co-operatively to the nest. In hunting for termites, a worker penetrates soil sheeting and captures and kills termites, which are then put into small caches on or alongside the trunk trail. Passing workers transport them to the nest which results in recruitment along the trail. The ant that has found the termites, meanwhile, lays a short-lived and very attractive trail from the poison gland between the food source and the trunk trail. This chemical stimulus alone is sufficient for rapid recruitment of ants from the trunk trail, and stimulates further digging activity, but in the absence of further prey is not re-inforced and recruitment rapidly ceases. The poison gland secretion appears not to function as a true alarm pheromone: it will attract workers away from the trail, but only a break in the continuity of the trunk trail provokes a fleeing response.

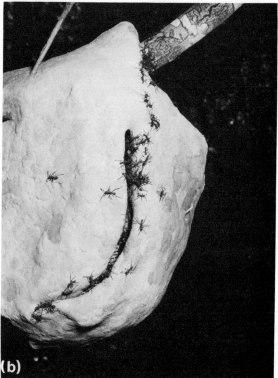

Fig. 15.13   (a) *Megaponera foetens* raiding party returning to the nest with termites (*Odontotermes* spp.) (photo (a) courtesy of Dr. W. A. Sands). (b) *Eciton* at the start of a raid on a nest of *Polybia singularis*. (c) near the end of the raid (photos (b) and (c) courtesy of Dr. W. A. Hamilton).

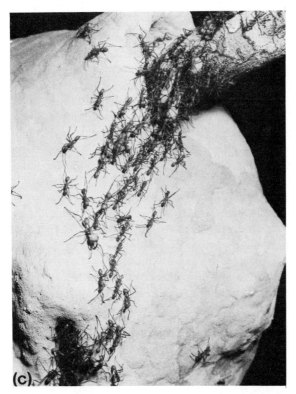

(c)

Nest moving is an important aspect of the biology of the *Leptogenys* species, and probably also of *Megaponera*. The species studied by Maschwitz and Mühlenberg live in natural cavities, e.g., termite mounds or hollow trees and do not construct nests themselves. (Indeed, *L. diminuta* make a bivouac of their own bodies over the brood, as do the doryline ants.) Emigration to a new nest site is frequent and rapid in *L. binghami*, occurring less frequently and at irregular intervals in the other two species. It is also frequent and rapid in *L. nitida*, *L. attenuata* and *L. stuhlmanni* (Fletcher, 1971, 1973). The poison gland trail is used in nest emigration, and this appears to be its sole usage in *L. binghami*.

The nest emigrations are not cyclic, as in doryline ants, but appear to be responses to the availability of food sources and to changing climatic conditions. The evolution of nomadic doryline raiding behavior has been a matter of some dispute, (see below, Fletcher, 1973; Maschwitz and Mühlenberg, 1975). Many of the elements of doryline behavior are present in the ponerines but may be seen as adaptations to deal with food sources that are rich but difficult to locate, and sometimes relatively large compared with the ants themselves. Maschwitz and Mühlenberg (1975) believe that nest emigration is more important than recruitment for coping with a rich but labile biotope, and that the recruitment trail of *Leptogenys* may hence have evolved first for nest moving and later become modified for food retrieval.

In general, group raiding depends upon speed of action, as prey can escape into galleries, burrows, etc., making a raiding party a wasteful exercise. The formation of caches allows rapid attacks to be carried out on insects such as termites. *Megaponera foetens* is able to detect termites without coming into contact with them. Certain other termite predators studied in Nigeria by Long-hurst *et al.* (1979) divided into two groups; those that could be detected by the termites (Macrotermitinae) evoking a rapid withdrawal response and alarm behavior, and those which evoked no response. Species of the tribes Tetra-moriini and Crematogastrini normally have mandibular gland alarm phero-mones containing aldehydes and ketones which are repellent, but specialized termite predators, such as *Decamorium uelense* and *Tetramorium termitobium* contain predominantly alcohols in their mandibular gland secretions, and are apparently undetected by the termites. This is one of the first reported examples of chemical crypsis and has, no doubt, increased the efficiency of the group raiding strategy. Another means of discovering and overcoming groups of social insects is by searching on a wide front and pillage of the whole colony by a superior force. This technique is aided by regular colony emigration into new areas, and is the one used by the doryline ants.

15.5.5 Dorylinae

Army ants (Ecitonini) and driver ants (Dorylini), have a fundamentally differ-ent way of life to the group-raiding ponerines in many respects. The army ants have regular cycles of alternating nomadic and stationary phases (Schneirla, 1971) mainly dependent upon the condition and food demands of the brood. Booty is transported to temporary caches before being carried back to the bivouac, in the statary phase, or carried with the emigration to a new site in the nomadic phase. Schneirla wrote that 'ants in the advance are not scouts in the human sense but temporary pioneers since trail pushing is done by any and all raiders that enter new ground'.

Two types of raiding pattern occur. In swarm raiding species, the ants fan out at some point on the trail and comb the area for prey: the swarm raid of *Eciton burchelli* may be 12–15 m wide with between 100 000 and 500 000 ants in it (Schneirla, 1971). In column raiders, the ants form branching trails on a tree-like pattern. The swarm raiders take a wide-range of prey, while the column raiders take mainly soft-bodied prey. *Neivamyrmex nigrescens* is a column raider that feeds commonly on *Camponotus* ants. When there is no detectable trail pheromone on the ground, workers move slowly and press themselves against objects on the ground, such as twigs and stones. After creeping forwards for a few centimetres, the ant turns and runs back along its path, until it reaches its starting point, when it reverts to moving slowly. These ants also run along artificial hind-gut trails, and by inference, are probably laying their own trails when creeping forward (Topoff and Lawson, 1979). This behavior is repeated numerous times at the head of a foraging column, resulting in net forwards movement, although many of the ants in the vanguard are moving in the opposite direction.

Recruitment to large food sources normally takes place as a branch from these exploratory trails. *N. nigrescens* scouts leave the main group and, on finding food, lay a chemical trail, which differs in quality from the exploratory trail (Topoff *et al.*, 1980). Ants which they encounter are stimulated to follow the branch trail by a motor display, but the trail itself is excitatory, i.e., this is a form of mass recruitment. Ants encountering the branch trail also perform the same display, causing a rapid onset of recruitment similar to that observed in raiding ponerines (see above). In addition to trail pheromones, contact between workers is maintained by tactile interactions and a volatile chemical 'body odour' (Topoff and Mirenda, 1975).

In the southern USA, LaMon and Topoff (1981) found that *Camponotus festivellus* has a striking evasion response. Contact with one worker of *Neivamyrmex nigrescens* is sufficient to provoke evacuation of the nest in which the workers carry the brood to the tops of grass stalks where the army ants do not normally forage. The recognition by the *Camponotus* appears to depend mainly upon tactile stimuli, although Schneirla (1971) reports that *Camponotus* and other ants will give startle reactions to trails of army ants at 1–2 cm distance.

Doryline ants are very thorough and efficient predators, and may completely destroy the social insect colonies they raid. Bodot (1961) reported the demise of 60 *Macrotermes bellicosus* colonies on the Ivory Coast site of 24 hectares due to attacks of subterranean *Dorylus* species. This represents a turnover of about one third of the colonies per year and illustrates the importance of nomadism to enlarge the trophic field. The predation of *Megaponera foetens* in the Mokwa area of Nigeria is much higher, representing the total annual standing crop of *M. bellicosus*, but the cropping technique does not kill the colonies or the reproductive pair.

Nomadism in army ants has been considered an extension of the group raiding pattern of ponerines (Wilson, 1958, 1971). However, we have seen that nest emigration and group raiding already exist together in ponerine ants, and that emigration may be the habit that evolved first. Furthermore, relocation of the bivouac in *Neivamyrmex nigrescens* is determined primarily by the availability of suitable nest sites and is not necessarily related to areas where the most successful raids occur, (Topoff and Mirenda, 1980). The direction and frequency of emigrations can, however, be influenced by availability and location of food.

Among the few predators of army ants, is the blind snake, *Leptotyphlops dulcis*, which attacks *Neivamyrmex nigrescens* and *N. opacithorax* (Watkins *et al.*, 1969). It is able to locate army ants by following their trails. If attacked by the ants, it releases a cloacal sac secretion containing fatty acids in a slippery glucoprotein (Blum *et al.*, 1971) which is highly repellent to the ants and makes it difficult for them to grasp the snake's integument. The secretion also repels other ophiophagous and insectivorous snakes and acts as a sex pheromone for the *Leptotyphlops*. Skatole, produced by cephalic glands of the worker *Neivamyrmex*, also attracts the *Leptotypholops*, but repels other predatory snakes and ants. The interactions are summarized in Fig. 15.14.

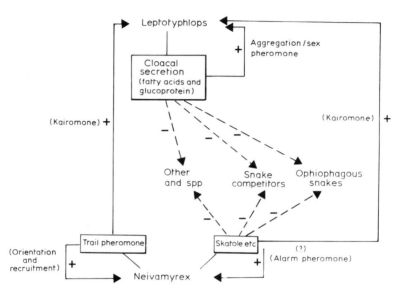

Fig. 15.14   Diagram to show the chemical stimuli involved in reactions between the blind snake *Leptotyphlops dulcis* and its prey, *Neivamyrmex nigrescens*, according to hypotheses of Watkins *et al.* (1969) and Blum *et al.* (1971). Plus signs indicate attraction (+); repellency (−).

15.5.6 Multiple recruitment systems in *Oecophylla longinoda*

In many species, the trail pheromone is used in two distinct contexts, namely food retrieval and nest emigration. Hölldobler and Wilson (1978) have distinguished no less than five types of recruitment in the African weaver ant, *O. longinoda*. The alarm and territorial pheromones of this species have already been discussed, and these, together with the recruitment systems, constitute the most complex use of chemical communication yet discovered in ants.

Recruitment to new food sources is directed by odour trails produced by the rectal gland. These trails are not active in recruitment, but ants which are laying trails frequently contact other workers and open their mandibles as if offering food, although actual regurgitation does not often occur. The trail back to the nest is built up over a period of time, as there is a strong tendency for trail-layers to reverse their direction following a contact. Thus, starting at the food, a multiple trail is deposited, which gradually approaches the nest, but will only reach it if the food-source is sufficiently large to warrant recruitment on a large scale. The trail is further re-inforced by recruited workers, even before they have contacted the food. Thus in this species, flexibility in response is apparently achieved by the length of the trail, rather than by the trail-laying activity of workers responding to the quality of the food, as in *Atta cephalotes*.

The rectal gland secretion is also used in three other contexts, which are recruitment to new terrain, to new nest-sites, and long-range recruitment to intruders. The first two involve similar recruitment displays by trail-layers towards nest-mates which consist of antennation, and a jerking motion which is often 'answered' by the same display in the responding worker. Mandible opening and regurgitation are not observed. Application of fresh territorial markers to the new terrain may help to indicate the context to recruited ants. Long-range recruitment to intruders is characterized by a much more intense jerking display, which elicits both trail-laying and further jerking displays in the recruited ants. In these examples, the same orientation stimulus takes on different meanings when modified by visual and tactile displays. Short-range recruitment to intruders involves a separate chemical signal, the sternal gland secretion. This secretion is emitted by dragging the last abdominal sternite over the substratum, when major workers encounter alien ants of roughly their own size or larger, which would require more than one or two *Oecophylla* workers to hold them down. Small groups of major workers form around places where the sternal gland pheromone has been deposited, even when the intruding ant and the original depositor of the secretion, have moved on. These groups of workers are then available to immobilize intruders much more quickly than can single individuals. This short-range recruitment also accompanies the long-range recruitment to intruders, described above.

15.5.7 Leaf-marking pheromone in *Atta cephalotes*

Although many myrmicines use their Dufour's gland secretion for trail-laying, *A. cephalotes* use this secretion to mark leaves before and during transport back to the nest. While they are cutting leaves, the workers brush the tip of their abdomens across the surface of the part to be removed, and after cutting, hold the leaf in their mandibles and rotate the tip of the gaster forwards to touch its edge. Marked leaves are more attractive than unmarked, and are more quickly transported back to the nest (Bradshaw *et al.*, 1984). The secretion which releases both these types of behavior, and also elicits leaf-marking itself, is that of Dufour's gland, although part of the attractiveness may be due to deliberate or incidental application of poison gland secretion as well. In this species, Dufour's gland contains hydrocarbons, but only in small amounts (Evershed and Morgan, 1980, 1981). Of these, *n*-tridecane is the main attractant, and (*Z*)-9-nonadecene releases pickup of leaves and re-marking (Bradshaw *et al.*, 1983). Moreover, leaves marked with the latter chemical are carried for longer periods than unmarked leaves or those marked with other components of the secretion. Since the signal which releases this extended carrying is usually produced by the same ant which responds, this is an 'autostimulatory' pheromone.

This marking of food, or rather the substrate for the fungus that produces the larval food, has not been found yet outside leaf-cutting ants, and may be unique to that group. However, chain transport, the relaying of food from one worker to another over long distances, is quite common (Wilson, 1971), and may involve similar types of communication.

15.6 SEX PHEROMONES

Compared with aggressive and recruitment behavior, literature on the involvement of chemical signals in sexual behavior and reproduction in ants is scarce. This is probably because sexual activity in ants is normally restricted to a few days in the year, and is not easily induced in the laboratory. Two aspects will be discussed here, the bringing together of the sexes prior to mating, and the control of reproductive activities by queens.

15.6.1 Sex attractants

Rather few sex attractants have been demonstrated in ants, and in no cases have they been characterized chemically. Many species mate on the wing, and are difficult to observe. Recent reviews of various aspects include those by Blum (1981a) and Buschinger (1975). Males of *Lasius neoniger*, *L. alienus* and *Acanthomyops claviger* discharge terpenes and indoles from their mandibular glands during the nuptial flight, but the function of these compounds is unknown (Law *et al.*, 1965). Males of *Camponotus herculeanus* discharge their mandibular gland secretions to stimulate flight in the females, although the secretion does not attract them (Hölldobler and Maschwitz, 1965). Although many of the constituent compounds have been identified (Brand *et al.*, 1973), the one(s) which stimulate swarming are apparently not among these.

Males of *Pogonomyrmex* species also discharge their mandibular glands, but away from the nest at species-specific, perennial mating sites, or 'leks'. The first males arriving at the site discharge the secretion and other alates, both male and female, approach from downwind, apparently attracted by the odour. In four sympatric species in Arizona, nuptial flights often occur on the same days, and separation of species is partially achieved by the time of day at which leks form, and partially by the type of site chosen, whether on the ground or in trees. Once the females have landed, their poison gland secretions attract males for mating, and there is probably also a species-specific surface pheromone which prevents mating between members of different species (Hölldobler, 1977). Once they have mated several times, they indicate their unwillingness to mate again by stridulation (Markl *et al.*, 1977).

In some species, females 'call' males from positions on the ground, in much the same way as in some other Hymenoptera (e.g., *Cephalcia lariciphila*, Borden *et al.*, 1978). Females of *Formica montana* and *F. pergandei* emerge singly from their nests, and walk or fly to nearby vegetation, where they release their sex pheromone. Males fly in zig-zags across wind until they detect the pheromone, when they are stimulated to fly upwind, locate the female, and mate (Kannowski and Johnson, 1969). In female worker reproductives or 'ergatoids' of the ponerine *Rhytidoponera metallica*, the pheromone originates in the tergal gland (Hölldobler and Haskins, 1977) and in the myrmicines *Harpagoxenus sublaevis* and *H. canadensis*, the poison gland is the source (Buschinger and Alloway, 1979). It is not known whether similar chemical attractants are used by females which mate on the wing.

Males of the ponerine *Megaponera foetens* employ a totally different strategy to locate females. The females of this species are wingless ergatoids, and mating normally takes place inside the nests. The male leaves its parent nest, climbs up any vertical surface, and flies, weakly, for up to 100 metres. On landing, it moves in circles, palpating the ground with its antennae, until it locates a trail laid by workers which raided a termite nest earlier that day. The male then follows this trail to the nest (Longhurst and Howse, 1979a).

15.6.2 Queen pheromones

The use of pheromones by ants to control the behavior and reproduction of workers has not been investigated so successfully as the analogous pheromones in honey bees and termites (Wilson, 1971). Most ant queens are highly attractive to their workers; the source of the attractant in *Solenopsis invicta* has been identified as the poison sac (Vandermeer *et al.*, 1980). In the same species, Fletcher and Blum (1981) have demonstrated the presence of a non-volatile pheromone in the queen, which is transferred around the colony by the workers, and inhibits dealation and oögenesis in the virgin females in the same nest.

15.7 INTER-RELATIONSHIPS AND ASSOCIATIONS

In the previous sections, we have described the chemical ecology of ants as it relates to other ants, and to termites. Knowledge of the chemical inter-relationships between ants and other organisms is fragmented, and only a few examples have been studied in detail. These include the symbiosis between leaf-cutting ants (Attini) and their fungus-gardens (see Howse and Bradshaw, 1977), and the relationships between ants and myrmecophiles, insects which live more or less as the guests of ants (Hölldobler, 1971b, 1978). Some idea of the possible complexity of such interactions can be given by the following examples.

Many larvae of lycaenid butterflies produce attractants and food for ants, although only a few species live inside ant nests. The ecological significance of these secretions has been investigated for the larvae of *Glaucopsyche lygdamus* (Pierce and Mead, 1981). The larvae of this species feed on flowers, and are attended by a retinue of ants, species of *Formica*. The larvae have two types of gland, one secreting attractants for the ants, and the other, Newcomer's organ, producing honeydew. The ants defend the larvae from braconid and tachinid parasites (*Apanteles cyaniridis* and *Apolyma theclarum* respectively), and the relationship is therefore mutualistic, although some other lycaenids that are attended by ants do not have honeydew glands. It has been suggested that the females of the ithomiine butterflies *Mechanitis* and *Melinaea* that follow army ant swarms do so by means of the swarm odour. These butterflies feed on the droppings of antbirds, family Formicariidae, which themselves follow the army ants, feeding off insects flushed from the ground by the ant column (Ray and Andrews, 1980).

Several species of ants have close mutualisms with particular plants, called myrmecophytes; the plants provide nest sites, and often food from special organs, in return for defense against herbivores (Wilson, 1971). Recent evidence suggests that production of food may only occur in the presence of the ants (Risch and Rickson, 1981), and it is likely that some aspects of the mutualism are influenced by chemical factors. Many other species of ant rely on honeydew from homopterans as a source of food, and there is evidence here for the involvement of chemical factors, at least between aphids and *Formica* ants (Kleinjan and Mittler, 1975; Nault *et al.*, 1976; Nault and Phelan, Chapter 9).

However, these few examples represent only a small fraction of the complex chemical relationships that must exist between ants and other organisms. Considering the arthropods alone, the presence of the same volatile chemicals in large numbers of distantly related organisms (Blum, 1981b) hints at shared components of a chemical language which we are only beginning to appreciate.

REFERENCES

Adams, E. S. and Traniello, J. F. A. (1981) Chemical interference competition by *Monomorium minimum* (Hymenoptera: Formicidae). *Oecologia*, **51**, 265–70.
Alloway, T. M. (1979) Raiding behaviour of two species of slave-making ants, *Harpagoxenus americanus* (Emery) and *Leptothorax duloticus* (Wesson). *Anim. Behav.*, **27**, 202–10.
Bazire-Bénazet, M. and Zylberberg, L. (1979) An integumentary gland secreting a territorial marking pheromone in *Atta* sp.: detailed structure and histochemistry. *J. Insect Physiol.*, **25**, 751–65.
Blum, M. S. (1977) Ecological and social aspects of the individual behaviour of social insects. *Proc. VIII Int. Cong., IUSSI, Wageningen* pp. 54–9.
Blum, M. S. (1981a) Sex pheromones in social insects: chemotaxonomic potential. In: *Biosystematics of Social Insects* (Howse, P. E. and Clément, J.-L., eds) pp. 163–74. *Systematics Association Special Volume 19*.
Blum, M. S. (1981b) *Chemical Defenses of Arthropods*. Academic Press, New York & London.

Blum, M. S., Padovani, F. and Amante, E. (1968) Alkanones and terpenes in the mandibular glands of *Atta species*. *Comp. Biochem. Physiol.*, **26**, 291–9.

Blum, M. S., Byrd, J. B., Travis, J. R. Watkins II, J. F. and Gehlbach, F. R. (1971) Chemistry of the cloacal sac secretion of the blind snake *Leptotyphlops dulcis*. *Comp. Biochem. Physiol.*, **38B**, 103–7.

Blum, M. S. and Hermann, H. R. (1978a) Venoms and venom apparatuses of the Formicidae: Myrmeciinae, Ponerinae, Dorylinae, Pseudomyrmecinae, Myrmicinae and Formicinae. In: *Arthropod Venoms* (Bettini, S., ed.) Springer, Berlin.

Blum, M. S. and Hermann, H. R. (1978b) Venoms and venom apparatuses of the Formicidae: Dolichoderinae and Aneuretinae. In: *Anthropod Venoms* (Bettini, S., ed.) Springer, Berlin.

Blum, M. S., Jones, T. H., Hölldobler, B., Fales, H. M. and Jaouni, T. (1980) Alkaloidal venom mace: offensive use by a thief ant. *Naturwissenschaften*, **67**, 144–5.

Bodot, P. (1961) La destruction des termitières de *Bellicositermes natalensis* par une fourmi: *Dorylus (Typhlopone) dentifrons* Wasm. *C.R. Acad. Sci. Paris*, **253**, 3053.

Borden, J. H., Billany, D. J., Bradshaw, J. W. S., Edwards, M., Baker, R. and Evans, D. A. (1978) Pheromone response and sexual behaviour of *Cephalcia lariciphila* Wachtl (Hymenoptera: Pamphiliidae). *Ecol. Ent.*, **3**, 13–23.

Bossert, W. H. and Wilson, E. O. (1963) The analysis of olfactory communication among animals. *J. Theoret. Biol.*, **5**, 443–69.

Bradshaw, J. W. S. (1981) The physicochemical transmission of two components of a multiple chemical signal in the African weaver ant, (*Oecophylla longinoda*). *Anim. Behav.*, **29**, 581–5.

Bradshaw, J. W. S., Baker, R. and Howse, P. E. (1975) Multicomponent alarm pheromones of the weaver ant. *Nature*, **258**, 230–1.

Bradshaw, J. W. S., Baker, R. and Howse, P. E. (1979a) Multicomponent alarm pheromones in the mandibular glands of major workers of the African weaver ant, *Oecophylla longinoda*. *Physiol. Ent.*, **4**, 15–25.

Bradshaw, J. W. S., Baker, R., Howse, P. E. and Higgs, M. D. (1979b) Caste and colony variations in the chemical composition of the cephalic secretions of the African weaver ant, *Oecophylla longinoda*. *Physiol. Ent.*, **4**, 27–38.

Bradshaw, J. W. S., Baker, R. and Howse, P. E. (1979c) Chemical composition of the poison apparatus secretions of the African weaver ant, *Oecophylla longinoda*, and their role in behaviour. *Physiol. Ent.*, **4**, 39–46.

Bradshaw, J. W. S., Howse, P. E. and Baker, R. (1984) A novel pheromone regulating chain transport of leaves in *Atta cephalotes*. **Animal Behaviour** (in press).

Brand, J. M., Duffield, R. M., MacConnell, J. G., Blum, M. S. and Fales, H. M. (1973) Caste-specific compounds in male carpenter ants. *Science*, **179**, 388–9.

Brand, J. M., Blum, M. S., Lloyd, H. A. and Fletcher, D. J. C. (1974) Monoterpene hydrocarbons in the poison gland secretion of the ant *Myrmicaria natalensis* (Hymenoptera: Formicidae). *Ann. ent. Soc. Am.*, **67**, 525–6.

Brough, E. J. (1978) The multifunctional role of the mandibular gland secretion of an Australian desert ant. *Z. Tierpsychol.*, **46**, 279–97.

Brown, W. L. (1968) An hypothesis concerning the function of the metapleural gland in ants. *Am. Nat.*, **102**, 188–91.

Bruinsma, O. H. (1979) An analysis of building behaviour of the termite *Macrotermes subhyalinus*. Doctoral thesis, Agricultural University, Wegenigen, Holland.

Buschinger, A. (1975) Sexual pheromones in ants. In: *Pheromones and Defensive Secretions in Social Insects, Symposium of the International Union for the Study of Social Insects* (Noirot, C., Howse, P. E. and LeMasne, G., eds) pp. 225–34. University of Dijon, Dijon.

Buschinger, A. and Alloway, T. M. (1979) Sexual behaviour in the slave-making ant,

Harpagoxenus canadensis M. R. Smith, and sexual pheromone experiments with *H. canadensis, H. americanus* (Emery) and *H. sublaevis* (Nylander) (Hymenoptera; Formicidae). *Z. Tierpsychol.*, **49**, 113–19.

Buschinger, A. and Winter, U. (1977) Rekrutierung von Nestgenossen mittels Tandemlaufen bei sklavenraubzügen der dulotischen Ameise *Harpagoxenus sublaevis* (Nyl.). *Insectes Sociaux*, **24**, 183–90.

Buschinger, A., Ehrhardt, W. and Winter, U. (1980) The organisation of slave raids in dulotic ants – a comparative study (Hymenoptera: Formicidae). *Z. Tierpsychol.*, **53**, 245–64.

Cammaerts, M.-C., Morgan, E. D. and Tyler, R. (1977) Territorial marking in the ant *Myrmica rubra L. (Formicidae). Biol. Behav.*, **2**, 263–72.

Cammaerts, M.-C., Evershed, R. P. and Morgan, E. D. (1981) Comparative study of the Dufour gland secretions of workers in four species of *Myrmica* ants. *J. Insect Physiol.*, **27**, 59–65.

Cammaerts-Tricot, M. C. (1974a) Production and perception of attractive pheromones by differently aged workers of *Myrmica rubra* (Hymenoptera: Formicidae). *Insectes Sociaux*, **21**, 235–47.

Cammaerts-Tricot, M.-C. (1974b) Piste et pheromone attractive chez la fourmi *Myrmica rubra. J. comp. Physiol.*, **88**, 373–82.

Casnati, G., Ricca, A. and Pavan, M. (1967) Sulla secrezione difensiva delle glandole mandibolari *Paltothyreus tarsatus* (Fabr.). *Chimica Indica (Milano)*, **49**, 57–61.

Cole, B. J. (1981) Dominance heirarchies in *Leptothorax* ants. *Science*, **212**, 83–4.

Collart, A. (1927) Notes sur la biologie des fourmis Congolaises. *Rev. Zool. Afric.*, **14**, 249–53.

Crewe, R. M. (1973) An examination of biochemical polymorphism in ants. *Proc. VII Int. Cong., IUSSI*, pp. 77–83. London.

Crewe, R. M. and Blum, M. S. (1972) Alarm pheromones of the Attini: their phylogenetic significance. *J. Insect Physiol.*, **18**, 31–42.

Cross, J. H., Byler, R. C., Ravid, U., Silverstein, R. M., Robinson, S. W., Baker, P. M., Sabino de Oliveira, J., Jutsum, A. R. and Cherrett, J. M. (1979) The major component of the trail pheromone of the leaf-cutting ant, *Atta sexdens rubropilosa* Forel: 3-Ethyl-2, 5-dimethylpyrazine. *J. Chem. Ecol.*, **5**, 187–203.

Evershed, R. P. and Morgan, E. D. (1980) A chemical study of the Dufour glands of two attine ants. *Insect Biochem.*, **10**, 81–6.

Evershed, R. P. and Morgan, E. D. (1981) Chemical investigations of the Dufour gland contents of attine ants. *Insect Biochem.*, **11**, 343–51.

Evershed, R. P., Morgan, E. D. and Cammaerts, M. C. (1981) Identification of the trail pheromone of the ant *Myrmica rubra* L., and related species. *Naturwiss.*, **68**, 374–5.

Fletcher, D. J. C. (1971) The glandular source and social function of the trail pheromones of two species of ants (*Leptogenys*). *J. Ent. (A)*, **46**, 27–37.

Fletcher, D. J. C. (1973) Army ant behaviour in the Ponerinae: a reassessment. *Proc. VII Int. Cong., IUSSI*, pp. 116–20. London.

Fletcher, D. J. C. and Blum, M. S. (1981) Pheromonal control of dealation and oögenesis in virgin queen fire ants. *Science*, **212**, 73–5.

Hangartner, W. (1967) Spezifität und Inaktivierung des Spurpheromons von *Lasius fuliginosus* Latr. und Orientierung der Arbeiterinnen im Duftfeld. *Z. vergl. Physiol.*, **57**, 103–36.

Hefetz, A. and Blum, M. S. (1978) Biosynthesis and accumulation of formic acid in the poison gland of the carpenter ant *Camponotus pennsylvanicus. Science*, **201**, 454–5.

Hölldobler, B. (1971a) Recruitment behaviour in *Camponotus socius* (Hym. Formicidae). *Z. vergl. Physiol.*, **75**, 123–42.

Hölldobler, B. (1971b) Communication between ants and their guests. *Sci. Am.*, **224**, (3) 86–93.

Hölldobler, B. (1973) Chemische Strategie beim Nahrungserwerb der Diebsameise (*Solenopsis fugax* Latr.) und der Pharaoameise (*Monomorium pharaonis* L.). *Oecologia*, **11**, 371−80.

Hölldobler, B. (1976) Recruitment behavior, home range orientation and territoriality in harvester ants, *Pogonomyrmex*. *Behav. Ecol. Sociobiol.*, **1**, 3−44.

Hölldobler, B. (1977) The behavioural ecology of mating in harvester ants (Hymenoptera: Formicidae: *Pogonomyrmex*). *Behav. Ecol. Sociobiol.*, **1**, 405−23.

Hölldobler, B. (1978) Ethological aspects of chemical communication in ants. *Adv. Behav.*, **8**, 75−115.

Hölldobler, B. (1979a) Territories of the African weaver ant (*Oecophylla longinoda* (Latreille). *Z. Tierpsychol.*, **51**, 201−3.

Hölldobler, B. (1979b) Territoriality in ants. *Proc. Am. Philosoph. Soc.*, **123**, 211−18.

Hölldobler, B. and Maschwitz, U. (1965) Der Hochzeitsschwarm der Rossameise *Componotus herculeanus* L. (Hym. Formicidae). *Z. vergl. Physiol.*, **50**, 551−68.

Hölldobler, B., Möglich, M. and Maschwitz, U. (1974) Communication by tandem-running in the ant *Camponotus sericeus*. *J. comp. Physiol.*, **90**, 105−27.

Hölldobler, B., Stanton, R. and Engel, H. (1976) A new exocrine gland in *Novomessor* (Hymenoptera: Formicidae) and its possible significance as a taxonomic character. *Psyche*, **83**, 32−41.

Hölldobler, B. and Haskins, C. P. (1977) Sexual calling behavior in primitive ants. *Science*, **195**, 793−4.

Hölldobler, B. and Wilson, E. O. (1977a) Colony-specific territorial pheromone in the African weaver ant *Oecophylla longinoda* (Latreille). *Proc. Nat. Acad. Sci., USA*, **74**, 2072−5.

Hölldobler, B. and Wilson, E. O. (1977b) The number of queens: an important trait in ant evolution. *Naturwissenschaften*, **64**, 8−15.

Hölldobler, B. and Engel, H. (1978) Tergal and sternal glands in ants. *Psyche*, **85**, 285−330.

Hölldobler, B., Stanton, R. C. and Markl, H. (1978) Recruitment and food-retrieving behaviour in *Novomessor* (Formicidae, Hymenoptera). I. Chemical signals. *Behav. Ecol. Sociobiol.*, **4**, 163−81.

Hölldobler, B. and Wilson, E. O. (1978) The multiple recruitment systems of the African weaver ant *Oecophylla longinoda* (Latreille) (Hymenoptera: Formicidae). *Behav. Ecol. Sociobiol.*, **3**, 19−60.

Hölldobler, B. and Lumsden, C. J. (1980) Territorial strategies in ants. *Science*, **210**, 732−9.

Hölldobler, B. and Möglich, M. (1980) The foraging system of *Pheidole militicida* (Hymenoptera: Formicidae). *Insectes Sociaux*, **27**, 237−64.

Hölldobler, B. and Traniello, J. F. A. (1980a) The pygidial gland and chemical recruitment communication in *Pachycondyla* (= *Termitopone*) *laevigata*. *J. Chem. Ecol.*, **6**, 883−93.

Hölldobler, B. and Traniello, J. (1980b) Tandem running pheromone in ponerine ants. *Naturwissenschaften*, **67**, 360.

Howse, P. E. and Bradshaw, J. W. S. (1977) Some aspects of the chemistry and biology of leaf-cutting ants. *Outlook on Agriculture*, **9**, 160−6.

Howse, P. E. and Bradshaw, J. W. S. (1980) Chemical systematics of social insects with particular reference to ants and termites. In: *Systematics Association Special Vol. 16, Chemosystematics: Principles and Practice* (Bisby, F. A., Vaughan, J. G. and Wright, C. A., eds) pp. 71−90. Academic Press, London and New York.

Huwyler, S., Grob, K. and Visconti, M. (1975) The trail pheromone of the ant, *Lasius fuliginosus*: identification of six components. *J. Insect Physiol.*, **21**, 299−304.

Jaffé, K., Bazire-Benazét, M. and Howse, P. E. (1979) An integumentary pheromone-secreting gland in *Atta* sp.: territorial marking with a colony-specific pheromone in *Atta cephalotes*. *J. Insect Physiol.*, **25**, 833–9.

Jaffé, K. and Howse, P. E. (1979) The mass recruitment system of the leaf-cutting ant *Atta cephalotes*. *Anim. Behav.*, **27**, 930–9.

Jaisson, P. (1975) L'impregnation dans l'ontogenèse du comportement de soins aux cocons chez la jeune fourmi rousse (*Formica polyctena* Först). *Behaviour*, **52**, 1–37.

Jaisson, P. (1980) Environmental preference induced experimentally in ants (Hymenoptera: Formicidae). *Nature*, **286**, 388–9.

Jander, R. and Jander, U. (1979) An exact field test for the fade-out time of the odour trails of the Asian weaver ants *Oecophylla smaragdina*. *Insectes Sociaux*, **26**, 165–9.

Jessen, K., Maschwitz, U. and Hahn, M. (1979) Neue Abdominaldrüsen bei Ameisen. I. Ponerini (Formicidae: Ponerinae). *Zoomorphologie*, **94**, 49–66.

Jutsum, A. R. (1979) Interspecific aggression in leaf-cutting ants. *Anim. Behav.*, **27**, 833–8.

Jutsum, A. R., Saunders, T. S. and Cherrett, J. M. (1979) Intraspecific aggression in the leaf-cutting ant *Acromyrmex octospinosus*. *Anim. Behav.*, **27**, 839–44.

Kannowski, P. B. and Johnson, R. L. (1969) Male patrolling behaviour and sex attraction in ants of the genus *Formica*. *Anim. Behav.*, **17**, 425–9.

Kleinjan, J. E. and Mittler, T. E. (1975) A chemical influence of ants on wing development in aphids. *Ent. exp. appl.*, **18**, 384–8.

Kugler, C. (1978) Pygidial glands in the Myrmicine ants (Hymenoptera: Formicidae). *Insectes Sociaux*, **25**, 267–74.

Kugler, C. (1979) Alarm and defense: a function for the pygidial gland of the myrmicine ant *Pheidole biconstricta*. *Ann. ent. Soc. Am.*, **72**, 532–6.

LaMon, B. and Topoff, H. (1981) Avoiding predation by army ants: defensive behaviour of three ant species of the genus *Camponotus*. *Anim. Behav.*, **29**, 1070–81.

Law, J. H., Wilson, E. O. and McCloskey, J. A. (1965) Biochemical Polymorphism in ants. *Science*, **149**, 544–6.

Le Roux, A. M. and le Roux, G. (1979) Activité et agressivité chez les ouvrières de *Myrmica laevinodis* Nyl. (Hymenoptere, Formicides). Modification en fonction du groupement et de l'expérience individuelle. *Insectes Sociaux*, **26**, 354–63.

Leston, D. (1978) A neotropical ant mosaic. *Ann. ent. Soc. Am.*, **71**, 649–53.

Leuthold, R. H. and Schlunegger, U. (1973) The alarm behaviour from the mandibular gland secretion in the ant *Crematogaster scutellaris*. *Insectes Sociaux*, **20**, 205–14.

Levieux, J. (1966) Note préliminaire sur les colonnes de chasse de *Megaponera foetens* (Fab.) (Hymenoptera: Formicidae). *Insectes Sociaux*, **13**, 117–26.

Lofqvist, J. (1977) Toxic properties of the chemical defence systems in the competitive ants *Formica rufa* and *F. sanguinea*. *Oikos*, **28**, 137–51.

Longhurst, C., Baker, R., Howse, P. E. and Speed, W. S. (1978) Alkyl pyrazines in ponerine ants: their presence in three genera, and caste-specific behavioural responses to them in *Odontomachus troglodytes*. *J. Insect Physiol.*, **24**, 833–7.

Longhurst, C. and Howse, P. E. (1978) The use of kairomones by *Megaponera foetens* (Fab.) (Hymenoptera: Formicidae) in the detection of its termite prey. *Anim. Behav.*, **26**, 1213–18.

Longhurst, C., Baker, R. and Howse, P. E. (1979a) Termite predation by *Megaponera foetens* (Fab.) (Hymenoptera: Formicidae): co-ordination of raids by glandular secretions. *J. Chem. Ecol.*, **5**, 703–21.

Longhurst, C., Baker, R. and Howse, P. E. (1979b) Chemical crypsis in predatory ants. *Experientia*, **35**, 870–1.

Longhurst, C. and Howse, P. E. (1979a) Some aspects of the biology of the males of *Megaponera foetens* (Fab.) (Hymenoptera: Formicidae). *Insectes Sociaux*, **26**, 85–91.

Longhurst, C. and Howse, P. E. (1979b) Foraging, recruitment and emigration in

Megaponera foetens (Fab.) (Hymenoptera: Formicidae) from the Nigerian guinea savanna. *Insectes Sociaux*, **26**, 204–15.

Longhurst, C., Baker, R. and Howse, P. E. (1980) A multicomponent mandibular gland secretion in the ponerine ant *Bothroponera soror*. *J. Insect Physiol.*, **26**, 551–5.

Longhurst, C., Bolwell, S., Bradshaw, J. W. S., Howse, P. E. and Evans, D. A. (1983). Multicomponent alarm pheromones from the poison gland secretion of *Myrmicaria eumenoides* and *M. striata* (Hymenoptera: Formicidae) (in preparation).

Markl, H., Hölldobler, B. and Hölldobler, T. (1977) Mating behaviour and sound production in harvester ants (*Pogonomyrmex*, Formicidae). *Insectes Sociaux*, **24**, 191–212.

Markl, H. and Hölldobler, B. (1978) Recruitment and food-retrieving behaviour in *Novomessor* (Formicidae: Hymenoptera). II. Vibration signals. *Behav. Ecol. Sociobiol.*, **4**, 183–216.

Maschwitz, U. (1974) Vergleichende Untersuchungen zur Funktion der Ameisenmetathorakaldrüse. *Oecologia*, **16**, 303–10.

Maschwitz, U., Hölldobler, B. and Möglich, M. (1974) Tandemlaufen als Rekrutierungsverhalten bei *Bothroponera tesserinoda* Forel (Formicidae: Ponerinae). *Z. Tierpsychol.*, **35**, 113–23.

Maschwitz, U. and Mühlenberg, M. (1975) Zur Jagdstrategie einiger orientalischer *Leptogenys*-Arten (Formicidae: Ponerinae). *Oecologia*, **20**, 65–83.

Maschwitz, U., Hahn, M. and Schönegge, P. (1979) Paralysis of prey in ponerine ants. *Naturwissenschaften*, **66**, 213–14.

Maschwitz, U., Jessen, K. and Maschwitz, E. (1981). Foaming in *Pachycondyla*: a new defense mechanism in ants. *Behav. Ecol. Sociobiol.*, **9**, 79–81.

Möglich, M. (1979) Tandem calling pheromones in the genus *Leptothorax* (Hymenoptera: Formicidae): Behavioural analysis of specificity. *J. Chem. Ecol.*, **5**, 35–52.

Möglich, M., Maschwitz, U. and Hölldobler, B. (1974) Tandem calling: a new signal in ant communication. *Science*, **186**, 1046–7.

Möglich, M. and Hölldobler, B. (1975) Communication and orientation during foraging and emigration in the ant *Formica fusca*. *J. comp. Physiol.*, **101**, 275–88.

Moser, J. C., Brownlee, R. G. and Silverstein, R. (1968) Alarm pheromones of the ant *Atta texana*. *J. Insect Physiol.*, **14**, 529–35.

Nault, L. R., Montgomery, M. E. and Bowers, W. S. (1976) Ant–aphid association: role of aphid alarm pheromone. *Science*, **192**, 1349–51.

Olubajo, O., Duffield, R. M. and Wheeler, J. W. (1980) 4-Heptanone in the mandibular gland secretion of the nearctic ant, *Zacryptocerus varians* (Hymenoptera: Formicidae). *Ann. ent. Soc. Am.*, **73**, 93–4.

Oster, G. and Wilson, E. O. (1978) *Caste and Ecology in the Social Insects*. Princeton University Press, Princeton, New Jersey.

Pasteels, J. M. (1975) Some aspects of the behavioural methodology in the study of social insects pheromones. In: *Pheromones and Defensive Secretions of Social Insects*, *Symp.*, *IUSSI*, Dijon. pp. 105–21.

Pierce, N. E. and Mead, P. S. (1981) Parasitoids as selective agents in the symbiosis between Lycaenid butterfly larvae and ants. *Science*, **211**, 1185–7.

Ray, T. S. and Andrews, C. C. (1980) Antbutterflies: butterflies that follow army ants to feed on antbird droppings. *Science*, **210**, 1147–8.

Regnier, T. E. and Wilson, E. O. (1971) Chemical communication and 'propaganda' in slave-making ants. *Science*, **172**, 267–9.

Riley, R. G., Silverstein, R. M. and Moser, J. C. (1974a) Isolation, identification, synthesis and biological activity of volatile compounds from the heads of *Atta* ants. *J. Insect Physiol.*, **20**, 1629–37.

Riley, R. G., Silverstein, R. M., Carroll, B. and Carrol, R. (1974b) Methyl 4-methylpyrrole-2-carboxylate; a volatile trail pheromone from the leaf-cutting ant, *Atta cephalotes*. *J. Insect Physiol.*, **20**, 651–4.

Risch, S. J. and Rickson, F. R. (1981) Mutualism in which ants must be present before plants produce food bodies. *Nature*, **291**, 149–50.

Ritter, F. J., Bruggeman-Rotgans, I. E. M., Verweil, P. E. J., Persoons, C. J. and Talman, E. (1977) Trail pheromones of the Pharaoh's ant, *Monomorium pharaonis*: isolation and identification of faranal, a terpenoid related to juvenile hormone II. *Tetrahedron Letters*, 2617–18.

Rosengren, R. and Cherix, D. (1981) The pupa-carrying test as a taxonomic tool in the *Formica rufa* group. In: *Biosystematics of Social Insects* (Howse, P. E. and Clément, J.-L., eds) pp. 263–82. Academic Press, London and New York.

Schildknecht, H. (1976) Chemical ecology – a chapter of modern natural products chemistry. *Angewandte Chemie, Int. edn*, **15**, 214–22.

Schildknecht, H. and Koob, K. (1971) Myrmicacin, the first insect herbicide. Defensive Compounds in Arthropods, Part 50. *Angewandte Chemie, International Edition*, **10**, 124–5.

Schneirla, T. C. (1971) *Army Ants: a Study in Social Organisation* (Topoff, H. R., ed.) Freeman, San Francisco.

Talbot, M. (1967) Slave-raids of the ant *Polyergus lucidus* Mayr. *Psyche*, **74**, 299–313,

Taylor, R. W. (1978) *Nothomyrmecia macrops*: a living-fossil ant rediscovered. *Science*, **201**, 979–85.

Topoff, H. and Mirenda, J. (1975) Trail-following by the army ant *Neivamyrmex nigrescens*: responses by workers to volatile odors. *Ann. ent. Soc. Am.*, **68**, 1044–6.

Topoff, H. and Lawson, K. (1979) Orientation of the army ant *Neivamyrmex nigrescens*: integration of chemical and tactile information. *Anim. Behav.*, **27**, 429–33.

Topoff, H. and Mirenda, J. (1980) Army ants do not eat and run: influence of food supply on emigration behaviour in *Neivamyrmex nigrescens*. *Anim. Behav.*, **28**, 1040–6.

Topoff, H., Mirenda, J., Droval, R. and Herrick, S. (1980) Behavioural ecology of mass recruitment in the army ant *Neivamyrmex nigrescens*. *Anim. Behav.*, **28**, 87–103.

Traniello, J. F. A. (1977) Recruitment behaviour, orientation and the organisation of foraging in the carpenter ant *Camponotus pennsylvanicus* De Geer (Hymenoptera: Formicidae). *Behav. Ecol. Sociolbiol.*, **2**, 61–79.

Traniello, J. F. A. (1980) Colony specificity in the trail pheromone of an ant. *Naturwissenschaften*, **67**, 361–2.

Tumlinson, J. H., Moser, J. C., Silverstein, R. M., Brownlee, R. G. and Ruth, J. M. (1972) A volatile pheromone of the leaf-cutting ant, *Atta texana*. *J. Insect Physiol.*, **18**, 809–14.

Vandermeer, R. K., Glancey, B. M., Lofgren, C. S., Glover, A., Tumlinson, J. H. and Rocca, J. (1980) The poison sac of red imported fire ant queens: source of a pheromone attractant. *Ann. ent. Soc. Am.*, **73**, 609–12.

Vandermeer, R. K., Williams, F. D. and Lofgren, C. S. (1981) Hydrocarbon components of the trail pheromone of the red imported fire ant, *Solenopsis invicta*. *Tetrahedron Letters*, **22**, 1651–4.

Vinson, S. B., Phillips, S. A. and Williams, H. J. (1980) The function of the postpharyngeal glands of the red imported fire ant, *Solenopsis invicta* Buren. *J. Insect Physiol.*, **26**, 645–50.

de Vroey, C. (1979) Aggression and Gause's law in ants. *Physiol. Ent.*, **4**, 217–22.

de Vroey, C. and Pasteels, J. M. (1978) Agonistic behaviour of *Myrmica rubra* L. *Insectes Sociaux*, **25**, 247–65.

Watkins II, J. R., Gehlbach, F. R. and Kroll, J. C. (1969) Attractant-repellent secretions of blind snakes (*Leptotyphlops dulcis*) and their army ant prey (*Neivamyrmex nigrescens*). *Ecology*, **50**, 1098–102.

Weber, N. A. (1949) The functional significance of dimorphism in the African ant, *Oecophylla*. *Ecology*, **30**, 397–400.

Wheeler, J. W., Evans, S. L., Blum, M. S. and Torgerson, R. L. (1975). Cyclo-pentyl ketones: identification and function in *Azteca* ants. *Science*, **187**, 254–5.

Wheeler, W. M. (1936) Ecological relations of ponerine and other ants to termites. *Proc. Am. Acad. Arts Sci.*, **71**, 159–243.

Williams, H. J., Strand, M. R. and Vinson, S. B. (1981) Trail pheromone of the red imported fire ant *Solenopsis invicta* (Buren). *Experientia*, **37**, 1159–60.

Williams, W. G., Kennedy, G. G., Yamamoto, R. T., Thacker, J. D. and Bordner, J. (1980). 2-Tridecanone: a naturally occurring insecticide from the wild tomato *Licopersicon hirsutum* f. *glabratum*. *Science*, **207**, 888–9.

Wilson, E. O. (1958) The beginnings of nomadic and group-predatory behavior in the ponerine ants. *Evolution*, **12**, 24–36.

Wilson, E. O. (1962) Chemical communication among workers of the fire ant *Solenopsis saevissima* (Fr. Smith) 1. The organisation of mass-foraging. *Anim. Behav.*, **10**, 134–47.

Wilson, E. O. (1971) *The Insect Societies*. Harvard University Press, Cambridge, Massachusetts.

Wilson, E. O. (1975) Enemy specification in the alarm-recruitment system of an ant. *Science*, **190**, 798–800.

Wilson, E. O. and Regnier, F. E. (1971) The evolution of the alarm-defense system in the Formicine ants. *Am Nat.*, **105**, 279–89.

16

Sociochemicals of Termites

P. E. Howse

16.1 INTRODUCTION

The study of termite chemistry has been slow to evolve compared with that of
ants. One undoubted reason is the scarcity of competent taxonomists working
in the field, and another is the difficulty of maintaining many species in laboratory cultures or indeed of obtaining them in the first place. The termites are
predominantly a tropical and subtropical group. For example, in Europe they
are mainly confined to areas with a mediterranean climate and are not found in
northern France or Britain, while in north America they go little further north
than the Canadian border. Those species best known are those sympatric with
researchers in the USA and Europe. These termites are mostly members of the
Kalotermitidae (dry-wood termites) Termopsidae (damp-wood termites) and
Rhinotermitidae (subterranean termites). Members of these families, together
with the morphologically primitive single-genus Australian family Mastotermitidae, are essentially wood feeders, living within their food, and relying upon
internal symbionts (flagellate protozoa) for breakdown of cellulose. In many of
the species concerned, recruitment to food sources scarcely occurs as the colony
simply extends its galleries further, colony defense depends upon the activity of
soldiers with powerful jaws, and caste development has not extended to the
possession of a morphologically fixed worker caste.

The greatest adaptive radiation is found in the remaining families which
move outside their nests for food. The Hodotermitidae, which are harvesting
termites with functional compound eyes, cut grasses foraging on open above-
ground trails, and take them back to the nest. The Termitidae comprises about
three-quarters of all known species in three subfamilies. Many species build

Chemical Ecology of Insects. Edited by William J. Bell and Ring T. Cardé
© 1984 Chapman and Hall Ltd.

complex mound, arboreal or underground nests which are bases for the colony, from which foraging parties exploit neighbouring sources of wood and dead vegetation. The subfamily Macrotermitinae are fungus-growers, often building mound nests of a spectacular size. The Nasutitermitinae have soldiers with greatly reduced mandibles and a beak-like tubular projection of the head capsule (nasus) used for dispensing a chemical defensive secretion. Amitermitinae have both a nasus and sabre-like mandibles and commonly construct mound or tree nests.

From this extremely brief conspectus it may be clear that the foraging strategies, caste regulation mechanisms, and defensive strategies, all of which may be dependent to a greater or lesser extent on chemical secretions according to the species concerned, show great diversity.

When we compare ants and termites, many of the differences can be said to stem from the feeding habits of the latter. Termites are entirely herbivorous and have a mainly cryptic existence. They retain a soft cuticle which allows them to burrow and excavate in constricted spaces. They do not have a sting: indeed, a sting requires firm well-articulated abdominal segments to be wielded effectively. The weapons of defence are, instead, located in the head, either as powerful jaws or as glandular openings secreting toxic or sticky chemicals. The location of defence in the head immediately implies constraints, because animals with jaws specialized for crushing, cutting, slicing and snapping cannot use them very effectively for feeding. Specialization of function has resulted, with the defensive capacity vested primarily in a soldier caste forming a small percentage of the total colony number. This caste must be fed by the workers, which provide stomodeal, or occasionally protodeal food. The soldier caste would be less effective if it had the same thresholds for behavioral responses as the workers, and normally the behavior and responsiveness of this caste differ markedly from that of other castes. The reasons for this must be sought in hormonal changes that may also determine development. The colony will only invest what natural selection has judged that it can afford in production of soldiers or reproductives, and therefore mechanisms of regulation of caste numbers are essential to colony function.

The main predators of termites throughout the world are ants. Many species of ponerine ants are obligate predators in the tropics (see Chapter 15), and many ant species will take termites if they have the opportunity. As we shall see, termites rely on a combined defence of soldiers, nest structure, and worker building behavior. Away from the nest, or if the nest is breached by larger animals, the defence provided by the soldiers alone is rarely effective for long.

The general biology of termites has been reviewed by Harris (1971), Lee and Wood (1971), Howse (1970), and in Krishna and Weesner (1969). Termite defensive behavior has been reviewed by Prestwich (1979a, b, c) and Deligne *et al.* (1980).

16.2 MASTOTERMITIDAE

The only extant member of this family is *Mastotermes darwiniensis* from Australia. This species is morphologically primitive, and has a number of characters reminiscent of supposed blattoid ancestors, such as an anal lobe in the hindwing of the adults, and eggs glued together in two rows, as in an ootheca (Gay and Calaby, 1970). *M. darwiniensis*, is, however, a highly successful species and the most destructive in Australia. It is able to construct chambers, from hard, semi-digested ligneous faecal material or *carton*, in or under tree trunks or logs, and to forage in subterranean galleries over 30 meters from the nest (Gay and Calaby, 1970). Its foraging behavior is evidently aided by a trail pheromone from the sternal glands. There are three sternal glands (Fig. 16.2, on abdominal segments 3, 4 and 5) in *Mastotermes*, which sets it apart from all other termite genera which have a single gland (Ampion and Quennedey, 1981).

The proportion of soldiers is low (1–2% according to Gay and Calaby, 1970). They have well-developed mandibles, but also produce a highly odorous chemical secretion from the mouth which is initially colourless and fluid, but soon sets to a dark rubbery material which may immobilize enemies (Moore, 1968). This consists of *p*-benzoquinone and toluquinone, which tan proteins present in the saliva turning them into dark, insoluble sclerotin.

16.3 TERMOPSIDAE AND KALOTERMITIDAE

The Termopsidae are a morphologically primitive group of damp-wood termites. The genus which has been most intensively studied is *Zootermopsis*, which contains three species found in western North America living on fallen or damaged pine.

The Kalotermitidae are dry-wood termites. They are considered here with the Termopsidae because their caste structure and mode of life are essentially similar. Both families have soldiers with powerful biting mandibles (Fig. 16.9), there is no fixed worker caste, and the colony gains its main protection from the tree, stump or log in which it excavates galleries. The European species *Kalotermes flavicollis* has been used extensively in the study of caste determination mechanisms.

Caste determination and regulation in the so-called 'lower termites' has been reviewed by Light (1942–3), Miller (1969), Lüscher (1974, 1977) and others. Light (1942–3) showed that any larva of *Zootermopsis* is capable of developing into either a soldier, winged adult, or supplementary reproductive, and that extracts of the head and thorax were capable of inhibiting the formation of supplementary reproductives (Light, 1944).

The presence of reproductives has been shown to inhibit the production of replacement reproductives in *Z. angusticollis* (Light, 1942–3) while production of winged adults is inhibited by the presence of functional reproductives in *Z. angusticollis* and *Z. nevadensis* (Lüscher, 1974). Further, soldiers inhibit pre-soldier formation (Light, 1942–3), but stimulate the production of replacement reproductives (Lüscher, 1974). This work clearly implicates pheromones as factors in the control of social polymorphism in *Zootermopsis*.

Caste regulation in *Kalotermes* appears to be very similar to that in *Zootermopsis*, and the various possibilities of development are shown in Fig. 16.1. The fully grown larva is known as a 'pseudergate', and the pseudergates are the main source of replacement reproductives and (via nymphal stages with wing pads) of winged primary reproductives. Removal of the king and queen from a *Kalotermes flavicollis* colony usually results in several pseudergates moulting to form replacement reproductives, but the surplus is killed by fighting among reproductives leaving only one pair (Rüppli, 1969). Lüscher (1952) was able to demonstrate the role of pheromones in barrier experiments in which a sexual pair was separated from the colony by a gauze screen through which only the antennae could pass. Replacement reproductives were produced, as in the total absence of the sexuals. When a queen was fixed in a solid screen such that her abdomen protruded into a colony with reproductives, there was inhibition of reproductive production, even if her tergal and sternal glands and genital opening were covered with varnish, but if the anus was blocked, inhibition ceased. Complete inhibition, however, occurs only if the king is present with the queen. Such a synergism between the sexes has also been found in the Australian species *Porotermes adamsoni* (Mensa-Bonsu, 1976). Lüscher (1974) has some evidence that pseudergates inhibit the production of female replacement reproductives and vice versa.

In damp and dry-wood termites, the mechanisms that control production and elimination of members of the various castes are clearly very complex.

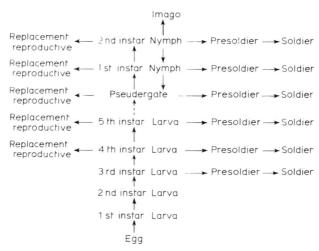

Fig. 16.1   Developmental possibilities in *Kalotermes flavicollis* (after Lüscher, 1976).

They lead to a kind of dynamic equilibrium in which a single reproductive pair and a small percentage of soldiers are maintained in the colony. The king and queen are mobile within the nest, and their chances of being taken by predators must be greater than in species with fortified nests that have a central well-defended queen cell. In the absence of either or both of the pair, regulatory systems act rapidly to provide replacements, and if soldiers are lacking more soldiers are produced.

In *Kalotermes*, inhibition of reproductive production by other reproductives is apparently absolute in small colonies. However, it is incomplete in *Zootermopsis* and *Porotermes*. Mensa-Bonsu (1976) notes that replacement reproductives are very common in field colonies of *P. adamsoni*, and nests (as in *Zootermopsis*) may ramify and divide into separate sub-units each with a nursery area. With such a nesting habit, toleration of more than one pair of functional reproductives in the colony may increase the chances of survival of the colony as a whole.

Little is known of the chemical nature of the pheromones referred to above. It is well-established, however, that implantation of corpora allata into pseudergates results in soldier-formation at the next moult (Lüscher and Springhetti, 1960; Lebrun, 1967), but feeding or applying juvenile hormone (JH) or JH analogues to *Zootermopsis* gave rise to high proportions of soldier–worker or nymph–alate intercastes which are rare in nature (Lüscher, 1977). Hence, it is unlikely that JH is the actual pheromone although it must reach a high titre in the individual to make soldier production likely (Lüscher, 1977), but it appears that the reproductive pheromones stimulate the corpora allata to an increased but critical level of activity which makes production of inter-castes unlikely.

Trophallactic exchange plays a crucial part in the social life of termites, as in all social insects. In the lower termites this serves to perpetuate the gut fauna, which is lost after each moult, and also to spread caste pheromones throughout the colony.

Communication in colonies of damp- and dry-wood termites is facilitated by a trail substance from the sternal gland, found in all castes, the structure and distribution of which (Fig. 16.2) has been reviewed by Ampion and Quennedey (1981). In the Termopsidae, there is a single gland on the fourth sternite, containing three cell types in the glandular epithelium. In the Kalotermitidae, the gland is on the fifth sternite and contains only two cell types; a class of cells with cuticular ducts opening to the exterior is lacking. In *Zootermopsis*, the fourth sternite overlaps the fifth posteriorly, forming a reservoir in which the pheromone may be stored (Stuart, 1964). Lüscher and Müller (1960) showed that *Z. nevadensis* workers recruited others to food (moist sawdust) using a pheromone trail, and followed faithfully an extract of the gland, even if they were separated from direct contact with the trail by a gauze platform. Several components have been identified from extracts of the gland by Hummel and Karlson (1968), of which hexonoic acid was the most attractive in bioassays. Geraniol and citronellol were less attractive on artificial trails. The sternal gland

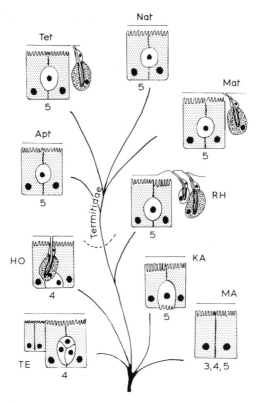

Fig. 16.2   Variations in sternal gland morphology among termite families. Type 1 cells lightly stippled; Type 2 cells heavily stippled; Type 3 cells without stippling. TE = Termopsidae, HO = Hodotermitidae, MA = Mastotermitidae, KA = Kalotermitidae, RH = Rhinotermitidae. Mat = Macrotermitinae, Nat = Nasutitermitinae, Tet = Termitinae, Apt = Apicotermitinae. The numbers refer to the segments on which the glands are found (from Quennedey, 1975; Ampion and Quennedey, 1981).

secretion of *Z. nevadensis* winged adults is very attractive to individuals of the opposite sex (Pasteels, 1971).

Communication of alarm in *Zootermopsis angusticollis* involves an interplay between pheromonal and acoustic mechanisms (Howse, 1964, 1970). Tapping sounds are produced by alarmed termites knocking their heads against the roof and/or floor of the galleries. Under the influence of substrate vibrations, which act as alerting signals, termites withdraw from light and air movements and follow odour trails that lead in a downwind direction or toward the centre of the nest. Trail laying may also occur in recruitment to a stimulus that provokes alarm, or recruitment to a site where building occurs to seal the nest or cover an intruding stimulus such as a dead ant or a drop of phenol (Stuart, 1967).

The variety of chemical signals in damp- and dry-wood termites may be related to their nesting habits. The colony needs to be mobile because it consumes its own nest. The king and queen cannot be protected in a central fortified cell and so may run a higher risk of loss to predators than in other families of termite. As the nest extends, the reproductives must remain mobile or, as an additional strategy, supplementary reproductives must be produced to service the outer parts of a ramifying nest structure. There is a complex system for regulation of caste numbers, especially for the production of replacement reproductives, which can be produced within a few days if circumstances demand.

For defense, the insects rely upon soldiers with large heads and powerful jaws which can operate well within gallery systems and block them easily (Fig. 16.9) (some Kalotermitidae have soldiers with phragmotic heads). There is a rapid response to air movement (Howse, 1966) which results in sealing of any gaps to the outside air, thereby denying entry to predators, and this depends in part on recruitment using a trail pheromone from the sternal gland.

16.4 HODOTERMITIDAE

The Hodotermitidae are harvester termites. They mostly feed on grasses and seeds, constructing subterranean nests in dry savanna areas in which they store the harvested material. One of the best studied species is *Hodotermes mossambicus*, which is distributed in drier regions from South Africa to Ethiopia. This species swarms one day after rain in Kenya (Leuthold, 1977). As an apparent adaptation to colonization of arid regions the winged forms have large salivary reservoirs which are water stores (Watson *et al.*, 1971). After flight, both sexes shed their wings and the males expose a very large sternal gland with the abdomen held high (Leuthold, 1977). Females approach upwind running in a zig-zag pattern typical of anemotactic orientation, although males are detected at distances of only 6–10 cm. After contact with the female the male ceases calling and the pair begin to excavate a chamber in the soil. Leuthold believes that this rapid mechanism for pairing allows the insects to swarm in sunny conditions after rain when the ground is moist but not inundated, while avoiding prolonged exposure to natural predators which are more active in daylight.

Hodotermes workers are darkly pigmented and have well-developed compound eyes. They leave the nest through long underground galleries which come to the surface near foraging areas. The insects then establish a trail, up to a maximum of 3 m in length, to suitable sites where they cut grass and transport it back to the nest hole. Leuthold *et al.* (1976) studied orientation experimentally on a turntable with a foraging hole in the centre. In the dark or in diffuse light, the insects relied on a pheromone trail laid from the sternal gland. They appear to follow the trail klinotactically, weaving from left to right, unlike ants which follow trails tropotactically with little deviation. Normally, with the sun or moon visible, the workers orientate menotactically. Orientation is then more precise, and the insects respond to optical cues rather than pheromone trails when the two are in competition in turntable experiments. Pheromonal cues are more important to workers leaving the nest, which first follow a trail and then switch to optical orientation. Chemical markers near the entrance hole are also important, as workers have difficulty in finding the hole if the dust around it is removed. Memory of distance travelled is a supplementary orientation mechanism: termites displaced while feeding walk for a distance approximating to the distance of the feeding site from the nest hole.

The foraging behavior of *H. mossambicus* thus bears interesting comparisons with that of ants. Exposure of the insects above ground is, however, relatively brief. The sub-spherical subterranean nests or 'hives' are linked by long galleries. Some of the hives do not have reproductives and are apparently used as storage 'granaries' (Coaton, 1946), and the whole nest system may extend over 200 m in one direction and exploit grass over more than a third of a hectare.

16.5 RHINOTERMITIDAE

The Rhinotermitidae are small subterranean wood-eating termites. In this family, and in the Termitidae, the winged adults and the soldiers possess an unpaired cephalic gland known as the frontal gland, which opens through a frontal pore, or fontanelle, situated on the dorsal surface of the head.

Reticulitermes is a genus with a small frontal gland, which has a biology similar to the Termopsidae. These small termites are destructive to timber used in construction of buildings in many parts of the world, including the USA, Southern Europe, and the Far East.

The nests of *Reticulitermes* consist of a network of underground galleries and chambers of carton, often centring on moist or fungus-infected wood. Above-ground covered galleries are commonly built, linking subterranean galleries with above-ground timber. The brood may be dispersed in several centres (Noirot, 1970) interconnected by a network of galleries, and many supplementary reproductives are often found in nests. *Reticulitermes* species are frequently found in wood infected with the fungus *Gloeophyllum trabeum* (= *Lenzites trabea*) and are strongly attracted towards (Z,Z,E)-3,6,8-dodecatrienl-lo isolated from wood infected with the fungus and also from *Reticulitermes virginicus* (Matsumara *et al.*, 1968). Howard *et al.* (1976) showed that *R. virginicus* did not distinguish a trail made with the extract of its own body from a trail made with the fungus extract, but three other *Reticulitermes* species responded preferentially to their own trails. Hence, it seems likely that the dodecatrienol is the only component of the trail pheromone in *Reticulitermes*.

The sternal gland in Rhinotermitidae (Fig. 16.2) is situated on the fifth abdominal sternite and is bilobed, with a large extracellular reservoir between the cuticle and the glandular cells (Ampion and Quennedey, 1981). The cuticle in this region has hypertrophied pore canals. The gland is non-functional in the three larval instars, but becomes bilobed in the subsequent worker and soldier castes (Quennedey, 1977). The anterior part of the gland, which contains cells abutting on to the cuticle (class I of Noirot and Quennedey 1974) and cells not directly in contact with the cuticle (class 2), hypertrophies in the sexual forms, becoming largest of all in the female which has a predominance of class 2 cells. Class 3 cells, with a cuticular duct, are present in both parts of the gland, but generally, more are present in the winged adults. Quennedey (1977) suggests that the sex pheromone is secreted from class 1 and 2 cells in the anterior part of the gland, and the trail pheromone is secreted by class 3 cells from the posterior part.

Courtship in *Reticulitermes* begins with 'calling' in the female, in which the abdomen is lifted and the sternal gland exposed (Buchli, 1960; Clément, 1982). The males then approach and form tandems, in which the palps of the male are placed on the abdominal tergites of the female. The pair then moves off and excavates a suitable cavity for the brood. Buchli (1960) found that the calling female of *R. lucifugus* would stop calling and run when she was touched by the antennae of the male, but if he lost contact she would stop and call again. He also found that he could induce males to respond to glass models first placed in contact with the abdomen of a calling female.

Clément (1977, 1978a, b, 1981, 1982) has made a particularly thorough study of geographical variation and isolating mechanisms in the European species, *R. santonensis* and *R. lucifugus*. *R. santonensis* is found in a circumscribed zone in Saintonge (SW France), while *R. lucifugus* occurs in S. France, Spain, Italy, Corsica, and other areas in the mediterranean basin. On morphological grounds, *lucifugus* can be separated into several subspecies including *R. l. grassei* in SW France and W. Spain and *R. l. banyulensis* from the eastern end of the Pyrenees southward into Spain (Fig. 16.3). Adults that had flown and lost their wings were tested in an olfactometer which had air issuing into a chamber from

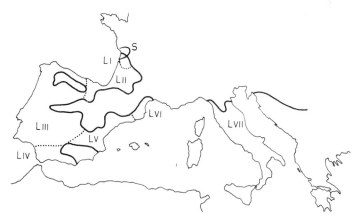

Fig. 16.3 Distribution of *Reticulitermes* species in Europe. S = *R. santonensis*. L = *R. lucifugus*. Populations LI and LII = *R. l. grassei*, LVI = *R. l. banyulensis* (from Parton *et al.*, 1981).

four pipettes, each charged with a sternal gland extract on filter paper. A clear preference was expressed in approach of the termites to air currents containing extracts of their own species (Fig. 16.4) although the pheromones of females were also attractive to other females, and that of males to other males, and there was significant attraction of male *santonensis* to female *grassei* and of male *grassei* to female *santonensis*.

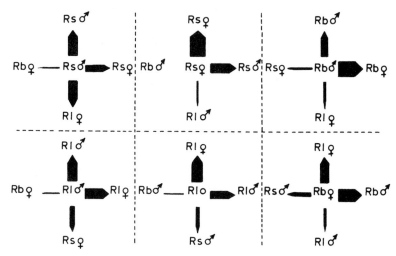

Fig. 16.4   Results of olfactometer tests with sternal gland extracts of *Reticulitermes*. The thickness of arrows represents proportions of sexuals responding to each of four air inlets with extracts of *R. santonensis* (Rs), *R. lucifugus grassei* (Rl) and *R. l. banylensis* (Rb) (from Clément, in Parton *et al.*, 1981).

When adults of *santonensis* and *grassei* were mixed and released together in an experimental arena 50 cm square, 88% of the tandems formed were conspecific, but they were all heterosexual. There are thus two kinds of chemical signal, one acting at a distance and stimulating upwind orientation, and another acting at short range when individuals come into contact. Apart from extracts of the sternal gland, dodecantriene-1-ol and various terpenes and alcohols are also attractive to adults in olfactometer experiments, and these may help to guide the female in tandem to a suitable site for nest excavation.

The chemical nature of the sternal gland secretion of adults is unconfirmed, but two major volatile components have been found in differing proportions in *R. santonensis*, *R. lucifugus grassei* and *R. l. banyulensis* (Fig. 16.5). They occur in the females and in much lower quantities in the males (Parton *et al.*, 1981). Their existence could explain the interspecific responsiveness to extracts and the responsiveness of females to males in olfactometers (Fig. 16.4). The limited attraction that occurs between *santonensis* and *grassei* is countered by other factors in nature. In the sympatric zone, the two species swarm at different times, *R. santonensis* in mid-April, and *R. l. grassei* during the first

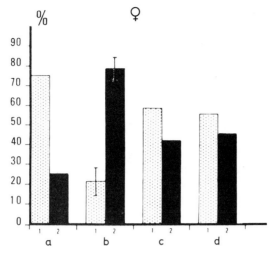

Fig. 16.5 Relative proportions of the two major volatile components of the sternal gland of female sexuals of *Reticulitermes santonensis* (a), *R. lucifugus grassei* (b), and *R. l. banyulensis* of population LVI in Fig. 16.3. (c) and LIV (d) (from Parton *et al.*, 1981).

three weeks of May (Clément, 1977). Aggression between workers of colonies in the sympatric zone is higher than elsewhere (Clément, 1978b), but very low between *santonensis* and *lucifugus grassei*. It is also very high between *grassei* from SW France and *banyulensis* from the Catalan region.

Aggressive behavior results when workers of two different species (or sub-species) of *Reticulitermes* come into contact (Clément, 1981). There are then reciprocal movements of the antennae and palps in which the antennae are placed against the cuticular surfaces of the antennae of the partner, the labrum, the palps, and the pleura. Aggression follows rupture of the movement sequence, suggesting that contact pheromones on the cuticle are involved in recognition. It is an open question whether these are glandular secretions spread over the cuticle, or whether the particular blend of hydrocarbons in the lipid layer of the cuticle represents the pheromone. Howard *et al.* (1978) have examined the cuticular hydrocarbons of the N. American species *Reticulitermes flavipes*. All castes contained the same blend of species-specific hydrocarbons, but the relative proportions of branched to unbranched components and saturated to non-saturated components was characteristic of each caste. In addition, soldiers had twenty times more hydrocarbon than workers. Broadly similar caste differences were found in *Zootermopsis angusticollis* (Blomquist *et al.*, 1979), and evidence was found for biosynthesis of dimethylalkanes by the insects themselves (Blomquist *et al.*, 1978). The relatively involatile hydrocarbons would make suitable species and caste-recognition pheromones. Volatile compounds would saturate the enclosed spaces in which termites live and they would need a mechanism for coping with sensory adaptation (Blomquist *et al.*, 1978).

The host-specific staphylinid beetle *Trichopsenius frosti* has a blend of cuticular hydrocarbons that are identical to those of its host, *Reticulitermes flavipes* (Howard *et al.*, 1978). The beetle grooms workers and rides on the back of female reproductives. This appears to be a case of chemical mimicry of the species-recognition pheromones of the host.

Aggressivity in *Reticulitermes santonensis* and *R. lucifugus* varies with the ecological conditions in which the colonies are found, and with the season (Clément, 1978a, 1981). Colonies in humid zones with plenty of available food sources tend to reproduce by 'budding' from a ramifying gallery system. Here, genetic uniformity of the populations is lower than in colonies from predominantly arid areas which reproduce only by swarming. In the latter, intercolony aggression is more intense. In *R. lucifugus*, intercolony aggression is high during the winter months, co-inciding with a high level of production of soldiers and nymphs in the pre-imaginal stage with long wing pads. In the summer, after swarming, aggressivity is low and there is the possibility that colonies could mix. The chemical basis of this seasonal variation in aggressivity has not yet been investigated.

Regional differences also exist in the nature of the frontal gland secretion employed by the European *Reticulitermes* species (Parton *et al.*, 1981). The secretion is emitted from the frontal pore of soldiers and runs forward on to the mandibles. Ants (*Atta cephalotes*) bitten by the soldier termites develop motor ataxia (personal observations). The effects are usually only temporary, but death occasionally results. *R. santonensis* soldiers produce α-pinene, β-pinene limonene, and a sesquiterpene, none of which occur in *R. lucifugus* (Parton *et al.*, 1981). Soldiers of the latter commonly contain the sesquiterpenes, germacrene A and β-farnesene, an acyclic diterpene alcohol (geranyl linalool) (Baker *et al.*, 1982) and alkanes and alkenes (Fig. 16.11). A complete range of all the compounds identified was never found in individuals of a given colony. No volatile components were detected in the subspecies *banyulensis* from around Banyuls-sur-Mer (Roussillon). Propionates and acetates were found only in the secretions of colonies from around Valencia in Spain, in which sesquiterpenes were absent.

Repeated analyses of the same colony gave the same results. Colonies from the same locality but from different woods had similar secretions, and analyses of spring and autumn samples of the same colony gave similar results. The polychemism thus appears to be genetically based and variations in the quantities of components present can be correlated with aspects of enzyme polymorphism and with geographical position on a line from SW France running parallel to the coast through Spain and Portugal to the eastern end of the Pyrenees.

The soldier secretions of the North American species, *Reticulitermes flavipes* and *R. virginicus* contain none of the above compounds, but two sesquiterpenes, γ-cadinene and its aldehyde (Fig. 16.11), have been isolated from both species (Zalkow *et al.*, 1981). These compounds, or attack by soldiers, apparently had no irritant or toxic effect on the ant *Solenopsis geminata*, and the authors conclude that the sesquiterpenes have no role in defence. It is frequently observed, however, that termite defensive secretions do not have an all-or-nothing effect (see below), but give rise to temporary paralysis in varying degrees, which is easier to observe in large ants than in small ants such as *Solenopsis*. Further, the susceptibility of ants may be expected to vary with the species, and certainly does so.

The variation that occurs in the nature of the defensive secretions of *Reticulitermes lucifugus* has parallels among nasute termites (see below). The correlation that occurs between this variation, the distribution of esterase alleles and geographical distribution could be explained as an adaptation to a different spectrum of principal predators which occur in the different ecosystems or climatic zones. We may find that, because of co-evolutionary pressures, specialized ant predators are immune to the secretions, and opportunistic predators are most affected. The development of chemical ecology of termites thus depends upon the study of the comparative effects of defensive secretions on a range of predatory species.

Antagonism also exists between different colonies of the Australian species *Coptotermes acinaciformis* collected from the field (Howick and Greffield, 1980). Parts of a colony kept separately for four weeks showed no antagonism when rejoined, but different colonies showed undiminished antagonism after they were kept separately for four weeks in the laboratory. The perturbation of collection and installation in the laboratory does not affect intraspecies aggression, suggesting that the colony recognition factors are very stable.

The soldiers of *Coptotermes* have a large frontal pore (Fig. 16.6), from a very large frontal gland which extends backwards into the abdomen, displacing the viscera and giving the soldiers an opaque whitish-yellow appearance. Moore (1968) showed that the secretion of *Coptotermes lacteus* soldiers is a milky fluid that dries rapidly to form a colourless, rubbery film. It can be reconstituted by addition of water, and consists of a suspension of lipids in aqueous mucopolysaccharide. The secretion appears to have no toxic action, but immobilizes other arthropods. Bugnion (1927) reports that the secretion of *C. ceylonicus* appears as a droplet and then rolls down towards the mandibles. It sticks the mandibles and antennae of ants together, but the termite itself may often be put out of action by its own secretion. This is compensated for by high soldiers numbers. Harverty (1979) has reported that experimental laboratory colonies of *C. formosanus* regulate the proportions of soldiers within broad limits, killing off excess numbers. In this species, the soldier ratio of field colonies is between 4.7 and 10% (*in lit.*).

Soldiers of *Reticulitermes* differ from those of *Coptotermes* by the possession of a shallow gutter running from the frontal pore towards the labrum (Quennedey and Deligne, 1975). This groove is also present in *Prorhinotermes*, considered by some to be close to the ancestral line of the genera *Parrhinotermes*, *Schedorhinotermes*, *Rhinotermes*, *Dolichorhinotermes* and *Acorhinotermes*,

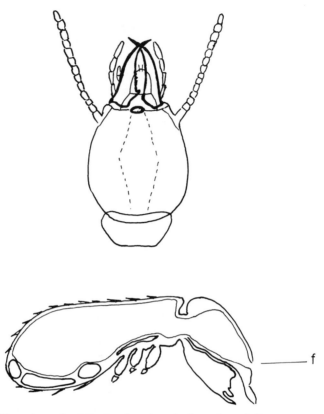

Fig. 16.6 Plan view of soldier head and sagittal section of *Coptotermes ceylonicus* showing the frontal pore (f) and the extent of the frontal gland (after Bugnion, 1927).

in which the frontal gutter is well-marked and extended to the tip of the elongate labrum which bears brush-like cuticular spines (Fig. 16.7) (Quennedey and Deligne 1975). Such a brush is found in *Prorhinotermes simplex* and *Schedorhinotermes*. In the latter, and in *Dolichorhinotermes* and *Rhinotermes*, the soldiers are dimorphic, large soldiers having well-developed mandibles and a relatively small labral brush, and small soldiers having reduced mandibles and a long narrow labral gutter (Fig. 16.7). In *Acorhinotermes*, the soldiers are secondarily monomorphic.

The defence secretions of the species with labral brushes are lipophilic contact poisons (Prestwich, 1979, Prestwich and Collins, 1982). The major component of the soldier secretion of *Prorhinotermes simplex* is a nitrogenous compound, 1-nitro-*trans*-pentadecene (Vrkoč and Ubic, 1974). *Schedorhinotermes putorius* produces three ketones, 1-tetradecen-4-one, 1-hexadecen-3-one, and 2-tridecanone (Quennedey *et al.*, 1973). Similar ketones have been found in *S. lamanianus* (Prestwich *et al.*, 1975), the main component in both

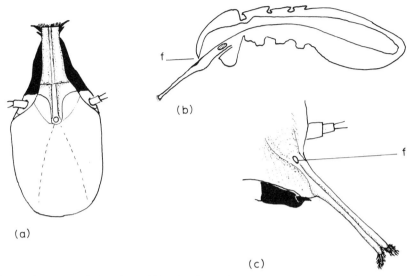

Fig. 16.7 (a) Head of major soldier of *Schedorhinotermes putorius*, showing the frontal pore (f) and frontal groove leading to the labral brush. Mandibles in black (drawn from stereoscan photographs of Quennedey *et al.*, 1973); (b) Sagittal section of *Rhinotermes nasutus* minor worker showing frontal pore (f) and the front gland (from Mill, 1982); (c) Anterior head region of minor soldier of *Rhinotermes marginalis* showing reduced mandibles (in black) and frontal pore (f) opening to narrow labral gutter ending in a daubing brush (drawn from stereoscan photographs of Quennedey and Deligne, 1975).

major and minor soldiers being 1-tetradecen-3-one. Minor soldiers of *Rhino-termes hispidus* and *R. marginalis* produce 3-keto-13-tetradecenal and 3-ketotetradecanal (Prestwich and Collins, 1982), which are lacking in the sparse secretion of the major soldier. The monomorphic soldiers of *Acorhinotermes subfusciceps* contain 3-keto-(*Z*)-9-hexadecenal and (*Z*)-8-pentadecen-2-one.

The increasingly specialized development of the labial brush from *Prorhino-termes* and continuing along the rhinotermitine line (Fig. 16.8), according to the hypothesis of Prestwich and Collins (1982), may be correlated with increased toxicity and chemical reactivity of the contact poisons used. Quennedey *et al.* (1973) found that exposure of various European ant species to the vapour of the vinyl ketones of *Schedorhinotermes putorius*, or topical application of the compounds, was sufficient to kill them in a short time. The principal component of the *Prorhinotermes simplex* secretion is markedly toxic to houseflies (Hrdý *et al.*, 1977). The β-ketoaldehydes that occur in the more specialized rhinotermitine soldiers are more highly reactive with a greater affinity for peptide groups (Prestwich and Collins, 1982).

Although it is known that the Rhinotermitidae with a labral brush use this for smearing or daubing their adversaries with the frontal gland secretion, remarkably little is known of the efficiency of this method of defence in natural

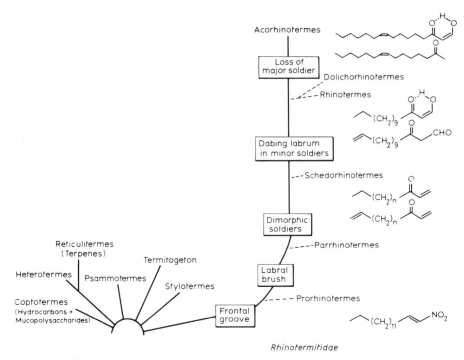

Fig. 16.8 Evolutionary tree of some rhinotermitids based on stereoscan micrographs (Quennedey and Deligne, 1975) and chemical data (Prestwich and Collins, 1980, 1981, 1982; Prestwich *et al.*, 1975; Vrkoč and Ubic, 1974).

conditions. The Rhinotermitinae mainly live in rotting wood on the forest floor. *Schedorhinotermes* constructs a carton nest, within tree trunks, with a central queen cell and forages over tree trunks under covered galleries. Many Australian *Coptotermes* species build mound nests, the outer walls consisting of clay material, and the inner chambers of carton. Prestwich and Collins (1982) report an unusual kind of defensive behavior in *Acorhinotermes subfusciceps*, in which, during breaking open of the nest, the minor soldiers carry eggs and larvae in their mandibles with 'an effective defensive shield (the labrum) hanging over the surface of their otherwise defenseless charges'. This behavior was also noted by Holmgren (1906), writing about *Rhinotermes nasutus*, and the importance of this strategy as a means of colony defence remains to be investigated.

The development of insecticidal secretions in social insects means there must be concomitant pressure for the co-evolution of immunity for the individuals that produce the secretion (Howse, 1975). This immunity must extend at least to resistance of the glandular linings and cuticle over which the substance is discharged, and may depend upon specializations of individual behavior that minimize the chances of self-contamination. Spanton and Prestwich (1981) found that *Schedorhinotermes lamanianus* and *Prorhinotermes simplex* workers detoxify secretions of their soldiers by means of substrate-specific alkene reductases which catalyse conversion to the saturated compounds. These are then converted to the acetates and taken up again by the soldiers for recirculation. The authors exposed three species of termite to the principal component of the defensive secretions of *S. lamanianus* and *P. simplex* and the two saturated compounds resulting from detoxification (Table 16.1). It appears that *P. simplex* is very resistant to its own secretion and detoxification product, but very susceptible to the *S. lamanianus* secretion, and *vice versa. Reticulitermes flavipes* is very susceptible to all the compounds.

Table 16.1 Median lethal doses (at 48 h) of components of *Prorhinotermes simplex* and *Schedorhinotermes lamanianus* defensive secretions, and their detoxification products. (From Spanton and Prestwich, 1981)

	R. flavipes	*P. simplex*	*S. lamanianus*
P. simplex nitroalkene	0.45	>10	1.2
Detoxification product (nitroalkane)	0.75	>10	1.5
S. lamanianus vinyl ketone	0.05	0.11	0.60
Detoxification product (ethyl ketone)	0.65	0.65	>5

Another possible means of chemical defence in Rhinotermitidae has been described by Bacchus (1979), who found the apparent pore-like openings of glands on the distal ends of the tibiae and on the ventral surfaces of the first and second tarsal segments. These structures are more numerous in workers than soldiers, and may serve to repel small myrmicine ants which tend to attack workers by grabbing their legs.

16.6 TERMITIDAE

The Termitidae is the largest of the termite families, containing about three-quarters of all known species (Krishna, 1970). It is divided into four sub-families, the Termitinae, Nasutitermitinae, Apicotermitinae and Macrotermitinae (Sands, 1972). Of these, the Nasutitermitinae and Macrotermitinae (fungus-growers) have been most intensively studied from ecological, physiological and chemical aspects.

16.6.1 Termitinae

The Termitinae includes a group of species in which defense of the colony depends mainly upon specializations of the soldier mandibles. A classification of methods by which the mandibles are employed is provided by Deligne (1971), who describes biters, cutters, symmetrical snappers and asymmetrical snappers (Fig. 16.9). Soldiers with biting and cutting mandibles sometimes use these in conjunction with a chemical secretion which dribbles on to the mandibles and may be introduced into a wound. Snapping mandibles depend upon the storage of energy that occurs when the slightly elastic mandibles are adpressed. This energy is released suddenly, as when someone snaps their fingers, resulting in a powerful blow either side of the insect. In soldiers with asymmetrical blade-like mandibles, such as *Pericapritermes* and *Neocapritermes*, the blow is unilateral. As a defensive technique, the use of snapping mandibles is effective only in nest galleries and similar enclosed spaces. In the open air operation of the snapping mechanism often results in the soldiers being thrown into the air. Species with

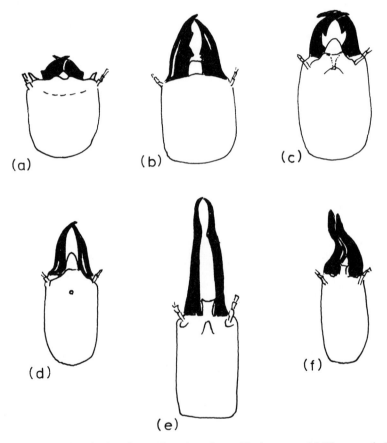

Fig. 16.9   Soldier heads showing a diversity of mandibular types (a) Phragmotic head (*Cryptotermes domesticus*); (b) biting-crushing mandibles (*Zootermopsis angusticollis*); (c) biting-cutting, with frontal gland secretion (*Amitermes messinae*); (d) cutting, with frontal gland secretion (*Reticulitermes lucifugus*); (e) symmetrical snapping (*Termes odontomachus*); (f) asymmetrical snapping (*Pericaptritermes dumicola*) (redrawn from Harris, 1971).

snapping soldiers rarely have a chemical defence mechanism as well, but an exception to this is *Orthognathotermes*. *O. gibberorum* studied by the writer in Brazil (unpublished) has soldiers with slender elastic symmetrical mandibles of the snapping type. The soldiers are sluggish in their movements, and when grasped by the legs with forceps contract antagonistic abdominal muscles which result in rupture of the body wall in the region of the ventral prothorax. A slightly milky secretion, the contents of the hypertrophied labral gland which has a large reservoir in the abdomen, issues in a large droplet, and may also issue from the mouth (Fig. 16.10). This secretion rapidly becomes tacky.

Fig. 16.10 *Orthognathotermes gibberorum* soldier (a) before and (b) following abdominal dehiscence.

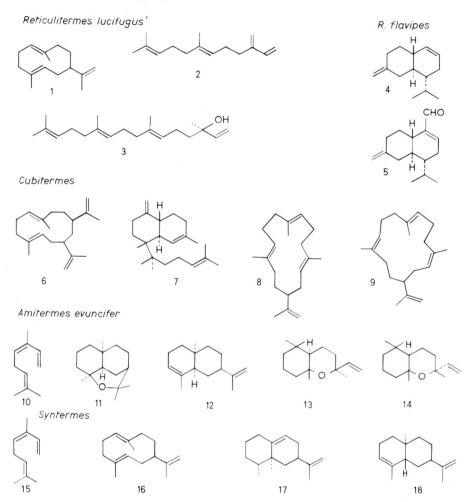

Reticulitermes lucifugus[']

R. flavipes

Cubitermes

Amitermes evuncifer

Syntermes

Camponotine ants smeared with the secretion in the field were either killed within a few seconds or immobilized for several minutes. This sacrificial behavior is common among soldierless species of the Apicotermitinae, and will be discussed further, but among the Termitinae it also occurs in *Globitermes*.

G. *sulphureus*, so-named because of the yellow colour of the soldiers imparted by an abdominal glandular reservoir, is able to emit the sticky secretion over cuts, but may also rupture in the process (Noirot, 1969a). G. *radaensis* also holds intruders with its curved mandibles and coats them with a secretion from the mouth, which immobilizes and often kills them (Kemner, in Roonwal, 1970).

A more common method of chemical defense in the Termitinae is the use of biting or cutting mandibles which are contaminated with a frontal gland secretion, such as occurs in *Cubitermes* and *Amitermes* species.

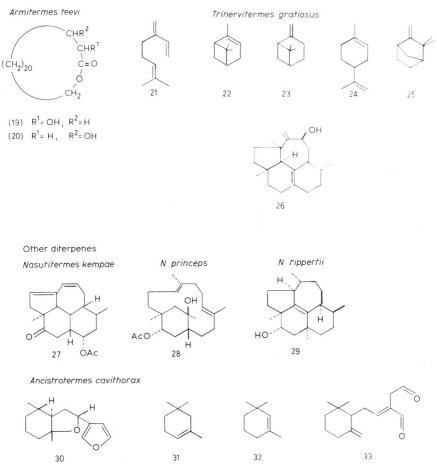

Fig. 16.11 Some components of termite defense secretions (*1*) Germacrene A, (*2*) β-farnesene, (*3*) geranyllinalool, (*4*) γ-cadinene, (*5*) γ-cadinene aldehyde, (*6*) cubitene, (*7*) biflora-4,10,(19), 15-triene, (*8*) cembrene A, (*9*) 3-*Z*-cembrene A, (*10*) *cis*-β-ocimene, (*11*) 4,11-epoxy-*cis*-eudesmane, (*12*) 10-epi-eudesma-3,11-diene, (*13*) 8-epi-cararrapi oxide, (*14*) cararrapi oxide, (*15*) *cis*-β-ocimene, (*16*) Germacrene A, (*17*) aristolochene, (*18*) epi-α-selenene, (*19*, *20*) macrocyclic lactones, (*21*) myrcene, (*22*) α-pinene, (*23*) β-pinene, (*24*) limonene, (*25*) camphene, (*26*) a trinervitene, (*27*) a kempene, (*28*) a secotrinervitene, (*29*) a rippertane, (*30*) ancistrofuran, (*31*) α-cyclogeraniolene, (*32*) β-cyclogeraniolene, (*33*) ancistrodial (cf. Fig. 16.8; see text for references).

Cubitermes species are soil feeders which do not usually venture to the surface. They construct hard earthen nests, which are common in African savannas, and usually take the form of small tumuli or are mushroom-shaped. The cap of the mushroom-shaped nest of *C. sankurensis* contains 58% of the free space of the nest but a minority of the colony (Bouillon and Lekie, 1964). The cap may help in insulating the stem of the nest from high daytime temperatures caused by insolation. This type of earthen nest is also an efficient defense against opportunist ant predators unless broken open and the colony can also survive attacks of the subterranean doryline ant, *Dorylus kohli* (Williams, 1959). Compounds of remarkable complexity have been identified in the soldier secretion of *Cubitermes umbratus*, including cubitene (Fig. 16.11) which has a skeleton with a novel 12-membered ring arising from irregular joining of isoprene units (Meinwald *et al.*, 1978; Prestwich, 1979b). Other compounds identified include biflora-4,10,(19),15-triene, which has a resemblance to the antibiotic biflorins (Weimer *et al.*, 1980), cembrene-A and 3,Z-cembrene-A. (Fig. 16.11). The diterpenes and several others are of widespread occurrence in East and West African *Cubitermes* species (Prestwich, 1979b, c). The secretion is oily and non-toxic according to Prestwich *et al.* (1978), who imply its function may be as an anti-coagulant.

African species of *Amitermes* build hard small mound nests of wood carton and earthen material, often based on a rotting tree stump. The frontal gland secretions of the *Amitermes* species are, however, totally different from those of *Cubitermes*. The West African species *A. evuncifer* contains a sesquiterpene ether, 4,11-epoxy-*cis*-eudesmane (>90%) (Wadhams *et al.*, 1974) with 10-epi-endesma-3,11-diene, 8-epi-cararrapi oxide, cararrapi oxide, and *cis*-β-ocimene as minor components (Fig. 16.11) (Baker *et al.*, 1978). The secretion is toxic to some ants on topical application, causing motor ataxia, frequently followed by death. Two potential predators, the ants *Odontomachus haematodus* and *Oecophylla longinoda* are affected in this way.

The secretion of the East African species *A. messinae* contains the eudesmane but the minor component is limonene (Meinwald *et al.*, 1978). Of two sympatric species, *A. unidentatus* produces mainly methyl ketones, principally 2-tridecanone, and *A. lonnbergianus* has no hexane-extractable material. The North American species, *A. wheeleri* produces a sesquiterpene hydrocarbon only (Prestwich, 1979b), while the Australian species, *A. herbertensis* and *A. laurensis* contain monoterpene hydrocarbons, the former terpinolene (>98%) and α-phellendrene and the latter limonene and α-pinene (Moore, 1968).

The diversity of frontal gland secretions in members of the genus *Amitermes* so far explored hardly suggests that chemotaxonomy has a promising future in termites, although, in the Nasutitermitinae, which we will now consider, there have been attempts to discern evolutionary trends on the basis of frontal gland chemistry.

16.6.2 Nasutitermitinae

The Nasutitermitinae is a subfamily in which chemical defence mechanisms are developed to an extraordinary degree allowing dense populations to exist in many areas of the tropics. Some species are so well-protected that they are able to forage in the open air forming long trails (e.g., *Hospitalitermes*, *Trinervitermes* – Harvester termites).

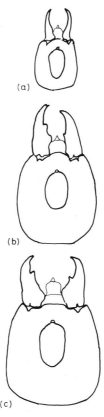

Fig. 16.12 Soldier heads of (a) *Syntermes molestus*, (b) *S. dirus* and (c) *S. grandis* showing the position and relative size of the frontal gland (from Coles, 1980).

Species regarded as morphologically primitive, such as the S. American genus *Syntermes*, have soldiers with a short tubular opening to the frontal pore and well-developed cutting mandibles, while the more 'advanced' species have a long syringe-like nasus and mandibles are vestigial or absent (Figs 16.12 and 16.13).

(a) Syntermes

Ecological studies on various *Syntermes* species in the *cerrado* vegetation zone around Brasilia have shown that foraging activity and the degree of chemical defense shown by a species can be correlated (Coles, 1980; Coles and Howse, 1982). *S. grandis*, *S. dirus* and *S. molestus*, all build hard, earthen, broadly domed mound nests. They forage above-ground at night or in the early morning for grasses which they cut and transport back to the nest. The foraging workers on trails are protected by soldiers in flanking positions. Trails are longest in *S. molestus* (mean 42 cm), shortest in *S. grandis* (Mean 12 cm), with *S. dirus* in an intermediate position (24 cm). The number of soldiers in the trail increases with time, but differs only slightly among the three species (Table 16.2).

In the three species the balance between mechanical and chemical defense differs (Fig. 16.12 and Table 16.2). *S. molestus*, which appears to be the most efficient forager has soldiers which produce a droplet of frontal gland secretion rich in the monoterpene *cis-β*-ocimene (Baker *et al.*, 1981a) with smaller quantities of the sesquiterpenes germacrene A, aristolochene, *epi-α*-selinene and *β*-elemene (a re-arrangement product of germacrene A), (Fig. 16.11). *S. molestus* and three other *Syntermes* species also produce *cis-β*-ocimene and aristolochene, but *S. grandis* soldiers have a secretion that is totally different, consisting only of $C_{19}-C_{22}$ hydrocarbons. Soldiers of this latter species, unlike the other two, retire into the mound when it is opened, and rarely emit a droplet of secretion.

Cis-β-ocimene applied topically to *Camponotus* ants induces paralysis, but the action of the sesquiterpenes is not known. *Cis-β*-ocimene which has a fairly pungent odour noticeable when a *Syntermes* nest is opened, has also been assayed as a feeding inhibitor to the giant anteater, *Myrmecophaga tridactyla*, which is an important mammalian termite predator in S. America. The time captive anteaters spent feeding from a box containing meat, bread and milk with a sample of synthetic *cis-β*-ocimene was significantly less than from a box without the chemical (Coles, 1980).

The hydrocarbon secretion of *S. grandis* also caused ataxia in Brazilian *Camponotus*, although only in about 20% of treated ants. It remains possible that the secretion is anticoagulant in action, as in *Macrotermes* (see below), increasing the severity of wounds caused by the large cutting mandibles in this species.

Table 16.2   Length of foraging trails and amount of soldiers and defensive secretion available to foraging parties of three *Syntermes* species in the Brazilian Federal District (From Coles, 1980)

	S. grandis	*S. dirus*	*S. molestus*
Length of foraging trails (cm)	12.4($\pm$1.1)	24.2($\pm$2.1)	42($\pm$5.9)
Soldier biomass (mg) per 100 workers on trail	1625($\pm$162)	1391($\pm$114)	1278($\pm$50)
Workers per soldier on trail	12.1($\pm$1.7)	9.7($\pm$0.8)	1.6($\pm$0.06)
Potential volume of secretion available to 100 workers (frontal gland vol. in mm^2 × no. soldiers)	22.8($\pm$2.1)	26.5($\pm$2.2)	68.9($\pm$2.7)
Frontal gland vol.: head vol. of soldier	0.022	0.037	0.122

(b) Armitermes

Species such as *Procornitermes* and *Cornitermes* construct hard mound nests in South America, and the soldiers have a frontal pore on a forwardly projecting

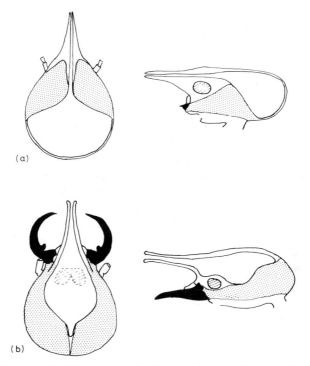

Fig. 16.13   Diagram of soldier heads of a *Nasutitermes* species (a) and *Armitermes holmgreni* (b). Muscle is shown stippled mandibles in black, and brain hatched (from Mill, 1982).

tube. The frontal gland secretion collects as a droplet (Fig. 16.14), which can be daubed into a wound as it is made by the cutting mandibles. In the genus *Armitermes*, the nasus is even more prolonged (Fig. 16.13) extending in some species to the tip of the hooked mandibles or beyond. The secretion of *A. euamignathus*, which has a slight toxic effect on ants, contains large quantities of tri-, tetra-, and pentadecene. These substances are also common in the alarm pheromones of camponotine and formicine ants (Blum, 1981), and may act as 'propaganda pheromones' (allomones) alarming such ants and reducing their raiding efficiency. Three other species from Guyana, *A. holmgreni, A. neotenicus* and *A. teevani* have a soldier secretion with macrocyclic lactones with 22–36 carbon atoms (Prestwich and Collins, 1982), the significance of which is unknown, although they may function as antihealants.

Nests of *Armitermes* species are mainly small and rounded, made from carton and sand particles, with a central attachment to the ground leaving a shallow recess between the bottom of the nest and the soil. In Brazil, invertebrates such as spiders and centipedes, as well as small snakes, use the recess as a refuge and may provide some protection to the nest inhabitants against predators.

Fig. 16.14   Soldiers of (a) *Procornitermes* and (b) *Constrictotermes*.

(c) 'Nasute' species

In the so-called nasutes, the mandibles have undergone considerable regression and the nasus is elongated to form a syringe capable of squirting a liquid glue over distances of up to ten cm (Fig. 16.14). The glue is sequestered in the frontal gland reservoir which occupies the greater part of the head capsule (Fig. 16.13). The secretion dries rapidly in air, becoming sticky and entangling and immolizing arthropods which are attacked.

Curvitermes strictinasus, from Brazil, nests in mounds made by other termite species. When the nest is opened, the soldiers which have small but functional mandibles and a long nasus, come to the breached galleries and daub a droplet of frontal gland secretion on to the substratum. An intense lemony odour is perceptible near the open galleries, which may in itself be a repellent to both invertebrate and vertebrate predators. Workers rapidly rebuild to close the open ends of the galleries under the defensive shield of the soldiers. The secretion contains limonene and terpinolene, with *p*-cymen-8-ol, tridecen-2-one, tridecen-2-one, *cis*,*trans*-farnesal and *trans*,*trans*-farnesal (Baker *et al.*, 1981b).

The secretion of *Curvitermes* is unusual among nasute soldiers in containing no diterpene glue. It appears that soldiers with vestigial mandibles always have a mixture of monoterpenes and diterpenes. Major soldiers of the African harvester species, *Trinervitermes gratiosus*, for example, produce α-pinene with various novel oxygenated tricyclic diterpenes known as '*trinervitenes*' (Prestwich, 1978; Prestwich *et al.*, 1976). Nasute soldiers have since formed a happy hunting ground for chemists searching for diterpenes with novel skeletons among mundane monoterpene hydrocarbons. Tetracyclic 'kempanes' have been found in the arboreal nesting *Nasutitermes kempae* from East Africa (see Prestwich, 1979), bicyclic 'secotrinervitanes' in *Nasutitermes princeps* from New Guinea (Braekman *et al.*, 1980), tetracyclic 'rippertanes' in *N. rippertii* and *N. ephratae* from the neotropics (Prestwich *et al.*, 1980).

A range of monoterpenes has been identified from *Nasutitermes* species (Prestwich, 1979; Baker and Walmsley, 1982) of which α-pinene, β-pinene, limonene, terpinolene, and myrcene are the most common. In *Trinervitermes gratiosus*, intraspecific variations occur (Prestwich, 1978) in allopatric populations from Kenya. Major soldiers produce α-pinene. Minor soldiers from two populations produce α-pinene, β-pinene, camphene, and limonene, but only α-pinene in the other group. More pronounced variations occur in the diterpenes, and populations of this species and *T. bettonianus* (Prestwich and Chen, 1981) are easily distinguished on the basis of the diterpene profiles of major and minor soldiers (Fig. 16.15). These population differences are not due to diet, and injection of labelled precursors (Prestwich *et al.*, 1981) into *Nasutitermes octopilis* has demonstrated that the soldiers are capable of synthesizing the mono- and diterpenoid components of their secretion.

The efficiency of the nasute secretion as a toxic and immobilizing material for other arthropods is beyond question. For example, Bugnion (1927) described

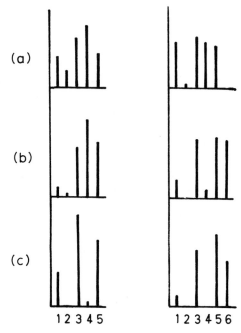

Fig. 16.15   Relative proportions of six diterpenes in the frontal gland secretions of *Trinervitermes gratiosus* soldier frontal glands from mounds in Kenya at Kaekrorok (a) Voi (b) and Kibwezi (c), 40 and 20 km apart, respectively, in a straight line (adapted from Prestwich, 1978).

attacks of the tree ant *Oecophylla smaragdina* on *Hospitalitermes monoceros* in Sri Lanka. After attack by a soldier, an ant falls almost immediately from the tree. 'Henceforth it is vanquished, and incapable of resisting . . . if we follow the ant which has fallen to the ground, we see her busied for a long time in rubbing her mouthpieces against stones, roots, etc., in the effort to free herself from the liquid which limes them.' Eisner *et al.* (1976) have made a detailed study of the defensive behavior of the Australian species *Nasutitermes exitiosus*, and described a similar fate for *Iridomyrmex purpureus* attacked by nasute soldiers in the field.

The possible roles of the monoterpene hydrocarbons in nasute secretions are indicated by the following. (i) Experiments with cockroaches and houseflies (Eisner *et al.*, 1976), showed that α- and β-pinene act as irritants, inducing preening and scratching movements that help to spread a secretion further over a contaminated insect. (ii) Scanning electron micrographs show that the *N. exitiosus* secretion can block and cover spiracles and sensilla (Eisner *et al.*, 1976). (iii) They may act as alarm pheromones (Moore, 1969). Eisner *et al.* (1976) found that contaminated targets attracted other soldiers over a distance of up to 30 mm although they were unable to identify the chemical components involved. Vrkoč *et al.* (1978) measured alarm behavior chromatographically, recording the increase in concentration of volatile defence secretion released in laboratory colonies in response to various stimuli. They found α-pinene to be most effective for *N. rippertii*, and 3-carene for *N. costalis*, but in each case the natural secretion was more effective than any of the individual components. (iv) The monoterpene components α- and β-pinene, and other monoterpene hydrocarbons, are toxic to *Formica rufa*, *Reticulitermes flavipes* and *Myrmicaria eumenoides* on topical application ($LD_{50} = 0.05-0.91 \mu g$ per insect for *F. rufa*) (Howse, 1975). They are hydrophobic and dissolve easily into the insect cuticle. Similarly, synthetic samples of limonene, α-phellendrene, α-pinene, β-pinene, α-thujene and *cis-β-ocimene* applied to indigenous species of *Camponotus* and *Acromyrmex* in Brazil proved to be toxic or to cause ataxia in doses of $0.15 \mu l$ per ant (Coles, 1980; Coles and Howse, 1982). (v) Feeding tests with captive giant anteaters (*Myrmecophaga tridactyla*) showed that limonene, β-pinene and *cis-β-ocimene* acted as feeding deterrents, and also that species with nasute soldiers were eaten less readily than those with mandibulate soldiers in choice tests (Coles and Howse, 1982). Similarly, Lubin and Montgomery (1981) have found that an anteater (*Tamandua mexicana*) in Panama rejected nasute soldiers in feeding tests while accepting workers and reproductives. *T. mexicana* and *T. tetradactyla* frequently withdrew from *Nasutitermes* when they began to inspect them in the field, often brushing the nose and sneezing. Lubin and Montgomery attribute this to the irritant effect of pinenes on vertebrate mucous membranes. There are also various observations in the literature (e.g., Eisner *et al.*, 1976) which suggest that the monoterpenes are repellent at a distance to ants. The writer has also sprinkled nasutes in the path of army ants (*Eciton* spp.) in Brazil, and found that the ants diverted their trail within a few minutes to circumvent the termite soldiers, which they did not attack. (vi) Monoterpenes act as efficient solvents for the higher molecular weight diterpene glues (Prestwich, 1979), and at the same time the diterpenes can provide a slow-release formulation, prolonging the effect of the monoterpenes. (vii) The possibility remains that insects that synthesize a mixture of volatile compounds can hide under the 'smoke-screen' generated by activation of receptors of different kinds, including generalist receptors, in insect predators (Blum, 1981). This is pure speculation, but it may be noted that some nasutes produce up to 10 different monoterpene hydrocarbons (Prestwich, 1979a).

The diterpenes are characterized by a domed molecular shape (Prestwich, 1979a) with hydrophilic and hydrophobic portions. Prestwich argues that a mixture of different diterpenes will make a more viscous glue in which the possibility of crystallization is removed, and a predator would have great difficulty in developing immunity to a multicomponent system.

The toxicology of diterpenes requires investigation. Hrdý *et al.* (1977) showed that the major diterpene of *Nasutitermes rippertii* is as toxic to *Musca domestica* larvae as many common monoterpenes.

The way in which the secretions are used may be related to nest structure (see below). If a nest is opened, nasute soldiers typically swarm out in large numbers and cover the area around the breach. They eject their secretion with an oscillation of the body that serves to spread the sticky thread in a series of loops covering a greater area (Nutting *et al.*, 1974; Eisner *et al.*, 1976). It is curious that the large soldiers of the Australian mound-builder *Nasutitermes exitiosus* are non-aggressive and retreat into the mound (McMahan, 1974), and are also repelled by the frontal gland secretion. It is suggested (Kriston *et al.*, 1977) that they act as messengers, transmitting alarm in the nest, but their role in foraging parties has not been fully investigated.

Relatively little is known of trail pheromones in the Nasutitermitinae. Jander and Daumer (1974) showed that *Hospitalitermes sharpi*, which makes long (*ca.* 100 m) and highly populated trails in the open in East Asia, does not use visual cues in orientation but relies almost completely on the chemical trail. Outgoing workers are on the outside of the trail and returning ones on the inside. This does not depend on directional information in the trail, but on the pattern of flow that is established, so that termites moving against the stream turn into it as a result of the collisions they suffer. Soldiers flank the trail, facing outwards, and head an advancing trail but form a rearguard to a retreating trail.

Sands (1961) in a study of five sympatric species of *Trinervitermes* in W. Africa showed that they occupy distinct ecological niches with different kinds of trailing behavior. All the species forage from small mound nests for grasses. *T. geminatus* forages extensively in long columns at night, taking longer grass than the others. *T. trinervius* forages under covered mud runways constructed over the surface of the soil and lives in more moist conditions than *T. geminatus*. *T. suspensus* takes smaller and finer grasses, but most nests are found in the nests of other species. The two remaining species differ from those above in not storing grass in the nest. *T. occidentalis* occurs in more sheltered habitats. It can use ligneous material and constructs mud runways up tree trunks, and forages by day in the shade. *T. oeconomus* does not make long foraging trails, but insects spread out in the vegetation close to the mound, and often continue their foraging into the earlier part of the day. *T. oeconomus* and *T. occidentalis* forage throughout the year, while the 'true harvesters' which store grasses restrict their foraging activity, avoiding the wettest and driest months.

Polymorphism of the sternal gland has been shown in *Trinervitermes bettonianus* and *T. gratiosus* (Quennedey and Leuthold, 1978; Leuthold and Lüscher, 1974), but not in *T. trinervius*. The glands are found on the fifth abdominal sternite in the Termitidae. In *T. bettonianus* the gland of the freshly dalate female is 1200 times greater than that of the worker, and that of the male 70 times greater. The soldiers, which do not appear to lay trails, have vestigial glands. Pasteels (1965) found a polymorphism within the worker castes of *Nasutitermes lujae*. The glands were largest in second-stage minor workers, and second- and third-stage major workers which are found on trails, and smallest in the first-stage workers which do not forage.

The sternal gland polymorphism of *T. bettonianus* is related to the courtship behavior of the winged forms (Fig. 16.16) (Leuthold, 1977). Females call from

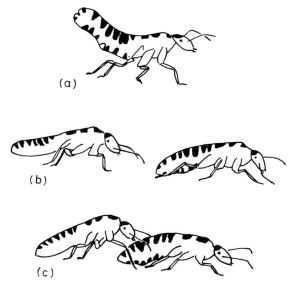

Fig. 16.16   Post-flight behavior of *Trinervitermes bettonianus*. (a) Female calling behavior, showing tergal glands and sternal gland exposed; (b) Male following strong pheromone trail of female; (c) Tandem run, with male in attennal contact with female (after Leuthold, 1975).

elevated points and also lay a powerful pheromone trail while walking. At 10–20 cm distance in still air the female's tergal gland releases searching behavior in the male. If the male encounters a fresh (<10 s old) sternal gland trail it is able to follow the trail, presumably up the concentration gradient. At 3 cm, the male orientates by the aerial odour trail. Tandem running occurs when he contacts the female, and running of the pair is released by the contact, maintained by the combined action of the sternal and tergal glands on the male.

The only termitid trail pheromone known is neocembrene A (Birch *et al.*, 1972), a diterpene hydrocarbon, which is found in the Australian species *Nasutitermes exitiosus*, *N. walkeri* and *N. graveolus*. This compound, also present in some conifers, is closely related to cembrenes, believed to be intermediates in the biosynthesis of nasute defensive secretion diterpenes (Prestwich, 1979a).

(d) Macrotermitinae

The Macrotermitinae are fungus-growers. They do not exist in the New World, their place being taken by the Attini (leaf-cutting ants) which have a similar fungal symbiosis.

A consequence of the fungus-growing habit is that the combined metabolism of fungus and insects in enclosed nests produces excess heat and carbon dioxide. The Macrotermitinae have a variety of nesting habits that provide solutions to the problems generated by the fungus. *Microtermes* builds single combs. Species such as *Syncanthotermes* have dispersed calies of fungus gardens, which are interconnected by galleries. Some S. African *Odonotermes* species have above-ground chimney-like structures with cavities leading down to fungus gardens. Warm air ascends from below, although the ducts may be occluded at certain times of the year.

The structure of *Macrotermes* mounds shows great diversity. That of *M. bellicosus* (Figs 16.17 and 16.18) consists of a central 'hive' containing fungus gardens, queen cell, brood and the majority of the colony at any one time. The hive stands on conical pillars and is separated from the thick outer wall by a shallow air-space. On the basis of measurements made on mounds in the Ivory Coast, Lüscher (1955) showed that there was an air circulation system within the mound. Warm humid air, rich in CO_2 but depleted in oxygen, rises to a space above the hive and is then distributed peripherally in ducts which narrow and ramify close to the surface of the mound (Fig. 16.18). Here air exchange

Fig. 16.17   Mound nest of *M. bellicosus* from near Mokwa, Nigeria.

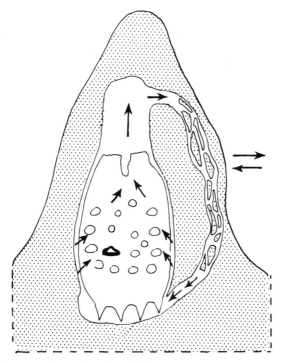

Fig. 16.18   Diagrammatic section of *M. bellicosus* mound showing system of air circulation (arrowed) from central hive with queen cell and fungus gardens to the top and periphery of the mound where air exchange takes place (after Lüscher, 1955).

takes place. Cooler, denser air, enriched in oxygen then descends through ducts to the base of the hive.

Even within the species *bellicosus* there is much diversity of nest structure. In Kinshasa, Ruelle (1964) found that there was no regular thermosiphon effect in nests, but air movements in the large ducts were greatly influenced by wind, local insolation and other factors. Mounds in the Mokwa region of Nigeria reach a height of up to 10 m. The hive is mainly below ground level and has a curious base-plate, with a downwardly projecting spiral ridge, that nests on a central pillar (Collins, 1979). Lüscher (1955) also described closed-type mounds in Uganda, in which air ascends into thin-walled cavities near the surface of the mound, and in which there is no air circulation as in Ivory Coast mounds. Harris (1971) described open-type mounds from E. Africa (?) with central chimneys.

Both closed and open mounds of *Macrotermes* exist in the Kajiado district of Kenya (van der Werff, 1981). Those of the open type are *M. subhyalinus*, while the closed type are now referred to as *M. michaelseni*. Along with morphometric data, details of nest structure, and behavior, the composition of the soldier frontal gland secretion has been used to separate the species. *M. subhyalinus*

Table 16.3   Responses of *Megaponera foetens* workers on foraging trails to filter papers on which soldier termites had been freshly squashed. Tests in Guinea savanna near Mokwa, Nigeria (data from Longhurst and Howse, *in preparation*)

	Alerting	Alarm	Stridulating	Repellency	Approach and withdrawal	Rapid approach	Bite	Sting	Sometimes carry away paper	Importance as prey
Macrotermes bellicosus	+					+	+	+	+	Preferred
Odontotermes smeathmani	+					+	+	+	+	Frequent
Macrotermes subhyalinus	+				+		+	+		Rare
Microtermes sp. A	+		+		+		+	+		Rare
Ancistrotermes cavithorax	+	+ (low)			+ (and inspection)					Rare
Trinervitermes geminatus and *togoensis*	+	+ (high)	+	+	+		+			Never
α-pinene + β-pinene + sabinene	+	+ (high)	+	+						(Present in *T. geminatus* secretion)

contains long-chain saturated and mono-unsaturated hydrocarbons, (C_{25}–C_{29}), including *n*-tricosane, *n*-pentacosane, 3- and 5-methylpentacosane, 5-methyl-heptacosane, (Z)-9-heptacosene and (Z)-9-nonacosene (Prestwich *et al.*, 1977). *M. michaelseni* contains mainly C_{27}–C_{35} alkanes and alkenes.

The major soldiers of *M. subhyalinus*, which have powerful cutting mandibles, contain about 500 times more frontal gland secretion than the minors, and are mainly found within the nest, while the minor soldiers are found with foragers. Prestwich *et al.* (1977), tested the crude soldier secretion against the obligate termite predator *Megaponera foetens* known to take *M. subhyalinus* as prey, and found that ants were killed only when the secretion was applied to a wound. The reason for this was claimed to be excess loss of haemolymph compared with controls.

Presentation of filter papers with the 'juice' of crushed soldiers of a Nigerian population of *M. subhyalinus* to *Megaponera foetens* foraging columns in the field (Longhurst and Howse, unpublished) resulted in alerting of ants followed by an oscillating process of orientation toward the termite and retreat. The initial approach was to within 50 mm of the odour source, but after 40 s, the ant bit and stung the filter paper. Filter papers on which *Macrotermes hellicosus* and *Odontotermes smeathmani* had been crushed were detected by foragers at 150–200 mm, after which an ant rapidly approached, bit and stung the paper. The reactions to these and other species are summarized in Table 16.3. It appears that the secretions of the commonest prey in the Mokwa area may act as kairomones. In contrast, the secretions of the *Trinervitermes* species (and some of the monoterpene hydrocarbons they contain) are highly repellent and alarming.

It seems that *Megaponera foetens* has developed resistance to the *Macrotermes subhyalinus* secretion and therefore the bioassays of Prestwich *et al.* (1977) must be treated with some caution. It is of interest that Löfquist (1977) found saturated alkenes (tridecane and undecane) to be more toxic to *Formica rufa* and *F. sanguinea* than formic acid. These compounds penetrated through the tracheae. It may be wrong to assume that higher-molecular-weight alkanes and alkenes are non-toxic to all species of ant. Longhust, Briner and others (unpublished) found that *M. subhyalinus* populations around Mokwa were resistant to *Megaponera* predation, while those from riverine areas of Zugurma and Rabba nearby were heavily predated. This may be due to a high level of a C_{16} diene found in the Mokwa populations that is present in only a low level in soldiers from the other areas.

Odontotermes badius and *O. stercivorus* from E. Africa produce a mixture of benzoquinone and protein, which rapidly tans becoming glutinous (Wood *et al.*, 1975). Some *Odontotermes*, *Hypotermes*, *Macrotermes* and *Microtermes* species from S.E. Asia produce secretions containing benzoquinone and toluquinone, which are expelled from hypertrophied labial glands (Fig. 16.19) (Maschwitz and Tho, 1974). *Macrotermes bellicosus* produces toluquinone (Baker, Briner, Evans and Howse, unpublished) and *Ancistrotermes cavithorax*

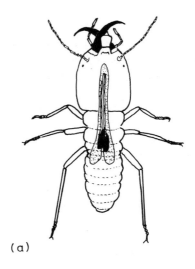

(a)

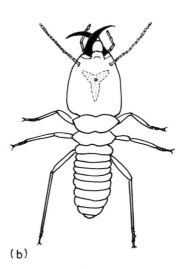

(b)

Fig. 16.19   (a) Soldier of *Macrotermes carbonarius* showing position of the hyper-tophied labial gland (black) and reservoir (stippled) (after Maschwitz *et al.*, 1972); (b) Soldier of *Macrotermes subhyalinus* showing the position of the frontal gland (stippled) and fontanelle (after Prestwich *et al.*, 1977).

major soldiers the furanosesquiterpene ancistrofuran, with α- and β-cyclo-geranolenes and toluene (Baker *et al.*, 1978), while the minor soldiers produce the oxygenated sesquiterpene ancistrodial.

It appears, therefore, that *Megaponera foetens* may have evolved general resistance to quinoid and alkane and alkene secretions, but remains susceptible to terpenoids.

Noirot (1969b) has shown that the development of castes in the Termitidae is largely determined by sexual mechanisms. Thus, in *Macrotermes bellicosus* male larvae give rise to major workers and female larvae to minor workers, and major or minor soldiers. Sexual forms develop through six nymphal stages.

There is indirect evidence that pheromones act in the production of nymphs and soldiers. Nymphs develop in *M. bellicosus* if the royal couple are removed (Bordereau, 1975). In *M. subhyalinus*, there is a large increase in the volume of the corpora allata in the presoldier instar, but the volume remains low in soldiers (Okot-Kotber, 1977). High but variable juvenile hormone titres also exist in eggs of *M. subhyalinus* (Lüscher, 1976). The titres are lowest before the seasonal appearance of nymphs in *M. subhyalinus*. Titres of JH in the queen haemolymph also show a seasonal fluctuation. It appears likely that first-instar larvae tend to develop into nymphs when the JH titre is low. When it is high, after swarming or shortly after colony foundation, production of nymphs is inhibited. A high JH level can also lead to soldier production starting from the third-stage larvae. Lüscher (1976) believes that pheromones from the soldiers inhibit corpora allata activity and pheromone from the reproductives stimulate corpora allata activity.

The queen of *M. subhyalinus* plays another important role in the society by controlling building activities. Bruinsma (1979) showed that workers respond to an isolated physogastric queen by building a replacement royal cell around her. A chemical stimulus, shown by bioassays to originate from the fat body, is responsible for the organization of this building behavior. A queen with a varnished abdomen only elicited building if the spiracles were uncovered. It appears that the building pheromone diffuses through the spiracles and spreads on to the cuticle, where it stimulates building a short distance away. It may consist of various unsaturated fatty acids. The trail pheromone also plays a part in facilitating the building activity of workers, guiding them to specific locations. A 'cement pheromone' from the worker salivary gland also focuses the activity of workers. In the building of pillars and galleries, the edge of a chemical trail stimulates deposition of particles, and the width and height of galleries is determined by the distribution of the trail pheromone. The cement pheromone controls the building of arches between neighbouring pillars.

(e) Apicotermitinae

The Apicotermitinae includes soldierless genera and genera with regressed mandibles. Soldiers are rare and apparently redundant in colonies of *Speculitermes*, but the workers can whip their abdomens round and deposit a faecal droplet on the head of a predator (Sands, 1972). Similar behavior has been described in *Skatitermes psammophilus*, which has a very mobile abdomen (Coaton, 1971), and also occurs in *Grigiotermes metoecus* from South America, although this species must first turn to face away from the target (Mill, 1982).

Abdominal dehiscence is common among the soldierless African species (Sands, 1972) in which the abdominal wall fractures, as a result of contraction of abdominal muscles, and the gut contents are expelled. Grassé and Noirot (1951) described columns of *Anoplotermes*, flanked by workers facing outward in a soldier-like manner, which were attacked by ants and killed, but subsequently rejected and were still untouched 20 h later. This suggests that their bodies contain substances noxious to ants.

Abdominal dehiscence occurs in *Ruptitermes* with emission of a sticky fluid from the enlarged salivary glands which have abdominal reservoirs (Mathews, 1977). The fluid may also be emitted through the mouth. Camponotine ants are sometimes killed instantly by the secretion (personal observations) or are immobilized and die some hours later. Mill (1982) found that an *Ectatomma* species often found preying on termites and *Pseudomyrmex termitarius*, a scavenging and sometimes opportunistic termite predator, were killed by 1–4 termite loads (LD_{50}) of three *Nasutitermes* species, *Ruptitermes* sp. and *Armitermes teevani*. A 'termite load' is the full defensive secretion load of one termite. *Ruptitermes proratus* secretion was effective in 5–6 loads for *P. trinervius* and 8–9 for *Ectatomma*. Ant species differed considerably in their susceptibility to *R. proratus* secretion, a *Pheidole* and another *Pseudomyrmex* species requiring only about 1 load (LD_{50}), and *Camponotus abdominalis*, 2–3.

16.7 TOWARDS A CHEMICAL ECOLOGY OF TERMITES

Chemical secretions, as we have seen, play an important role in termite foraging, recruitment, caste determination, and nest-building, as they do in social insects in general. However, in termites, we are faced with an excellent opportunity to explain inter-relationships between species in chemical terms because of their frequent reliance on chemical defense, which is linked with other aspects of their biology, such as foraging, feeding habits, nest structure and building behavior.

Mill (1982) estimates that 96% of all termite species in Brazil that forage in the open air or under covered galleries have soldiers with chemical defenses. On the other hand, about half the species that are subterranean or forage under litter, have mandibute soldiers. It appears then that mechanical defence alone is inadequate against most above-ground foragers. Ant predation on undefended insects in the tropics is especially high. Jeanne (1979), in experiments with wasp larva baits, found that rates of ant predation were significantly higher and baits were found sooner in the neotropics than in the USA. Ant species richness, as well as the overall predation rate, also increased towards the equator.

Species which rely upon mechanical defense tend to be those which build hard mounds. The soldiers withdraw and defend within the mound if it is broken open, where they can effectively block galleries with their large heads, or crush intruding ants against the walls. Conversely, species with well-developed chemical defense are able to form long foraging trails and exploit a wider area in their search for food. Their nests are more fragile. A *Nasutitermes* nest is easily opened − it has only a thin brittle lamella for an outer wall − and soldiers then swarm over the area around the breach. If they defended within the nest they would probably become entangled in their own glue, make the galleries hazardous for workers, and bring the concentration of toxic volatiles within the nest to a high level.

The behavior of workers is also important: they seal opened nest galleries more rapidly in hard mounds of *Cornitermes*, *Procornitermes araujoi* and *Armitermes euamignathus* than in nests of species with nasute soldiers such as *Velocitermes paucipilis* (Coles, 1980). The soldierless termite, *Grigiotermes metoecus*, also builds very hard mounds and repairs openings very rapidly with sticky faecal material. Dehiscent species are well-defended both inside and outside the nest. Inside, they can block galleries with their faeces or by self-immolation. Outside they are not dependant on the presence of soldiers for defense of a foraging party.

The importance of vertebrate predators should not be underestimated. Feeding tests with giant anteaters (p. 498) show that monoterpene hydrocarbons can act as feeding deterrents, and that mandibulate termite species are preferred as food to nasute species (Coles and Howse, 1982). Tamandua anteaters are apparently repelled by nasutes (Lubin and Montgomery, 1981). Similarly, analysis of the aardwolf (*Proteles cristatus*) diet showed that *Trinervitermes bettonianus*, was the main item in the Serengeti (Kruuk and Sands, 1972) but a high concentration of soldiers at the feeding site appears to inhibit feeding.

The role of sesquiterpenes and diterpenes in soldier secretions is little understood. Possibly sesquiterpenes, which are also common toxins in many poisonous plants, may sometimes act as feeding deterrents or inhibitors to vertebrate termite predators. Aristolochene, found in various *Syntermes* species, has a very high mammalian toxicity. The total secretion also has a powerful 'gamey' smell, and an association between odour of a mound and palatability of the inhabitants could be easily formed.

Multiple occupation of old mounds by a variety of termite species and also ants is common in S. America and elsewhere in the tropics. A statistical analysis by Coles (1980) showed significant associations occurring in mounds on a study site near Brasilia. Soldierless species were commonly found at the top of the mound, humivores with short mandibles in the centre, species with snapping mandibles at the base, and nasutes around the edges. An invader of such a mound therefore has to face a changing sequence of weaponry, analogous to a herbivore facing plant defence guilds (Atsatt and O'Dowd, 1976).

The association between termite defensive chemicals and their action on potential predators will clearly repay further study. The evidence so far suggests that predatory ants and termites have co-evolved and that a predator is successful because it is more or less immune to the secretions of the prey. The variations in defensive chemistry of termites may well reflect co-evolutionary struggles in which defense must be maintained against specialized predatory ant species and against a spectrum of opportunists that may differ according to geographical locality and microhabitat.

REFERENCES

Ampion, M. and Quennedey, A. (1981) The abdominal epidermal glands of termites and their phylogenetic significance. In: *Biosystematics of Social Insects* (Howse, P. E. and Clément, J.-L., eds). Academic Press, London and New York.

Atsatt, P. R. and O'Dowd, D. J. (1976) Plant defense guilds. *Science*, **193**, 24–9.

Bacchus, S. (1979) New exocrine gland on the legs of some Rhinotermitidae (Isoptera). *Int. J. Insect Morph. Embryol.*, **8**, 135–42.

Baker, R., Briner, P. H. and Evans, D. A. (1978) Chemical defence in the termite *Ancistrotermes cavithorax*: ancistrodial and ancistrofuran. *Chem. Commun.*, 410–11.

Baker, R., Coles, H. R., Edwards, M., Evans, D. A., Howse, P. E. and Walmsley, S. (1981a) Chemical composition of the frontal gland secretion of *Syntermes* soldiers (Isoptera, Termitidae). *J. Chem. Ecol.*, **7**, 135–45.

Baker, R., Edwards, M., Evans, D. A. and Walmsley, S. (1981b) Soldier-specific chemicals of the termite *Curvitermes strictinasus* Mathews (Isoptera, Nasutitermitinae). *J. Chem. Ecol.*, **7**, 127–33.

Baker, R., Parton, A. H. and Howse, P. E. (1982) Identification of an acyclic diterpene alcohol in the defense secretion of soldiers of *Reticulitermes lucifugus*. *Experientia*, **38**, 297.

Baker, R. and Walmsley, S. (1982) Soldier defense secretions of the South American termites *Cortaritermes silvestri*, *Nasutitermes* sp.n.D. and *Nasutitermes kemneri*. *Tetrahedron*, **38**, 1899–910.

Birch, A. J., Brown, W. V., Corrie, J. E. T. and Moore, B. P. (1972) Neocembrene A, a termite trail pheromone. *J. Chem. Soc.*, *Perkin Trans.*, **1**, 2653–8.

Blomquist, G. J., Howard, R. W. and McDaniel, C. A. (1978) Biosynthesis of the cuticular hydrocarbons of the termite *Zootermopsis angusticollis* (Hagen). Incorporation of propionate into dimethylalkanes. *Insect Biochem.*, **9**, 371–4.

Blomquist, G. J., Howard, R. W. and McDaniel, C. A. (1979) Structures of the cuticular hydrocarbons of the termite *Zootermopsis angusticollis* (Hagen). *Insect Biochem.*, **9**, 365–70.

Blum, M. S. (1981) *Chemical Defences of Arthropods*. Academic Press, New York.

Bordereau, C. (1975) Determinisme des castes chez les termites supérieurs: mise en évidence d'un contrôle royal dans la formation de la caste sexuée chez *Macrotermes bellicosus* Smeathman (Isoptera, Termitidae). *Insectes Sociaux*, **22**, 363–74.

Bouillon, A. and Lekie, R. (1964) Populations, rhythme d'activité diurne et cycle de croissance au nid de *Cubitermes sankurensis* Wasmann (Isoptera, Termitinae). In: *Etudes sur les Termites africains* (Bouillon, A., ed.) pp. 197–213. Masson, Paris.

Braekman, J. C., Daloze, D., Dupont, A., Pasteels, J., Tursch, B., Declerq, J. P., Germain, G. and Van Meerssche, M. (1980) Secotrinervitane, a novel bicyclic diterpene skeleton from a termite soldier. *Tetrahedron Letters*, **21**, 2761–2.

Bruinsma, O. M. (1979) *An Analysis of Building Behaviour of the Termite Macrotermes subhyalinus*. Doctoral thesis, Agricultural University, Wageningen, Holland.

Buchli, H. (1960) Les tropismes lors de la pariade des imagos de *Reticulitermes lucifugus* R. *Vie et Milieu*, **6**, 308–15.

Bugnion, E. (1927) The origin of instinct. A study of the war between the ants and the termites. *Psyche Monogr.*, **1**, 1–44.

Clément, J.-L. (1977) Écologie des *Reticulitermes* (Holmgren) français (Isoptères). Position systematique des populations. *Bull. Soc. Zool. France*, **102**, 169–85.

Clément, J.-L. (1978a) Nouveaux critères taxonomiques dans le genre *Reticulitermes* (Holmgren) (Isoptera). Description de nouveaux taxons français. *Ann. Soc. ent. France*, **14**, 131–9.

Clément, J.-L. (1978b) L'agression interspécifique et intraspécifique des espèces français du genre *Reticulitermes* (Isoptère). *CR Acad. Sci. Paris*, 351–4.

Clément, J.-L. (1981) *Spéciation des Reticulitermes Européens (Isoptères)*. Doctoral thesis, Université Pierre et Marie Curie, p. 244.

Clément, J.-L. (1982) Pheromones d'attraction sexuelle des termites Européens du genre *Reticulitermes*. Mechanismes comportementaux et isolements spécifiques. *Biol. Comportement* (in press).

Coaton, W. G. H. (1946) The harvester termite problem in South Africa. *S. Af. Dept. Agric. Sci. Bull.*, **292**.

Coaton, W. G. (1971) Five new termite genera from West Africa (Isoptera: Termitidae). *Cimbebasia* (A), **2**, 1–34.

Coles, H. M. (1980) *Defensive Strategies of the Ecology of Neotropical Termites*. Unpublished PhD thesis, University of Southampton.

Coles, H. M. and Howse, P. E. (1982) Termite defensive secretions: Ecological aspects. In: *Social Insects in the Tropics* (International Union for the Study of Social Insects Symposium) (Jaisson, P., ed.) Université Paris Nord.

Collins, H. M. (1979) The nests of *Macrotermes bellicosus* (Smeathman) from Mokwa, Nigeria. *Insectes Sociaux*, **26**, 240–6.

Deligne, J. (1971) Mechanics of the fighting behaviour of the termite soldiers (Insecta: Isoptera). *Forma et Function*, **4**, 176–87.

Deligne, J., Quennedey, A. and Blum, M. S. (1980) The enemies and defense mechanisms of termites. In: *Social Insects*, Vol. II (Hermann, H. R., ed.) pp. 1–76. Academic Press, New York.

Eisner, T., Kriston, I. and Aneshansley, D. J. (1976) Defensive behaviour of a termite (*Nasutitermes exitiosus*). *Behav. Ecol. Sociobiol.*, **1**, 1–126.

Gay, F. J. and Calaby, J. H. (1970) Termites of the Australian Region. In: *Biology of Termites*, Vol. II (Krishna, K. and Weesner, F. M., eds) pp. 393–447. Academic Press, New York.

Grassé, P.-P. and Noirot, C. (1951) La sociotomie: migration et fragmentation de la termitière chez les *Anoplotermes* et les *Trinervitermes*. *Behaviour*, **3**, 146–66.

Harris, W. V. (1971) *Termites, Their Recognition and Control*. Longmans, London.

Haverty, M. I. (1979) Soldier production and maintenance of soldier proportions in laboratory experimental groups of *Coptotermes formosanus* Shiraki. *Insectes Sociaux*, **26**, 69–84.

Holmgren, N. (1906) Studien über südamerikanische Termiten. *Zool. Jahrb.*, *Abt. Systematik*, **23**, 521–676.

Howard, R., Matsumura, F. and Coppel, H. C. (1976) Trail-following pheromones of the Rhinotermitidae: approaches to their authentication and specificity. *J. Chem. Ecol.*, **2**, 147–66.

Howard, R. W. and McDaniel, C. A. (1981) Chemical mimicry as an integrating mechanism: cuticular hydrocarbons of a termitophile and its host. *Science*, **210**, 431–3.

Howard, R. W., McDaniel, C. A. and Blomquist, G. J. (1978) Cuticular hydrocarbons of the Eastern Subterranean termite *Reticulitermes flavipes* (Kollar) Isoptera: Rhinotermitidae. *J. Chem. Ecol.*, **4**, 233–45.

Howick, C. D. and Greffield, J. W. (1980) Intraspecific antagonism in *Coptotermes acinaciformis* (Froggatt) (Isoptera: rhinotermitidae). *Bull. Ent. Res.*, **70**, 17–23.

Howse, P. E. (1964) The significance of the sounds produced by the termite *Zootermopsis angusticollis* (Hagen). *Anim. Behav.*, **12**, 284–300.

Howse, P. E. (1966) Air movement and termite behaviour. *Nature*, **210**, 967–8.

Howse, P. E. (1970) *Termites: A Study in Social Behaviour*. Hutchinson University Library, London.

Howse, P. E. (1975) Chemical defence of ants, termites and other insects: some outstanding questions. In: *Pheromones and Defensive Secretions in Social Insects* (Noirot, C., Howse, P. E. and Le Masne, G., eds) pp. 23–40. Université de Dijon.

Hrdý, I., Kreček, J. and Vrkoč, J. (1977) Biological activity of soldiers secretions in the termites: *Nasutitermes rippertii*, *N. costalis* and *Prorhinotermes simplex*. *Proc. VIII Int. Cong. IUSSI*. Pudoc, Wageningen. pp. 303–4.

Hummel, H. and Karlson, P. (1968) Hexansäure als Bestandteil des Spurpheromons der Termite *Zootermopsis nevadensis* Hagen. *Hoppe Seyler's Z. Physiol. Chem.*, **349**, 725–7.

Jander, R. and Daumer, K. (1974) Guide-line and gravity orientation of blind termites foraging in the open (Termitidae: Macrotermitidae). *Insectes Sociaux*, **21**, 45–69.

Jeanne, R. L. (1979) A latitudinal gradient in rates of ant predation. *Ecology*, **60**, 1211–24.

Krishna, K. (1970) Taxonomy, phylogeny and distribution of termites. In: *Biology of Termites*, Vol. II (Krishna, K. and Wessner, F. M., eds). Academic Press, New York.

Krishna, K. and Weesner, F. M. (eds) (1969) *The Biology of Termites*, Vols I and II. Academic Press, New York.

Kriston, I., Watson, J. A. L. and Eisner, T. (1977) Non-combative behaviour of large soldiers of *Nasutitermes exitiosus* (Hill): an analytical study. *Insectes Sociaux*, **24**, 103–11.

Kruuk, H. and Sands, W. A. (1972) The aardwolf (*Proteles cristatus* Sparrman, 1783) as a predator of termites. *East Af. Wildlife J.*, **10**, 211–27.

Lee, K. E. and Wood, T. G. (1971) *Termites and Soils*. Academic Press, New York.

Lebrun, D. (1967) Implications hormonales dans la morphogenèse des castes du termite *Kalotermes flavicollis* Fabr. *Bull. Soc. Zool. France*, **103**, 351–8.

Leuthold, R. H. (1977) Postflight communication in two termite species, *Trinervitermes bettonianus* and *Hodotermes mossambicus*. *Proc. VIII Int. Cong. IUSSI*. Pudoc, Wageningen. pp. 62–4.

Leuthold, R. H. and Lüscher, M. (1974) An unusual caste polymorphism of the sternal gland and its trail pheromone production in the termite *Trinervitermes bettonianus*. *Insectes Sociaux*, **21**, 335–42.

Leuthold, R. H., Bruinsma, O. and Van Huis, A. (1976) Optical and pheromonal orientation and memory for homing distance in the harvester termite, *Hodotermes mossambicus* (Hagen). *Behav. Ecol. Sociobiol.*, **1**, 127–39.

Light, S. F. (1942–3) The determination of castes of social insects. *Q. Rev. Biol.*, **17**, 312–26; **18**, 42–63.

Light, S. F. (1944) Experimental studies on ectohormonal control of the development of supplementary reproductives in the termite genus *Zootermopsis* (formerly *Termopsis*). *Univ. Calif. Pub. Zool.*, **43**, 413–54.

Lofquist, J. (1977) Toxic properties of the chemical defense systems in the competitive ants *Formica rufa* and *F. sanguinea*. *Oekos*, **28**, 137–51.

Lubin, Y. D. and Montgomery, G. G. (1981) Defense of *Nasutitermes* termites (Isoptera, Termitidae) against *Tamandua* anteaters (Edentata, Myrmocophagidae). *Bioptropica*, **13**, 66–76.

Lüscher, M. (1952) Die Produktion und Elimination von Ersatzgeschlechtstieren bei der Termite *Kalotermes flavicollis* Fabr. *Z. vergl. Physiol.*, **34**, 123–41.

Lüscher, M. (1955) Der Sauerstoffverbrauch bei Termiten und die Ventilation des Nestes bei *Macrotermes natalensis* (Haviland). *Acta Tropica*, **12**, 289–307.

Lüscher, M. (1974) Kasten und Kastendifferenzierung bei niederen Termiten. In: *Sozial-polymorphismus bei Insekten*. pp. 694–739. Wissenschaftliches Verlagsgesellschaft, Stuttgart.

Lüscher, M. (1976) Evidence for an endocrine control of caste determination in higher termites. In: *Phase and Caste Determination in Insects*. (Lüscher, M., ed.) pp. 91–103. Pergamon Press, Oxford and New York.

Lüscher, M. (1977) Queen dominance in termites. *Proc. VIII Int. Cong. IUSSI*, Pudoc, Wageningen. pp. 238–42.

Lüscher, M. and Müller, B. (1960) Ein spurbildendes Sekret bei Termiten. *Naturwissenschaften*, **27**, 503.

Lüscher, M. and Springhetti, A. (1960) Untersuchungen über die Bedeutung der Corpora Allata für die Differenzierung der Kasten bei der Termite *Kalotermes flavicollis* Fabr. *J. Insect Physiol.*, **5**, 190–212.

McMahan, E. A. (1974) Non-aggressive behaviour in the large soldier of *Nasutitermes exitiosus* (Hill) (Isoptera: Termitidae). *Insectes Sociaux*, **21**, 95–106.

Maschwitz, U., Jander, R. and Burkhardt, D. (1972) Wehrsubstanzen und Wehrverhalten der Termite *Macrotermes carbonarius*. *J. Insect Physiol.*, **18**, 1715–20.

Maschwitz, U. and Tho, Y. P. (1974) Chinone als Wehrsubstanzen bei einiger orientalischen Macrotermitinen. *Insectes Sociaux*, **21**, 231–3.

Mathews, A. G. A. (1977) *Studies on Termites from the Mato Grosso State, Brazil*. Academia Brasileira de Ciencias, Rio de Janeiro.

Matsumara, F., Coppel, H. C. and Tai, A. (1968) Isolation and identification of a termite trail-following pheromone. *Nature*, **219**, 963–4.

Meinwald, J., Prestwich, G., Nakanish, K. and Kubo, I. (1978) Chemical ecology: results from East Africa. *Science*, **199**, 1167–73.

Mensa-Bonsu, A. (1976) The production and elimination of supplementary reproductives in *Porotermes adamsoni* (Froggatt) (Isoptera, Hodotermitidae). *Insectes Sociaux*, **23**, 133–53.

Mill, A. E. (1982) Foraging and defence in Neotropical termites (Insecta, Isoptera). Unpublished PhD thesis, University of Southampton.

Miller, E. M. (1969) Caste differentiation in the lower termites. In: *Biology of Termites*, Vol. I (Krishna, K. and Weesner, F. M., eds) pp. 283–310. Academic Press, New York.

Moore, B. P. (1968) Studies on the chemical composition and function of the cephalic gland secretion in Australian termites. *J. Insect Physiol.*, **14**, 33–9.

Moore, B. P. (1969) Biochemical studies in termites. In: *Biology of Termites*, Vol. I (Krishna, K. and Weesner, F. M., eds) pp. 407–32. Academic Press, New York.

Noirot, C. (1969a) Glands and secretions. In: *Biology of Termites*, Vol. I (Krishna, K. and Weesner, F. M., eds) pp. 223–82. Academic Press, New York.

Noirot, C. (1969b) Formation of castes in the higher termites. In: *Biology of Termites*, Vol. I (Krishna, K. and Weesner, F. M., eds) pp. 89–123. Academic Press, New York.

Noirot, C. (1970) The nests of termites. In: *Biology of Termites*, Vol. II (Krishna, K. and Weesner, F. M., eds) pp. 223–82. Academic Press, New York.

Noirot, C. and Quennedey, A. (1974) Fine structure of insect epidermal glands. *Ann. Rev. Ent.*, **19**, 61–80.

Nutting, W. L., Blum, M. S. and Fales, H. M. (1974) Behaviour of the North American termite, *Tenuirostritermes tenuirostris*, with special reference to the soldier frontal gland secretion, its chemical composition and its use in defense. *Psyche*, **18**, 167–77.

Okot-Kotber, B. M. (1977) Changes in corpora allata volume during development in relation to caste differentiation in *Macrotermes subhyalinus*. *Proc. VIII Int. Cong. IUSSI*, Pudoc, Wageningen, pp. 262–4.

Parton, A. H., Howse, P. E., Baker, R. and Clément, J.-L. (1981) Variation in the chemistry of the frontal gland secretion of European *Reticulitermes* species. In: *Biosystematics of Social Insects* (Howse, P. E. and Clément, J.-L., eds). Academic Press, New York and London.

Pasteels, J. M. (1965) Polyéthisme chez les ouvriers de *Nasutitermes lujae* (Termitidae, Isoptères). *Biol. Gabon.*, **1**, 191–205.

Pasteels, J. M. (1971) Sex-specific pheromones in a termite. *Experientia*, **28**, 105–6.

Prestwich, G. D. (1978) Isotrinervi-2β-ol. Structural isomers in the defense secretions of allopatric populations of the termite *Trinervitermes gratiosus*. *Experientia*, **34**, 682–3.

Prestwich, G. D. (1979a) Interspecific variation in the defence secretions of *Nasutitermes* soldiers. *Biochem. Syst. Ecol.*, **7**, 211–21.

Prestwich, G. D. (1979b) Chemical defense by termite soldiers. *J. Chem. Ecol.*, **5**, 459–80.

Prestwich, G. D. (1979c) Termite chemical defense: new natural products and chemo-systematics. *Sociobiol.*, **4**, 127–38.

Prestwich, G. D., Bierl, B. A., Devilbiss, E. D. and Chaudhury, M. F. B. (1977) Soldier frontal glands of the termite *Macrotermes subhyalinus*: Morphology, chemical composition and use in defense. *J. Chem. Ecol.*, **3**, 579–90.

Prestwich, G. D. and Chen, D. (1981) Soldier defense secretions of *Trinervitermes bettonianus* (Isoptera, Nasutitermitinae): chemical variation in allopatric populations. *J. Chem. Ecol.*, **7**, 147–57.

Prestwich, G. D., Collins, M. S. (1980) A novel enolic β-ketoaldehyde in the defense secretion of the termite *Rhinotermes hispidus*. *Tetrahedron Lett.*, **21**, 5001–2.

Prestwich, G. D., Collins, M. S. (1981) Chemotaxonomy of *Subulitermes* and *Nasutitermes* termite soldier defense secretions. Evidence against the hypothesis of diphyletic evolution of the Nasutitermitinae. *Biochem. Syst. Ecol.*, **9**, 83–8.

Prestwich, G. D. and Collins, M. S. (1982) Chemical defense secretions of the termite soldiers of *Acorhinotermes* and *Rhinotermes* (Isoptera, Rhinotermitidae): ketones, vinyl ketones and β-ketoaldehydes derived from fatty acids. *J. Chem. Ecol.*, **8**, 147–61.

Prestwich, G. D., Jones, R. W. and Collins, M. S. (1981) Terpene biosynthesis by nasute termite soldiers (Isoptera: Nasutitermitinae). *Insect Biochem.*, **11**, 331–6.

Prestwich, G. D., Kaib, M., Wood, W. F. and Meinwald, J. (1975) 11,13-Tetra-decadien-3-one and homologs: New natural products isolated from *Schedorhino-termes* soldiers. *Tetrahedron Letters*, 4701–4.

Prestwich, G. D., Spanton, S. G., Lauher, J. W., Vrkoč, J. (1980) Structure of 3α-hydroxy-15-rippertene. Evidence for 1,2-methyl migration during biosynthesis of a tetracyclic diterpene in termites. *J. Am. Chem. Soc.*, **102**, 6825–8.

Prestwich, G. D., Tanis, S. P., Pilkiewicz, F., Miura, I., Nakanishi, K. (1976) Nasute termite frontal gland secretions II. Structures of trinervitane congeners from *Trinervitermes* soldiers. *J. Am. Chem. Soc.*, **98**, 6062–4.

Prestwich, G. D., Wiemer, D. F., Meinwald, J., Clardy, J. (1978) Cubitene, an irregular twelve-membered-ring diterpene from a termite soldier. *J. Am. Chem. Soc.*, **100**, 2560–61.

Quennedy, A. (1975) Morphology of exocrine glands producing pheromones and defensive substances in subsocial and social insects. In: *Pheromones and Defensive Secretions in Social Insects* (Noirot, C, Howse, P. E. and Le Masne, G., eds) pp. 1–21. Université de Dijon.

Quennedy, A. (1977) An ultrastructural study of the polymorphic sternal gland in *Reticulitermes santonensis* (Isoptera, Rhinotermitidae); another way of looking at the true termite trail-pheromone. *Proc. VIII Int. Cong. IUSSI*, Pudoc, Wageningen. pp. 48–9.

Quennedy, A., Baulé, G., Rigaud, J., Dubois, P. and Brossut, R. (1973) La glande frontale des soldats de *Schedorhinotermes putorius* (Isoptera): analyse chemique et fonctionnement. *Insect Biochem.*, **3**, 61–7.

Quennedey, A. and Deligne, J. (1975) L'arme frontale des soldats de termites. I. Rhinotermitidae. *Insectes Sociaux*, **22**, 243–67.

Quennedey, A. and Leuthold, R. H. (1978) Fine structure and pheromonal properties of the polymorphic sternal gland in *Trinervitermes bettonianus* (Isoptera). *Insectes Sociaux*, **25**, 153–62.

Roonwal, M. L. (1970) Termites of the oriental region. In: *Biology of Termites*, Vol. I (Krishna, K. and Weesner, F. M., eds) pp. 315–91. Academic Press, New York.

Ruelle, J. E. (1964) L'architecture du nid de *Macrotermes natalensis* et son sens fonctionnel. In: *Études sur les Termites africains* (Bouillon, A., ed.) pp. 327–63. Masson, Paris.

Rüppli, E. (1969) Die Elimination überzähliger Ersatzgeschlechtstiere bei der Termite *Kalotermes flavicollis* (Fabr.). *Insectes Sociaux*, **16**, 235–48.

Sands, W. A. (1961) Foraging behaviour and feeding in five species of *Trinervitermes* in West Africa. *Ent. exp. appl.*, **4**, 277–88.

Sands, W. A. (1972) The soldierless termites of Africa (Isoptera: Termitidae). *Bull. Brit. Mus. Nat. Hist. (Entomology)*, Supplement **18**, 1–243.

Spanton, S. G. and Prestwich, G. D. (1981) Chemical self-defense by termite workers: prevention of autotoxication in two rhinotermitids. *Science*, **214**, 1363–4.

Stuart, A. M. (1964) Morphological and functional aspects of an insect epidermal gland. *J. Cell Biol.*, **36**, 527–49.

Stuart, A. M. (1967) Alarm, defense, and construction behaviour in termites (Isoptera). *Science*, **156**, 1123–5.

Van der Werff, P. A. (1981) Two mound types of *Macrotermes* near Kajiado (Kenya): intraspecific variation or interspecific divergence? In: *Biosystematics of Social Insects* (Howse, P. E. and Clément, J.-L., eds) pp. 231–48. Academic Press, New York.

Vrkoč, J., Kreček, J. and Hrdý, J. (1978) Monoterpenic alarm pheromones in two *Nasutitermes* species. *Acta ent. Bohem.*, **75**, 1–8.

Vrkoč, J. and Ubic, K. (1974) 1-Nitro-trans-1-pentadecene as the defensive compound of termites. *Tetrahedron Letters*, **15**, 1463–4.

Wadhams, L. J., Baker, R. and Howse, P. E. (1974) 4,11-Epoxy-*cis*-eudesmane, a novel oxygenated sesquiterpene in the frontal gland secretion of the termite *Amitermes evuncifer*. *Tetrahedron Letters*, 1697–1700.

Watson, J. A. L., Hewitt, P. H. and Nel, J. J. C. (1971) The water-sacs of *Hodotermes mossambicus*. *J. Insect Physiol.*, **17**, 1705–9.

Weimer, D. F., Meinwald, J., Prestwich, G. D., Solheim, B. A. and Clardy, J. (1980) Bilflora-4, 10(19), 15-triene: a new diterpene from a termite soldier (Isoptera Termitidae Termitinae). *J. Org. Chem.*, **45**, 191–2.

Williams, R. M. C. (1959) Colony development in *Cubitermes ugandensis* Fuller (Isoptera, Termitidae). *Insectes Sociaux*, **6**, 291–304.

Wood, W. F., Truckenbrodt, W. and Meinwald, J. (1975) Chemistry of the defensive secretion from the African termite *Odontotermes badius*. *Ann. ent. Soc. Am.*, **68**, 359–60.

Zalkow, L. H., Howard, R. W., Gelbaum, L. T., Gordon, M. M., Deutsch, H. M. and Blum, M. S. (1981) Chemical ecology of *Reticulitermes flavipes* (Kollar) and *R. virginicus* (Banks) (Rhinotermitidae): chemistry of the soldier cephalic secretions. *J. Chem. Ecol.*, **7**, 717–32.

Index